GW00492773

ADVANCES IN MEDICINE AND BIOLOGY

ADVANCES IN MEDICINE AND BIOLOGY. VOLUME 20

ADVANCES IN MEDICINE AND BIOLOGY

Additional books in this series can be found on Nova's website under the Series tab.

Additional E-books in this series can be found on Nova's website under the E-books tab.

ADVANCES IN MEDICINE AND BIOLOGY

ADVANCES IN MEDICINE AND BIOLOGY. VOLUME 20

LEON V. BERHARDT
EDITOR

Nova Science Publishers, Inc.
New York

For permission to use material from this book please contact us:
Telephone 631-231-7269; Fax 631-231-8175
Web Site: http://www.novapublishers.com

LIBRARY OF CONGRESS CATALOGING-IN-PUBLICATION DATA

ISSN: 2157-5398

ISBN 978-1-61209-135-8

Published by Nova Science Publishers, Inc. † New York

CONTENTS

PREFACE

This continuing series gathers and presents original research results on the leading edge of medicine and biology. Each article has been carefully selected in an attempt to present substantial topical data across a broad spectrum. Topics discussed in this compilation include the cell cycle alteration in down syndrome; TLR agonists as immune adjuvants; heart rate variability in eating disorders; current trends in glaucoma; pharmacological aspects of borates; phage display technology and the quantitative determination of oxytocin.

Chapter 1 - More sensitive and rapid detection magnetic immunoassay methods are emerging based on magnetic immunoassay and its utility in *in vivo* imaging is emerging to pinpoint the location of TroponinT release after damage in cardiac tissue. The possibility is growing of imaging the cardiac muscle focal damage due to acute myocardial infarction as rapid noninvasive imaging technique. *Hypothesis:* The paramagnetic nanoparticle antitroponin complexes can be detected by magnetic resonance imaging. *Materials and Methods:* The nanoparticle complex was prepared by coprecipitation method. The iron oxide-avidin microimulsion was used followed by polymer coating and coupling with biotinylated antitroponin. The prototype phantom of nanocomplexes were imaged by magnetic resonance imaging. Excised mouse heart tissue was imaged and compared with histology. The image processing of heart images was used for shape analysis. *Results:* The nanoparticle complex made by co-precipitation method measured in the range of 30-100 nm size. The effect of avidin and biotin binding with iron oxide was effective to generate change in relaxation time constant and MRI visibility of cardiac tissue. The troponin T concentration variation in phantom sample was proportionate with variation of relaxation constants of nanoparticles and varied with concentration of iron oxide in the nanoparticle complex. The MALDI method evidenced to image the location of cardiac proteins. The MRI images of excised mouse heart showed muscle fiber orientation associated with troponin and image segmentation predicted the heart shape and muscle fiber orientation. *Discussion:*. TroponinT have shown significance in assessment of acute myocardial infarction. The present study presents an innovative idea to explore the possibility of: 1.Antitroponin bound magnetic particle in polymer coat preparation as specific nanospheres for magnetic resonance imaging technique; 2. A new approach of antitroponin and biotin bound with fluorescent marker. The troponinT specific antitroponin nanoparticle sensitive MRI signal of the cardiac tissue sample predicted the mouse heart shape by image segmentation. The image correlation with muscle fiber histology showed enhanced tissue morphometry accuracy. *Conclusion:* The emerging new nanoparticle methods of TroponinT detection are useful in rapid MRI imaging the myocardial infarction damage.

Chapter 2 - Troponin molecules and their T,C and I subtypes have shown significance in early point-of-care assessment of acute myocardial infarction. More sensitive and rapid detection magnetic immunoassay methods are emerging based on polymer bound enzymes for lab use and early point-of-care. *Hypothesis:* The magnetic particles in immunoassay enhance the utility of immunoassay by their ability to facilitate the separation of the targeted compounds and their high sensitivity of analyte. The basic principle of antitroponin-magnetic immunoassay was based on formation of sandwitch complex of two different antibodies at two binding sites on same target troponinT antigen (one antibody site was attached with magnetic nanoparticles, other antibody binding site attached with glucose in equal proportion for its detection by glucose oxidase-peroxidase enzyme measurement. *Materials and Methods:* The concentration of troponinT in the blood sample was correlated with proportional concentration of glucose measured by glucometer. The present report presents an innovative idea to explore the newer possibility of: 1. Troponin subtyping by MALDI mass spectrometry technique; 2. Magnetic immunoassay in serum sandwitch magnetic immunoassay for both solid phase Enzyme Linked Immunosorbent Assay (ELISA) in clinical chemistry lab and point-of-care miniature pencil (POCM) in quick emergency. *Results and Discussion:* The application of nanotechnology in troponin analysis minimized the artifacts of troponin analysis in AMI. Using site-specific agent such as a magnetic nanosphere is novel choice to estimate 1 pico gm/L. In addition, the magnetic nanoparticles enhance the analyte detectibility, sensitivity to analyte with optical reaction speed. The nanospheres could be extracted out by using external magnet. The technique provides simple point-of-care method with a method to use proteomics and magnetic nanoparticles to isolate and quantify AMI marker troponin in the blood sample with use of miniature glucometer. The MALDI mass spectrometry technique showed potential of exploring AMI sensitive Troponins with other possible proteins. The concentration of troponinT in blood sample was correlated with proportional concentration of glucose measured by pencil miniature glucometer. Conclusion: The magnetic immunoassay of point-of-care method shows potential as rapid reliable cost-effective and efficient as cardiac protection device. The technique is simple, user friendly and cost-effective. The nanotechnology based glucometer may provide portable miniature point-of-care testing at low cost.

Chapter 3 - Lead is a toxic and cumulative metal that exerts its harmful action in virtually all body organs and systems. Exposure to this metal may result in a wide variety of biological effects depending on the level and duration of the exposure. Despite the progressive reduction in its applications, there are many industry sectors in which it is still present. In addition, lead has been used by humans since ancient times so, since it is a non-biodegradable element, environmental pollution caused is persistent and widespread, affecting the population at large.

The International Agency for Research on Cancer has classified lead and its inorganic compounds as probable human carcinogens (group 2A). Nowadays there is broad evidence of the relationship between the interaction with genetic material and cancer development, and thus genotoxicity tests are applied as biomarkers of the early effects of most carcinogen agents.

Although exposure to lead, both in environmental and occupational settings, has been frequently associated with an increase in genotoxic damage in humans, controversy is nurtured by several works reporting conflicting results.

This chapter conducts a review of the studies on genotoxicity of exposure to lead, compiling *in vitro*, *in vivo* and epidemiological studies using different DNA damage parameters from mutagenicity to chromosomal alterations.

Chapter 4 - Trisomy 21 or Down syndrome is determined by the triplication of human chromosome 21 and the overdosage of genes on this chromosome in all cells. This genetic overdosage determines transcriptional alteration of most of the chromosome 21 genes, and trisomic cells suffer of dosage imbalance of a lot of proteins that can potentially impair all molecular pathways, including cell cycle control. Thus, trisomy 21 provides a genetic model for the role of aneuploidy in neurogenesis, leukemia, tumors and cancer. An example of chromosome 21 gene involved in alteration of cell cycle in neuronal cell, determining abnormal neurogenesis, is Dyrk1A, the ortholog of the *Drosophila minibrain* (MNB) gene which is involved in the regulation of neural proliferation. Three oncogenes mapping on chromosome 21, AML1, ERG and ETS2, highly related to the genesis of leukemias, are of the most interest to contribute to the general understanding of the oncogenic mechanisms of chromosomal aneuploidies, the most common abnormalities in cancer. The chromosome 21 gene TIAM1 (T-cell lymphoma invasion and metastasis 1 gene) has important role in the progression of epithelial cancers, especially carcinomas of the breast and colon. The direct involvement of the human chromosome 21 SIM2 (Single minded 2) in the inhibition of gene expression, growth inhibition, inhibition of tumor growth as well as induction of apoptosis that could involve a block of cell cycle, establish SIM2 gene as a molecular target for cancer therapeutics and have additional important implications for the future diagnosis and treatment of specific solid tumors as well as for further understanding the cancer risk in Down syndrome patients.

Chapter 5 - Eating disorders, which include anorexia nervosa (AN) and bulimia nervosa (BN), are characterized by significant alterations of many systems maintaining body homeostasis. With respect to the cardiovascular system, a series of studies have been performed to assess whether heart rate variability (HRV), which is an indicator of cardiac autonomic regulation, changes in patients affected by eating disorders in comparison to healthy subjects. Even if some discrepancies exist, possibly due to different methods of investigation and/or criteria for patient selection, most results show the existence of a predominance of vagal over sympathetic activity in the modulation of HRV in both AN and BN. This cardiac sympathovagal imbalance may be the pathophysiological basis for the higher risk of sudden cardiac death observed in eating disorders. The pathological modifications of HRV and cardiac autonomic modulation might be secondary to the endocrine abnormalities associated with AN and, to a lesser extent, BN. Indeed, alterations of HRV have been described in classical endocrine diseases, like diabetes mellitus or acromegaly. In particular, hormones controlling caloric intake and body fat composition may play a primary role. In the second part of this chapter, after a comprehensive review of the available data about HRV in eating disorders, the possible relationships between HRV and hormonal changes are discussed.

Chapter 6 - *Introduction.* Methicillin- and also vancomycin (glycopeptide)-resistant Gram-positive organisms have emerged as an increasingly problematic cause of hospital-acquired infections, and are also spreading into the community. Vancomycin (glycopeptide) resistance has emerged primarily among Enterococci, but the MIC values of vancomycin are also increasing for the entire *Staphylococcus* species, worldwide.

Materials and Methods. The aim of our review is to evaluate the efficacy and tolerability of newer antibiotics with activity against methicillin-resistant and glycopeptide-resistant Gram-positive cocci, on the grounds of our experience at a tertiary care metropolitan Hospital, and the most recent literature evidence in this field.

Results. Quinupristin-dalfopristin, linezolid, daptomycin, and tigecycline show an excellent *in vitro* activity, comparable to that of vancomycin and teicoplanin for methicillin-resistant staphylococci, and superior to that of vancomycin for vancomycin-resistant isolates. Dalbavancin, televancin and oritavancin are new glycopeptide agents with excellent activity against Gram-positive cocci, and have superior pharmacodynamics properties compared to vancomycin. We review the bacterial spectrum, clinical indications and practical use, pharmacologic properties, and expected adverse events and contraindications associated with each of these novel antimicrobial agents, compared with the present standard of care.

Discussion. Quinupristin-dalfopristin is the drug of choice for vancomycin-resistant *Enterococcus faecium* infections, but it does not retain effective activity against *Enterococcus faecalis*. Linezolid activity is substantially comparable to that of vancomycin in patients with methicillin-resistant *Staphylococcus aureus* (MRSA) pneumonia, although its penetration into the respiratory tract is exceptionally elevated. Tigecycline has activity against both *Enterococus* species and MRSA; it is also active against a broad spectrum of Enterobacteriaceae and anaerobes, which allows its use for intra-abdominal, diabetic foot, and surgical infections. Daptomycin has a rapid bactericidal activity for *Staphylococcus aureus* and it is approved in severe complications, like bacteremia and right-sided endocarditis. It cannot be used to treat pneumonia and respiratory diseases, due to its inactivation in the presence of pulmonary surfactant.

Chapter 7 - Toll like receptors (TLRs) are part of the innate immune system, and they belong to the pattern recognition receptors (PRR) family, which is designed to recognize and bind certain molecules that are restricted to pathogens, like LPS and CpG. Different TLR signals converge through few common adapter proteins to relay their signals, a process that results in the activation of several genes essential for mounting an immune response. The wide distribution of TLRs on hematopoietic and non-hematopoietic cells, and their high potential for activating the host immune system makes TLR ligands great adjuvant candidates. They will elicit their function on a wide variety of cells, and stimulate both the innate and adaptive immune systems. This review will describe some of the TLR agonists that are considered as major targets for the development of agonists that could serve as adjuvants and agents of immunotherapy. Several of these TLR agonists, including TLR2, TLR4, TLR7 and TLR9 agonists are being developed and tested in clinical trials.

Chapter 8 - Neuroprotection was initially investigated for disorders of the central nervous system such as amyotrophic lateral sclerosis, Alzheimer's disease, Parkinson and head trauma, but only few therapies have been approved (nemantine for Azheimer's dementia).

Neuroprotection is a process that attempts to preserve the remaining cells that are still vulnerable to damage. The main aim of neuroprotective therapy is to apply pharmacologic or other means to attenuate the hostility of the environment surrounding the degenerating cells, or to supply the cells with the tools to deal with this aggression, providing resilience to the insult. By definition, glaucoma neuroprotection must be considered independent of intraocular pressure lowering, and the target neurons should be in the central visual pathway, including retinal ganglion cells.

Several agents have been reported neuroprotective in glaucoma, both in clinical studies, such as Ca2+ channel blockers, and in experimental studies, such as betaxolol, brimonidine, NMDA antagonists, Nitric Oxide Synthase inhibitors, neurotrophins and ginkgo biloba extract. However, to establish neuroprotective drugs for glaucoma, well-designed clinical trials are required, specially randomized clinical trials comparing neuroprotective treatments to placebo, and although there is laboratory evidence for glaucoma neuroprotection by several drugs, we still evidence from randomized clinical trials.

The ideal anti-glaucoma drug would be one that when applied topically, reduces IOP, but also probes to reach the retina in appropriate amounts, and activates specific receptors in the retina to attenuate retinal ganglion cell death.

Chapter 9 - Boric acid and sodium borate are pharmaceutical necessities and are used as anti-infective agents and as a buffer in pharmaceutical formulations. Boric acid is used as an antimicrobial agent for the treatment of Candida, Aspergillus and Trichomonas infections. It is employed in the treatment of psoriasis, acute eczema, dermatophytoses, prostrate cancer, deep wounds and ear infections. Borates are toxic in nature and produce symptoms of poisoning on exposure to mineral dust, oral ingestion and topical use. Boric acid induces developmental and reproductive toxicity in mice, rats and rabbits. It has been shown to produce cytotoxic, embryotoxic, genotoxic, ototoxic and phytotoxic effects. Borates are considered to be completely absorbed by the oral route. They are also absorbed through denuded or irritated skin. Boric acid causes changes in lipid metabolism and in acid-base equilibrium of the blood. Boric acid appears to modulate certain inflammatory mediators and to regulate inflammatory processes. It takes part in the reversible or irreversible inhibition of certain enzyme systems. Borates are widely used as a preservative and insecticide.

Chapter 10 - Synthetic biology is potentially one of the most powerful emerging technologies today. It is the art of synthesizing and engineering new biological systems that are not generally found in nature, and also of redesigning existing biological molecules, structures and organisms so as to understand their underlying mechanisms. The molecules used in these systems might be naturally occurring or artificially synthesized for a variety of nanotechnological objectives.

Protein evolution in vitro technologies can provide the tools needed to synthetic biology both to develop “nanobiotechnology” in a more systematic manner and to expand the scope of what it might achieve. Proteins designed and synthesized from the scratch –the so-called “never born proteins” (NBPs)– and those endowed with novel functions can be adapted for nano-technological uses.

The phage display technique has been used to produce very large libraries of proteins having no homology with known proteins, and being selected for binding to specific targets. The peptides are expressed in the protein coats of bacteriophage, which provides both a vector for the recognition sequences and a marker that signals binding to the respective target. Specific ligands for virtually any target of interest can be isolated from highly diverse peptide libraries, and many of them selected up till now have shown considerable potential for nanotechnological applications. These involve drug discovery, “epitope discovery”, design of DNA-binding proteins, source of new materials, antibody phage and recombinant phage probes, next-generation nano-electronics, targeted therapy and phage-display vaccination.

All this can be regarded as a kind of synthetic biology in that it involves the reshaping and redirecting of natural molecular systems, phage, typically using the tools of protein and genetic engineering.

Chapter 11 - Oxytocin is a clinically important nonapeptide that is used for the induction and/or augmentation of labor and is normally administered as a slow intravenous infusion diluted with normal saline or Ringer's lactate solution. Oxytocin is also indicated for use in the prevention and treatment of post partum hemorrhage and may be administered via either the intramuscular or intravenous routes in order to increase uterine tone and/or reduce bleeding. The analysis of oxytocin in different media has evolved over the past 30 years with the result that more sophisticated, selective and sensitive techniques are used for the determination of the compound. A variety of techniques have been applied to the determination of oxytocin in different matrices ranging from simple paper chromatography to hyphenated liquid chromatographic such as liquid chromatography coupled with mass-spectrometry. Additionally enzyme linked immuno-sorbent assays (ELISA) and radio immuno-assays (RIA) are used for the determination of low concentrations of oxytocin in biological matrices. This manuscript provides a systematic survey of the analytical methods that have been reported for isolation and quantitation of oxytocin in different matrices.

Chapter 12 - Proteomics has become an important contributor to the knowledge of plant cell wall structure and function by allowing the identification of proteins present in cell walls. This chapter will give an overview on recent development in the cell wall proteomic field. Results from proteomics show some discrepancies when compared to results from transcriptomics obtained on the same organ. It suggests that post-transcriptional regulatory steps involve an important proportion of genes encoding cell wall proteins (CWPs). Proteomics thus complements transcriptomic. The cell wall proteome of *Arabidopsis thaliana* is the most completely described at the moment with about one third of expected CWPs identified. CWPs were grouped in functional classes according to the presence of predicted functional domains to allow a better understanding of main functions in cell walls. The second best-described cell wall proteome is that of *Oryza sativa*. Same functional classes were found, with different compositions reflecting the differences in polysaccharide structure between dicot and monocot cell walls. All these proteomic data were collected in a new publicly accessible database called *WallProtDB* (http://www.polebio.scsv.ups-tlse.fr/WallProtDB/). In conclusion, some perspectives in plant cell wall proteomics are discussed.

Chapter 13 - Currently, the primary technique for preserving isolated human organs for transplant is maintaining them at low temperature, but the time limit of this method is from 4 hr to 24 hr depending on the type of organs. If organs could be preserved long-term like blood cells or microorganisms, then the problem of organ shortage could be considerably alleviated. New techniques for the long-term preservation of organs still have to be developed and they remain eagerly awaited.

Seki *et al.* (1998) focused on cryptobiosis which enables living things to adapt themselves to extreme environment such as severe desiccation or very low temperature by decreasing the water content of the organism, which leads to a slow down of the metabolism. In this context, Seki *et al.* succeeded in resuscitating tardigrades placed under extremely high pressure (600 MPa). Since 1998, using liquid perfluorocarbon (PFC) for organ preservation, Seki *et al.* decreased the water contained in isolated rat hearts and preserved them for resuscitation. This occasionally led to good results, but the reproducibility was poor. Taking advantage of the anesthetic and metabolic inhibitory actions of carbon dioxide (CO_2) gas on organisms, Seki *et al.* (1998) repeatedly conducted experiments on rat hearts whose water content was considerably decreased and were exposed to high partial pressure of PCO_2=200 hPa during preservation and good results were obtained with reproducibility [21]. After having thus established organ preservation method by means of desiccation with CO_2, Yoshida *et al.* successfully extended the preservation time of isolated rat hearts to 72 hr by supplying CO_2 (PCO_2=200 hPa) to liquid PFC in the form of bubbles.

Various combinations of gas mixtures were tried. For example, in place of the usual CO_2, CO which is in reversible relation with oxygen (O_2) was introduced. Isolated rat hearts were placed in a hyperbaric environment of 2 ATA where high partial pressure between 200 hPa and 1000 hPa was chosen for the CO gas and the remaining pressure was complemented by the partial pressure of O_2. The rat hearts preserved 24 hr in this hyperbaric environment were heterotopically transplanted and were resuscitated. This successfully demonstrated the usefulness of CO for the preservation of organs and in search of longer preservation period, the partial pressure of CO was varied. For example, when the predetermined partial pressure of CO was between 1000 hPa and 5000 hPa, O_2 gas was supplied until the total pressure reached 7 ATA. After 48 hr of heart preservation by desiccation, heterotopic transplantation of the rat heart was performed and it was resuscitated. The reproducibility of this series of experiments was good.

In addition, CO_2 was introduced into the hyperbaric environment of 7 ATA. At the same time, to prevent CO_2 from causing decompression sickness, helium (He), an inert gas, was used to create a hyperbaric environment of 7 ATA. Consequently in the hyperbaric environment of PCO=400 hPa + PCO_2=100 hPa + PO_2=900 hPa + PHe=5600 hPa, isolated rat hearts were desiccated and preserved for 72 hr and were heterotopically transplanted and resuscitated successfully. Significant reproducibility was obtained. By using a mixture of CO_2, CO, O_2 and He, reproducible techniques have been developed allowing the desiccation and preservation of isolated rat hearts for a maximum of 120 hr, followed by their heterotopical transplantation and resuscitation. Therefore, the new scientific field of semibiology had been created.

Chapter 14 - A group of bacterial system specific chaperones are involved with the maturation pathway of redox enzymes that utilize the twin-arginine protein translocation (Tat) system. These chaperones are referred collectively as REMPs (*R*edox *E*nzyme *M*aturation *P*rotein). They are proteins involved in the assembly of a complex redox enzyme which itself does not constitute part of the final holoenzyme. These proteins have been implicated in coordinating the folding, cofactor insertion, subunit assembly, protease protection and targeting of these complex enzymes to their sites of physiological function. The substrates of REMPs include respiratory enzymes such as N- and S-oxide oxidoreductases, nitrate reductases, and formate dehydrogenases, which contain at least one of a range of redox-active cofactors including molybdopterin (MoPt), iron sulfur [Fe-S] clusters, and b- and c-type haems. REMPS from *Escherichia coli* include TorD, DmsD, NarJ/W, NapD, FdhD/E, HyaE, HybE and the homologue YcdY. The biochemical, structural and functional information on these REMPs are reviewed in detail here.

Chapter 15 - Vancomycin treatment failures clinically account for 23-40% of patients with *S. aureus* infections; this occurs in the absence of laboratory-documented vancomycin resistance. Resistance of methicillin-resistant *S. aureus* (MRSA) clinical isolates to vancomycin can be phenotypically increased by exposure to dehydroepiandosterone (DHEA), an androgen with a chemical structure analogous to that of cholesterol. This study describes a phenotypic increase in resistance to the host cationic defense peptide, β-1 defensin, as well as vancomycin and other antibiotics that have a positive charge, in response to DHEA. The

DHEA-mediated alteration in cell surface architecture appears to correlate with increased resistance to vancomycin. DHEA-mediated cell surface changes include alterations in: cell surface charge, surface hydrophobicity, capsule production, and carotenoid production. In addition, exposure to DHEA results in decreased resistance to lysis by Triton X-100 and lysozyme, indicating activation of murien hydrolase acivity. We propose that DHEA is an interspecies quorum-like signal that triggers innate phenotypic host survival strategies in *S. aureus* including but not limited to increased carotenoid production. A side effect of this phenotypic change is increased vancomycin resistance. Furthermore this DHEA-mediated survival system may share the cholesterol-squalene pathway shown to be statin sensitive.

Versions of these chapters were also published in *International Journal of Medical and Biological Frontiers,* Volume 16, Numbers 1-12, edited by Tsisana Shartava, published by Nova Science Publishers, Inc. They were submitted for appropriate modifications in an effort to encourage wider dissemination of research.

In: Advances in Medicine and Biology. Volume 20
Editor: Leon V. Berhardt

ISBN 978-1-61209-135-8

Chapter 1

TROPONIN T: A SEARCH OF SUPERPARAMAGNETIC IRON-OXIDE BOUND ANTITROPONIN NANOPARTICLE FOR MAGNETIC RESONANCE IMAGING

Rakesh Sharma[*1,2]
[1] Center of Nanomagnetics and Biotechnology,
Forida State University, Tallahassee, Florida 32310 , U.S.A.
[2] University of North Carolina, Greensboro, NC., U.S.A.

Abstract

More sensitive and rapid detection magnetic immunoassay methods are emerging based on magnetic immunoassay and its utility in *in vivo* imaging is emerging to pinpoint the location of TroponinT release after damage in cardiac tissue. The possibility is growing of imaging the cardiac muscle focal damage due to acute myocardial infarction as rapid noninvasive imaging technique. *Hypothesis:* The paramagnetic nanoparticle antitroponin complexes can be detected by magnetic resonance imaging. *Materials and Methods:* The nanoparticle complex was prepared by coprecipitation method. The iron oxide-avidin microimulsion was used followed by polymer coating and coupling with biotinylated antitroponin. The prototype phantom of nanocomplexes were imaged by magnetic resonance imaging. Excised mouse heart tissue was imaged and compared with histology. The image processing of heart images was used for shape analysis. *Results:* The nanoparticle complex made by co-precipitation method measured in the range of 30-100 nm size. The effect of avidin and biotin binding with iron oxide was effective to generate change in relaxation time constant and MRI visibility of cardiac tissue. The troponin T concentration variation in phantom sample was proportionate with variation of

[*] Email: rksz2004@yahoo.com

relaxation constants of nanoparticles and varied with concentration of iron oxide in the nanoparticle complex. The MALDI method evidenced to image the location of cardiac proteins. The MRI images of excised mouse heart showed muscle fiber orientation associated with troponin and image segmentation predicted the heart shape and muscle fiber orientation. *Discussion:*. TroponinT have shown significance in assessment of acute myocardial infarction. The present study presents an innovative idea to explore the possibility of: 1.Antitroponin bound magnetic particle in polymer coat preparation as specific nanospheres for magnetic resonance imaging technique; 2. A new approach of antitroponin and biotin bound with fluorescent marker. The troponinT specific antitroponin nanoparticle sensitive MRI signal of the cardiac tissue sample predicted the mouse heart shape by image segmentation. The image correlation with muscle fiber histology showed enhanced tissue morphometry accuracy. *Conclusion:* The emerging new nanoparticle methods of TroponinT detection are useful in rapid MRI imaging the myocardial infarction damage.

Keywords: Troponin, antitroponin, acute myocardial infarction, ELISA, imaging, cardiac muscle.

Introduction

Myocardial Infarction is caused due to myocardial necrosis and ischemia. It is confirmed by measurement of biomarkers such as troponin T as a cause of sudden death, percutaneous coronary disturbances, coronary artery bypass recovery of patients and stent thrombosis. Recently, the nanomaterials promised a great advantage in monitoring myocardial infarction and cardiac architectural changes during disease. Three major advances were reported to monitor the disease recovery progress. First, troponinT rapid immunoassays, second troponin proteomics typing and third, imaging of troponinT leakage sites by fluorescent or magnetic behavior of nanoparticles [1]. Troponin subtypes T/C/I play a distinct role in recovery of myocardial infarction associated with calcium regulation and conformational changes in Lys amino acid moieties of troponinT molecule. Antitroponin human antibody raised in mouse provides a unique opportunity to measure the troponin in the cardiac tissue and blood [2]. The present study focuses on imaging of troponin T leakage.

Ultrasmall superparamagnetic iron oxide nanoparticles have been widely used during the past decade as MR intravascular contrast agents in the study of animal heart models. Such agents enhance both T1 and T2/T2* relaxation constants of tissue protons. The strong microscopic intravascular susceptibility effect of nanoparticles enables mapping the local blood volume distribution in myocardium. The long half-life of these nanoparticles in blood generates high spatial resolution and sensitivity in presence of anesthetic chemicals. This capability has been utilized to study the cerebrovascular blood volume distributions and their changes in normal, activated, pathologic and pharmacologically or genetically modified states, particularly in rodent animal heart models [3]. It has also been applied to study blood volume changes in myocardium. The relaxation rate shifts delta R2 and delta R2* induced by iron oxide agents may differ depending on certain morphological characteristics of the microvascular myocardial network, and sensitive delta R_2 and delta R_2* mapping can provide, in addition to blood volume, measurement of other important microvascular parameters such as blood vessel density and infarct size. On other hand, iron-oxide (maghemite) as nanomaterial offers distinct MRI signal properties based on its dephasing effect and MRI signal loss [4]. In other report, the dephasing effect in MRI was multifactorial and was dependent on ferritin protein physiology. In the presence of oxygen and at circumneutral pH conditions, ferrous iron is quickly oxidized to Fe(III) and precipitates as iron oxides [5]. Ferritin protein consist of 24 protein subunits that self-assemble into a cage-like architecture in which a hydrated ferric oxide or ferrous phosphate is mineralized. The role of ferritin *in vivo* is to sequester Fe as a hydrated form of iron oxide (or phosphate), predominantly as the mineral ferrihydrite. Demineralized ferritin (apoferritin) is a hollow, spherical protein shell homogeneously dispersed in aqueous media. The ferrimagnetic iron oxide phase Fe_3O_4 synthesized within ferritin protein makes 'magnetoferritin'[5]. Magnetoferritin cores get oxidized to make maghemite (γ-Fe_2O_3) cage-like structure to entrap molecules as reported distinct by electron microscopy[5]. In our initial experiment, a ferritin cage was disassembled under low-pH conditions and subsequently reassembled at almost neutral pH in the presence of the Gd-chelate complex GdDOTP (H8DOTP = 1, 4, 7, 10-tetra-kis(methylenephosphonic acid)-1,4,7,10-tetra-azacyclo-dodecane). This results in the entrapment of about 10 Gd-chelate complexes within the protein cage as shown in figure 1. This material has been shown to exhibit high magnetic resonance imaging proton relaxivities, making it a promising candidate for applications in magnetic resonance imaging [6]. Other contrast enhancing paramagnetic material is gadolinium. The evaluation, development and use of gadolinium and ferritin protein require the investigation of Gd-chelate contrast chemistry and use in MR imaging techniques. The fundamental issues were magnetic contrast agent development, rational drug delivery, MR molecular imaging.[7]. The strength of magnetic field along with nanoparticle size and its concentration in tissue determine the MRI image quality and its contrast properties. [8]. Other recent approach of troponin imaging was based on Fluorescent Ca^{2+} indicator proteins (FCIPs). The investigators described transgenic mouse lines expressing a troponin C (TnC)-based biosensor. The biosensor was widely expressed in neurons and has improved Ca^{2+} sensitivity both *in vitro* and *in vivo*[9]. Still technique is in infancy and in evaluation phase.

In present study, we proposed a new superparamagnetic antitroponin-iron oxide as potential cardiac nanoparticle for its preparation and application in magnetic resonance imaging and microimaging. The new nonoparticle-cardiac protein based MRI technique is described to visualize 3D anatomy and dynamic architecture to visualize possible troponin leakage sites. In future, troponin mapping may be extended in cardiac muscle fiber directional mapping as time-series monitoring tool usable in pharmacotherapy, anatomy atlas, functional imaging. The novelty of this technique is to localize very specific sites of troponinT leakage from cardiac tissue in pure noninvasive way.

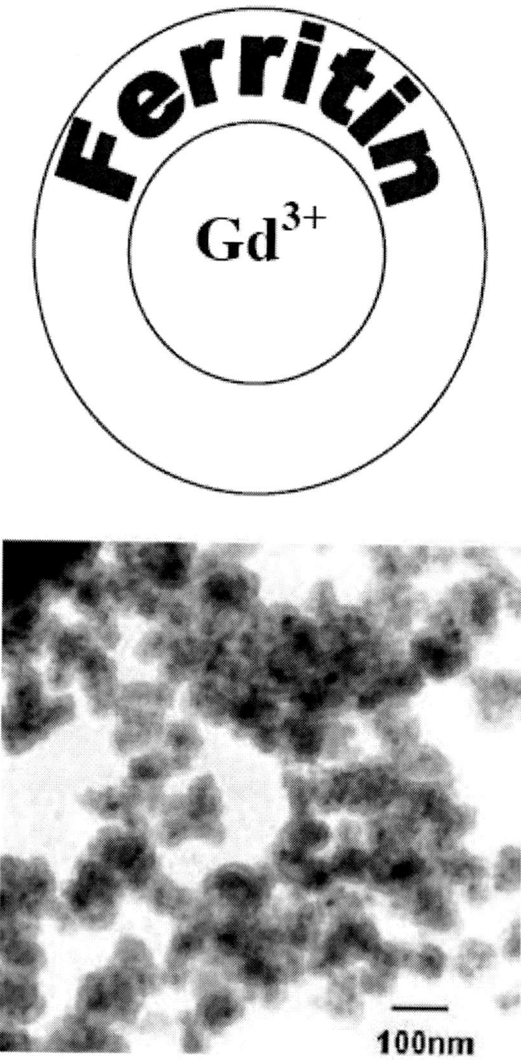

Figure 2. Nanoparticle complex having paramagnetic molecule like iron oxide or gadolinium in center coated with avidin-polymer and coupled with biotinylated antitroponin antibody lebel at outermost layer. The specificity of monoclonal antibody to cardiac troponinT offers the accurate localization of cardiac damage.

2. Materials and Methods

2.1. Preparation and Characterization of Nonoparticle Complex

The nanoparticles were prepared by chemical and physical co-precipitation methods for antitroponin-biotin bound iron-oxide conjugated with avidin and encapsulated within polyethylene. Alternate wet chemical method was also reported to produce nanoparticles with better production scalability and uniform nature [10]. Precipitation of a solid nanoparticles from a solution was a preferred technique for the synthesis of fine 10-100 nm sized nanoparticles [10]. The preparation was performed in three steps:

1. The general procedure involved the reactions in aqueous or non aqueous solutions containing the soluble or suspended salts. Once the solution becomes supersaturated with the product, the precipitate was formed by either homogeneous or heterogeneous nucleation. The formation of nuclei after formation proceeded by diffusion in which case concentration gradients and reaction temperatures were very important in determining the growth rate of the particles, for example, to form monodispersed particles.
2. Ultrasonication was done to have homogeneous sized particles in nanometer range (10-100 nm). In order to minimize the oxidation effect, these particles were left overnight under a continuous flow of oxygen and nitrogen. These particles were further coated with polymer by conventional precipitation process and were designed for biomedical applications. Polymer coated particles are best for their capacity to bound with proteins and many other biological molecules.
3. The preparation of antitroponin biotinylated complex consists the following steps as we described elsewhere:

a. Anti-troponin antibody (cTnT) human recombinant antibody containing immunoglobin G1 (with 2 epitopes) in Freund's adjuvant from cardiac muscle tissue (source: T6077 Sigma) with molecular homogeneity of purified antigens by SDS-PAGE, OD 0.706 IU at 280 nm wave length) was used to raise monoclonal antibody in mouse against cTnT by fusion of myocytes with splenocytes from Balb/C mouse.
b. The troponin antibodies were screened by Western Blot (or MALDI spectroscopy) as previously done [11].
c. The biotinylated complex of said troponin antibody (antiTrT-biotin) was prepared by use of biotin cross-linked with N-hydroxysuccinimide. The complex can be used for ELISA[1] or as contrast agent[2].
d. Coating of biotinylated antibody complex on polystyrene or polyethylene polymer surface was done by external tagging of antibody in balance between biotinylated antibody and conjugated avidin [12]

Here in present study, the antibody was used to make nanoparticle complex for imaging experiments.

4. The central core of iron oxide of nanoparticle sphere contributes magnetic resonance relaxation constants and image contrast. The avidin provides affinity of paramagnetic central core with rest of the sphere. The peripheral biotin-polymer coat offers a

[1] For ELISA, complex coupled with alkaline phosphtase enzyme in incubation buffer (0.1 mol/liter sodium citrate, 0.47 mol/liter Na2HPO4, 0.1% bovine serum albumin at pH 6.8). The biotinylated troponin T antibody (1B10) was precoated to a streptavidin-covered microtiter well plate. Fifty microliters of sample was added with 200 μl of a second cardiospecific troponin T antibody (M7) covalently linked to enzyme alkaline phosphatase in incubation buffer (pH 6.8) to measure cTnT as described elsewhere[11].

[2] The nanomaterial complex is made of iron-oxide core surrounded with avidin and coupled with polymer –biotin-antitroponin.

binding between antibody and biocompatible polymer. The monoclonal antitroponin antibody serves as specific biosensor to troponin sites in cardiac tissue.

2.2. Gadolinium Enhanced T1 Weighted Imaging of Apoferritin-Gadolinium Complex

The samples of apoferritin (alone), GdII-apoferritin and 100 mM Gd-HPDO$_3$A solutions (Prohance) were imaged in glass tubes by proton 11.7 Tesla MRI imaging. The longitudinal T1 relaxation time constants of these solutions were measured by inversion recovery using fast spin echo method [6].

2.3. Characterization of Cardiac Troponin T/I/C Subtypes and MALDI Imaging

The Matrix Assisted Laser Desorption Ionization (MALDI) technique was used by using troponinT as standard, HABA, CHCHA as standard calibration proteins to compare fixed on steel MALDI plate [13]. The slice of mice heart was cryo-fixed on MALDI steel plate and its mass spectroscopy was done as described elsewhere [13].

2.4. Development of Antitroponin Based Nanoparticles as MRI Imaging Contrast Agents

The superparamagnetic iron-oxide antimyoglobin capsulated nanoparticle complex was placed in capillary tubes and imaged at high resolution MRI imager.

Selection of TE and TR for Optimizing Contrast

The prototype capillaries loaded with nanoparticles were discriminated on MRI image by varying TE or TR scan parameters at a time and keeping constant TE 200 msec and TR 500, 750, 1000, 1500, 2000 msec (for T2 weighted) or TR 500 msec and TE 5, 10, 15, 20, 25 msec (for T1 weighted) value. T2-w imaging generated good contrast between antitroponin loaded with nanoparticles tube and free troponin loaded tube.

2.5. MRI Microimaging of Heart and Prototype Tube

The superparamagnetic contrast agent was injected into animal tail. For prototype, a polymeric tube that prototyped a vascular vein, free troponin T was injected in a different tube under the same condition. The tube cap was tightened down to exactly 11 mm so as to fit snugly in the imaging probe head of a 11.7 T superconducting magnet (35 mm bore). Transverse slices of 1 mm thickness was obtained using a select gradient of 4.65 G/cm and readout gradient of 2.40 G/cm. In order to standardize troponin T encapsulated superparamagnetic image contrast, scan parameters were typically optimized at TE=30 ms; TR=500 ms; a 512 x 256 matrix zero-filled to 2048 x 2048 [6].

2.6. Characterization, Detection of and Measurement of Cardiac Muscle Components

After MRI images were acquired these images were analyzed to visualize different cardiac features. Edge detection and 4D segmentation techniques were used to demarcate the cardiac tributaries. Delineation of cardiac feature mass and cardiac features was extracted out by manual delineation methods of edge detection or thresholding. The measurement of deformity and curves of cardiac structures were analyzed by texture analysis [14-16].

3. Results

By physical method, morphology and sizes of all the particles were determined by scanning transmission electron microscopy. Fig 1 shows the size of typical nanoparticles by TEM micrograph. The particles were measured mostly in the size range of 50-200 nm. The magnetoferritin and gadolinium complex is shown in figure 2. The synthesized complexes at different stages are shown in figure 3. The magnetoferritin and gadolinium contribute in contrast enhancement as shown in figure 4. Earlier we synthesized iron oxides particles and particles were functionalized with antimyoglobin as mouse heart tissue contrast agent [17]. Fig. 5 shows MRI image of capillary filled with iron oxide and mice heart image without contrast agent. The particles served to enhance the contrast of vascular wall region after the injection of superparamagnetic iron oxide contrast agent.

Encapsulated superparamagnetic particles have been reported for biomedical applications in the areas of immunology, cell separation and drug delivery. From our lab, earlier reports developed a line of superparamagnetic micro and nanoparticles coated with albumin, polyethylene, polypropylene or polystyrene for biomedical applications [18].

3.1. The Polymer Coated Superparamagnetic Biotinylated Antitroponint Microspheres as Image Contrast

The microspheres appear as multilayered tiny complex by electron microscopy. The inner most layer is iron oxide coated with biotinylated avidin-polymer coupled with antitroponin on outermost surface as shown in Fig 3. The surface of the polymer coating was functionalized with a site-specific troponinT antibody to target the cardiac muscle specific troponinT in a heterogeneous heart tissue. The magnetic character of iron oxide generates magnetic moment in high magnetic field and dephases the proton spins in heart tissues. The dephased proton MRI signal loss creates image contrast to target trponinT leakage from heart. However, the nanoparticle complex surface antibody binds with specific troponinT from the cardiac tissue but quantification of troponinT leakage and interpretation is still remains a challenge.

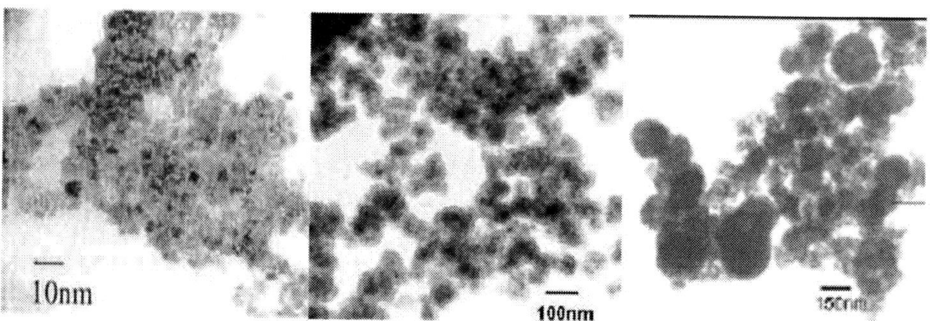

Figure 3. The figure represents iron oxide particles (panel on left); biotinylated avidin coupled polymer complex(panel in center); antibody labeled polymer surface as multilayered complex microsphere.

3.2. Gadolinium Enhanced T1 Weighted Imaging of Apoferritin-Gadolinium Complex

In preliminary experiments, pH drop and monomerization was associated with marked loss of turbidity of aopferritin solution. After pH elevation the solution again became turbid. Following dialysis to remove $GdHPDO_3A$ that was free in solution, the apoferritin was introduced to an MR phantom. Figure 4 shows the T1 maps of apoferritin (alone) (1030 $\pm$ 20 ms), GdII-apoferritin (75.5 $\pm$ 1.3 ms) and 100 mM $GdHPDO_3A$ (17.5 $\pm$ 2.1 ms). The gadolinium generates T1 signal loss as source of enhanced contrast as shown in figure 4.

If the apoferritin proves to be useful contrast vehicle, it may be synthesized through recombinant techniques and furthermore to incorporate a fluorescent tag to aid histological analysis.

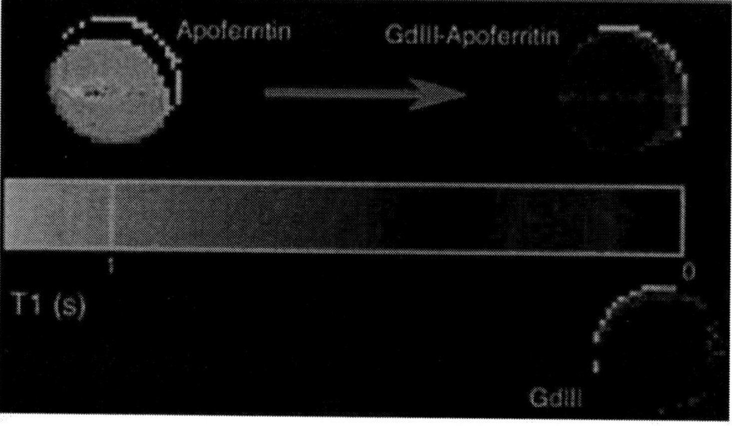

Figure 4. Quantitative T1 maps with pseudocolor obtained from solutions imaged in glass MR tubes. The tubes demonstrate a marked T1 shortening effect by loading GdHPDO3A (Prohance) into apoferritin. The samples on lower right contains 100 mM Prohance only. After loading, the apoferritin was extensively dialysed to remove free Prohance from solution, ensuring that the observed increase in relaxivity was due to GDIII-apoferritin complexes.

3.3. Characterization of Cardiac Troponin T/I/C Subtypes and MALDI Imaging

To validate the human troponin constitution, MALDI spectroscopy and Western blotting showed distinct troponin T/I/C subunits in the human cardiac troponin with specific protein mass/charge groups(data not shown). Troponin isotypes T,C and I were distinct by 2D electrophoresis (Western Blotting) and showed three distinct regions on MS spectroscopy [19]. These proteins suggested strongly their role in calcium mediated tropomyosin regulation by active peptides rich in Lys and Asp amino acids in troponin. However, antitroponin antibody light fragments showed compatibility to these peptides but it is not understood or information not available. The MALDI-MS technique showed cardiac tissue slice image as shown in figure 5. However, troponin rich sites could not be confirmed and mass spectroscopy visualized protein peaks in range (data not shown). NMR spectroscopy may elucidate these intricacies to some extent.

3.4. High Resolution Magnetic Resonance Imaging of Nanoparticles and Post-Injected Mouse Heart

For microimaging experiments at 11.7 Tesla magnetic field strength, the pixel resolution was achieved up to 25 micron as shown in figure 5. The concentration and inversion time of 180 degree pulse in inversion recovery method were determinant factors of T1 weighted MR signal intensity of nanoparticle images as shown in figure 5. The iron oxide caused MR signal loss due to dephasing while gadolinium enhanced the signal by shortening T1 delay time as shown in figure 5. To validate the MR signal characteristics of nanoparticle complexes, the prototype tubes filled with paramagnetic iron oxide bound and gadolinium bound polymer coated biotinylated antitroponin complexes were imaged as shown in figure 5.

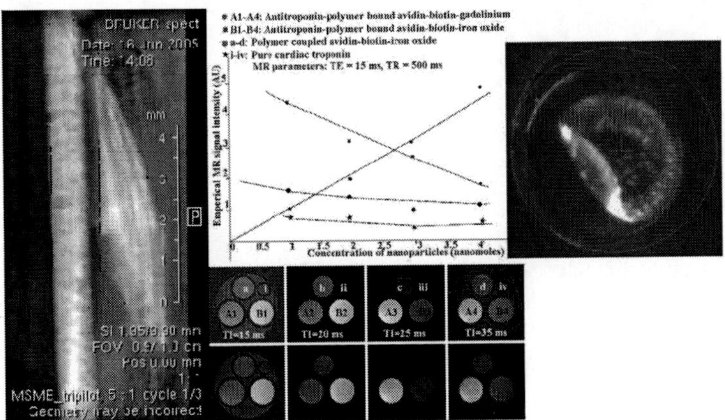

Figure 5. (Panel on left) An T1 weighted MR image of capillary filled with nanoparticle complex. (Panel in center) The different concentrations of nanoparticles filled in capillaries display MR signal intensities. Notice the Gadolinium bound complex showed signal enhancement (A1-A4) while iron oxide showed dephased MR signal loss at increased concentrations (B1-B4). (Panel on right) The heart slice was placed on MALDI plate and heart image was acquired during MS spectroscopy experiments. The MS spectroscopy data of proteins are reported in elsewhere.

The microimaging of injected nanoparticles in mouse distinguished the heart and vascular territories as shown in figure 6. Magnetic resonance imaging had the ability to quantify cardiac tributaries in vivo to measure the wall thickness at any angle. The heart image predicts the different segments at different levels as shown in figure 6 with marked levels. The troponin leakage sites of tissue were likely to show distinct MR signal changes caused by accumulation of antitroponin nanocomplexes at troponin rich regions. At troponin rich regions iron-oxide or gadolinium in the center of nanocomplex caused dephasing or enhancement of MR signal. However, limitations of poor signal-to-noise ratio precluded the use of in vivo techniques directly to an in vivo setting. To overcome the poor signal, the present study showed the signal enhancement by nanoparticles with better contrast against molecular and cellular cardiac targets.

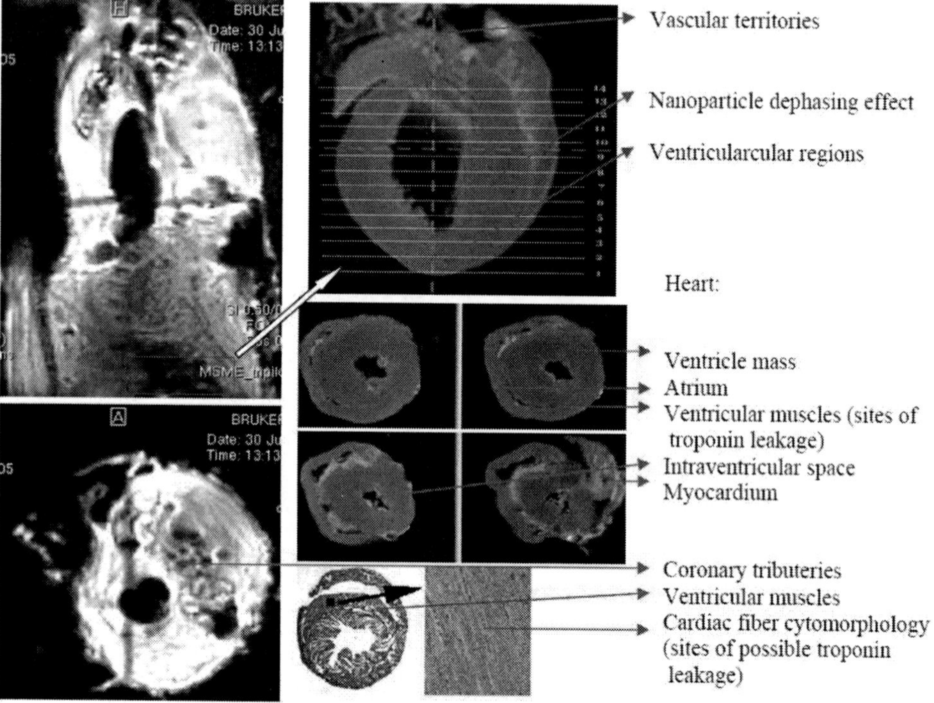

Figure 6. (on left) The figure shows nanoparticle injected in vivo animal heart (shown with arrow) with distinct tributaries on sagittal plane (on top) and axial plane (on bottom). In axial plane, these tributaries show nanoparticle dephased blood as dark and walls as brighter. (on right) a axial series of mice heart images is shown with distinct ventricular wall as heavy musculature mass. The lower panels show the matched histology details. Notice the measurable cardiac wall of left ventricle (on left) and details of muscular mass shown with arrow in the insert on right.

3.5. Characterization, Detection of and Measurement of Cardiac Muscle Components

To validate the measurement accuracy of MRI visible cardiac features on images, the image features were matched with excised tissue histology as shown in figure 6. Histological-

MRI feature comparison by applying coregistration of histologic digital images with MRI images using fiduciary markers showed prominent features visible on both histology and MRI images as shown in figure 6. By using pixel-by-pixel match of different regions in cardiac territories, cardiac mass showed extracted out shapes of cardiac features as shown in figure 7. The texture analysis[3] measured the delineation of margin as possible wall deformity or subtle curvatures by using occurrence matrix (vector of two voxel intensities) to evaluate contrast, correlation, homogeneity and the entropy. This matrix specified the scale and orientation in texture anisotropy analysis. Other approach of gradient density matrix by convolution calculated the intensity gradient vector in cylindrical polar coordinates as shown in figure 7.

The shape analysis of cardiac features showed the cardiac tissue shapes as determined by intuitive measures using hypothesis of compactness, eccentricity and rectangularity. The alternate statistical shape analysis by spatial configuration variation and deformation analysis by volumetric variation also predicted the in shape variations in position such as geometry based transformation. The shape = surface area/volume$^{2/3}$.

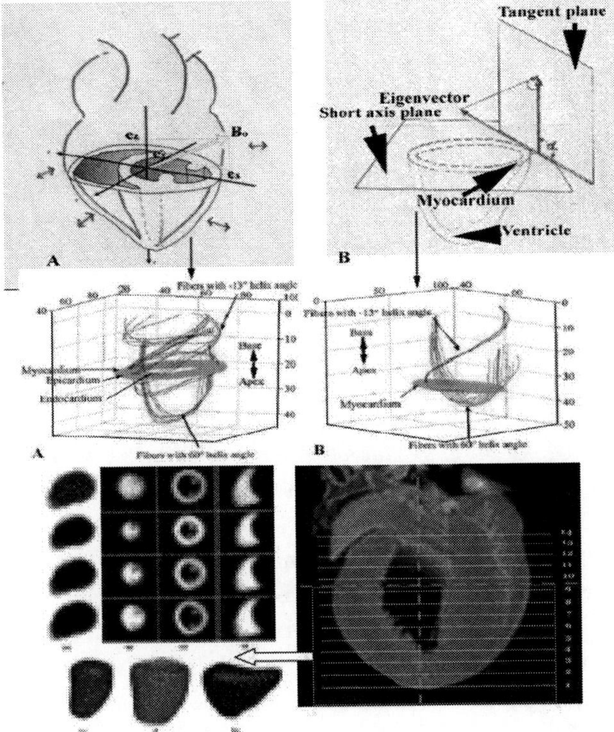

Figure 7. (on top row) The sketch of heart is shown to visualize cardiac wall evaluation method and cardiac fiber orientation analysis (panel A). The eigenvector of myocardium are displayed in tangent and axis planes (panel B). (in mid row) The cardiac fiber orientation is shown to calculate fiber arrangement (panels A and B). Modified from ref 3. (at bottom row on left) The calculated shapes of heart are shown with wall at different planes. The mouse excised heart T1 weighted image is shown at vertical plane at different levels. Modified from ref 14. (at bottom on right) The MRI visualization of heart wall muscles improved arrangement by using antitroponin paramagnetic particles at different planes. Modified from ref 14.

[3] The image analysis is adopted from our other data on rat heart image data analysis.

4. Discussion

The biotin protein serves as the bridging link between the magnetic nanospheres with antibodies and the marker enzyme for developing ELISA technique. The biotin-avidin system for conjugation of molecular labels to antibodies offers a convenient economic use. In our lab, we chose antibodies as two complimentary clones of monoclonal mouse antihuman cardiac troponinT (T6077 Sigma) [20]. The avidin was coupled with superparamagentic microspheres and these microspheres were also used to develop ELISA assay using streptavidin coupled enzyme marker by biotinylation of the antibodies and incubate them with the avidin and streptavidin coupled labels [21].

Strepatavidin is a terameric protein with four binding sites that is similar to Avidin in its molecular structure and has a similar binding capacity. (Ka = 10^{15}/M). The antibody biotinylation was done using Sulfo-NHS-LC-biotinylation kit (Pierce Chemicals). Pierce offered two biotinylated kits: Biotin with long chain arm (22 A spacer arm between primary amine and biotin moiety) and standard Biotin [21]. The long chain arm biotin couples with much larger paramagnetic microspheres and enhances the sensitivity. After biotinylation of antibodies and microspheres coupled with avidin, the antibody provides troponin sensitive biomarker labels on surface. For ELISA, the addition of human serum albumin (HSA) to the antibody-label conjugates to block all other possible binding sites on biotin and/or Avidin in both the microsphere-antibody conjugate and on the Alkaline Phosphatase-Antibody conjugate. Our focus was microspheres in imaging application.

Two major objectives were accomplished: development of antitroponin based nanoparticles as MRI imaging contrast agents; and characterization, detection of and measurement of cardiac muscle components. First aim was to synthesize magnetic antitroponin nanoparticles with size distribution less than 25 nm as imaging contrast agent. Antitroponin based magnetic nanoparticles had the advantage of being less expensive and more specific to cardiac muscle and easily characterized.

Production of antitroponin nanoscale magnetic particles with a distribution of less than 25 nm. The following parameters of the magnetic nanoparticles are critical: (a) Specificity of Troponin T with cardiac muscle and their particle size (small as possible to improve tissular diffusion, and to have long sedimentation times and high effective surface areas), (b) Troponin specific surface characteristics (easy encapsulation of the magnetic nanoparticles protects them from degradation and endows biocompatibility), and (c) rapid and easily transportable nanoparticles at the target site after injection.

The major challenge in the synthesis of nanoparticles was to control of its size. The difficulty arises as a result of the high surface energy of these particles. The interfacial tension acts as the driving force for spontaneously reducing the particle surface area by growing it during the initial steps of the precipitation, and during aging of particle [22]. Therefore, the synthetic routes are able to produce magnetic troponin T nanoparticles of desired size without particle aggregation as MRI contrast agents.

For the imaging application superparamagnetic particles of uniform shape were needed. The superparamagnetic nanoparticles usually are made of maghemite Fe_2O_3. These particles have shown a superparamagnetic behavior when synthesized in sizes less than 25 nm. Production of uniform sized particles was necessary for constant reaction kinetics between the spheres and the desired molecules. The size of particles ensured the rapid and efficient physical or chemical reactions in a solution. Various reports suggested that true spherical magnetic nanoparticles minimize the chemical agglutination and offered the efficient use of the magnetic particles' surface for functionalization [22].

There has been a substantial interest in preparing monodisperse polymer particle preparation of monodispersed polystyrene particles[23]. The micron sized particles were generated by homogenization of monomer droplet in the presence of a stabilizer and subsequent polymerization of the monomer. However, the monodispersity of particles was poor because of inherent size distribution due to mechanical homogenization. Ugelstad et al [24] derived a two-step swelling method to achieve monodispersity. In this method polymer seed particles were synthesized by surfactant free emulsion polymerization and subsequent swelling by emulsion of low molecular weight compound, which diffuses from the emulsion droplet to the polymer particles. The particle size reaches up to 50 μm and the process was also tedious. Major factors to affect the particle size were the monomer concentrations, the initiator and surfactant used, the stirring speed, and the polymerization period. A second approach of dispersion polymerization was proposed to achieve a very narrow distribution of particle size. This process involved the polymerization of a monomer dissolved in a solvent in presence of the graft copolymer (or its precursor). It acts as a steric stabilizer and prevents the aggregation of the formed insoluble polymer particles. The particles can be prepared in the range of micron and submicron sizes.

By using a microemulsion technique it was possible to have particles in uniform size range in nanometer range (10 nm-100 nm) [25]. In this process the monomer was dispersed in water using a surfactant (e.g. SDS, CTAB) and a co-surfactant and then subsequently polymerized by means of water or oil soluble initiator to produce stable latex. The droplet size was thermodynamically controlled by the amount and properties of surfactant. We propose to produce polymer composite particles with a size distribution less than 100 nm using inverse microemulsion polymerization [25] and also using emulsion and precipitation polymerization [26] and ultrasonication. Ultrasonication and surface treatment with a surfactant generate the particles of uniform sizes. In addition, it has been shown that by adjusting the pH and the ionic strength of the precipitation medium, it was possible to control the mean size of the particles over one order of magnitude (from 15 to 20 nm) [27].

Encapsulation of the magnetic particles in an application-suitable surfactant is significant due to their potential toxicity or to functionalize the surface for facilitating the uploading of the particles into the nanoparticle complex. At present, HSA-encapsulated particles have been functionalized for cell separation and diagnosis in our lab. Polymers (e.g. polystyrene, polyethylene, polymethyl-metacrylate, polyvinyl-tolouene, hydroxy-propyl-cellulose) and copolymers have been used by the investigators but never been functionalized. Other materials could also be used to produce the spheres such as silica and glass, both of which offer high temperature stability and minimum swelling in solution. In addition, porous glass spheres are sometimes used to create a large surface area for adsorption or reaction. The surface of the surfactant is modified to suit the target compound. A large number of functional groups have been introduced to obtain a variety of linkages for attachments. Particle coating is based on either covalent bonding or physical surface adsorption of a protein or ligand onto the surface of the particle. This process required the amount of protein or ligands needed for coating based on the particle size [28]. Most covalent coupling applications start with a monolayer of protein coated on the nanoparticles to ensure correct spatial orientation and decrease in the likelihood of nonspecific binding. This monolayer amount can be derived from the following equation

$$S = \frac{6C}{\rho D}$$

where S is the amount of coating protein needed to achieve surface saturation (mg protein/g of nanoparticle), C is the capacity of nanoparticles surface for a given protein which will depend on the size and molecular weight of the protein to be coupled (mg protein/m^2 of polymer surface), 6/ρD is the surface area/mass(m^2 /g) for nanoparticles of a given diameter, ρ is the density of the nanoparticle, and D is the diameter of the particle. The avidin-coated beads were able to capture biotin moiety via avidin-biotin binding. To achieve reproducible results, uniform size and shape were required to provide consistent and uniform physical and chemical properties.

Magnetic Resonance Imaging on a prototyped tissue troponin offers a new Magnetic Resonance Imaging (MRI) technology of nanoparticles based imaging modality of non invasive cardiac intracellular events. It generated MRI signal based on the difference of magnetization of surrounding protons due to dephasing by nanoparticles in high magnetic fields subjected to a specific radiofrequency pulse sequence. The levels of magnetizations at different gradient levels create MRI image were reported with different contrasts [29]. Achievement of high and steady magnetic fields has enabled the accurate MRI tissue characterization of pathological areas at high resolution. Different types of superparamagnetic particles are shown as contrast agents to enhance the MRI signal. The iron oxide and gadolinium as imaging contrast agents in nanoparticles display reasonable specificity. Utilizing the cardiac specific troponin T antibody as contrast agent carrier improved the troponin specificity. A model capillary made out of polymeric tubing was utilized in the feasibility MRI evaluation. Both antitroponin T loaded solution with nanoparticles and troponin free solution were injected into the polymeric tubing in the presence of buffer solution. Multiple complementary methods were used to pursue the imaging.

Selection of TE and TR for optimizing contrast in prototype capillaries loaded with nanoparticles was dependent on TE or TR scan parameters. At constant TE 200 msec and variable TR 500, 750, 1000, 1500, 2000 msec (for T2 weighted) or TR 500 msec and TE 5, 10, 15, 20, 25 msec (for T1 weighted) generated good contrast between antitroponin loaded with nanoparticles tube and free troponin loaded tube. It corroborates with MRI principles of image contrast [30].

Magnetic Resonance Microimaging (mMRI) technique at optimal scan parameters, number of averages=128 for proton weighted images was optimal to get enough MRI T1-wt signal. Images were rendered using VNMR software and an Epson printer. Acquisition of images *ex vivo* at 11.7 T utilized a special lab built holder. The pixel resolution enhancement was achieved 256 x 256 matrix size with slice thickness 0.4 mm to get 20 microns high resolution and 0.5 mm^3 volume of interest comparable with earlier report [31].

Characterization, detection and measurement of cardiac troponin rich regions was done using prototyped tissue troponin. The analysis was done for segmentation by gray level thresholding, troponin orientation and fiber movement by filtered magnitude and directional assessment of intensity gradients. The approach agrees with earlier reports [32, 33]. The fiber orientation was performed by gradient vector providing a 3x3 matrix of each voxel. The resulting vector represented the outcome of filtered troponin motion components using Guassian karnel appearing as beam like fiber orientation in tube while direction corresponds to eigen values in earlier reports [32, 33].

5. Limitations and Challenges of the Study

The superparamagnetic particle synthesis is a challenge to keep functionality on polymer surface with active antibody specific to troponinT. Another limitation was avidin-biotin binding with iron-oxide as size of iron oxide is very crucial to generate dephasing effect and MRI contrast. It is not always necessary that iron-oxide size and antitroponinT specificity will represent the linear correlation as shown in present study. It needs a careful evaluation of magnetic behavior of nanoparticle in imaging application.

6. The Futuristic View of Art

The nanoparticle complex delivery is also an issue of clinical safety. The safe delivery without change in antitroponin lebels on polymer coupled avidin-biotin gadolinium or iron oxide complex is crucial. In future, several biocompatible polymer or alternate safe coating will enhance the possibility of antitroponin delivery at target. The present state of art using troponin leakage as cause of cardiac damage and its MRI interpretation in defining disease is emerging as clinical modality. However, lot of information remains unconfirmed and it needs further investigations as emphasized in recent reports [34-36].

7. Conclusion

The troponin T is major cardiac functional protein biomarker. The nanoparticles iron-oxide and gadolinium based microimaging methods provide better details of tissue for proteins and fiber orientation. MALDI imaging is a new method of protein localization and confirmation of damage. The heart image analysis is significant to predict cardiac wall damage and chamber shape.

Acknowledgments

The author acknowledges the facility of FSU center of nanomagnetics and biotechnology through cornerstone grant award to Dean CJ Chen through years 2002-2006.

Author acknowledges the facility support of NMR microimaging at Chemical-Biomedical Engineering Department of FAMU-FSU College of Engineering, Tallahassee.

References

[1] Sharma R. Search of Troponins as Smart Biosensors for Point-of-care Detection of Acute Myocardial Infarction. Proceedings of NSTI 2007 meeting at Boston, MA.Abstract 1222. http://www.nsti.org/Nanotech2007/showabstract.html?absno=1222

[2] Latini R, Masson S Diagnostic and prognostic value of troponin T in patients with chronic and symptomatic heart failure. *Recenti. Prog. Med.* 2008;99(10):505-8.

[3] Wu EX, Tang H, Jensen JH. Applications of ultrasmall superparamagnetic iron oxide contrast agents in the MR study of animal models. *NMR* 2004;17(7):478-83.

[4] Smirnov P, Gazeau F, Beloeil JC, Doan BT, Wilhelm C, Gillet B Single-cell detection by gradient echo 9.4 T MRI: a parametric study. *Contrast Media Mol. Imaging.* 2006;1(4):165-74

[5] Bulte JW, Douglas T, Mann S, Frankel RB, Moskowitz BM, Brooks RA, Baumgarner CD, Vymazal J, Strub MP, Frank JA. Magnetoferritin: characterization of a novel superparamagnetic MR contrast agent. *J. Magn. Reson. Imaging.* 1994 May-Jun;4(3):497-505.

[6] Sharma R, Kwon S. New Applications of Nanoparticles in Cardiovascular Imaging. *J. Exp. Neurosci.* 2007; 2(1 & 2): 115-126.

[7] Lanza GM, Winter P, Caruthers S, Schmeider A, Crowder K, Morawski A, Zhang H, Scott MJ, Wickline SA. Novel paramagnetic contrast agents for molecular imaging and targeted drug delivery. *Curr. Pharm. Biotechnol.* 2004; 5(6):495-507.

[8] Farrar CT, Dai G, Novikov M, Rosenzweig A, Weissleder R, Rosen BR, Sosnovik DE. Impact of field strength and iron oxide nanoparticle concentration on the linearity and diagnostic accuracy of off-resonance imaging. *NMR Biomed.* 2007; 21(5):453 – 463.

[9] Heim N, Garaschuk O, Friedrich MW._Improved calcium imaging in transgenic mice expressing a troponin C–based biosensor. *Nature Methods.* 2007, 4: 127 – 129.

[10] Zhang S, Bian Z, Gu C, Zhang Y, He S, Ning Gu N, Zhang J. Preparation *Colloids Surf. B. Biointerfaces.* 2007, 55(2):143-148.

[11] Dengler TJ, Zimmermann R, Braun K. Müller-Bardorff M, Zehelein J, Sack FU, Schnabel PA, Kübler W, Katus HA. Elevated serum concentrations of cardiac troponin T in acute allograft rejection after human heart transplantation. *J. Am. Coll. Cardiol.* 1998; 32:405-412.

[12] Farina L, Iacobello C, Orlandini S, Ius A, Albertini A. The effect of the concentration ratio of avidin and biotin on a single step sandwich enzyme immunoassay. *Ann. Clin. Biochem.* 1993;30 (1):87-9.

[13] Labugger R, Simpson JA, Quick M, Brown HA, Collier CE, Neverova I, Van Eyk JE. Strategy for analyais of cardiac troponins in biological samples with a combination of affinity chromatography and mass spectrometry. *Clin. Chem.* 2003;49(6):873-879.

[14] Feintuch A, Zhu Y, Bishop J, Davidson L, Dazai J, Bruneau BG, Henkelman RM. 4D cardiac MRI *NMR Biomed.* 2007 May;20(3):360-5.

[15] Säring D, Ehrhardt J, Stork A, Bansmann MP, Lund GK, Handels H._Computer-assisted analysis of 4D cardiac MR image *Methods Inf. Med.* 2006;45(4):377-83.

[16] Lorenzo-Valdés M, Sanchez-Ortiz GI, Elkington AG, Mohiaddin RH, Rueckert D. Segmentation of 4D cardiac MR images *Med. Image Anal.* 2004; 8(3):255-65.

[17] Sharma R, Haik Y, Chen CJ. Superparamagnetic iron oxide-myoglobin as potential nanoparticle: Iron Oxide magnetic resonance imaging marker in mouse imaging. *Journal of Experimental Nanoscience.*2007; 2(2) 127-138.

[18] Chatterjee J, Haik Y, Chen CJ. Modification and characterization of polystyrene based magnetic microspheres and its comparison with albumin based magnetic microspheres. *J. Mag. Mag. Mat.* 2001, 225:21-29.

[19] Labugger R, McDonough JL, Neverova I, Van Eyk JE. Solubilization, two-dimensional separation and detection of the cardiac myofilament protein troponin T. *Proteomics.* 2002, 2, 673–678.

[20] Kajandar E, Aho K., Ciftioglu N. Detection of calcifying nanoparticles and associated proteins thereon and correlation to disease. World Intellectual Property Organization. WO2007070021.http://www.wipo.int/pctdb/en/wo.jsp?IA=WO2007070021&wo=2007070021&DISPLAY=DESC

[21] Dutra RF, Kubota LT. An SPR immunosensor for human cardiac troponin T using specific binding avidin to biotin at carboxymethyldextran-modified gold chip. *Clin. Chim. Acta.* 2007;376(1-2):114-20.

[22] Shmilovich H, Danon A, Binah O, Roth A, Chen G, Wexler D, Keren G, George J. Autoantibodies to cardiac troponin I in patients with idiopathic dilated and ischemic cardiomyopathy. *Int. J. Cardiol.* 2007;117(2):198-203.

[23] Kobayashi S, Uyama H, Lee SW, Matsumoto Y. Preparation of micron-size monodisperse polymer particles by dispersion copolymerization of styrene with poly(2-oxazoline) macromonomer. *Journal of Polymer Science Part A: Polymer Chemistry.* 2003; 31(12):3133 – 3139.

[24] Ugelstad J, Berge A, Ellingsen T, Schmid R, Nilsen TN, Mørk PC, Stenstad P, Hornes E, Olsvik O,. Preparation and application of new monosized polymer particles. *Prog. Polym. Sci.* 1992; 17, 87 - 161.

[25] López-Quintela MA. Synthesis of nanomaterials in microemulsions: formation mechanisms and growth control. *Current Opinion in Colloid & Interface Science.* 2003;8(2): 137-144.

[26] Chatterjee J, Haik Y, Chen CJ. Polyethylene magnetic nanoparticle: a new magnetic material for biomedical applications. *Journal of Magnetism and Magnetic Materials.* 2002;246(3): 382-391.

[27] Shi G, Rouabhia M, Wang Z, Dao LH, Zhang Z. A novel electrically conductive and biodegradable composite made of polypyrrole nanoparticles and polylactide. *Biomaterials.* 2004; 25(130:2477-2788.

[28] Ye A, Flanagan I, Singh H. Formation of stable nanoparticle via electrostatic complexation between sodium caseinate and gum Arabic. *Biopolymers.* 2006;82(2):121-123.masse S, Laurent G,

[29] Chubaru F, Cadiou C, Dechamp I, Coradin T. Modification of the strober pocess by a polyaza macrocycle leading to unusual core shell silica nanoparticles. *Langmuir.* 2008; 24(8): 4026-31.

[30] Cho JH, Nyugen FT, Barone PW, Heller DA, Moll AE, Patel D, Boppart SA, Strano MS. Multimodal biomedical imaging with asymmetric single walled CNT/iron oxide nanoparticle complexes. *Nano Latters.* 2007;7(4):861-867.

[31] Hadjipanayis CG, Bonder MJ, Balakrishna S, Xiaoxia Wang X, Hui Mao H, Hadjipanayis GC. *Metallic Iron Nanoparticles for MRI Contrast Enhancement and Local Hyperthermia.* 2008;_4 (11):1925 – 1929.

[32] Sharma R. MRI microimaging of rat skin at 900 MHz. *Magnetic Resonance Imaging.* 26(9):1-13.

[33] Arai AE. False positive or true positive troponin in patients presenting with chest pain but 'normal' coronary arteries: lessons from cardiac. *MRI. EHJ.* 2007 28: 1175-1177.

[34] Selvanayagam JB, Pigott D, Balacumaraswami L, Petersen SE, neubauer S, Taggart DP. Relationship of irreversible myocardial injury to troponin I and creatine kinase-MB elevation after coronary artery bypass surgery: Insights from cardiovascular magnetic resonance imaging. *J. Am. Coll. Cardiol.* 2005; 45:629-631.

[35] Schwaiger M. Diagnostic value of contrast-enhanced magnetic resonance imaging and single-photon emission computed tomography for detection of myocardial necrosis early after acute myocardial infarction. J. Am. Coll. Cardio. 2007;49(2):208-216.

[36] Gallegos R, Swingen C, Xu X, Wang X, Bianco R, Jerosch-Herold M, Bolman R. Infarct extent by MRI corelates with peak serum troponin level in the canine model12. Journal of Surgical Research. Volume 120 , Issue 2 , Pages 266 – 271.

[37] Sommer T, Naehle CP, Yang A,Zeijlemaker V, Hackenbroch M, Schmiedel A, Meyer C, Strach K, Skowasch D, Vahlhaus C, Litt H, Schild H. Strategy for safe performance of extrathoracic magnetic resonance imaging at 1.5 Tesla in the presence of cardiac pacemakers in non-pacemaker-dependent patients. A prospective study with 115 Examinations. *Circulation.* 2006;114:1285-1292.

In: Advances in Medicine and Biology. Volume 20
Editor: Leon V. Berhardt
ISBN 978-1-61209-135-8

Chapter 2

TROPONIN T: NEWER MAGNETIC IMMUNOASSAY METHOD OF TROPONINS AS POINT-OF-CARE DETECTION OF ACUTE MYOCARDIAL INFARCTION

Rakesh Sharma*

[1] Center of Nanomagnetics and Biotechnology,
Florida State University, Tallahassee, FL 32310 USA

Abstract

Troponin molecules and their T,C and I subtypes have shown significance in early point-of-care assessment of acute myocardial infarction. More sensitive and rapid detection magnetic immunoassay methods are emerging based on polymer bound enzymes for lab use and early point-of-care. *Hypothesis:* The magnetic particles in immunoassay enhance the utility of immunoassay by their ability to facilitate the separation of the targeted compounds and their high sensitivity of analyte. The basic principle of antitroponin-magnetic immunoassay was based on formation of sandwitch complex of two different antibodies at two binding sites on same target troponinT antigen (one antibody site was attached with magnetic nanoparticles, other antibody binding site attached with glucose in equal proportion for its detection by glucose oxidase-peroxidase enzyme measurement. *Materials and Methods:* The concentration of troponinT in the blood sample was correlated with proportional concentration of glucose measured by glucometer. The present report presents an innovative idea to explore the newer possibility of: 1. Troponin subtyping by MALDI mass spectrometry technique; 2. Magnetic immunoassay in serum sandwitch magnetic immunoassay for both solid phase Enzyme Linked Immunosorbent Assay (ELISA) in clinical chemistry lab and point-of-care miniature pencil (POCM) in quick emergency. *Results and Discussion:* The

* Email: rksz2004@yahoo.com

application of nanotechnology in troponin analysis minimized the artifacts of troponin analysis in AMI. Using site-specific agent such as a magnetic nanosphere is novel choice to estimate 1 pico gm/L. In addition, the magnetic nanoparticles enhance the analyte detectibility, sensitivity to analyte with optical reaction speed. The nanospheres could be extracted out by using external magnet. The technique provides simple point-of-care method with a method to use proteomics and magnetic nanoparticles to isolate and quantify AMI marker troponin in the blood sample with use of miniature glucometer. The MALDI mass spectrometry technique showed potential of exploring AMI sensitive Troponins with other possible proteins. The concentration of troponinT in blood sample was correlated with proportional concentration of glucose measured by pencil miniature glucometer. Conclusion: The magnetic immunoassay of point-of-care method shows potential as rapid reliable cost-effective and efficient as cardiac protection device. The technique is simple, user friendly and cost-effective. The nanotechnology based glucometer may provide portable miniature point-of-care testing at low cost.

Keywords: Troponin, antitroponin, acute myocardial infarction, MRI, magnetic immunoassay, imaging, cardiac muscle.

1. Introduction

Troponin-T (TpT) release was first reported as quickest within hours and known as most sensitive marker of early diagnosis of AMI and myocardial necrosis [1]. Other slowly released proteins are Creatinf Kinase-MB, fatty acid-binding protein and myoglobin. Troponin has different subtypes T, I and possible other subtypes [2]. Conventional antibody-antigen Enzyme Linked Immunosorbent Assay (ELISA) immunoassays do measure these markers and need laboratory setting [3]. So, these lab methods are not suitable for point-of-care rapid detection in emergency. The magnetic particles in immunoassay enhance both the utility and specificity of immunoassay. The magnetic particles facilitate the separation of the targeted troponin bound complexes in serum and the particle concentrations are highly sensitive to analyte activity (magnetic particle bound biomarker enzyme). In research lab, two analytical approaches enhanced the sensitivity: first, troponin subtype proteomics by MALDI mass spectrometry technique; second, developing solid-phase Enzyme linked Immunosorbent Assay (ELISA) based on formation of "sandwich" complex of two different antibodies at two binding sites on same target troponin antigen [4]. In this complex, one antibody is attached with magnetic nanoparticles at one binding site and other antibody with enzyme or fluorescent at other binding site. The magnetic assays were reported to measure concentration of troponinT in the blood sample based on the glucose oxidase-peroxidase enzyme lebel attached. TroponinT can be correlated with concentration of glucose measured by miniature glucometer[5]. We reported that technique may provide simple point-of-care method to improve detection of silent MI. The troponinT marker assay at point-of-care miniature (POCM) was reported as an accurate and timely evaluation of patients in ambulance or emergency room [6, 7]. Current belief suggests the ability of troponinT and other markers to determine the AMI soon after infarction.

The troponinT (TpT) is a single peptide chain (17kDa) as a small cardiac protein that shows the elevated serum levels soon after the infarction. The levels of plasma troponinT rise about 1.5-2.5 μg/L. Still it is believed that TpT is most specific to cardiac injury while other troponin C and I subtypes are also indicators of myocardial damage [8, 9]. Later emphasis was diverted to monitor troponin release by ELISA at different time intervals with high accuracy. Recent studies suggested that troponinT measurement allows quick discrimination of myocardial tissue damage within 4-6 hours [10, 11]. Recently, immunoassays have emerged as potential quantitative methods to measure troponin in tissues based on specific binding of troponin antigen with homologous antibodies [11]. The commonly used methods for cardiac markers are radioimmunoassay, latex agglutination and two-site immunoassay techniques [12, 13]. However, these methods are time consuming 3-4 hours and require special analytical instruments that limit their usefulness in point-of-care small facility in emergencies. Other factor is cost-effective rapid decision-making with efficient evaluation of disease progress and rapid treatment of AMI in emergency rooms of hospitals. Point-of-care devices allow diagnostic assays immediate result at the site of patient care delivery in emergency departments or intensive care unit [14, 15, 16]. Most of the commercial assays use tubes, wells or plastic beads as immobilized solid phase for either antigen or antibodies with less specificity [17, 18]. The use of magnetic particles in immunological assays has grown considerably. The magnetic properties of nanoparticles permit the easy separation and concentration of these particles in large serum volumes. It allows more accurate and faster troponin assays with improved sensitivity over currently available clinical methods [19, 20, 21].

The use of polymer coated magnetic particles for cell and protein separation has been established at our Center of Nanomagnetics and Biotechnology lab since year 1998. The coated magnetic particles were classified according to their shape, size, composition and surface coating. In biomedical applications the shape of nanoparticles was observed almost exclusively spherical. The size of the beads ranges from nano to micrometers in size of 10 nm- 500 nm. Our laboratory has developed magnetic particles made of iron oxide in the range of 6 nm to 30 nm. These particles exhibited superparamagnetic behavior [6]. The surface properties of these magnetic particles were manipulated by coating them with avidin and biotin protein, polymer, silica. In addition, a large number of functional groups of amines, carboxylic acid, hydroxyl, apoxy, amide, aldehyde, ketone, chloromethyl, sulphate, hydrazide were reported to provide a variety of linkage for attachment on surface of spheres [22].

A novel sandwich magnetic immunoassay technology was developed at center of nanomagnetics and biotechnology to detect minute concentration of cardiac markers in blood [23]. The sandwich magnetic immunoassay used solid-phase enzyme linked immunoassay (ELISA). The sandwich was formed by attachment of two different antibodies to different epitopes on the same target antigen (cardiac marker in this case). One antibody was attached to a solid surface used for the separation of the antigen (cardiac marker) from the blood sample. The second antibody, attached to enzyme molecules, to measure the relative concentration of glucose proportional to the cardiac troponinT marker in blood stream. Attaching glucose molecules at the end of the anti-cardiac marker antibody facilitates the detection of AMI. The glucose is easily measured by photometer at wave length of 560 nm as distinct.

1.2. Synthesis of Nanomagnetic Particles for Biomedical Applications

Encapsulated superparamagnetic iron oxide and gadolinium particles have found novel biomedical applications in the imaging and drug delivery. Research conducted by the research team at our lab has resulted in the development of a line of superparamagnetic microspheres and nanospheres such as polystyrene, polyethylene coated magnetic spheres attached to a red blood cell [24]. To form conjugate, 76 mg avidin coupled iron oxide microsphere solution (10 nanomoles of avidin or 10 mg/mL solution) was mixed with 1 ml biotinylated antibody solution (0.1 mg/mL) or each of 76 mg avidin protein molecules coupled microspheres represent 10 nanomoles of avidin. The ratio of biotin molecules binding on troponin antibody clone (12 biotin groups per antibody needs 10 nanomoles of avidin to couple with 0.1 mg troponin antibody molecules) was calibrated using HABA method. The final solution of troponin antibody and avidin coupled microspheres were incubated by constant mixing at $37^{\circ}C$ for 1.5 hours. Initially avidin and biotin proteins were used as model ligands due to their strong bond-forming ability with the various ligands used in immunoassays. Later, their magnetization measurements displayed the characteristic superparamagnetic behavior of the composite particles for imaging application. Typically, a 40 nM biotin-succinimide derived ester [6-(6-biotinoyl)-amino-hexanoyl-amino-hexanoic acid] or succinimidyl ester was added in dimethylformamide slowly with mixing to an antibody (anti-Troponin solution S-51761 or T6277 Sigma) at 2 mg/ml in the mixture of 50 mM potassium borate and 150 mM sodium chloride at pH 8.2, (TBS) to achieve a final molar ratio of 20:1 biotin-ester: antibody. This solution was incubated at room temperature for 2 hours. Later the solution was dialyzed at 4°C for overnight or 12 hours with 1 ml volume of iron oxide maghemite Fe_3O_4 particles or Gadolinium DTPA particles (10% solid mixture). After dialysis, subsequently the particles were reacted with 9 ml of 0.55 mg/ml avidin-HS protein (Scripps Laboratories, San Diego, Calif.) in 50 mM Tris hydrochloride, 150 mM sodium chloride at pH 7.5. The mixture of nanoparticle:biotin-ester: antibody solution was incubated at 45°C for 2 hours. The mixture was washed 3 times, each with 10 ml TBS, and suspended again in 10 ml TBS. In next step, anti-troponin T antibody was used for two purposes. First, fluorescence labeled with a Cy5 pigment for bioimaging similar to the method reported recently [25]. For another experiment, maghemite nanoparticles (MNPs) were synthesized by chemical coprecipitation and coated with meso-2,3-dimercaptosuccinic acid (HOOC-CH(SH)-CH(SH)-COOH or DMSA).

1.3. Characterization of Antitroponin Based Nanoparticle Complex by Sandwich ELISA

Using a standard protocol developed at our lab, a superparamagnetic nonoparticle complex was developed using iron-oxide in center coated with streptavidin and biotin-antitroponin human antibody IgG1 raised in mouse. The biotinylation was done by synthesizing 40 nM biotin-succinimide derived ester [6-(6-biotinoyl)-amino-hexanoyl-amino-hexanoic acid]. Later, followed by polymer polyethylene encapsulation. The morphology and properties of the nanoparticles were characterized by TEM, XRD, Zeta Potential Analyzer and VSM. Subsequentially, the anti-human cardiac troponin I (cTnI) immunomagnetic nanoparticles (IMNPs) were prepared by grafting anti-human cTnI antibodies on the surface of DMSA-coated MNPs using the linker of EDC (1-ethyl-3-[3-dimethylaminopropyl] carbodiimide hydrochloride). The conjugation amount of the antibodies with protein bound activity of IMNPs was evaluated by enzyme linked immunosorbent assay (ELISA) and Western blotting as described earlier [26]. At center of Nanobioscience, paramagnetic particles using antibody-conjugated contrast agents were prepared.

The results showed that the physical and chemical adsorption occurred at the same time, but the physical adsorption was unstable and showed quick desorption. The maximum conjugation amount of antibody was measured about 96 micrograms on the 0.1 mg MNPs by covalent bond. The stability showed that after 300 days the antibodies on the IMNPs retained the significant biological activity.

1.4. Principle of Nanoparticle in Magnetic Assay

The gadolinium or iron oxide are magnetic metals and can be separated by magnet. The nanoparticles generate magnetic moments in the magnetic field. These complex gadolinium or iron oxide nanoparticles (NP) are bound with biocompatible protein such as avidin or biotin (BP) and coated in polymers such as polyethylene, polystyrene (P). The polymer act as sandwich lebeled with antibody sites: one site for biomarker enzyme such as glucose oxidase-peroxidase(GOD-POD) or horse redox peroxidase (E) and other site for antitroponin as shown in figure 1. The enzyme glucose oxidase-proxidase activity was used in glucose measurement to measure the proportionate concentration of TroponinT in serum.

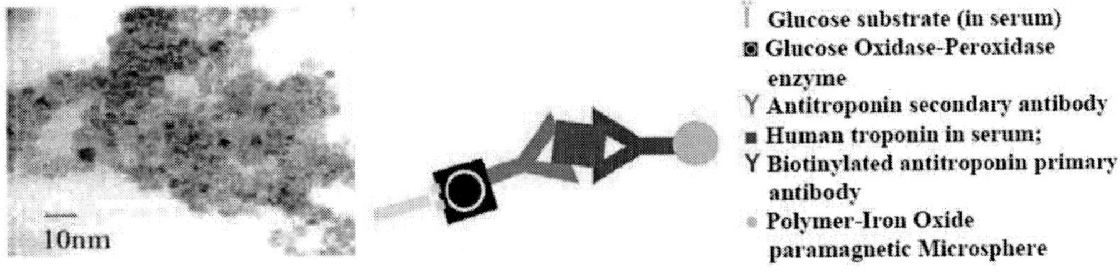

Figure 1. Figure shows the nanoparticles (leftmost panel). The sketch of nanoparticle complex components is represented as “sandwich sites” on antitroponin molecule for biotin and superparamagenetic iron oxide or gadolinium chelating metal (Modified from ASME IMEC Seattle meeting 2008: Presenter: Dr CJ Chen IMECE2007-43617).

2. AMI Detectable Device

1. The proposed detection device composed of glucometer with modified strip design to incorporate the nanoparticles in the glucostrip. The phase I experiments of the development were concentrated on developing reliable experimental data that can be used to develop the testing device.

2.1. The POCM Device

The device meets the American College of Cardiology/American Heart Association guidelines to perform lab assays within 30 minutes. The choice of nanomagnetic particles into

POC device for AMI was based on: (a) The reactants mix thoroughly and interact with each other. (b) Both reactants and products on the spheres are not diluted by buffer or other reagents. Therefore, both the detection limit and the reaction speed are maintained at a level corresponding to the concentration of analytes. (c) Unlike a plug of solution, magnetic spheres are easy to localize so that compounds on the sphere do not diffuse away. (d) it is easy to separate the product on the spheres from undesired materials by wash cycles. (e) The use of nano particles allows the detection of very small amounts of proteins and compounds in the heterogeneous solution. For more information of rapid troponin testing, readers are referred to the recent reports [31-35].

The device utilized nonmagnetic particles to isolate cardiac markers from whole blood thereafter labeling these cardiac markers with a glucose molecule to allow for a glucometer to measure the concentration of the Troponin cardiac marker in the sample. Blood glucose concentration was measured by glucose oxidase-peroxidase system. In this system, the reagent strips contained test area, which was impregnated with enzyme and two metal electrodes. The monitor in which strip was inserted applies a voltage across the electrode, causing the oxidation of mediator. This reaction generates a current proportional to the amount of glucose present and is translated by the meter to the numeric glucose conversion. The proposed device meets almost nine requirements of CLIA and POC diagnostic assays: Validity, reliability, quick and less time consuming, specimen accessibility, simple technique, high efficiency, disposable, and low cost.

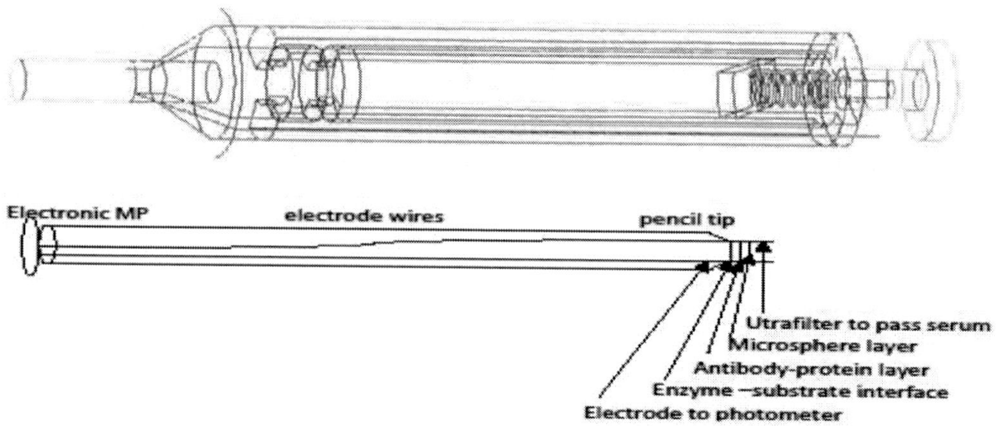

Figure 2. (on top) A magnetic pen, consisting of a ferrite tip that is magnetized after it is connected to a permanent magnet. Notice that magnetic effect of permanent magnet is used to separate out paramagnetic spheres bound with TroponinT conjugates for magnetic immunoassay by fluorescence or colorimetric methods. (on bottom) A sketch of magnetic pen is shown with arrangement of different layers at the tip and microprocessor to read the electronic signal to measure the troponinT.

2.2. Device for Acute Myocardial Infarction Detection and Troponin Screening

The proposed detection device works on the principle of glucometer with modified dry chemistry strip design to allow for incorporating the nanoparticles on the strip. Initially, the phase I study of its development will concentrate on acquiring reliable experimental data that can be used to calibrate and develop a 'pen' shaped testing device as shown in the figure 2. The device was used using the following steps:

The device was suitable to measure troponinT using the following steps:

2. A sample of known volume was placed in a buffer solution that has predetermined amount of magnetic nanoparticles in excess with antitroponin antibodies on their surface.
3. The magnetic pen developed for this project was used to mix the magnetic nanoparticles in excess with blood sample to capture essentially all the marker troponin molecules. Then the magnetic pen was pressed to magnetize its tip. Once magnetized the tip, it will collect the magnetic nanoparticles from the mixed buffer solution and blood sample. At this time all the TroponinT molecules are assumed to be linked to antibodies on the nanoparticles' surface.
4. The magnetic particles were transferred to another buffer to wash excess blood components while the nanoparticles are still attached to the magnetic tip.
5. The particles were transferred to a vial containing antiTroponin antibody coupled with glucose.
6. Another wash removed the excess glucose. A known amount of (20 μg) was placed on the glucose strip and placed in the glucometer for proportionate troponin measurement.

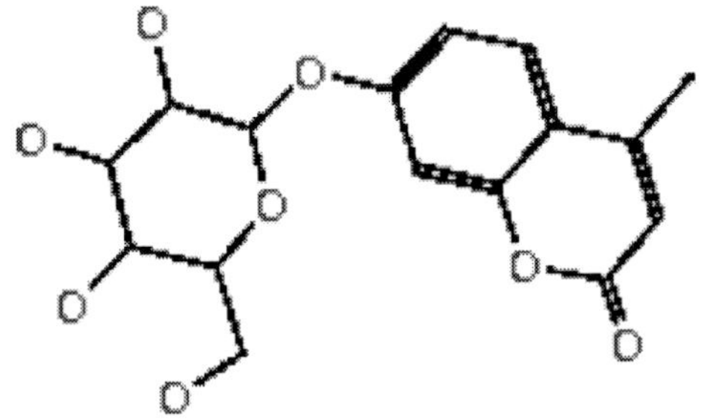

Figure 3. The structure of methylumbelliferyl α-D-glucose used in fluorescent measurement to measure glucose.

Twice washes will remove the excess glucose. A known amount (20 μg) of antiTroponin bound methylumbelliferyl glucose complex will be placed on the glucose strip and placed on the strip platform of glucometer for glucose measurement. The glucose concentration will measure proportional amount of Troponin T in blood.

Still the art of rapid point-of-care device is in infancy and measurement of troponin T and the role of other troponin subunits I and C is controversial for its specificity and statistical accuracy as reported recently [37-42].

2.3. Principle of Dry Chemistry Glucodetection in Troponin Estimation

Attaching glucose molecules at the end of the anticardiac-marker antibody facilitated the detection of AMI by a glucometer.

Phase I study: Terminal glucose was measured by one of three methods: electrical, fluorometric, and acidity measurements using the following reaction.

Glucose + O_2 → Gluconic acid + H_2O_2
H_2O_2 + reduced chromogen(uncolored) → H_2O + Oxidized chromogen(colored) or
H2O2 + methylumbelliferyl- (reduced)→H20 + Oxidised (fluorescent)

Glucose assay methods:

Electrical: measures electrical resistance on strips
Colorimetric: measures color intensity or fluorescence by spectrophotometer
Acidity: measures pH change by pH meter

The glucose concentration was be measured by "amperometric" system for point of care determination. The method utilizes glucose dehydrogenase enzyme impregnated on test area of reagent strip. The strip was embedded with two metal strips (as electrodes). The enzyme action on glucose liberates an electron, which reacts with a mediator on strip to result its oxidation and current generation proportional to the glucose linked conjugate. The measurement of current measures the numeric glucose concentration. Other alternate method was using glucose oxidase (GOD) enzyme instead of glucose dehydrogenase.

In colorimetric approach, glucose reacts with glucose oxidase to form gluconic acid and hydrogen peroxide. It reacts with a chromogen to form pink-colored compound, which can be monitored by reflectance photometer [38]. The fluorescence of oxidized compound will be monitored by change in fluorescence at wave length.

In the pH meter approach, microelectrode detects the acidity resulting from the reaction between glucose and its enzyme to produce gluconic acid. The acidity of glutonic acid was detected by regular pH meter where more acidic solution indicates more glucose attached to the conjugate.

2.4. Calibration of Glucose Measurement

We have conducted preliminary experiments to test the concept using electrical glucometer to measure electrical resistance that generates due to the reaction that occurred between glucose oxidase (GOD) enzyme and its substrate, glucose was measured using two electrodes connected to a voltameter model (Fluke 23). The readings showed that the electrical resistance generated in strips was inversely related to the different glucose concentrations; the highest concentration had the lowest resistance as shown in figure 4.

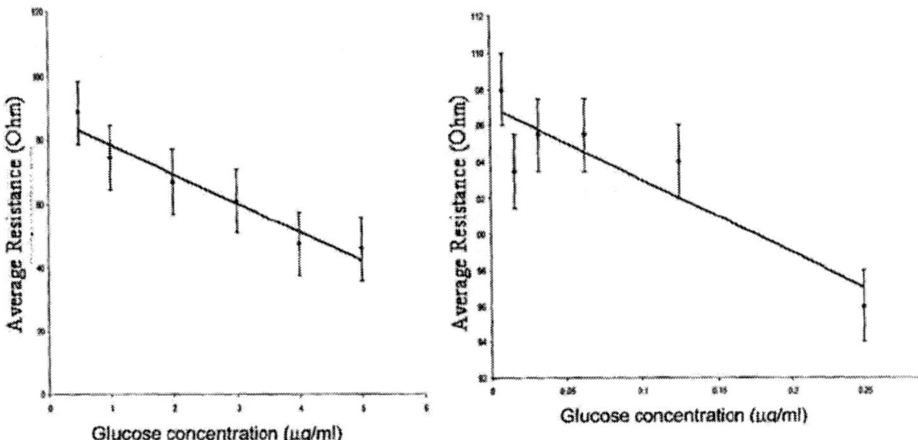

Figure 4. The relationship is shown between glucose concentration (higher concentrations on left panel and lower concentrations on right in µgm/ml) and its average resistance.

In colorimetric assay, a volume of 500 µL of glucose oxidase reagent solution was warmed up for 5 minutes in water bath under 37 °C and mixed with 50 µL of sample conjugate of each concentration of troponin and incubated for 10 minutes in a water bath at 37 °C. The samples were measured using a spectrophotometer model (Turner SP-830) and the resulted solution had different intensities of pink color depicting the amount of glucose in each sample as shown in figure 5.

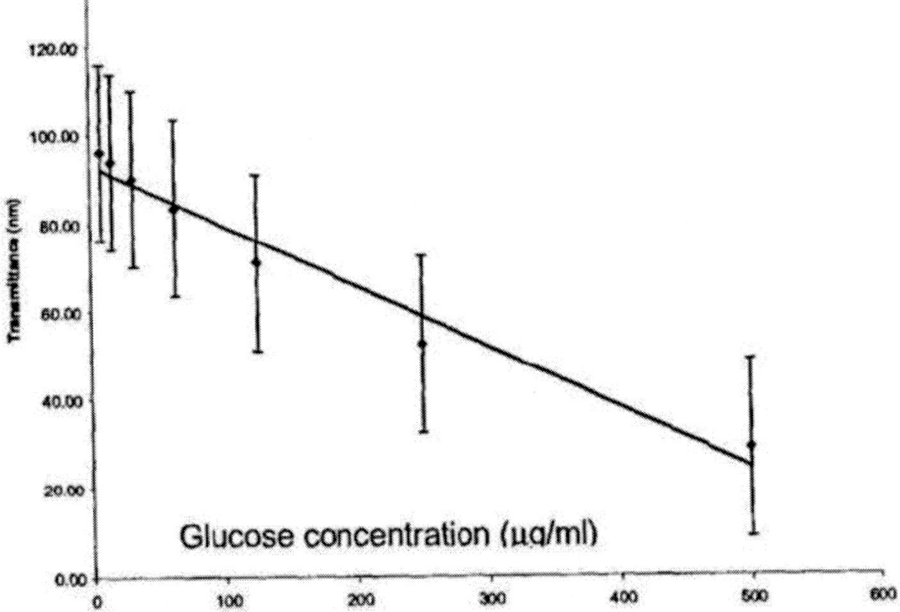

Figure 5. The relationship is shown between each glucose concentration (µgm/ml) and its average transmittance level or inverse log of optical density. The linearity is best at lower concentrations.

The enzymatic action on glucose liberated electron that reacts with strip mediator as shown below:

Glucose + O_2 → Gluconic acid + H_2O_2; H_2O_2 + reduced chromogen(uncolored) → H_2O + Oxidized chromogen(colored). Glucose assay methods use the following principle: 1.Electrical: measures electrical resistance on strips; 2.Colorimetric: measures color intensity by spectrophotometer; 3. Acidity: measures pH change by pH meter.

For calibration and comparison a colorimetric assay was used. In figure 6, troponin at different concentrations showed relationship with absorbance curve of glucose assay. A volume of 500 μL of glucose oxidase reagent solution was warmed up for 5 minutes in water bath under 37°C and mixed with 50 μL of sample conjugate of each concentration of troponin and incubated for 10 minutes in a water bath at 37°C. The samples were measured using a spectrophotometer model 9Turner SP-830) and the resulted solution had different intensities of pink color as described earlier [36].

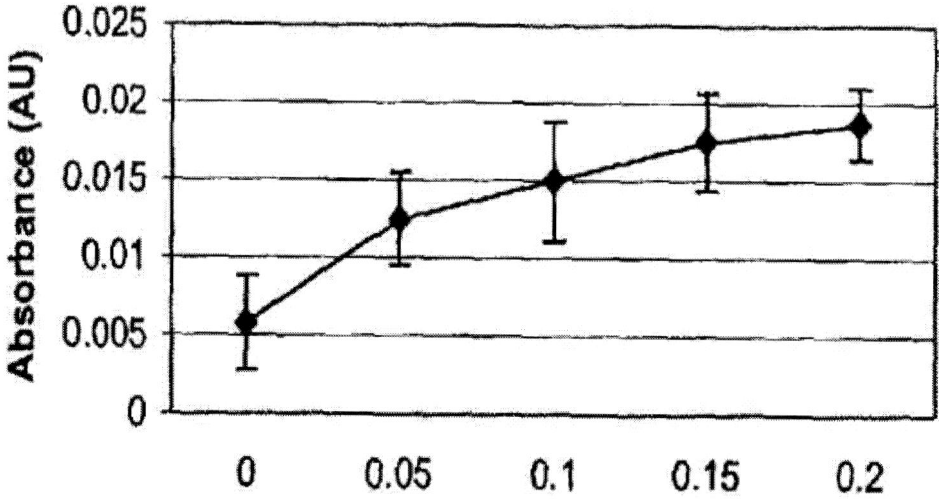

Figure 6. The concentration curve is shown for average reading of 5 different Troponin concentration samples (μgm on x axis) based on the reaction of glucose with glucose oxidase-peroxidase enzyme following coupling with labeled antitroponin antibodies and magnetic separation using magnetic immunoassay.

2.2. The Use of the Fluorescent Dry Chemistry in Glucodetection Device

Other technique incorporated the utilization of nanomagnetic particles to isolate Troponin cardiac markers from whole blood. Thereafter, labeling these cardiac markers with fluorogenic 4-methylumbelliferyl- alpha-D-glucose and/or 7-diethylamino-3-[4'malei-midylphenyl] -4-methylcoumarin (CPM) molecules allowed glucometer to measure the concentration of the Troponin cardiac marker as change in fluorescence in the blood sample on dry strip of glucometer [36, 37]. In this system, the reagent strips contain test area, which is impregnated with enzyme and two metal electrodes. The enzymatic action on glucose liberates electron that reacts with strip mediator. The monitor in which strip is inserted applies a voltage across the electrode, causing the oxidation of mediator. This reaction generates a current proportional to the amount of glucose present and is translated by the meter to the numeric glucose conversion.

3. Detection of Troponin T as Point-of-Care AMI Test

In our previous experiments, the magnetic assay of Troponin marker detection in sample utilized a routine solid-phase enzyme linked-immunoassay (ELISA), the sandwich forms by attaching two different antibodies to different epitopes on the same target cardiac troponin marker [22, 24]. One antibody was attached to a solid surface of magnetic microsphere, and the other was attached to a glucose molecule. The first antibody was used for the separation of troponinT from the blood sample whereas, the second antibody, attached to some glucose molecules. The double antibody labeled complex was used to measure the relative concentration of AMI troponinT marker in the sample. Attaching glucose molecules at the end of the antitroponinT-marker, the antibody facilitated the detection of AMI by using a glucometer.

4. Approach of Troponin MALDI Analysis

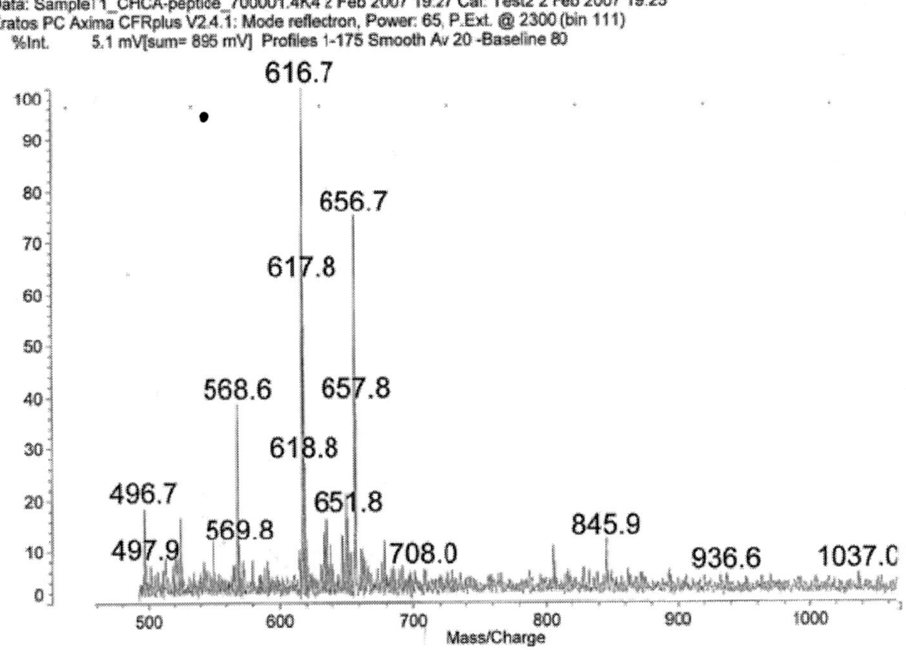

Figure 4. The typical MALDI TOF mass spectrometry analysis showed distinct peaks at mass/charge positions in the range of 300-1000 for small sized peptides. The proteins were concentrated in narrow range of 500-700. The detailed analysis matches these proteins with Lys and Asp aminoacids shown in table 1.

Affinity chromatography with antibodies specific for one cardiac troponin subunit facilitated the isolation of the entire cardiac troponin complex from myocardial tissue. The three different proteases were used for enzymatic digestion as decribed elsewhere [43]. The digestion increased the total protein amino acid sequence coverage by mass spectrometry for the three cardiac troponin subunits in the digested fractions. Combined amino acid sequence showed cardiac troponinI,T,andC (cTnI,cTnT,cTnC) subunits in proportion of 54%, 48%, and 40%, respectively. Peptide mixtures from protease digestion by matrix-assisted laser desorption/ionization time-of-flight (MALDI-TOF) masss pectrometry (VoyagerDEPro; Applied Biosystems) as described previously with cyano-4-hydroxycinnamic acid used as an energy-absorbing matrix. However, the matrix quality effects on cardiac troponin detection. In these experiments ignoring the matrix effects, total protein sequence coverages showed cTnI, cTnT, and cTnC in portion of 44%, 41%, and 19%, respectively. Although cTnI and cTnT in the first three dilutions (1:10,1:20,1:40) appeared visible, while cTnC only in the first two dilutions. At any dilution lower than 1:40, no cardiac troponins were visible. A typical MALDI spectrum shows the troponinT mass/charge ratio in the range of 700-2500.

5. MALDI-TOF Analysis in AMI Serum

The lower quantities of cTnI were expected in AMI patient serum. Using the binding capacity of affinity beads exceeded the amount of cTnI present in the serum sample from patient. Earlier, the affinity-bound proteins in serum were confirmed by Western blotting. The authors showed both cTnI with its degraded products and cTnC with weaker cTnT signals as shown in figure 1 [43]. After digestion, the mass spectrometric analysis revealed the presence of cTnI and cTnC with total protein sequence coverages of 42% and 19%, respectively [43]. However, cTnT was not detected by mass spectrometry. Obviously, author believes that TnI is present pre-dominantly as either free cTnI or as a binary cTnI-cTnC complex. In other set of analysis, investigator revealed the presence of cTnI, cTnT, and cTnC with total protein sequence coverages of 33%, 11%, and 28%, respectively probably due to presence of cTnI-cTnT or cTnI-cTnC-cTnT complexes [44].

Table 1. Total protein amino acid sequence coverage for cardiac troponins from human myocardial specimen as assessed by MALDI-TOF spectrometry. The yellow colored similarities are shown with standard Swiss Proteomics database. Reproduced from reference with permission[43]

Peptide	MALDI sequence of Troponin proteins*
TnC 40%	
MDDIYKAAVE QLTEEQKNEF KAAFDIFVLG AEDGCISTKE L GKVMRMLGQ NPTPEELQEM IDVDEDGSG TVDFDEFLVM MVRCMKDDSK GKSEEELSDL FRMFDKNADG YIDLEELKIM LQATGETITE DDIEELMKDG DKNNDGRIDY DEFLEFMKGV E	
TnI 54%	
ADGSSDAARE PRPAPAPIRR RSSNYRAYAT EPHAKKKSKI SASRKLQLKT LLLQIAKQEL EREAEERRGE KGRALSTRCQ PLELAGLGFA ELQDLCRQLH ARVDKVDEER YDIEAKVTKN ITEIADLTQK IFDLRGKFKR PTLRRVRISA DAMMQALLGA RAKESLDLRA HLKQVKKEDT EKENREVGDW RKNIDALSGM EGRKKKFES	
TnT 48%	
SDIEEVVEBY EEEEQEEAAV EEQEEAAEED AEAEAETEET RAEEDEEEEE AKEAEDGPME ESKPKPRSFM PNLVPPKIPD GERVDFDDIH RKRMEKDLNE LQALIEAHFE NRKKEEEELV SLKDRIERRR AERAEQQRIR NEREKERQNR LAEERARREB EENRRKAEDE ARKKKALSNM MIFCGYIQKQ AQTERKSGKR QTEREKKKKI LAERRKVLAI DHLNEDQLRE KAKELWQSIY NLEAEKFDLQ EKFKQQKYEI NVLRNRINDN QKVSKTRGKA KVTGRWK	

* Swiss Protdatabase (http://ca.expasy.org/sprot/). a, c TnT isoform 6 was used for PMF.

6. Limitations

The magnetic immunoassay method is capable of detecting small amounts of cardiac troponins in the serum of patients with AMI. The use of magnetic immunoassay in point-of-care detection is still in infancy. The MALDI-TOF mass spectrometry is not really quantitative and still remains as research tool. With technical advancement in near future, a fully automated system for high-throughput separation, using protein chips with surface-enhanced laser desorption, ionization mass spectrometry, is under development and could be an ext step in the analysis of cardiac troponins from biological samples.

Conclusion

Sandwich magnetic immunoassay is a commercial approach to detect separate out cardiac Troponin protein(s). However, present time, still it remains semi-quantitative while MALDI-TOF mass spectroscopy is further extension of protein subunit or peptide components precisely responsible for acute myocardial infarction process initially within 1-4 hours of disease progress.

Acknowledgments

The report was based on work done at Center of Nanomagnetics and Biotechnology, Florida State University, Tallahassee under supervision of Dr CJ Chen of grant funded by Cornerstone Foundation 2006. The work was presented at NSTI 2006 and NSTI 2007 annual meetings of Nanoscience and Technology Institute, Boston, MA

References

[1] Bodor GS, Porter S, Landt Y, Ladenson JH. Development of monoclonal antibodies for an assay of cardiac troponin-I and preliminary results in suspected cases of myocardial infarction. *Clin. Chem.* 1992;38(11):2203-14.

[2] Lin JC, Apple FS, Murakami MM, Luepker RV. Rates of positive cardiac troponin I and creatine kinase MB mass among patients hospitalized for suspected acute coronary syndromes. *Clin. Chem.* 2004;50(2):333-8.

[3] Lang K, Borner A, Figulla HR. Comparison of biochemical markers for the detection of minimal myocardial injury: superior sensitivity of cardiac troponin--T ELISA. *J. Intern. Med.* 2000;247(1):119-23.

[4] Suetomi K, Takahama K.A sandwich enzyme immunoassay for cardiac troponin *I. Nippon Hoigaku Zasshi.* 1995;49(1):26-32

[5] Apple FS, Anderson FP, Collinson P, Jesse RL, Kontos MC, Levitt MA, Miller EA, Murakami MM. Clinical evaluation of the first medical whole blood, point-of-care testing device for detection of myocardial infarction. *Clin. Chem.* 2000;46(10):1604-9.

[6] Sharma R Search of Troponins as Smart Biosensors for Point-of-care Detection of Acute Myocardial Infarction. *Proc. NSTI Nanotech.* 2007; Abstract 1222.

[7] Hsu LF, Koh TH, Lim YL.Cardiac marker point-of-care testing: evaluation of rapid on-site biochemical marker analysis for diagnosis of acute myocardial infarction. *Ann. Acad. Med. Singapore.* 2000;29(4):421-7.

[8] Newman DJ, Olabiran Y, Bedzyk WD, Chance S, Gorman EG, Price CP. Impact of antibody specificity and calibration material on the measure of agreement between methods for cardiac troponin I. *Clin. Chem.* 1999;45(6 Pt 1):822-8.

[9] Hetland O, Goransson L, Nilsen DW. Cardiac troponin T immunoassay on biotin--streptavidin-coated microplates: preliminary performance in acute myocardial infarction. *Scand. J. Clin. Lab. Invest.* 1995;55(8):701-13.

[10] Gao L, Zhuang J, Nie L, Zhang J, Zhang Y, Gu N, Wang T, Feng J, Yang D, Perrett S, & Yan X. Intrinsic peroxidase-like activity of ferromagnetic nanoparticles. *Nature Nanotech.* 2007;2(9):521-583.

[11] Zhang J. Directing cells to target tissues organs. USPTO Application #: 20070053839. http://www.freshpatents.com/Directing-cells-to-target-tissues-organs-dt20070308ptan20070053839.php

[12] Jossi S, Gordon SL, Legge MA, Armstrong GP. All troponins are not created equal. *Intern. Med. J.* 2006;36(5):325-7.

[13] Saadeddin SM, Habbab MA, Siddieg HH, Al Seeni MN, Tahery AB, Dafterdar RM.Evaluation of 6 cardiac troponin assays in patients with acute coronary syndrome. *Saudi Med. J.* 2003;24(10):1092-7.

[14] Kratz A, Januzzi JL, Lewandrowski KB, Lee-Lewandrowski E.Positive predictive value of a point-of-care testing strategy on first-draw specimens for the emergency department-based detection of acute coronary syndromes. *Arch. Pathol. Lab. Med.* 2002;126(12):1487-93.

[15] Zarich SW, Bradley K, Mayall ID, Bernstein LH.Minor elevations in troponin T values enhance risk assessment in emergency department patients with suspected myocardial ischemia: analysis of novel troponin T cut-off values. *Clin. Chim. Acta.* 2004;343(1-2):223-9.

[16] Wu AH, Smith A, Christenson RH, Murakami MM, Apple FS. Evaluation of a point-of-care assay for cardiac markers for patients suspected of acute myocardial infarction. *Clin. Chim. Acta.* 2004;346(2):211-9.

[17] Takahashi M, Lee L, Shi Q, Gawad Y, Jackowski G.Use of enzyme immunoassay for measurement of skeletal troponin-I utilizing isoform-specific monoclonal antibodies. *Clin. Biochem.* 1996;29(4):301-8.

[18] Larue C, Calzolari C, Bertinchant JP, Leclercq F, Grolleau R, Pau B.Cardiac-specific immunoenzymometric assay of troponin I in the early phase of acute myocardial infarction. *Clin. Chem.* 1993;39(6):972-9.

[19] Haug C, Bachem MG, Woehrle H, Hetzel M, Gruenert A.Evaluation of two modified cardiac troponin I enzyme immunoassays. *Clin. Chem. Lab. Med.* 2002;40(8):837-9.

[20] Wu AH, Valdes R Jr, Apple FS, Gornet T, Stone MA, Mayfield-Stokes S, Ingersoll-Stroubos AM, Wiler B.Cardiac troponin-T immunoassay for diagnosis of acute myocardial infarction. *Clin. Chem.* 1994;40(6):900-7.

[21] Lestin M, Hergert M, Lestin HG, Brinker C, Storm H, Kuhrt E, Kuhrt B, Lambrecht HG, Kuhnel W. Evaluation of the chemiluminescence immunoassays for the

measurement of troponin I, myoglobin and CK-MB using the IMMULITE system in comparison to other measuring systems. *Clin. Lab.* 2002;48(3-4):211-21.

[22] Chatterjee J, Haik Y, Chen CJ. Modification and characterization of polystyrene based magnetic microspheres and its comparison with albumin based magnetic microspheres. *J. Mag. Mag. Mat.* 2001, 225:21-29.

[23] Filatov VL, Katrukha AG, Bereznikova AV, Esakova TV, Bulargina TV, Kolosova OV, Severin ES, Gusev NB. Epitope mapping of anti-troponin I monoclonal antibodies. *Biochem. Mol. Biol. Int.* 1998;45(6):1179-87.

[24] Haik Y, Chatterjee J, Chen CJ. Synthesis and stabilization of Fe-Nd-B nanoparticles for biomedical applications. *J. Nanoparticle Res.*2005, 7:675-679.

[25] Song Zhang, Zhiping Bian, Chunrong Gu, Yu Zhang, Shiying He, Ning Gu, Jinan Zhang. Preparation Colloids Surf B Biointerfaces. 2006; http://www.wipo.int/pctdb/en/wo.jsp?WO=2007085411&IA=EP2007000537&DISPLAY=DESC

[26] Penttila I, Hirvonen K, Julkunen A, Penttila K, Rantanen T.Adaptation of the troponin T ELISA test to a microplate immunoassay reader. *Eur. J. Clin. Chem. Clin. Biochem.* 1995;33(1):59-63.

[27] Amit G, Gilutz H, Cafri C, Wolak A, Ilia R, Zahger D.What have the new definition of acute myocardial infarction and the introduction of troponin measurement done to the coronary care unit? Impacts on admission rate, length of stay, case mix and mortality. *Cardiology.* 2004;102(3):171-6.

[28] Apple FS, Chung AY, Kogut ME, Bubany S, Murakami MM. Decreased patient charges following implementation of point-of-care cardiac troponin monitoring in acute coronary syndrome patients in a community hospital cardiology unit. *Clin. Chim. Acta.* 2006; .

[29] Apple FS, Ler R, Chung AY, Berger MJ, Murakami MM.Point-of-care i-STAT cardiac troponin I for assessment of patients with symptoms suggestive of acute coronary syndrome. *Clin. Chem.* 2006;52(2):322-5.

[30] Apple FS, Quist HEDiagnostic and prognostic value of cardiac troponin I assays in patients admitted with symptoms suggestive of acute coronary syndrome. *Arch. Pathol. Lab. Med.* 2004;128(4):430-4..

[31] Ogawa M, Abe S, Saigo M, Kozono T, Yamaguchi K, Toda H, Lee S, Yamashita T, Atsuchi Y, Tateishi S, Tahara M, Torii H, Akimoto M, Mawatari K, Fukusaki M, Tei C.Usefulness of rapid bedside cardiac troponin T assay for the diagnosis of acute myocardial infarction]. *J. Cardiol.* 2000;35(3):157-64.

[32] Heeschen C, Goldmann BU, Moeller RH, Hamm CW.Analytical performance and clinical application of a new rapid bedside assay for the detection of serum cardiac troponin I. *Clin. Chem.* 1998;44(9):1925-30.

[33] Muller-Bardorff M, Freitag H, Scheffold T, Remppis A, Kubler W, Katus HA.Development and characterization of a rapid assay for bedside determinations of cardiac troponin T. *Circulation.* 1995;92(10):2869-75.

[34] Torabi F, Mobini Far HR, Danielsson B, Khayyami M.Development of a plasma panel test for detection of human myocardial proteins by capillary immunoassay. *Biosens Bioelectron.* 2006

[35] Hirschl MM, Lechleitner P, Friedrich G, Sint G, Sterz F, Binder M, Dienstl F, Laggner AN. Usefulness of a new rapid bedside troponin T assay in patients with chest pain. *Resuscitation.* 1996;32(3):193-8.

[36] Guo H, Yang D, Gu C, Bian Z, He N, Zhang J.Development of a low density colorimetric protein array for cardiac troponin I detection. *J. Nanosci. Nanotechnol.* 2005;5(12):2161-6.

[37] Meng QH, Zhu S, Booth C, Stevens L, Bertsch B, Qureshi M, Kalra J.Impact of the cardiac troponin testing algorithm on excessive and inappropriate troponin test requests. *Am. J. Clin. Pathol.* 2006;126(2):1-5.

[38] James SK, Lindahl B, Armstrong P, Califf R, Simoons ML, Venge P, Wallentin L; GUSTO-IV ACS Investigators.A rapid troponin I assay is not optimal for determination of troponin status and prediction of subsequent cardiac events at suspicion of unstable coronary syndromes. *Int. J. Cardiol.* 2004;93(2-3):113-20.

[39] Collinson PO, Gaze DC, Stubbs PJ, Swinburn J, Khan M, Senior R, Lahiri A.Diagnostic and prognostic role of cardiac troponin I (cTnI) measured on the DPC Immulite. *Clin. Biochem.* 2006.

[40] Babuin L, Jaffe AS.Troponin: the biomarker of choice for the detection of cardiac injury. *CMAJ.* 2005;173(10):1191-202.

[41] Apple FS, Murakami MM, Christenson RH, Campbell JL, Miller CJ, Hock KG, Scott MG. Analytical performance of the i-STAT cardiac troponin I assay. *Clin. Chim. Acta.* 2004;345(1-2):123-7.

[42] Apple FS, Murakami MM, Quist HH, Pearce LA, Wieczorek S, Wu AH.Prognostic value of the Ortho Vitros cardiac troponin I assay in patients with symptoms of myocardial ischemia. Risk stratification using European Society of Cardiology/American College of Cardiology recommended cutoff values. *Am. J. Clin. Pathol.* 2003;120(1):114-20.

[43] Labugger R, Simpson JA, Quick M, Brown HA, Collier CE, Neverova I, Van Eyk JE. Strategy for analyais of cardiac troponins in biological samples with a combination of affinity chromatography and mass spectrometry. *Clin. Chem.* 2003;49(6):873-879.

[44] Bizzarri M, Cavaliere C, Foglia P, Guarino C, Samperi R, Lagana A. A label free method based on MALDI TOF Mass spectrometry for the absolute quantitation of troponinT in mouse cardiac tissue. *Ana Bioana Chem.* 2008;391(5):1969-76.

In: Advances in Medicine and Biology. Volume 20
Editor: Leon V. Berhardt
ISBN 978-1-61209-135-8

Chapter 3

GENOTOXICITY EVALUATION OF EXPOSURE TO LEAD

***Julia García-Lestón*[1,2]*, Josefina Méndez*[2]*, Eduardo Pásaro*[1] *and Blanca Laffon*[*1]**

[1] Toxicology Unit, Dept. Psychobiology, University of A Coruña, Edificio de Servicios Centrales de Investigación, Campus Elviña s/n, 15071-A Coruña, Spain

[2]Dept. Cell and Molecular Biology, University of A Coruña, Faculty of Sciences, Campus A Zapateira s/n, 15071-A Coruña, Spain.

ABSTRACT

Lead is a toxic and cumulative metal that exerts its harmful action in virtually all body organs and systems. Exposure to this metal may result in a wide variety of biological effects depending on the level and duration of the exposure. Despite the progressive reduction in its applications, there are many industry sectors in which it is still present. In addition, lead has been used by humans since ancient times so, since it is a non-biodegradable element, environmental pollution caused is persistent and widespread, affecting the population at large.

The International Agency for Research on Cancer has classified lead and its inorganic compounds as probable human carcinogens (group 2A). Nowadays there is broad evidence of the relationship between the interaction with genetic material and cancer development, and thus genotoxicity tests are applied as biomarkers of the early effects of most carcinogen agents.

Although exposure to lead, both in environmental and occupational settings, has been frequently associated with an increase in genotoxic damage in humans, controversy is nurtured by several works reporting conflicting results.

[*] blaffon@udc.es

This chapter conducts a review of the studies on genotoxicity of exposure to lead, compiling *in vitro, in vivo* and epidemiological studies using different DNA damage parameters from mutagenicity to chromosomal alterations.

1. INTRODUCTION

Lead is a metal that is very ubiquitous in nature. Since antiquity, man has used this metal with many purposes. The Romans were the first to use lead on a large scale in the manufacture of pipes for water supply, manufacture of tableware and kitchen utensils or even as pigment. Lead acetate was used later as a sweetener for wine and cider as well as in medicine for treating several diseases. Lead poisoning was very important during the 16th to 19th centuries due to its widespread use in pottery, pipes, boat building, manufacture of windows, arms industry, pigments and printing of books. Many of these uses declined or disappeared throughout the 19th century, but also introduced new ones, as its application for improving the octane rating of gasoline by the addition of tetraethyl lead, its use in glass containers for cooking or the use of paint with lead compounds (Hernberg, 2000; Moreno, 2003).

At present, though many of the uses of lead have disappeared, exposure to this metal is still present in many industrial activities such as car repair, manufacturing and recycling of batteries, lead paint removal, demolition, refining and smeltery. It is also used for maintenance of structures found in the open air as bridges or water towers, in solders of cans of food or beverages, glazed ceramic, and can also be present in drinking water or in tobacco smoke (Patrick, 2006; Spivey, 2007).

In addition to occupational exposure, given that lead has been used since ancient times and that it is a non-biodegradable element, environmental pollution caused is persistent and widespread, affecting the population at large.

1.1. Lead Modes of Action

Although lead toxicity on different biological systems and functions has been well documented (Goyer and Rhyne, 1973; Gerber *et al.*, 1980; Todd *et al.*, 1996), there are still conflicting data on its genotoxic and carcinogenic properties. The International Agency for Research on Cancer (IARC) classified lead as possible human carcinogen (group 2B) (IARC, 1987) and inorganic lead compounds as probable human carcinogens (group 2A) (IARC, 2006) on the basis of sufficient evidence for carcinogenicity in experimental animals but inadequate evidence for carcinogenicity in humans. Organic lead compounds can not be classified as carcinogenic for humans (group 3) (IARC, 2006). However, in some epidemiological studies exposure to lead has been linked to an increased incidence of some cancers such as stomach, lung and bladder cancers (Fu and Boffeta, 1995). There are several proposed mechanisms to better understand the carcinogenic properties of lead and the conditions required for this purpose. These mechanisms include mitogenesis, alterations in gene transcription, oxidative damage and several indirect genotoxicity mechanisms (Hartwig, 1994; Silbergeld, 2003). It has been seen that the role of lead in carcinogenesis is more permissive than causal, that is, lead can increase the risk of cancer by reducing the ability of cells to protect or repair DNA damaged by other exposures, rather than causing alterations in DNA directly (Silbergeld *et al.*, 2000).

As it has been previously pointed out, lead affects virtually all systems of the body. Several studies support the idea that lead toxicity is due to its interference with different enzymatic systems (WHO, 2001). This interference can be performed through interaction with proteins by binding to their sulphydryl groups or because of its ability to replace essential divalent cations in different molecular processes. In this way, lead inhibits essential cellular functions and is able to reach certain organs or organelles on which it exerts its effect (Silbergeld *et al.*, 2000).

In general, molecular processes by which it is believed that lead acts as a toxic agent can be divided into three types (Minozzo *et al.*, 2004). The first is related to lead binding to various electron donors, sulphydryl groups in particular. These bonds induce changes in protein structure and enzymatic function. The second type results from the biochemical and biophysical similarity of lead to essential polyvalent cations, which permits access of the metal to critical cellular organelles, such as mitochondria. The third type seems to affect nucleic acids by a mechanism not still understood, but it seems that production of free radicals may be involved.

Interaction with Sulphydryl Groups of Proteins

One of the main mechanisms of toxic action of lead is the interaction with proteins through their sulphydryl groups by which lead presents great affinity (Landrigan *et al.*, 2000). One of these interactions is by inhibiting the cytosolic enzyme δ-aminolevulinic acid dehydratase (ALAD), which is part of the route of synthesis of heme group. This enzyme is responsible for catalyzing the formation of porphobilinogen from two molecules of δ-aminolevulinic acid (ALA) (Onalaja and Claudio, 2000). Inhibition of ALAD is produced by the union of lead to the sulphydryl groups of the active centre of the enzyme, displacing zinc, which is normally linked to one of them (ATSDR, 1999). This union causes conformational changes that inhibit the enzyme activity and reduces the conversion of ALA to porphobilinogen, by preventing the association with the enzyme (Mijares *et al.*, 2006). Thus, it produces an accumulation of ALA in blood and urine, which is used as a biomarker of exposure to lead or as a marker of early effects of lead.

Another enzyme involved in the synthesis of heme group inhibited by lead is the ferrochelatase. This enzyme introduces iron into the protoporphyrin molecule to form the heme group. Lead acts by inhibiting the ferrochelatase in the same way as it does with ALAD, so that the iron is not incorporated in the protoporphyrin and this is accumulated in the tissues and body fluids, where it is spontaneously attacked by zinc forming the zinc-protoporphyrin (ZPP) (Onalaja and Claudio, 2000), which is also used as a biomarker of exposure to lead.

Supplanting of Essential Polyvalent Cations

Another well known mechanism of lead toxicity is its ability to substitute essential polyvalent cations in the organism cellular machinery, which is possible by an ionic structure that allows it to establish interactions with the groups that join the polyvalent cations in proteins, sometimes with more affinity than the supplanted ion. However, since lead has an unequal distribution of charges, the proteins that bind lead adopt a non-physiological conformation. Through this mechanism lead affects, among other molecular targets, metal transport proteins, ion channels, cell adhesion proteins, metabolic enzymes and DNA binding proteins (Garza *et al.*, 2005).

The main ion replaced by lead is zinc. This ion is indispensable in the activation of many proteins joining with their active centre and giving adequate spatial conformation to enable performing their function. There are a variety of proteins in which lead can replace zinc, including DNA binding proteins. Lead can interact with DNA protective proteins (such as histones or protamines) decreasing their affinity for DNA, and thus the genetic material becomes more vulnerable to damage by other toxic agents (Landrigan *et al.*, 2000).

Alterations in the interaction of DNA with the nuclear transcription factors in which zinc is substituted by lead cause changes in gene expression. The replacement of zinc by lead in certain proteins also plays an important role in the post-translational regulation. The tumour suppressor gene p53 is a zinc binding protein. If replacement of zinc by lead occurred in this protein, there would be a conformational change that would have similar consequences to a mutation in the p53 gene (Silbergeld, 2003). All these mechanisms of lead are facilitative or permissive for the development of processes that contribute to carcinogenesis.

Lead can also act as an agonist of calcium in many cellular processes. Lead can join to the calcium transporters in the cell membrane and use them as gateway to the cytoplasm. Thus, it is distributed inside the cell reaching different organelles such as mitochondria, where it reduces the energetic metabolism and favours the formation of free radicals, or the endoplasmic reticulum, where lead can inhibit the calcium ATPases leading to the accumulation of calcium in the cytoplasm. In addition, the abnormal functioning of intracellular regulatory proteins such as calmodulin or protein kinase C leads the toxic effects of lead to extend to a large segment of the molecular machinery of the cell, because these proteins are involved in intra- and intercellular signaling (Garza *et al.*, 2005).

Lead also interacts with the vitamin D receptor (VDR), which is involved in the absorption of calcium and mineralization of bone and is activated by binding of calcitrol (hormonal form of vitamin D that is transported in blood) (Weaver *et al.*, 2003). When the VDR is activated, it leads to the activation of genes that code for calcium binding proteins (Devlin, 1997). Since lead and calcium have a similar biochemical nature because they are divalent cations, they interact with the same biological systems interfering with each other in the absorption processes (Onalaja and Claudio, 2000). If there is a dietary deficiency of calcium, higher levels of calcitrol will be synthesized to stimulate the synthesis of calcium binding proteins to be able to absorb it as much as possible. If lead is present, it will join to the calcium binding proteins and will be transported to different parts of the body rich in calcium (Richardt *et al.*, 1986).

Free Radical Production

Another important mechanism for production of damage by lead is by producing free radicals through two events: depletion of cellular antioxidants and/or production of reactive oxygen species (ROS) which can induce oxidative stress of cells (Kasprzak, 1996). Oxidative stress occurs when production of ROS exceeds the antioxidant capacity of the target cell; therefore oxidative stress arises from the interaction of ROS with critical cellular macromolecules, such as proteins, lipids and DNA (Klaunig *et al.*, 1998). Lead can generate ROS and related compounds through their interference with detoxification enzyme systems.

Cellular glutathione is an antioxidant which reduces free radicals that can cause damage to DNA and it also functions as a scavenger as it reduces other antioxidants that have been oxidized. Lead forms a complex with this molecule and depletes it, so that the glutathione can not reduce ROS and they are accumulated in the body to produce, among other consequences, an increase in lipid peroxidation and DNA damage. Lead also interferes with detoxification systems causing alterations in the activity of enzymes involved in the Fenton reaction (mostly catalase and superoxide-dismutase), through which ROS are reduced.

1.2. Genotoxicity Tests

The integrity of cell genome is crucial not only for cell viability but also for the functional activity of cells. However, cellular genomes are constantly subjected to a variety of physical and chemical agents that can produce potentially harmful diseases on individuals because of their ability to induce mutations in essential genes of somatic cells (Albertini *et al.*, 1993; Olsen, 1996). The types of mutational events that may occur include point mutations, deletions, duplications, translocations, mitotic recombination, gene conversion, gene amplification and aneuploidy. As many of these mutational mechanisms are involved in the development of specific types of tumours in humans (Yunis, 1983; Varmus, 1984; Klein and Klein, 1985; Shtivelman *et al.*, 1985), measures of the frequency of different types of mutagenic events in human cells *in vivo* facilitate the study of the health risks associated with these events (Olsen *et al.*, 1996). To do this, several tests have been developed to determine the mutation frequency caused by mutagenic agents in somatic cells. Among them, the most frequently used are hypoxanthine-guanine phosphoribosyl-transferase (HPRT) and T cell receptor (TCR) mutation assays.

In addition, cytogenetic tests have been traditionally employed for the evaluation of lead genotoxic effects. Among them, the most representative ones are sister chromatid exchanges (SCE), micronucleus (MN) test, and the study of DNA lesions such as structural and numeric chromosome aberrations (CA). However, among all the lesions that toxic compounds may cause in the genetic material of the exposed organisms, DNA strand breaks are perhaps the most frequent, and also the easiest to detect (Bernstein and Bernstein, 1991). Many assays have been developed to evaluate these DNA alterations, such as the alkaline sucrose sedimentation (Humphrey *et al.*, 1968; Veatch and Okada, 1969; McBurney *et al.*, 1972) the alkaline unwinding (Rydberg, 1975; Taningher *et al.*, 1987), or the alkaline elution (Kohn *et al.*, 1976; Bihari *et al.*, 1990; Bolognesi *et al.*, 1992, 1996; Vukmirovic *et al.*, 1994); but since twenty years ago the single cell gel electrophoresis (comet) assay has being more and more used due to their sensitivity and simplicity (Singh *et al.*, 1988; Olive *et al.*, 1990, 1991; Steinert *et al.*, 1998; Wilson *et al.*, 1998).

The following sections are focused on the description of each one of the most commonly used techniques to evaluate the genotoxicity of lead and summarize the studies that use these techniques compiling *in vitro*, *in vivo*, and epidemiological studies.

2. Hypoxanthine-Guanine Phosphoribosyl-Transferase Gene Mutation Assay

Hypoxanthine-guanine phosphoribosyl-transferase (HPRT) is a non-essential enzyme that phosphoribosylates free guanine or hypoxanthine for the synthesis of nucleic acids. The gene is located on the X chromosome, thus, a single inactivation of the gene results in a mutant phenotype of somatic cells even for women, since only one of the X chromosomes are actively transcribed (Nakamura *et al.*, 1991). In normal cells the base analogue 6-thioguanine (TG) is phosphoribosylated by HPRT transforming it into a cytotoxic compound that is introduced in the DNA and paralyzes replication resulting in cell death. Cells lacking HPRT would be unaffected by TG since they will not form this toxic compound. Therefore, cells carrying mutations that inactivate *hprt* gene are able to grow in the presence of toxic concentrations of TG, providing a good medium for their selection by direct cloning (Albertini *et al.*, 1993). The *hprt* gene is capable of reflecting a wide variety of genetic alterations such as DNA base pair substitutions, large and small deletions, inversions and heterologous chromosome recombinations (Albertini and Hayes, 1981; Cole and Skopek, 1994).

When applied to human populations, this test requires a large amount of blood (10 ml) due to the relatively low frequency of mutation that occurs in this gene (from 10^{-5} to 10^{-6}). Time of culture is very long (2 weeks). The test involves direct cloning of peripheral blood lymphocytes, therefore, it is essential to provide the culture medium with a good source of human interleukin 2 (IL-2). Moreover, it seems that a strong tendency toward negative selection of mutants *in vivo* exists, so the test does not appear to be practical for a long-term study (Nakamura *et al.*, 1991). A big advantage of this test is the location of the *hprt* locus on the X chromosome. This makes the *hprt* gene hemicygous in males and functionally hemicygous in females. Because of this, *hprt* mutations can be analyzed in any cell line without regard to gender (Glickman *et al.*, 1994). In addition, the location of the gene on the X chromosome is advantageous to ignore dominance-recessive considerations, but disadvantageous in the aspect that genotoxic major events such as somatic recombination can not be detected (Albertini *et al.*, 1993). In addition, the test has the advantage of allowing the isolation of the clones for culture and subsequent analysis of molecular mutations at the *hprt* gene (Olsen *et al.*, 1996).

Another short-term method to detect mutations at the *hprt* gene involves autoradiography or immunofluorescence to detect tritiated thymidine or 5-bromo-2'-deoxyuridine incorporation, respectively, in mutant cells that are resistant to TG inhibition of first-round phytohaemagglutinin stimulated DNA synthesis *in vitro* (Albertini *et al.*, 1993). These short term assays are simple, relatively inexpensive, and have the potential for automation.

Several studies used this assay to evaluate the genotoxic effects of lead *in vitro*. Most of them did not show any increase in the mutation frequency at the *hprt* locus (Patierno *et al.*, 1988; Patierno and Landolph, 1989; Hartwig *et al.*, 1990; Hwua and Yang, 1998). Only two studies (Zelikoff *et al.*, 1988; Yang *et al.*, 1996) found positive results.

Yang *et al.* (1996) evaluated the mutagenicity of lead acetate at the *hprt* gene of CHO K1 cells. Their results showed that lead acetate induced a dose-dependent mutant frequency at doses lower than the LD_{50}. To determine the mutational specificity induced by lead cDNA and genomic DNA sequences were characterized. Sequencing analysis showed lead-induced single-base substitutions, small sequence alterations, splicing defects, and gene deletions. Although lead produced a broad mutational spectrum similar to that occurring spontaneously, the positions and kinds of base substitutions at the *hprt* gene induced by lead were different. The predominant mutations were G·C→C·G transversions in low-mutant-frequency populations and G·C→A·T transitions in high-mutant-frequency populations. These observations indicate that different lead mutagenic mechanisms may exist for inducing low and high mutant frequencies. On the contrary, Hwua and Yang (1998) did not find any increase in the mutation frequency at the *hprt* gene in diploid human skin fibroblasts (HFW) at similar cytotoxic dosages of lead acetate to those used by Yang *et al.* (1996). The fact that lead mutagenicity observed in CHO K1 cells, but not in human fibroblasts, may be attributable, according to the authors, to the different defence mechanisms against lead genotoxicity in HFW and CHO K1 cells. In fact, the basal metallothionein (the most important metal-binding protein) level is approximately 5 μg/mg protein in HFW (Lin *et al.*, 1995), but undetectable in CHO K1, and the catalase level is ≈ 50-fold higher in HFW than in CHO K1 (Lee and Ho, 1995; Yang *et al.*, 1996; Hwua and Yang, 1998). This comparison suggests that metal mutagenicity is positively correlated to the cellular oxidative stress, i.e. mutation may be induced in cells having limited antioxidant machinery (Hwua and Yang, 1998).

3. T Cell Receptor Mutation Assay

T-cell receptor (TCR) is an heterodimer consisting of two chains, α and β, which is expressed in the cell surface of the vast majority of peripheral T lymphocytes $CD4^+$ and $CD8^+$. The genes encoding TCR α and β chains are located on chromosomes 14 and 7, respectively, and they are phenotypically hemizygous. It is believed that only one of the TCR alleles is actively expressed as a result of an allelic exclusion mechanism similar to that observed in the immunoglobulin genes of B cells. Therefore, it is expected that a single mutation in any of the TCR chains results in the absence of phenotypic expression of TCR on the cell surface (Kronenberg *et al.*, 1986). In addition, TCR can only be expressed in the cell surface when it forms a complex with CD3. TCR and CD3 play an important role in antigen recognition and in signal transduction, and the union of the two components is essential for these functions (Clevers *et al.*, 1988). If the expression of any of the TCR α or β chains is inactivated, the complex TCRαβ/CD3 cannot be transported to the surface of the cell membrane and the altered complexes are accumulated in the cytoplasm (Akiyama *et al.*, 1995). Thus, mutations in any of the TCR chain genes can be detected by means of flow cytometry using antibodies that recognize the CD3 molecule. This technique allows to identify and to enumerate the TCRαβ mutants in the T helper lymphocytes population that express CD4 (Akiyama *et al.*, 1995). The frequency of mutant cells $CD3^-CD4^+$ is measured within the T $CD4^+$ normal cells.

This methodology considers the total mutations in TCR chain genes, thus it is not able to differentiate those mutations occurred in the α chain gene from those occurred in the β chain gene. It uses non fixed lymphocytes, so the analysis takes only few hours and it is only necessary 1 ml of fresh blood since the mutation frequencies are of the order of 10^{-4}. There are no restrictions on the donor genotype and the mutant clones of T cells can be isolated by a cell sorter to be grown *in vitro* and carry out the molecular analysis. This mutation test uses commercially available antibodies and flow cytometry, so it can be widely used in many laboratories (Nakamura *et al.*, 1991; Kyoizumi *et al.*, 1992). Furthermore, it seems that the life span of the mutant lymphocytes in the blood is about 2-3 years. As most of the lymphocytes in peripheral blood are in G0 phase and rarely come into mitosis, mutant cells take some time to lose the expression of pre-existing TCR/CD3 complex. Therefore, it is not considered an appropriate method to estimate the biological dose immediately after exposure to genotoxic agents (Ishioka *et al.*, 1997).

To date there are only two studies which used the T-cell receptor (TCR) mutation assay to evaluate the potential genotoxic effects of human exposure to lead. In our laboratory we evaluated 30 workers from two different factories engaged in the production of lead-acid batteries and glass chips, respectively (García-Lestón *et al.*, 2008). We found statistically significant differences between exposed and controls, showing the exposed group higher TCR mutation frequency than the control group. On the contrary, Chen *et al.* (2006) did not observe significant differences between exposed and controls when they evaluated the genotoxic effects of lead exposure in individuals from a workplace producing storage battery. However, in this study the population size was smaller (25 exposed *vs.* 25 controls). Moreover, since the composition of both populations was statistically similar in terms of sex and tobacco smoking, the authors did not assess the impact of these factors on the results achieved, and they only focused on the evaluation of the differences between exposed and controls. Nevertheless, in the first study the populations were different with regard to tobacco consumption, but not with regard to sex because all individuals were males, so the difference in the results could also be related to this fact.

4. Chromosome Aberrations

The chromosome aberrations (CA) test detects changes in the structure of chromosomes, which are visible under the optical microscope. These changes are breaks and rearrangements within a chromosome or between different chromosomes. Such reorganizations are produced mainly by substances that directly break the DNA strand (e.g. radiation) or distort the double helix of DNA (e.g. intercalating agents). The type of chromosomal structural alterations produced by physical or chemical agents depends on the lesions induced in the DNA and, therefore, on the mode of action of the genotoxic agent.

Structural aberrations can be classified as either stable or unstable, depending upon their ability to overcome cell division. Unstable aberrations (dicentric chromosomes, ring chromosomes, fragments and other asymmetric rearrangements) lead to cell death during mitotic cell division because the reorganizations produced do not allow the division of the cell or entail the loss of fragments of genetic material in daughter cells. Stable aberrations (balanced translocations, inversions and other symmetrical rearrangements) can be transmitted to offspring because they do not interfere with cell division. Furthermore, other classification distinguishes between chromosomal aberrations, that affect both chromatids since they occur before DNA replication, and chromatid aberrations, that affect only one chromatid because they are produced after DNA synthesis.

Numerical CA refers to changes in chromosome number due to abnormal cell division. Cells can be classified as aneuploid, when they contain up to a few more (hyperploid) or a few less (hypoploid) chromosomes than the normal complement; or as polyploid when they contain multiples of the normal complement. The mechanism by which aneuploidy can occur include damage to the mitotic spindle and associated elements, damage to chromosomal sub-structures, alterations in cellular physiology, and mechanical disruption (Oshimura and Barrett, 1986). The mechanisms by which polyploidy occurs are less clear (Mitchell *et al.*, 1995).

CA are usually determined and quantified in conventional Giemsa stained preparations. In combination with fluorescence in situ hybridization (FISH) technique, CA test allows the study of the mechanism of production of the aberrations by using region specific DNA probes as well as chromosome specific DNA libraries (Natarajan *et al.*, 1996).

Numerous studies (*in vitro*, *in vivo* and epidemiological studies) used the CA test to investigate the genotoxic effects induced by lead or its derivatives.

4.1. *In Vitro* Studies

Treatment of human leukocytes with lead acetate showed clearly elevated frequencies of achromatic lesions, chromatid breaks and isochromatid breaks in 72 h cultures, but not in 48 h cultures (Beek and Obe, 1975). Deknut and Deminatti (1978) observed that the most common aberration induced by lead acetate in human lymphocyte cultures was the occurrence of chromosome fragments. On the contrary, Schmid *et al.* (1972) and Gasiorek and Bauchinger (1981) did not observe effect of lead acetate on the frequency of CA in human lymphocytes. Bauchinger and Schmid (1972) obtained the same negative results for Chinese hamster cells.

Other studies with lead nitrate and lead glutamate were also negative (Lin *et al.*, 1994; Cai and Arenaz, 1998; Shaik *et al.*, 2006), except for Lerda (1992) who found an increase in the frequency of CA in *Allium cepa*. However, consistently positive results were obtained with lead chromate (Douglas *et al.*, 1980; Wise *et al.*, 1992; Xu *et al.*, 1992; Wise *et al.*, 1994, Wise *et al.*, 2003; Wise *et al.*, 2004; Xie *et al.*, 2005), which most of the authors related to the probable action of chromate.

4.2. *In Vivo* Studies

Significant increases in the CA rate were found in several mammalian studies: leukocytes from male and female mice fed with lead acetate (Muro and Goyer, 1969), in cynomolgus monkeys (*Macaca irus*) administered lead acetate in the diet (Jacquet and Tachon, 1981), in bone marrow cells of mice exposed daily to lead acetate (Huang *et al.*, 1985), in bone marrow cells of rats after intraperitoneal administration of lead acetate (Tachi *et al.*, 1985), in maternal bone marrow and foetal liver and lung cells of ICR Swiss Webster mice following maternal exposure to lead nitrate (Nayak *et al.*, 1989), and in Swiss albino male mice orally gavaged to leachates (Tewari *et al.*, 2005). Other authors did not find any increase in frequencies of CA in mice fed with lead acetate (Deknudt and Gerber, 1979).

Moreover, Ramsdorf *et al.* (2008) observed some little structural aberrations, such as gaps, in fishes (*Hoplias malabaricus*) treated with different doses of lead nitrate by intraperitoneal injections but the statistical analysis showed that this increase was not significant. Ferraro *et al.* (2004) also found in the same fishes treated orally with PbII a significant increase in the frequency of CA, but in this case the exposure time was higher than in Ramsdorf *et al.* (2008).

Other studies reported differential induction of several types of CA. Jacquet *et al.* (1977) carried out an experiment in which dietary lead at different dose levels was given to female C57B1 mice for periods up to three months. They found no severe chromosome or chromatid aberrations at any dose level, but the frequency of chromatid gaps increased significantly at the highest doses. Deknudt *et al.* (1977a) described that the type of CA induced by lead acetate in cynomolgus monkeys (*Macaca irus*) depended on the intake of calcium in the diet. The frequency of severe abnormalities (dicentrics, rings, translocations and exchanges) was significantly increased only in the group on a low calcium diet, whereas "light" abnormalities (gaps and fragments) increased with time in all groups receiving lead irrespective of the diet. Aboul-Ela (2002) found only structural aberrations like chromatid gaps, deletions and fragments in bone marrow cells of male Swiss mice after oral administration of lead acetate. Nehéz *et al.* (2000) investigated the possible genotoxic effects exerted by the pyrethroid cypermethrin and by either of the metals cadmium and lead alone or in combination, on bone marrow cells of outbred male Wistar rats. Treatment with lead acetate only increased significantly the number of aberrant cells and numerical aberrations but did not alter the number of structural aberrations. The combination of cypermethrin and lead caused a significant increase in aberrant cells and in structural aberrations but not in numerical aberrations. The most frequently structural aberrations observed were gaps and acentric fragments. These results agree with Lorencz *et al.* (1996) who found increases in numerical aberrations in Wistar rats treated with different doses of lead acetate.

4.3. Epidemiological Studies

Given the results observed in epidemiological studies for CA test, there is considerable controversy regarding the ability of lead to cause chromosomal damage on exposed individuals.

Several studies reported increases in the frequency of CA in human populations exposed to lead (Schwanitz *et al.*, 1970; Deknudt *et al.*,1973; Deknudt and Leonard, 1975; Schwanitz *et al.*, 1975; Forni *et al.*, 1976; Deknudt *et al.*, 1977b; Garza-Chapa *et al.*, 1977; Nordenson *et al.*, 1978; Sarto *et al.*, 1978; Hogstedt *et al.*, 1979; Forni *et al.*, 1980; Al-Hakkak *et al.*, 1986; Huang *et al.*, 1988; Piña-Calva *et al.*, 1991; Chen, 1992; Silva and Santos-Mello, 1996; Vaglenov *et al.*, 1997; Bilban, 1998; Pinto *et al.*, 2000; Testa *et al.* 2005; Madhavi *et al.*, 2008). However, other works found no effects of lead exposure in CA frequencies (Bauchinger and Schmid, 1970; Sperling *et al.*1970; Bauchinger *et al.*, 1972; Schmid *et al.*,1972; O'Riordan and Evans, 1974; Bijlsma and de France, 1976; Bauchinger *et al.*,1977; Horvat and Prpic-Majic, 1977; Maki-Paakkanen *et al.*,1981; Nordenson *et al.*, 1982; Pelclová *et al.*,1997). Moreover, Beckman *et al.* (1982) carried out a study in which CA frequencies in workers exposed to lead were evaluated at three different occasions (1976, 1978 and 1979). The results showed a significant increase of CA rates in lead-exposed workers in 1979, but not in 1976. But, after evaluating confounding factors, as altered smoking habits and simultaneous exposure to other toxic agents, the authors concluded that such variations in the frequency of CA could not be ascribed with certainty to changes in lead exposure.

Deknudt *et al.* (1977b) analyzed CA in cultured lymphocytes from two groups of lead-exposed people: workers from a smelting plant for storage battery in Lyon (France) and workers from a factory of tin dishes in Nerem (Belgium). They found an increased number of severe aberrations (rings and dicentrics) in people from Lyon, whereas no such aberrations but an increased number of chromosome fragments were observed in those from Nerem. Furthermore, Huang *et al.* (1988) obtained mainly "light" CA (gaps, breaks, deletions and fragments) when the frequencies of CA in lead-exposed workers from a battery factory were analysed, indicating minor lesions in the chromosomes.

Maki-Paakkanen *et al.* (1981) studied the frequency of CA in peripheral blood lymphocytes of workers exposed to lead in a smeltery. They found no significant differences in the CA rates between lead-exposed workers and unexposed controls. But similar to Forni *et al.* (1980), they observed significantly higher rates of total aberrations in the 72 h culture regarding to the 50 h cultures. These results favour the previously proposed hypothesis (Forni and Secchi, 1973; Forni *et al.*, 1976) that aberrations observed in lymphocyte cultures of lead-exposed subjects may be culture-born. This may be due to deficiency of repair functions in the presence of lead or some lead-induced metabolite which could accumulate with increasing culture time (Maki-Paakkanen *et al.*, 1981).

5. Sister Chromatid Exchanges

Sister chromatid exchanges (SCE) are reciprocal exchanges between chromatids of a chromosome that involve the breaking of both DNA strands, followed by a full-duplex exchange between sister chromatids. Such exchanges occur at apparently homologous *loci* and presumably involve DNA breakage and reunion during DNA replication. The interchange does not involve morphological alteration and is detected only by differential labelling of the sister chromatids. The currently used method for the detection of the SCE requires DNA replication in the presence of 5-bromo-2'-deoxyuridine (BrdU) for two consecutive cell cycles. BrdU closely resembles thymidine, and is efficiently incorporated into the elongating DNA strands during replication. After a second round of growth in BrdU medium, the two sister chromatids differ in the amount of BrdU present because of the semiconservative replication of DNA: one has two strands substituted with BrdU and the other one has only one substituted strand. BrdU has the effect of "bleaching" so that the chromatid with more BrdU is lighter in appearance after a normal Giemsa staining (Wilson III and Thompson, 2007).

SCE events are not lethal for the cell. Moreover they can not be considered by themselves as mutations because they do not produce any changes in the genetic information. However, it appears that the frequency of SCE increases when cells are exposed to known carcinogens and mutagens.

The SCE assay yields quantifiable data from every cell scored, which increases the efficiency of data collection and the identification of DNA damage resulting from exposure to genotoxic carcinogens compared to traditional CA analysis (Albertini *et al.*, 2000). However, there are some serious arguments against a general application of this method for screening weak mutagens (Gebhart, 1981): SCE test is only suitable to detect those weak mutagens which actually induce the primary DNA lesions finally leading to SCE formation. The limits of the resolving power of SCE test are documented by the rather large amount of evidently "false-negative" results, and also by the relatively large number of slightly positive results which are so difficult to interpret with substances which have yielded negative results in other mutagenic systems. Another disadvantage lies in the technique itself: because of the need of adding BrdU, in no case the activity of the test agent alone is measured, but always the co-activity of the test agent and the BrdU.

5.1. *In Vitro* Studies

The *in vitro* studies which evaluated alterations in SCE rates induced by lead report contradictory results. Increases in the frequency of SCE were found evaluating the genotoxic effects of different salts of lead. Douglas *et al.* (1980), evaluated the effect of lead chromate in CHO cells; Wulf (1980) tested lead sulphate in human lymphocytes; Sharma *et al.* (1985) and Poma *et al.* (2003) investigated the induction of SCE by lead acetate in pregnant mice and their foetuses and in human melanoma cells (B-Mel), respectively; and Lin *et al.* (1994) and Cai and Arenaz (1998) used lead nitrate in CHO cells. On the other hand, no effect of lead on the frequency of SCE was observed by Beek and Obe (1975) and Hartwig *et al.* (1990), who evaluated the genotoxic effect of lead acetate in human leukocytes and in Chinese hamster V79 cells, respectively, and by Zelikoff *et al.* (1988), who assessed the genotoxicity of lead sulphide and lead nitrate in V79 cells. Furthermore, Montaldi *et al.* (1985) analysed SCE induction by several salts of lead, among other heavy metals, in CHO cells and they observed that lead sulphate increased SCE frequency whereas lead acetate did not affect it.

Hartwig *et al.* (1990) investigated whether the genotoxicity of lead was due to indirect effects, such as interference with DNA repair processes, by means of SCE among other endpoints. They found that lead acetate alone did not induce sister chromatid exchanges in V79 Chinese hamster cells. However, lead ions enhanced the number of UV-induced SCE suggesting that lead interfered with the processing of UV-induced DNA lesions. The authors reached the conclusion that not only the DNA damage itself but also the genotoxic effects of other chemical or physical agents may be augmented when repair is inhibited, as illustrated by the enhancement of UV-induced SCE.

5.2. *In Vivo* Studies

Only three studies conducted in animals evaluated the influence of lead on the frequency of SCE. Willems *et al.* (1982) investigated the effect of lead on SCE rate in male rabbits after exposure to different doses of lead acetate subcutaneously injected three times a week during 14 weeks, each on a group of five rabbits. Statistical analysis of the number of SCE per metaphase in lymphocytes indicated no differences between the groups. Nayak *et al.* (1989) analysed the frequency of SCE in maternal bone marrow and foetal liver and lung cells of ICR Swiss Webster mice, following maternal exposure to lead nitrate. Lead caused a moderate, but statistically significant, increase in the frequency of SCE in maternal bone marrow cells. On the contrary, lead did not cause any effect on the frequency of SCE in liver or lung cells of foetus, although lead was shown to cross the placenta. Dhir *et al.* (1993) reported that intraperitoneal injection of low doses of lead nitrate caused a significant increase in SCE rate in bone marrow of male Swiss albino mice.

5.3. Epidemiological Studies

In most studies assessing the genotoxic effects of lead in exposed people an increase in the frequency of SCE was observed (Grandjean *et al.*, 1983; Huang *et al.,* 1988; Chen, 1992; Rajah and Ahuja, 1995; Bilban, 1998; Dönmez *et al.*, 1998; Pinto *et al.*, 2000; Duydu *et al.*, 2001; Wu *et al.*, 2002; Palus *et al.*, 2003; Wiwanitkit *et al.*, 2008). Moreover, in some of them a positive relationship between SCE frequency and blood lead levels (Grandjean *et al.*, 1983; Huang *et al.* 1988; Bilban, 1998; Duydu *et al.*, 2001; Wu *et al.*, 2002; Wiwanitkit *et al.* 2008) or between SCE rates and urinary lead concentrations (Chen, 1992) was found, whereas in other two works the first relationship could not be observed (Dönmez *et al.*, 1998; Pinto *et al.*, 2000).

Nevertheless, two works reported no increase in the frequency of SCE in occupationally lead-exposed workers. Maki-Paakanen *et al.* (1981) analysed the SCE frequency in individuals exposed to lead in a smeltery and they did not find significant differences between the control group and the exposed group. However, they observed an increase in the SCE rate in smoker exposed workers in comparison to smoker controls, the same as in Rajah and Ahuja (1995). This would indicate that tobacco smoke works as enhancer of the genotoxic effects of lead. Two other studies which assessed the effects of lead in children living in an extensively contaminated area (Dalpra *et al.*, 1983; Myelzynska *et al.*, 2006), , did not show effect of the exposure on the frequency of SCE.

6. Micronucleus Test

Micronuclei (MN) are formed by condensation of acentric chromosomal fragments or by whole chromosomes that are left behind (lagging chromosomes) during anaphase movements (Carrano and Natajaran, 1988). The formation of MN in dividing cells is the result of chromosome breakage due to unrepaired or mis-repaired DNA lesions, or chromosome malsegregation due to mitotic malfunction. These events may be induced by oxidative stress, exposure to clastogens or aneugens, genetic defects in cell cycle checkpoint and/or DNA repair genes, as well as deficiencies in nutrients required as co-factors in DNA metabolism and chromosome segregation machinery (Bonassi *et al.*, 2007).

This test can detect two types of effects: a) clastogenic, produced by compounds that generate chromosomal breaks and lead to structural changes, and b) aneugenic, produced by compounds that disrupt the normal migration of mitotic chromosomes during anaphase and lead to numerical alterations.

MN containing chromosomal fragments result from: (a) direct DNA breakage; (b) replication on a damaged DNA template; and (c) inhibition of DNA synthesis. MN containing whole chromosomes are primarily formed from failure of the mitotic spindle, kinetochore, or other parts of the mitotic apparatus, or by damage to chromosomal sub-structures, alterations in cellular physiology, and mechanical disruption. Thus, an increased frequency of micronucleated cells is a biomarker of genotoxic effects that can reflect exposure to agents with clastogenic or aneugenic modes of action. The most frequently applied methodology uses the cytokinesis-block MN technique (Fenech and Morley, 1985) in which scoring is limited to cells that have divided once since mitogen-stimulation (Albertini *et al.*, 2000). This method is based on the identification of cells that have completed one round of nuclear division and are therefore capable of expressing chromosome damage as MN. These cells are recognised by their binucleated appearance after blocking cytokinesis using cytochalasin-B, an inhibitor of microfilament assembly required for cytokinesis (Fenech, 1998).

The mechanistic origin of individual MN can be determined using the MN test in combination with FISH or CREST staining techniques. FISH uses a centromeric probe common to all human chromosomes, and CREST uses immunofluorescent antikinetochore antibodies, making it possible to distinguish between MN containing acentric chromosomal fragments (no signal) and MN containing whole chromosomes (signal). The MN test also allows the detection of early apoptosis in cells (with a nucleus divided into small pieces).

This test presents some advantages in comparison to other cytogenetic genotoxicity tests (Albertini and Kirsch-Volders, 1997; Miller *et al.*, 1997): potential of detecting clastogenic and aneugenic compounds; simplicity of the method; fast and inexpensive; higher statistical power (due to the fact that many cells are analyzed); potential for automation using image analysis; and possibility to obtain mechanistic information in combination with other techniques. However the assay has some inconvenients because it does not detect all the structural chromosomal aberrations, only acentric fragments, and it requires pre-division for the expression of MN.

6.1. *In Vitro* Studies

Poma *et al.* (2003) found that lead acetate induced MN in a dose-dependent manner when evaluated chromosomal damage induced in human melanoma cells (B-Mel) using the cytokinesis-blocked MN assay. Thier *et al.* (2003) and Bonacker *et al.* (2005) studied the genotoxic effects of inorganic lead salts in V79 Chinese hamster fibroblasts by means of MN test. They observed that lead chloride and lead acetate induced MN in a dose-dependent manner, and they determined by means of CREST assay that the effects of lead were predominantly aneugenic, which was consistent with the observed morphology of the MN. Moreover, Sandhu *et al.* (1989) assessed the clastogenic potential of various chemicals commonly found at industrial waste sites by means of MN test. They found positive results for lead in plant cuttings of *Tradescantia* clone 4430.

Only one MN *in vitro* study showed negative results for lead (Lin *et al.*, 1994). In this study the authors investigated the effects of cadmium nitrate and lead nitrate in CHO cells by means of a number of short-term assays, including the MN test. They found that lead nitrate did not significantly increase the frequencies of binucleated CHO cells with MN.

6.2. *In Vivo* Studies

Several studies which evaluated the genotoxic effects of lead acetate in rats by means of the MN test showed an increase in the frequency of MN (Tachi *et al.*, 1985; Celik *et al.*, 2005; Piao *et al.*, 2007). Alghazal *et al.* (2008) analysed the MN rate in bone marrow erythrocytes of male and female Wistar rats treated with lead acetate trihydrate. They found a significant increase in the total number of MN in polychromatic erythrocytes of both male and female rats with regard to the control group. Moreover, there was a decrease in the ratio of polychromatic to normochromatic erythrocytes in male rats, indicating both genotoxic and cytotoxic effects of lead acetate in male rats. Similarly, Jagetia and Aruna (1998) observed an increase in the frequency of MN in bone marrow cells of male and female mice treated with lead nitrate. The frequency of MN did not show a dose related increase but male mice were more sensitive to the induction of MN than female mice, evidenced by higher frequencies of micronucleated polychromatic erythrocytes.

In another work, Tewari *et al.* (2005) investigated the genotoxic potential of the leachate, a mixture of various organic chemicals and heavy metals among which lead is included, prepared from municipal sludge sample, in mouse bone marrow cells by MN tests. Male mice orally gavaged to leachates revealed also a significant increase in micronucleated polychromatic erythrocytes. In this case, the cause of genotoxic effects may be due to one component or to combination of toxicants in the leachate.

On the contrary, three studies carried out in mice (Jacquet *et al.*, 1977), rabbits (Willems *et al.*, 1982) and fishes (Ramsdorf *et al.*, 2008) did not find any increase in the MN frequency when compared the lead-exposed group to the control group. However, Ramsdorf *et al.* (2008) related their negative results to the low number of fishes analyzed and to the fact that the piscine MN assay may lack sensitivity, since it does not detect the mitotic disjunctions if they do not provoke chromosomal loss in the anaphases neither chromosome aberrations caused by rearrangement, such as translocations or inversions if these do not originate acentric fragments (Metcalfe, 1989).

6.3. Epidemiological Studies

Most epidemiological studies collected in the literature that use MN test to evaluate the potential genotoxic effects induced by exposure to lead are performed in individuals exposed in the workplace. In general, these studies could draw the conclusion that occupational exposure to lead causes genetic damage that, depending on the degree, could lead to cell unviability.

To date, the vast majority of works showed an increase in the frequency of MN in individuals occupationally exposed to lead compared with a control group. Minozzo *et al.* (2004) assessed the genetic damage in workers in the recycling of automotive batteries. They observed that both the concentrations of lead in blood and the frequency of MN in peripheral lymphocytes in the exposed group were significantly higher than in the control group. Moreover, the values of the nuclear division index (NDI) were significantly higher in the control group than in the exposed group, indicating a possible effect of lead on cell cycle. These results are consistent with other studies in individuals occupationally exposed to lead that showed an increase in the MN rate (Bilban, 1998; Vaglenov *et al.*, 1998; Pinto *et al.*,

2000; Hamurcu *et al.*, 2001; Vaglenov *et al.*, 2001; Martino-Roth *et al.*, 2003; Palus *et al.*, 2003; Chen *et al.*, 2006).

The only epidemiological study that did not find positive effect of lead on the genetic material using the MN test was conducted by Hoffmann *et al.* (1984). They investigated the genotoxic effects in a group of car repair and reconditioning radiators workers, and they observed no statistical differences between the exposed group and the control group. They had previously found a statistically significant correlation between "normal" blood lead values and CA in lymphocytes (Högstedt *et al.*, 1979, 1981). But, according to the authors, the difference between these results could be explained either by the different cytogenetic methods employed or, more likely, by the fact that the earlier findings were due to a confounding exposure to other chemical substances with known mutagenicity.

Moreover, there is one study in the literature that was conducted in children environmentally exposed to complex mixtures including polycyclic aromatic hydrocarbons and lead (Mielzynska *et al.*, 2006); in this work increases in the frequency of MN were obtained.

7. Single Cell Gel Electrophoresis (Comet) Assay

The single cell gel electrophoresis assay, also called the comet assay, is a simple, rapid and sensitive method for the analysis and quantification of DNA damage in virtually all cell types. This technique measures breaks in DNA chains and can thereby detect genotoxic effects induced by different physical or chemical agents. Cells are embedded in a thin layer of agarose on a glass slide or plastic film, and then lysed in a solution containing detergent and high concentration of salts. Thus membranes and soluble cell constituents, as well as histones, are removed leaving the DNA still supercoiled and attached to the nuclear matrix. Subsequent neutral or alkaline incubation and electrophoresis causes DNA loops containing breaks to move towards the anode, forming a "comet tail" when visualized by fluorescence microscopy with a suitable stain. The images resemble comets, and the relative content of DNA in the tail indicates the frequency of breaks.

DNA damage detected by the alkaline version of the assay can arise through various mechanisms, including DNA inter-strand cross-linking, DNA single-strand breaks, alkali-labile sites, and incompletely repaired excision sites present at the time of lysis. Compared with other genotoxicity tests the advantages of this technique include (Tice, 1995; Rojas *et al.*, 1999; Tice *et al.*, 2000; Olive and Banath, 2006): (1) data are collected at the level of individual cells, providing information on the intercellular distribution of damage and repair, (2) only a small number of cells is required, (3) almost any eukaryotic cell population can be used, (4) the assay is sensitive, simple and cost effective, (5) data can be obtained within a few hours of sampling, (6) a relatively short time period (a few days) is needed to complete an experiment, (7) the assay can evaluate DNA damage in non-proliferating cells, (8) cells do not need to be tagged with a radioisotope, thus allowing measurement of damage in any nucleated cell; (9) the method could be used to measure variations in response to DNA damaging agents between cells of the same exposed population, and (10) it has a demonstrated sensitivity for detecting low levels of DNA damage.

However, there are some limitations of this method (Olive and Banath, 2006): (1) the number of cells and samples that can be analyzed in each experiment is limited; (2) the assay requires a viable single-cell suspension, if samples contain predominantly necrotic or apoptotic cells, accurate information on the presence of specific lesions like strand breaks or base damage cannot be obtained; (3) it provides no information on DNA fragment size; (4) cells with actively replicating DNA behave differently during gel electrophoresis, and this can be confused with a difference in inherent sensitivity; and (5) interpretation of comet results is complicated by the fact that there is no simple relationship between the amount of DNA damage caused by a specific chemical and the biological impact of that damage. Each drug can differ in terms of the number of DNA breaks that are associated with a given biological effect (Olive and Johnston, 1997).

7.1. *In Vitro* Studies

Three studies were conducted *in vitro* to evaluate the genotoxicity of lead by means of the comet assay. In all of them a significant increase in DNA fragmentation was found when different cell types were exposed to lead salts. Robbiano *et al.* (1999) obtained significant dose-dependent increases of DNA strand breaks in primary rat and human kidney cells exposed to different concentrations of lead acetate. Wozniak and Blasiak (2003) evaluated the genotoxic effects of the same salt in human lymphocytes. They found an increase in the comet tail length due to the induction of DNA strand breaks and/or alkali labile sites. Lymphocytes exposed to the highest dose showed a decrease in the comet tail length caused by the formation of DNA-DNA and DNA-protein cross-links. In the same study, the neutral version of the assay revealed that lead acetate induced DNA double-strand breaks at all concentrations tested. Similarly, Shaik *et al.* (2006) found a significant increase in the comet tail length when they analysed human lymphocytes exposed to lead nitrate, and this increase was proportional to lead nitrate concentration.

7.2. *In Vivo* Studies

All studies conducted in animals with the comet assay showed a positive effect of lead in the induction of DNA damage in several tissues and organs. Ramsdorf *et al.* (2008) evaluated the effects of inorganic lead in fishes (*H. malabaricus*). They found a significant difference between control and contaminated groups. Although differences between the applied doses were not observed, blood cells showed a higher sensitivity than kidney cells, suggested to be caused by the acute contamination. There was one exception for kidney cells at the lowest dose probably due to the short exposure time and also to the low quantity of lead, as explained the authors. These results are in agreement with Ferraro *et al.* (2004) who found a significant increase of tailed nucleoids in fish erythrocytes (*H. malabaricus*) treated with PbII, showing that extended exposures to lead contaminants are capable of originating damages in the genetic material of fishes.

Valverde *et al.* (2002) used a lead inhalation model in mice in order to detect the induction of genotoxic damage as single-strand breaks and alkali-labile sites in several organs. They found a positive induction of DNA damage after a single inhalation only in the liver and the lung. In subsequent inhalations the response was positive in all organs tested except for the testicle. These results showed that lead acetate inhalations induced systemic DNA damage but some organs are special targets for this metal, such as lung and liver, depending in part on length of exposure. Devi *et al.* (2000) also found a significant increase in mean comet tail length at all time intervals after oral treatment of mice with lead nitrate when compared to controls. The same results were reported by Tewari *et al.* (2005), who observed a statistically significant DNA damage in bone marrow cells from Swiss albino male mice exposed to leachates.

Yuan and Tang (2001) studied the accumulation effect of lead on DNA damage and the protection offered by selenium in mice blood cells of three generations. A significant induction of DNA damage was observed in both sexes of the second and third generations, suggesting that the accumulation effect of lead was very significant starting from the second generation.

Valverde *et al.* (2001) explored the capacity of lead, cadmium or a mixture of both metals to interact with acellular DNA in cells from several organs of the CD-1 mice by employing a variant of the comet assay. By means of this modified assay, that uses an enriched-lysis solution with proteinase-K and was described by Kasamatsu *et al.* (1996), DNA is no longer held under the regulation of any metabolic pathway or membrane barrier. They obtained a negative response in the induction of DNA damage in cells derived from the liver, kidney and lung. However, they observed the production of lipid peroxidation and an increase in free radical levels in the different organs after inhalation of lead acetate, suggesting the induction of genotoxicity and carcinogenicity by indirect interactions, such as oxidative stress.

7.3. Epidemiological Studies

The comet assay has been used in many epidemiological studies as an important end-point to determine the possible induction of genotoxicity in individuals occupationally exposed to lead. Despite the controversy regarding the genotoxic properties of lead, all studies conducted to date in which damage was assessed using the comet assay, showed positive results.

Palus *et al.* (2003) assessed genotoxic damage in peripheral lymphocytes of workers from a Polish battery plant after high-level occupational exposure to lead and cadmium. The results of the comet assay showed a slightly but significantly increased rate in DNA migration compared to the control group. The same results were also reported in workers from battery plants by Restrepo *et al.* (2000) and Fracasso *et al.* (2002), in secondary smelter lead workers (Ye *et al.*, 1999; Steinmetz-Beck *et al.*, 2005), in workers producing storage battery (Martino-Roth *et al.*, 2003; Chen *et al.*, 2006), and in workers from a secondary lead recovery unit (Danadevi *et al.*, 2003), as well as in a study conducted in children environmentally exposed to lead (Méndez-Gómez *et al.*, 2008).

The fact that all studies using the comet assay showed positive results, while other tests to assess genotoxic effects provided such different results, could be due to the fact that the comet assay detects not only the induction of permanent DNA damage but DNA damage that is repaired quickly and also the DNA repair processes that are taking place at the time of the experiment and are reflected as strand breaks in the assay.

Concluding Remarks

Most studies reviewed in this article, *in vitro*, *in vivo*, as well as in human populations, showed positive results in the induction of CA, SCE, MN and DNA strand breaks by different chemical forms of lead. However, in those studies which evaluated the induction of CA by lead chromate, the authors related this fact to the toxic action of chromate and not to lead. In addition, several works did not report clear evidence on whether inorganic lead compounds exert clastogenic effects themselves, or whether they enhance CA induced by compounds that occur simultaneously or arise during cell culturing (Beck and Obe, 1975; Deknudt and Leonard, 1975; Forni *et al.*, 1980). Furthermore, some studies which did not find induction of SCE in workers occupationally exposed to lead showed increases in SCE rates in smoker exposed individuals indicating that tobacco smoke works as enhancer of the genotoxic effects of lead (Maki-Paakanen, 1981; Rajah and Ahuja, 1995).

This variability in findings could be due to the influence of some confounding factors, as duration and route of lead exposure, cell culturing time following the exposure, smoking habits and simultaneous exposure to other toxic agents that could modify the results. Moreover, the type of cell evaluated in each work also plays an important role in the interpretation of results. For example, most studies conducted in human populations involve the use of blood cells and they remain in an arrested G0 phase, so any lesions induced in DNA may persist for some time (for instance, until cells are stimulated to divide in culture) (Winder and Bonin, 1993). In addition, different types of cells could have different defence mechanisms against lead genotoxicity, as reported by Hwua and Yang (1998), suggesting that lead mutagenicity is positively correlated to the cellular oxidative stress.

On the other hand, it has been reported that genotoxicity of lead could be due to indirect mechanisms (Hartwig *et al.*, 1990; Hartwig, 1994; Landrigan *et al.*, 2000; Silbergeld *et al.*, 2003; Garza *et al.*, 2005). Lead can substitute calcium and/or zinc in processes and enzymes involved in DNA processing and repair leading to an inhibition of DNA repair and an enhancement in the genotoxicity when combined with other DNA damaging agents. Thus, confounding factors as well as the indirect mechanisms of genotoxicity must be considered when assessing the genotoxic potential of lead compounds.

Acknowledgments

This work was supported by a grant from the *Xunta de Galicia* (INCITE08PXIB106155PR).

References

Aboul-Ela, E.I. (2002) The protective effect of calcium against genotoxicity of lead acetate administration on bone marrow and spermatocyte cells of mice in vivo. *Mutat. Res.* 516: 1-9.

Akiyama, M., Kyoizumi, S., Hirai, Y., Kusunoki, Y., Iwamoto, K.S., Nakamura, N. (1995) Mutation frequency in human blood cells increases with age. *Mutat. Res.* 338: 141-149.

Albertini, R.J., Anderson, D., Douglas, G.R., Hagmar, L., Hemminki, K., Merlo, F., Natarajan, A.T., Norppa, H., Shuker, D.E.G., Tice, R., Waters, M.D., Aitio, A. (2000) IPCS guidelines for the monitoring of genotoxic effects of carcinogens in humans. *Mutat. Res.* 463: 111-172.

Albertini, R.J., Hayes, R.B. (1997) Somatic cell mutations in cancer epidemiology. In: Application of biomarkers in cancer epidemiology. IARC Scientific Publications 142. Toniolo, P., Boffeta, P., Shuker, D.E.G, Rothman, N., Hulka, B., Pearce, N. (Eds.). *International Agency for Research on Cancer, Lyon*. Pp. 159-184.

Albertini, R.J., Nicklas, J.A. y O'Neill, J.P. (1993) Somatic cell mutations in humans: biomarkers for genotoxicity. *Environ. Health Perspect.* 101 (Suppl. 3): 193-201.

Albertini, S., Kirsch-Volders, M. (1997) Summary and conclusions on the MNT in vitro and application on testing strategies. *Mutat. Res.* 392: 183-185.

Alghazal, M., Sutiakova, I., Kovalkovicova, N., Legath, J., Falis, M., Pistl, J., Sabo, R., Benova, K., Sabova, L. and Vaczi, P. (2008) Induction of micronuclei in rat bone marrow after chronic exposure to lead acetate trihydrate. *Toxicol. Ind. Health.* 24: 587-593.

Al-Hakkak, Z.S., Hamamy, H.A., Murad, A.M., Hussain, A.F. (1986) Chromosome aberrations in workers at a storage battery plant in Iraq. *Mutat. Res.* 171: 53-60.

ATSDR (Agency for Toxic Substances and Disease Registry) (1999) *Toxicological profile for lead (update)*. US Department of Health and Human Services. Agency for Toxic Substances and Disease Registry (Ed.), Atlanta.

Bauchinger, M., Dresp, J., Schmid, E., Englert, N., Krause, C. (1977) Chromosome analyses of children after ecological lead exposure. *Mutat. Res.* 56: 75-80.

Bauchinger, M., Schmid, E. (1970) Die cytogenetische wirkung von blei in menschlichen lymphozyten. *"Arbeitsgruppe Blei" der Kommission für Umveltgefahren des Bundesgesundheitsamtes. Berlin.*

Bauchinger, M., Schmid, E. (1972) Chromosome analysis of cultures of Chinese hamster cells after treatment with lead acetate. *Mutat. Res.* 14: 95-100.

Bauchinger, M., Schmid, E., Schmid, D. (1972) Chromosomenanalyse bei Verkehrspolizisten mit erhöhter Bleilast. *Mutat. Res.* 16: 407-412.

Beckman, L., Nordenson, I., Nordstrom, S. (1982) Occupational and environmental risks in and around a smelter in northern Sweden. VIII. Three-year follow-up of chromosomal aberrations in workers exposed to lead. *Hereditas.* 96: 261-264.

Beek, B., Obe, G. (1974) Effect of lead acetate on human leukocyte chromosomes in vitro. *Experientia.* 30: 1006-1007.

Beek, B., Obe, G. (1975) The human leukocyte test system. VI. The use of sister chromatid exchanges as possible indicators for mutagenic activities. *Humangenetik.* 29: 127-134.

Bernstein, C., Bernstein, H. (1991). *Aging, sex and DNA repair*. Academic Press USA, San Diego. Pp. 15-26.

Bihari, N., Batel, R., Zahn, R.K. (1990) DNA damage determination by the alkaline elution technique in the haemolymph of mussel *Mytilus galloprovincialis* treated with benzo(a)pyrene and 4-nitroquinoline-N-oxide. *Aquat. Toxicol.* 18: 13-22.

Bijlsma, J.B., de France, H.F. (1976) Cytogenetic investigations in volunteers ingesting inorganic lead. *Int. Arch. Occup. Environ. Health.* 38: 145-148.

Bilban, M. (1998) Influence of the work environment in a Pb-Zn mine on the incidence of cytogenetic damage in miners. *Am. J. Ind. Med.* 34: 455-463.

Bolognesi, C., Parrini, M., Roggieri, P., Ercolini, C., Pellegrino, C. (1992). Carcinogenic and mutagenic pollutants: impact on marine organisms. In: *Proceedings of the FAO/UNEP/IOC workshop on the biological effects of pollutants on marine organisms* (Malta, September 10-14, 1991). MAP Technical Reports Series. 69: 113-121.

Bolognesi, C., Rabboni, R., Roggieri, P. (1996) Genotoxicity biomarkers in *Mytilus galloprovincialis* as indicators of marine pollutants. *Comp. Biochem. Physiol.* 113: 319-323.

Bonacker, D., Stoiber, T., Bohm, K.J., Prots, I., Wang, M., Unger, E., Thier, R., Bolt, H.M., Degen, G.H. (2005) Genotoxicity of inorganic lead salts and disturbance of microtubule function. *Environ. Mol. Mutagen.* 45: 346-353.

Bonassi, S., Znaor, A., Ceppi, M., Lando, C., Chang, W.P., Holland, N., Kirsch-Volders, M., Zeiger, E., Ban, S., Barale, R., Bigatti, M.P., Bolognesi, C., Cebulska-Wasilewska, A., Fabianova, E., Fucic, A., Hagmar, L., Joksic, G., Martelli, A., Migliore, L., Mirkova, E., Scarfi, M.R., Zijno, A., Norppa, H., Fenech, M. (2007) An increased micronucleus frequency in peripheral blood lymphocytes predicts the risk of cancer in humans. *Carcinogenesis*. 28: 625-631.

Cai, M.Y., Arenaz, P. (1998) Antimutagenic effect of crown ethers on heavy metal-induced sister chromatid exchanges. *Mutagenesis.* 13: 27-32.

Carrano, A.V., Natarajan, A.T. (1988) Considerations for population monitoring using cytogenetic techniques. *Mutat. Res.* 204: 379-406.

Celik, A., Ogenler, O., Comelekoglu, U. (2005) The evaluation of micronucleus frequency by acridine orange fluorescent staining in peripheral blood of rats treated with lead acetate. *Mutagenesis.* 20: 411-415.

Chen, Q. (1992) Lead concentrations in urine correlated with cytogenetic damage in workers exposed to lead. *Zhonghua Yu Fang Yi Xue Za Zhi*. 26: 334-335.

Chen, Z., Lou, J., Chen, S., Zheng, W., Wu, W., Jin, L., Deng, H., He, J. (2006) Evaluating the genotoxic effects of workers exposed to lead using micronucleus assay, comet assay and TCR gene mutation test. *Toxicology.* 223: 219-226.

Clevers, H., Alarcon, B., Wileman, T., Terhorst, C. (1988) The T cell receptor/CD3 complex: a dynamic protein ensemble. *Ann. Rev. Immunol.* 6: 629-662.

Cole, J., Skopek, T.R. (1994) ICPEMC Working Paper No. 3. Somatic mutant frequency, mutation rates and mutational spectra in the human population in vivo. *Mutat. Res.* 304: 33-105.

Dalpra, L., Tibiletti, M.G., Nocera, G., Giulotto, P., Auriti, L., Carnelli, V., Simoni, G. (1983) SCE analysis in children exposed to lead emission from a smelting plant. *Mutat. Res.* 120: 249-256.

Danadevi, K., Rozati, R., Banu, B.S., Rao, P.H., Grover, P. (2003) DNA damage in workers exposed to lead using comet assay. *Toxicology.* 187: 183-193.

Deknudt, G., Colle, A., Gerber, G.B. (1977a) Chromosomal abnormalities in lymphocytes from monkeys poisoned with lead. *Mutat. Res.* 45: 77-83.

Deknudt, G., Deminatti, M. (1978) Chromosome studies in human lymphocytes after in vitro exposure to metal salts. *Toxicology.* 10: 67-75.

Deknudt, G., Gerber, G.B. (1979) Chromosomal aberrations in bone-marrow cells of mice given a normal or a calcium-deficient diet supplemented with various heavy metals. *Mutat. Res.* 68: 163-168.

Deknudt, G., Leonard, A. (1975) Cytogenetic investigations on leucocytes of workers from a cadmium plant. *Environ. Physiol. Biochem.* 5: 319-327.

Deknudt, G., Manuel, Y., Gerber, G.B. (1977b) Chromosomal aberrations in workers professionally exposed to lead. *J. Toxicol. Environ. Health.* 3: 885-891.

Deknudt, G.H., Leonard, A, Ivanov, B. (1973) Chromosome aberrations observed in male workers occupationally exposed to lead. *Environ. Physiol. Biochem.* 3: 132-138.

Devi, K.D., Banu, B.S., Grover, P., Jamil, K. (2000) Genotoxic effect of lead nitrate on mice using SCGE (comet assay). *Toxicology.* 145: 195-201.

Devlin, T.M. (1997) *Textbook of biochemistry. Fourth edition.* Wiley-Liss, New York. Pp. 906-907.

Dhir, H., Roy, A.K., Sharma, A. (1993) Relative efficiency of *Phyllanthus emblica* fruit extract and ascorbic acid in modifying lead and aluminium-induced sister-chromatid exchanges in mouse bone marrow. *Environ. Mol. Mutagen.* 21: 229-236.

Dönmez, H., Dursun, N., Ozkul, Y., Demirtas, H. (1998) Increased sister chromatid exchanges in workers exposed to occupational lead and zinc. *Biol. Trace Elem. Res.* 61: 105-109.

Douglas, G.R., Bell, R.D., Grant, C.E., Wytsma, J.M., Bora, K.C. (1980) Effect of lead chromate on chromosome aberration, sister-chromatid exchange and DNA damage in mammalian cells in vitro. *Mutat. Res.* 77: 157-163.

Duydu, Y., Suzen, H.S., Aydin, A., Cander, O., Uysal, H., Isimer, A., Vural, N. (2001) Correlation between lead exposure indicators and sister chromatid exchange (SCE) frequencies in lymphocytes from inorganic lead exposed workers. *Arch. Environ. Contam. Toxicol.* 41: 241-246.

Fenech, M. (1998) Important variables that influence base-line micronucleus frequency in cytokinesis-blocked lymphocytes - a biomarker for DNA damage in human populations. *Mutat. Res.* 404: 155-165.

Fenech, M., Morley, A.A. (1985) Measurement of micronuclei in lymphocytes. *Mutat. Res.* 147: 29-36.

Ferraro, M.V., Fenocchio, A.S., Mantovani, M.S., Oliveira-Ribeiro, C.A., Cestari, M.M. (2004) Mutagenic effects of lead (PbII) on the fish *H. malabaricus* as evaluated using the comet assay, piscine micronucleus and chromosome aberrations tests. *Genet. Mol. Biol.* 27: 103-107.

Forni, A., Cambiaghi, G., Secchi, G.C. (1976) Initial occupational exposure to lead: Chromosome and biochemical findings. *Arch. Environ. Health.* 31: 73-78.

Forni, A., Sciame, A., Bertazzi, P.A., Alessio, L. (1980) Chromosome and biochemical studies in women occupationally exposed to lead. *Arch. Environ. Health.* 35: 139-146.

Forni, A., Secchi, G.C. (1973) Chromosome changes in preclinical and clinical lead poisoning and correlation with biochemical findings. In: *International Symposium on Environmental Health Aspects of Lead* (Amsterdam). Barth, D., Berlin, A., Engel, R., Recht, P., Smeets, J. (Eds.). Commission of the European Communities, Luxembourg. Pp. 473-485.

Fracasso, M.E., Perbellini, L., Solda, S., Talamini, G., Franceschetti, P. (2002) Lead induced DNA strand breaks in lymphocytes of exposed workers: role of reactive oxygen species and protein kinase C. *Mutat. Res.* 515: 159-169.

Fu, H. y Boffetta, P. (1995) Cancer and occupational exposure to inorganic lead compounds: a meta-analysis of published data. *Occup. Environ. Med.* 52: 73-81.

García-Lestón, J., Laffon, B., Roma-Torres, J., Teixeira, J.P., Monteiro, S., Pásaro, E., Prista, J., Mayan, O., Méndez, J. (2008) Efectos genotóxicos e inmunotóxicos de la exposición laboral al plomo. *Arch. Prev. Riesgos. Labor.* 11: 124-130.

Garza, A., Chávez, H., Vega, R., Soto, E. (2005) Mecanismos celulares y moleculares de la neurotoxicidad por plomo. *Salud. Mental.* 28: 48-58.

Garza-Chapa, R., Leal-Garza, C.H., Molina-Ballesteros, G. (1977) Chromosome analysis in professional subjects exposed to lead contamination. *Arch. Invest. Med. (Mex).* 8: 11-20.

Gasiorek, K., Bauchinger, M. (1981) Chromosome changes in human lymphocytes after separate and combined treatment with divalent salts of lead, cadmium, and zinc. *Environ. Mutagen.* 3: 513-518.

Gebhart, E. (1981) Sister chromatid exchange (SCE) and structural chromosome aberration in mutagenicity testing. *Hum. Genet.* 58(3): 235-254.

Gerber, G.B., Léonard, A., Jacquet, P. (1980) Toxicity, mutagenicity and teratogenicity of lead. *Mutat. Res.* 76: 115-141.

Glickman, B.W., Saddi, V.A., Curry, J. (1994) Spontaneous mutation in mammalian cells. *Mutat. Res.* 304: 19-32.

Goyer, R.A., Rhyne, B. (1973) Pathological effects of lead. *Int. Rev. Exp. Pathol.* 12: 1-77

Grandjean, P., Wulf, H.C., Niebuhr, E. (1983) Sister chromatid exchange in response to variations in occupational lead exposure. *Environ. Res.* 32: 199-204.

Hamurcu, Z., Donmez, H., Saraymen, R., Demirtas, H. (2001) Micronucleus frequencies in workers exposed to lead, zinc, and cadmium. *Biol. Trace Elem. Res.* 83: 97-102.

Hartwig, A. (1994) Role of DNA repair inhibition in lead- and cadmium-induced genotoxicity: a review. *Environ. Health Perspect.* 102 Suppl 3: 45-50.

Hartwig, A., Schlepegrell, R., Beyersmann, D. (1990) Indirect mechanism of lead-induced genotoxicity in cultured mammalian cells. *Mutat. Res.* 241: 75-82.

Hernberg, S. (2000) Lead poisoning in a historical perspective. *Am. J. Ind. Med.* 38: 244-254.

Hoffmann, M., Hagberg, S., Karlsson, A., Nilsson, R., Ranstam, J., Hogstedt, B. (1984) Inorganic lead exposure does not effect lymphocyte micronuclei in car radiator repair workers. *Hereditas.* 101: 223-226.

Hogstedt, B., Gullberg, B., Mark-Vendel, E., Mitelman, F., Skerfving, S. (1981) Micronuclei and chromosome aberrations in bone marrow cells and lymphocytes of humans exposed mainly to petroleum vapors. *Hereditas.* 94: 179-187.

Hogstedt, B., Kolnig, A.M., Mitelman, F., Schutz, A. (1979) Correlation between blood-lead and chromosomal aberrations. *Lancet.* 2: 262.

Horvat, D.J., Prpic-Majic, D. (1977) Cytogenetic and biochemical study of lead exposed population. *Abstracts of the 2nd International Conference on Environmental Mutagen.* (Edinburgh, England). P.174.

Huang, X.P., Feng, Z.Y., Zhai, W.L., Xu, J.H. (1985) Chromosomal aberrations in the bone marrow cells of mice treated with lead acetic. *Occup. Health Chem. Ind.* 2: 7-9.

Huang, X.P., Feng, Z.Y., Zhai, W.L., Xu, J.H. (1988) Chromosomal aberrations and sister chromatid exchanges in workers exposed to lead. *Biomed. Environ. Sci.* 1: 382-387.

Humphrey, R.M., Steward, D.L., Sedita, B.A. (1968) DNA *Mutat. Res.* 6: 459-465.

Hwua, Y.S., Yang, J.L. (1998) Effect of 3-aminotriazole on anchorage independence and mutagenicity in cadmium- and lead-treated diploid human fibroblasts. *Carcinogenesis.* 19: 881-888.

IARC (International Agency for Research on Cancer) (1987). IARC Monographs on the evaluation of carcinogenic risk to humans. Volumes 1 to 42. (Suppl. 7). *Overall evaluations of carcinogenicity: an updating of IARC monographs*. IARC, Lyon.

IARC (International Agency for Research on Cancer) (2006) *Inorganic and organic lead compounds*. IARC monographs on the evaluation of carcinogenic risks to humans volume 87. IARC, Lyon.

Ishioka, N., Umeki, S., Hirai, Y., Akiyama, M., Kodama, T., Ohama, K., Kyoizumi, S. (1997) Stimulated rapid expression in vitro for early detection of in vivo T-cell receptor mutations induced by radiation exposure. *Mutat. Res.* 390: 269-282.

Jacquet, P., Leonard, A., Gerber, G.B. (1977) Cytogenetic investigations on mice treated with lead. *J. Toxicol. Environ. Health.* 2: 619-624.

Jacquet, P., Tachon, P. (1981) Effects of long-term lead exposure on monkey leukocyte chromosomes. *Toxicol. Lett.* 8: 165-169.

Jagetia, G.C., Aruna, R. (1998) Effect of various concentrations of lead nitrate on the induction of micronuclei in mouse bone marrow. *Mutat. Res.* 415: 131-137.

Kasamatsu, T., Kohda, K., Kawazoe, Y. (1996) Comparison of chemically induced DNA breakage in cellular and subcellular systems using the comet assay. *Mutat. Res.* 369: 1-6.

Kasprzak, K.S. (1996) Oxidative DNA damage in metal-induced carcinogenesis. In: *Toxicology of metals.* Chang, L., Magos,L., Suzuki,T. (Eds.). CRC Press, Boca Raton, Florida. Pp. 299-320.

Klaunig, J.E., Xu, Y., Isenberg, J.S., Bachowski, S., Kolaja, K.L., Jiang, J., Stevenson, D.E., Walborg, E.F., Jr. (1998) The role of oxidative stress in chemical carcinogenesis. *Environ. Health Perspect.* 106 (Suppl 1): 289-295.

Klein, G., Klein, E. (1985) Evolution of tumours and the impact of molecular oncology. *Nature.* 315: 190-195.

Kohn, K.W., Erickson, L.C., Ewig, R.A.G., Friedman, C.A. (1976) Fractionation of DNA from mammalian cells by alkaline elution. *Biochemistry.* 15: 4629-4637.

Kronenberg, M., Siu, G., Hood, L.E., Shastri, N. (1986) The molecular genetics of the T-cell antigen receptor and T-cell antigen recognition. *Annu. Rev. Immunol.* 4: 529-591.

Kyoizumi, S., Umeki, S., Akiyama, M., Hirai, Y., Kusunoki, Y., Nakamura, N., Endoh, K., Konishi, J., Sasaki, M.S., Mori, T., Fujita, S., Cologne, J.B. (1992) Frequency of mutant T lymphocytes defective in the expression of the T-cell antigen receptor gene among radiation-exposed people. *Mutat. Res.* 265: 173-180.

Landrigan, P.J., Boffetta, P., Apostoli, P. (2000) The reproductive toxicity and carcinogenicity of lead: A critical review. *Am. J. Ind. Med.* 38: 231-243.

Lee, T.C., Ho, I.C. (1995) Modulation of cellular antioxidant defense activities by sodium arsenite in human fibroblasts. *Arch. Toxicol.* 69: 498-504.

Lerda, D. (1992) The effect of lead on *Allium cepa* L. *Mutat. Res.* 281: 89-92.

Lin, C.J., Wu, K.H., Yew, F.H., Lee, T.C. (1995) Differential cytotoxicity of cadmium to rat embryonic fibroblasts and human skin fibroblasts. *Toxicol. Appl. Pharmacol.* 133: 20-26.

Lin, R.H., Lee, C.H., Chen, W.K., Lin-Shiau, S.Y. (1994) Studies on cytotoxic and genotoxic effects of cadmium nitrate and lead nitrate in Chinese hamster ovary cells. *Environ. Mol. Mutagen.* 23: 143-149.

Lorencz, R., Nehéz, M., Dési, I. (1996) Investigations on the mutagenic effects of cadmium and lead on the chromosomes of the bone marrow cells in subchronic experiments. In:

Abstracts of the XXVIIth Meeting of the Hungarian Society of Hygiene (Blatonföldvár, Hungary, September 25-27). P. 116.

Madhavi, D., Devi, K.R., Sowjanya, B.L. (2008) Increased frequency of chromosomal aberrations in industrial painters exposed to lead-based paints. *J. Environ. Pathol. Toxicol. Oncol.* 27: 53-59.

Maki-Paakkanen, J., Sorsa, M., Vainio, H. (1981) Chromosome aberrations and sister chromatid exchanges in lead-exposed workers. *Hereditas.* 94: 269-275.

Martino-Roth, M.G., Viegas, J., Roth, D.M. (2003) Occupational genotoxicity risk evaluation through the comet assay and the micronucleus test. *Genet. Mol. Res.* 2: 410-417.

McBurney, M.W., Graham, F.L., Whitmore, G.F. (1972) Sedimentation analysis of DNA from irradiated and unirradiated L-cells. *Biophys. J.* 12: 369-383.

Mendez-Gomez, J., Garcia-Vargas, G.G., Lopez-Carrillo, L., Calderon-Aranda, E.S., Gomez, A., Vera, E., Valverde, M., Cebrian, M.E., Rojas, E. (2008) Genotoxic effects of environmental exposure to arsenic and lead on children in region Lagunera, Mexico. *Ann. N. Y. Acad. Sci.* 1140: 358-367.

Metcalfe, C.D. (1989) Testes for predicting carcinogenicity in fish. *CRC Cr. Rev. Aquat. Sci.* 1: 111-129.

Mielzynska, D., Siwinska, E., Kapka, L., Szyfter, K., Knudsen, L.E., Merlo, D.F. (2006) The influence of environmental exposure to complex mixtures including PAHs and lead on genotoxic effects in children living in Upper Silesia, Poland. *Mutagenesis.* 21: 295-304.

Mijares, I.A., López, P., Rosado, J.L., Cebrián, A., Vera-Aguilar, E., Alatorre, J., Quintanilla-Vega, M.B., Rojas-García, A.E., Stoltzfus, R.J., Cebrián, M.E., García-Vargas, G.G. (2006) Delta-aminolevulinic acid dehydratase genotype and its relationship with blood lead and zinc protoporphyrin levels in lead-exposed children living in a smelter community in Northern Mexico. *Toxicol. Mech. Meth.* 16: 41-47.

Miller, B., Albertini, S., Locher, F., Thybaud, V., Lorge, E. (1997) Comparative evaluation of the in vitro micronucleus test and the in vitro chromosome aberration test: industrial experience. *Mutat. Res.* 392: 45-59.

Minozzo, R., Deimling, L.I., Gigante, L.P., Santos-Mello, R. (2004) Micronuclei in peripheral blood lymphocytes of workers exposed to lead. *Mutat. Res.* 565: 53-60.

Mitchell, I.D., Lambert, T.R., Burden, M., Sunderland, J., Porter, R.L., Carlton, J.B. (1995) Is polyploidy an important genotoxic lesion? *Mutagenesis.* 10: 79-83.

Montaldi, A., Zentilin, L., Venier, P., Gola, I., Bianchi, V., Paglialunga, S., Levis, A.G. (1985) Interaction of nitrilotriacetic acid with heavy metals in the induction of sister chromatid exchanges in cultured mammalian cells. *Environ. Mutagen.* 7: 381-390.

Moreno, M.D. (2003) Metales. In: *Toxicología Ambiental. Evaluación del riesgo para la salud humana*. García-Brage, A. (Ed.). McGraw-Hill, Madrid. Pp: 223-227.

Muro, L.A., Goyer, R.A. (1969) Chromosome damage in experimental lead poisoning. *Arch. Pathol.* 87: 660-663.

Nakamura, N., Umeki, S., Hirai, Y., Kyoizumi, S., Kushiro, J.-I., Kusunoki, Y., Akiyama, M. (1991) Evaluation of four somatic mutation assays for biological dosimetry of radiation-exposed people including atomic bomb survivors. In: *Progress in chemical and biological research.* Gledhill, B.L., Mauro, F. (Eds). Wiley/Liss, New York. Pp. 341-350.

Natarajan, A.T., Balajee, A.S., Boei, J.J.W.A., Darroudi, F., Dominguez, I., Hande, M.P., Meijers, M., Slijepcevic, P., Vermeulen, S., Xiao, Y. (1996) Mechanisms of induction of chromosomal aberrations and their detection by fluorescence in situ hybridization. *Mutat. Res.* 372: 247-258.

Nayak, B.N., Ray, M., Persaud, T.V., Nigli, M. (1989) Relationship of embryotoxicity to genotoxicity of lead nitrate in mice. *Exp. Pathol.* 36: 65-73.

Nehez, M., Lorencz, R., Desi, I. (2000) Simultaneous action of cypermethrin and two environmental pollutant metals, cadmium and lead, on bone marrow cell chromosomes of rats in subchronic administration. *Ecotoxicol. Environ. Saf.* 45: 55-60.

Nordenson, I., Beckman, G., Beckman, L., Nordstrom, S. (1978) Occupational and environmental risks in and around a smelter in northern Sweden. IV. Chromosomal aberrations in workers exposed to lead. *Hereditas.* 88: 263-267.

Nordenson, I., Nordstrom, S., Sweins, A., Beckman, L. (1982) Chromosomal aberrations in lead-exposed workers. *Hereditas.* 96: 265-268.

O'Riordan, M.L., Evans, H.J. (1974) Absence of significant chromosome damage in males occupationally exposed to lead. *Nature.* 247: 50-53.

Olive, P.L., Banáth, J.P. (2006) The comet assay: a method to measure DNA damage in individual cells. *Nature Protocols.* 1: 23-29.

Olive, P.L., Banath, J.P., Durand, R.E. (1990) Heterogeneity in radiation-induced DNA damage and repair in tumour and normal cells measured using the "comet" assay. *Radiat. Res.* 122: 86-94.

Olive, P.L., Johnston, P.J. (1997) DNA damage from oxidants: influence of lesion complexity and chromatin organization. *Oncol. Res.* 9: 287-294.

Olive, P.L., Wledeck, D., Banáth, J.P. (1991) DNA double-strand breaks measured in individual cells subjected to gel electrophoresis. *Cancer Res.* 51: 4671-4676.

Olsen, L.S., Nielsen, L.R., Nexo, B.A., Wassermann, K. (1996) Somatic mutation detection in human biomonitoring. *Pharmacol. Toxicol.* 78: 364-373.

Onalaja, A.O., Claudio, L. (2000) Genetic susceptibility to lead poisoning. *Environ. Health Perspect.* 108 Suppl 1: 23-28.

Oshimura, M., Barrett, J.C. (1986) Chemically induced aneuploidy in mammalian cells: mechanisms and biological significance in cancer. *Environ. Mutagen.* 8: 129-159.

Palus, J., Rydzynski, K., Dziubaltowska, E., Wyszynska, K., Natarajan, A.T., Nilsson, R. (2003) Genotoxic effects of occupational exposure to lead and cadmium. *Mutat. Res.* 540: 19-28.

Patierno, S.R., Banh, D., Landolph, J.R. (1988) Transformation of C3H/10T1/2 mouse embryo cells to focus formation and anchorage independence by insoluble lead chromate but not soluble calcium chromate: relationship to mutagenesis and internalization of lead chromate particles. *Cancer Res.* 48: 5280-5288.

Patierno, S.R., Landolph, J.R. (1989) Soluble vs insoluble hexavalent chromate. Relationship of mutation to in vitro transformation and particle uptake. *Biol. Trace Elem. Res.* 21: 469-474.

Patrick, L. (2006) Lead toxicity, a review of the literature. Part I: exposure, evaluation, and treatment. *Altern. Med. Rev.* 11: 2-22.

Pelclová, D., Picková, J, Patzelová, V. (1997) Chromosomal aberrations, hormone levels and oxidative phenotype (P450 2D6) in low occupational lead exposure. *Central Eur. J. Occup. Environ. Med.* 3: 314-322.

Piao, F., Cheng, F., Chen, H., Li, G., Sun, X., Liu, S., Yamauchi, T., Yokoyama, K. (2007) Effects of zinc coadministration on lead toxicities in rats. *Ind. Health.* 45: 546-551.

Piña-Calva, A., Madrigal-Bujaidar, E., Fuentes, M.V., Neria, P., Perez-Lucio, C. and Velez-Zamora, N.M. (1991) Increased frequency of chromosomal aberrations in railroad car painters. *Arch. Environ. Health.* 46: 335-339.

Pinto, D., Ceballos, J.M., Garcia, G., Guzman, P., Del Razo, L.M., Vera, E., Gomez, H., Garcia, A., Gonsebatt, M.E. (2000) Increased cytogenetic damage in outdoor painters. *Mutat. Res.* 467: 105-111.

Poma, A., Pittaluga, E., Tucci, A. (2003) Lead acetate genotoxicity on human melanoma cells in vitro. *Melanoma Res.* 13: 563-566.

Rajah, T., Ahuja, Y.R. (1995) In vivo genotoxic effects of smoking and occupational lead exposure in printing press workers. *Toxicol. Lett.* 76: 71-75.

Ramsdorf, W.A., Ferraro, M.V., Oliveira-Ribeiro, C.A., Costa, J.R., Cestari, M.M. (2008) Genotoxic evaluation of different doses of inorganic lead (PbII) in Hoplias malabaricus. *Environ. Monit. Assess.*

Restrepo, H.G., Sicard, D., Torres, M.M. (2000) DNA damage and repair in cells of lead exposed people. *Am. J. Ind. Med.* 38: 330-334.

Richardt, G., Federolf, G., Habermann, E. (1986) Affinity of heavy metal ions to intracellular Ca2+-binding proteins. *Biochem. Pharmacol.* 35: 1331-1335.

Robbiano, L., Carrozzino, R., Puglia, C.P., Corbu, C., Brambilla, G. (1999) Correlation between induction of DNA fragmentation and micronuclei formation in kidney cells from rats and humans and tissue-specific carcinogenic activity. *Toxicol. Appl. Pharmacol.* 161: 153-159.

Rojas, E., López, M.C., Valverde, M. (1999) Single cell gel electrophoresis assay: methodology and applications. *J. Chromatogr. B.* 722: 225-254.

Rydberg, B. (1975) The rate of strand separation in alkali of DNA of irradiated mammalian cells. *Radiat. Res.* 61: 274-287.

Sandhu, S.S., Ma, T.H., Peng, Y., Zhou, X.D. (1989) Clastogenicity evaluation of seven chemicals commonly found at hazardous industrial waste sites. *Mutat. Res.* 224: 437-445.

Sarto, F., Stella, M., Acqua, A. (1978) Cytogenetic study of a group of workers with increased lead absorption indices. *Med. Lav.* 69: 172-180.

Schmid, E., Bauchinger, M., Pietruck, S., Hall, G. (1972) Cytogenetic action of lead in human peripheral lymphocytes in vitro and in vivo. *Mutat. Res.* 16: 401-406.

Schwanitz, G., Gebhart, E., Rott, H.D., Schaller, K.H., Essing, H.G., Lauer, O., Prestele, H. (1975) Chromosome investigations in subjects with occupational lead exposure. *Dtsch. Med. Wochenschr.* 100: 1007-1011.

Schwanitz, G., Lehnert, G., Gebhart, E. (1970) Chromosome damage after occupational exposure to lead. *Dtsch. Med. Wochenschr.* 95: 1636-1641.

Shaik, A.P., Sankar, S., Reddy, S.C., Das, P.G., Jamil, K. (2006) Lead-induced genotoxicity in lymphocytes from peripheral blood samples of humans: *in vitro* studies. *Drug Chem. Toxicol.* 1: 111-124.

Sharma, R.K., Jacobson-Kram, D., Lemmon, M., Bakke, J., Galperin, I., Blazak, W.F. (1985) Sister-chromatid exchange and cell replication kinetics in fetal and maternal cells after treatment with chemical teratogens. *Mutat. Res.* 158: 217-231.

Shtivelman, E., Lifshitz, B., Gale, R.P., Canaani, E. (1985) Fused transcript of abl and bcr genes in chronic myelogenous leukaemia. *Nature.* 315: 550-554.

Silbergeld, E.K. (2003) Facilitative mechanisms of lead as a carcinogen. *Mutat. Res.* 533: 121-133.

Silbergeld, E.K., Waalkes, M., Rice, J.M. (2000) Lead as carcinogen: experimental evidence and mechanisms of action. *Am J Ind Med.* 38: 316-323.

Silva, J.M., Santos-Mello, R. (1996) Chromosomal aberrations in lymphocytes from car painters. *Mutat. Res.* 368: 21-25.

Singh, N.P., McCoy, T., Tice, R.R., Schneider, E.L. (1988) A simple technique for quantitation of low levels of DNA damage in individual cell. *Exp. Cell Res.* 175: 184-192.

Sperling, K., Weiss, G., Münzer, M., Obe, G. (1970) Cromosomenuntersuchung an bleiexponierten Arbeitern. *"Arbeitsgruppe Blei" der Kommission für Umveltgefahren des Bundesgesundheitsamtes, Berlin.*

Spivei, A. (2007) The weight of lead. Effects add up in adults. *Environ. Health Perspect.* 115: A31-A36.

Steinert, S.A., Streib-Moutee, R., Leather, J.M., Chadwick, D.B. (1998) DNA damage in mussels at sites in San Diego Bay. *Mutat. Res.* 399: 65-85.

Steinmetz-Beck, A., Szahidewicz-Krupska, E., Beck, B., Poreba, R., Andrzejak, R. (2005) Genotoxicity effect of chronic lead exposure assessed using the comet assay. *Med. Pr.* 56: 295-302.

Tachi, K., Nishimae, S., Saito, K. (1985) Cytogenetic effects of lead acetate on rat bone marrow cells. *Arch. Environ. Health.* 40: 144-147.

Taningher, M., Bordone, R., Russo, P., Santi, L., Carlone, S. (1987) Major discrepancies between results obtained with two different methods for evaluating DNA damage: Alkaline elution and alkaline unwinding. Posible explanations. *Anticancer Res.* 7: 669-680.

Testa, A., Festa, F., Ranaldi, R., Giachelia, M., Tirindelli, D., De Marco, A., Owczarek, M., Guidotti, M., Cozzi, R. (2005) A multi-biomarker analysis of DNA damage in automobile painters. *Environ. Mol. Mutagen.* 46: 182-188.

Tewari, A., Chauhan, L.K., Kumar, D., Gupta, S.K. (2005) Municipal sludge leachate-induced genotoxicity in mice – a subacute study. *Mutat. Res.* 587: 9-15.

Thier, R., Bonacker, D., Stoiber, T., Bohm, K.J., Wang, M., Unger, E., Bolt, H.M., Degen, G. (2003) Interaction of metal salts with cytoskeletal motor protein systems. *Toxicol. Lett.* 140-141: 75-81.

Tice, R.R. (1995) The single cell gel/comet assay: a microgel electrophoretic technique for the detection of DNA damage and repair in individual cells. In: *Environmental Mutagenesis*. Bios Scientific Publishers, Oxford. Pp. 315-339.

Tice, R.R., Agurell, E., Anderson, D., Burlinson, B., Hartmann, A., Kobayashi, H., Miyamae, Y., Rojas, E., Ryu, J.-C., Sasaki, Y.F. (2000) Single cell gel/comet assay: guidelines for in vitro and in vivo genetic toxicology testing. *Environ. Mol. Mutagen*. 35: 206-221.

Todd, A.C., Wtmur, J.G., Moline, J.M., Godbold, J.H., Levin S.H., Landrigan, P.J. (1996) Unraveling the chronic toxicity of lead: An essential priority for environmental health. *Environ. Health Perspect*. 104:141-146.

Vaglenov, A., Carbonell, E., Marcos, R. (1998) Biomonitoring of workers exposed to lead. Genotoxic effects, its modulation by polyvitamin treatment and evaluation of the induced radioresistance. *Mutat. Res.* 418: 79-92.

Vaglenov, A., Creus, A., Laltchev, S., Petkova, V., Pavlova, S., Marcos, R. (2001) Occupational exposure to lead and induction of genetic damage. *Environ. Health Perspect.* 109: 295-298.

Vaglenov, A.K., Laltchev, S.G., Nosko, M.S., Pavlova, S.P. (1997) Cytogenetic monitoring of workers exposed to lead. *Cent. Eur. J. Occup. Environ. Med.* 3: 298-308.

Valverde, M., Fortoul, T.I., Diaz-Barriga, F., Mejia, J., del Castillo, E.R. (2002) Genotoxicity induced in CD-1 mice by inhaled lead: differential organ response. *Mutagenesis.* 17: 55-61.

Valverde, M., Trejo, C., Rojas, E. (2001) Is the capacity of lead acetate and cadmium chloride to induce genotoxic damage due to direct DNA-metal interaction? *Mutagenesis.* 16: 265-270.

Varmus, H.E. (1984) The molecular genetics of cellular oncogenes. *Annu. Rev. Genet.* 18: 553-612.

Veatch, W., Okada, S. (1969) Radiation-induced breaks of DNA in cultured mammalian cells. *Biophys. J.* 9: 330-346.

Vukmirovic, M., Bihari, N., Zahn, R.K., Müller, W.E.G., Batel, R. (1994) DNA damage in marine mussel *Mytilus galloprovincialis* as a biomarker of environmental contamination. *Mar. Ecol. Prog. Ser.* 109: 165-171.

Weaver, V.M., Schwartz, B.S., Ahn, K.-D., Stewart, W.J., Kelsey, K.T., Todd, A.C., Wen, J., Simon, D.J., Lustberg, M.E., Parsons, P.J., Silbergeld, E.K., Lee, B.-K. (2003) Associations of renal function with polymorphisms in the δ-aminolevulinic acid dehydratase, vitamin D receptor, and nitric ocide synthase genes in Korean lead workers. *Environ. Health Perspect.* 111: 1613-1619.

WHO (World Health Organization) (2001) Lead. In: *Air quality guidelines – Second edition.* WHO Regional Office for Europe, Copenhagen.

Willems, M.I., de Schepper, G.G., Wibowo, A.A., Immel, H.R., Dietrich, A.J., Zielhuis, R.L. (1982) Absence of an effect of lead acetate on sperm morphology, sister chromatid exchanges or on micronuclei formation in rabbits. *Arch. Toxicol.* 50: 149-157.

Wilson III, D.M., Thompson, L.H. (2007) Molecular mechanisms of sister-chromatid exchange. *Mutat. Res.* 616: 11-23.

Wilson, J.T., Pascoe, P.L., Parry, J.M., Dixon, D.R., (1998) Evaluation of the comet assay as a method for the detection of DNA damage in the cells of a marine invertebrate *Mytilus edulis* L. (Mollusca: Pelecypoda). *Mutat. Res.* 399: 87-95.

Winder, C., Bonin, T. (1993) The genotoxicity of lead. *Mutat. Res.* 285: 117-124.

Wise, J.P., Leonard, J.C., Patierno, S.R. (1992) Clastogenicity of lead chromate particles in hamster and human cells. *Mutat. Res.* 278: 69-79.

Wise, J.P., Sr., Stearns, D.M., Wetterhahn, K.E., Patierno, S.R. (1994) Cell-enhanced dissolution of carcinogenic lead chromate particles: the role of individual dissolution products in clastogenesis. *Carcinogenesis.* 15: 2249-2254.

Wise, S.S., Schuler, J.H., Holmes, A.L., Katsifis, S.P., Ketterer, M.E., Hartsock, W.J., Zheng, T., Wise, J.P., Sr. (2004) Comparison of two particulate hexavalent chromium compounds: Barium chromate is more genotoxic than lead chromate in human lung cells. *Environ. Mol. Mutagen.* 44: 156-162.

Wise, S.S., Schuler, J.H., Katsifis, S.P., Wise, J.P., Sr. (2003) Barium chromate is cytotoxic and genotoxic to human lung cells. *Environ. Mol. Mutagen.* 42: 274-278.

Wiwanitkit, V., Suwansaksri, J., Soogarun, S. (2008) White blood cell sister chromatid exchange among a sample of Thai subjects exposed to lead: Lead-induced genotoxicity. *Toxicol. Environ. Chem.* 90: 765-768.

Wozniak, K., Blasiak, J. (2003) In vitro genotoxicity of lead acetate: induction of single and double DNA strand breaks and DNA-protein cross-links. *Mutat. Res.* 535: 127-139.

Wu, F.Y., Chang, P.W., Wu, C.C., Kuo, H.W. (2002) Correlations of blood lead with DNA-protein cross-links and sister chromatid exchanges in lead workers. *Cancer Epidemiol. Biomarkers Prev.* 11: 287-290.

Wulf, H.C. (1980) Sister chromatid exchanges in human lymphocytes exposed to nickel and lead. *Dan. Med. Bull.* 27: 40-42.

Xie, H., Wise, S.S., Holmes, A.L., Xu, B., Wakeman, T.P., Pelsue, S.C., Singh, N.P., Wise, J.P., Sr. (2005) Carcinogenic lead chromate induces DNA double-strand breaks in human lung cells. *Mutat. Res.* 586: 160-172.

Xu, J., Wise, J.P., Patierno, S.R. (1992) DNA damage induced by carcinogenic lead chromate particles in cultured mammalian cells. *Mutat. Res.* 280: 129-136.

Yang, J.L., Yeh, S.C. and Chang, C.Y. (1996) Lead acetate mutagenicity and mutational spectrum in the hypoxanthine guanine phosphoribosyltransferase gene of Chinese hamster ovary K1 cells. *Mol. Carcinog.* 17: 181-191.

Ye, X.B., Fu, H., Zhu, J.L., Ni, W.M., Lu, Y.W., Kuang, X.Y., Yang, S.L., Shu, B.X. (1999) A study on oxidative stress in lead-exposed workers. *J. Toxicol. Environ. Health. A.* 57: 161-172.

Yuan, X., Tang, C. (2001) The accumulation effect of lead on DNA damage in mice blood cells of three generations and the protection of selenium. *J. Environ. Sci. Health A Tox. Hazard Subst. Environ. Eng.* 36: 501-508.

Yunis, J.J. (1983) The chromosomal basis of human neoplasia. *Science.* 221: 227-236.

Zelikoff, J.T., Li, J.H., Hartwig, A., Wang, X.W., Costa, M., Rossman, T.G. (1988) Genetic toxicology of lead compounds. *Carcinogenesis.* 9: 1727-1732.

In: Advances in Medicine and Biology
Editor: Leon V. Berhardt

ISBN 978-1-61209-135-8

Chapter 4

CELL CYCLE ALTERATION IN DOWN SYNDROME

***Mohammed Rachidi**[*1,2], **Charles Tetaria**[1] and **Carmela Lopes**[1,3]*

[1] Molecular Genetics of Human Diseases (GMMH), BP 130228, F-98717 Punaauia, Tahiti, French Polynesia.
[2] Laboratory of Genetic dysregulation models: Trisomy 21 and Hyperhomocysteinemia, EA 3508, University Paris 7-Denis Diderot, Paris, France.
[3] University of French Polynesia, Dept of Sciences, Tahiti, French Polynesia.

ABSTRACT

Trisomy 21 or Down syndrome is determined by the triplication of human chromosome 21 and the overdosage of genes on this chromosome in all cells. This genetic overdosage determines transcriptional alteration of most of the chromosome 21 genes, and trisomic cells suffer of dosage imbalance of a lot of proteins that can potentially impair all molecular pathways, including cell cycle control. Thus, trisomy 21 provides a genetic model for the role of aneuploidy in neurogenesis, leukemia, tumors and cancer. An example of chromosome 21 gene involved in alteration of cell cycle in neuronal cell, determining abnormal neurogenesis, is Dyrk1A, the ortholog of the *Drosophila minibrain* (MNB) gene which is involved in the regulation of neural proliferation. Three oncogenes mapping on chromosome 21, AML1, ERG and ETS2, highly related to the genesis of leukemias, are of the most interest to contribute to the general understanding of the oncogenic mechanisms of chromosomal aneuploidies, the most common abnormalities in cancer. The chromosome 21 gene TIAM1 (T-cell

[*] Corresponding author: Mohammed RACHIDI; Molecular Genetics of Human Diseases (GMMH); BP 130228, F-98717 Punaauia, Tahiti; French Polynesia; mrachidi2@yahoo.fr

lymphoma invasion and metastasis 1 gene) has important role in the progression of epithelial cancers, especially carcinomas of the breast and colon. The direct involvement of the human chromosome 21 SIM2 (Single minded 2) in the inhibition of gene expression, growth inhibition, inhibition of tumor growth as well as induction of apoptosis that could involve a block of cell cycle, establish SIM2 gene as a molecular target for cancer therapeutics and have additional important implications for the future diagnosis and treatment of specific solid tumors as well as for further understanding the cancer risk in Down syndrome patients.

INTRODUCTION

Trisomy 21, the most common origin of Down syndrome, is an extra copy of chromosome 21 in all cells (LeJeune et al., 1959) generated by a chromosomal non-disjunction during meiosis which increases significantly with advancing maternal age (Connor and Ferguson-Smith, 1997).

This genetic disorder causes a complex phenotype, the main features of which are the morphological abnormalities of head and limbs, short stature, joint hyperlaxity, hypotonia, skeletal abnormalities, frequent occurrence of visceral malformation, particularly of the heart, immunological defects, increased risk of leukaemia, haematological and endocrinal alterations, early occurrence of an Alzheimer-like neuropathology and mental retardation (Rahmani et al., 1989; Delabar et al., 1993; Korenberg et al., 1994; Epstein, 1995). In addition to the specific mental retardation disease seen in Down syndrome patients, trisomy 21 is also a risk factor for a number of diseases. The incidence of congenital heart disease (Ferencz et al., 1989), the atherosclerosis (Murdoch et al., 1977; Yla-Herttuala et al., 1989), and the Hirschsprung disease are both significantly elevated in individuals with trisomy 21, and metabolic diseases are also observed (Puffenberger et al., 1994; Epstein, 1995; Cohen, 1999).

The genetic overdosage, caused by trisomy 21, determines alteration in transcriptional level of most of genes on chromosome 21. Most of the chromosome 21 genes are transcribed 1.5 fold, corresponding to the genetic over dosage; the other genes are over-expressed or down-expressed at variable levels. The chromosome 21 genes encode several transcription factors and other proteins controlling the transcription, and their dosage alterations determine transcriptional variations of several genes located on other chromosomes (Rachidi and Lopes, 2007). Thus, trisomic cells suffer of dosage imbalance of several proteins that can potentially impair all molecular pathways, including cell cycle control. Interestingly, some members of the phosphorylation pathways that control the cell cycle are encoded by chromosome 21 genes, suggesting the alteration in cell cycle control in Down syndrome individuals (Groet et al., 2000; Yang et al., 2000; Hasle, 2001).

Because, some specific Down syndrome phenotypes and some Down syndrome associated diseases correlate with an alteration of the cell cycle, we presented in this work the interest of the trisomy 21 studies, of the chromosome 21 genes and their contribution to elucidate some mechanisms of diseases related to cell cycle pathways and we discussed the possible implication of the cell cycle alterations in leukaemia, tumors and cancer in Down syndrome patients.

Cell Cycle and Neurogenesis Alterations in Down Syndrome

Down syndrome is considered mainly the most frequent genetic cause of mental retardation that contributes to about 30% of all moderate-to-severe cases of mental retardation (Lejeune, 1990; Stoll et al., 1990; Pulsifer, 1996), characterized by alterations in neurogenesis, neuronal differentiation, myelination, and synaptogenesis (Marin-Padilla, 1976; Takashima et al., 1981; Dalton and Crapper-McLachlan, 1986; Coyle et al., 1986; Becker et al., 1986; Wisniewski and Schmidt-Sidor, 1989; Schmidt-Sidor et al., 1990; Wisniewski, 1990; Becker et al., 1991; Wisniewski and Bobinski, 1991; Takashima et al., 1994; Bahn et al., 2002), smaller cerebellum (Jernigan et al., 1993; Raz et al., 1995; Aylward et al., 1997), larger parahippocampal gyrus, hypoplasia of the hippocampus and cortex (Weis et al., 1991; Wang et al., 1992; Raz et al., 1995; Epstein, 1995; Schneider et al., 1997; Risser et al., 1997; Teipel et al., 2003; Teipel and Hampel, 2006). Moreover, the most widely studied animal model of Down syndrome, Ts65Dn, (Davisson et al., 1990), is trisomic for major part of genes on human chromosome 21 and show similar phenotypes to those observed in Down syndrome patients, including reduced different brain areas (Baxter et al., 2000; Cooper et al., 2001) and cognitive impairments (Pennington et al., 2003).

In neuronal progenitors, as in other somatic cells, the proliferation and the growth arrest are regulated by a balance of signals that control entry, progression and exit from the cell cycle (Oppenheim et al., 1989; Cunningham and Roussel, 2001). Alteration in these mechanisms can determine the hypoplasic phenotype observed in Down syndrome patients and mouse models. In the postnatal Ts65Dn brain, a reduced mitotic index has been found in the external granule cells of the cerebellum (Roper et al., 2006), and proliferation of neural stem cells is reduced in the postnatal and adult dentate gyrus (Rueda et al., 2005; Clark et al., 2006; Contestabile et al., 2007). Thus, a neural cell proliferation phenotype is present in many cell types and central nervous system compartments. This proliferation defect persists after birth and likely continues to impact brain growth and function into adulthood in Down syndrome.

It has been demonstrated an altered proliferation of neuronal precursors in the hippocampus and dorsal neocortical wall in second trimester of Down syndrome brain (Contestabile et al., 2007). Comparison of normogenic versus trisomic subjects showed that Ts65Dn mice as well as Down syndrome fetuses had a larger percent number of cells in G2 phase of the cell cycle compared with controls. The opposite was true for the number of cells in the M phase (Contestabile et al., 2007). A longer time spent by proliferating cells in G2 may result in a longer cell cycle and, hence, a reduced proliferation rate. Similarly, reduced corticogenesis with elongation of cell cycle of neocortical progenitors has been demonstrated in the Ts16 mouse model for Down syndrome (Haydar et al., 2000). These results strongly suggest that alterations in the dynamics of cell cycle causally contribute to reduce cell proliferation potency in Down syndrome.

Interestingly, one member of the phosphorylation pathways that control the cell cycle is encoded by the chromosome 21 gene DYRK1A, within the Down Syndrome Critical Region, DSCR (Guimera et al., 1996), a genetically defined minimal region of chromosome 21 responsible for most of the relevant Down syndrome phenotypes, including mental retardation (Rahmani et al., 1989).

DYRK1A is a member of the DYRK (dual-specificity tyrosine[Y]-regulated kinase) subfamily that constitutes an emerging evolutionarily conserved subfamily of proline- or arginine-directed protein kinases belonging to the CMGC family of cyclin-dependent kinases (CDKs) (Becker and Joost, 1999; Kannan and Neuwald, 2004; Lochhead et al., 2005).

The involvement of Dyrk1A in the regulation of neural proliferation was initially suggested on the basis of neuronal alterations in the proliferation centres of the larval brain of Drosophila minibrain (Mnb) mutants (Tejedor et al., 1995). This supports the evolutionary conservation of this role as regulator of proliferation in neural progenitors (Hämmerle et al., 2008). This is also supported by the altered levels of cell cycle regulators found in transgenic mice overexpressing Dyrk1A (Branchi et al., 2004) and by the phosphorylation of cell cycle regulators by DYRK1A (de Graaf et al., 2004).

Homozygous null Dyrk1A mutant mouse embryos (Dyrk1A-/-) present delayed general growth with an overall reduction in organ growth, including neurogenesis decrease in the brain (Fotaki et al., 2004). The reduction in brain size and behavioral defects observed provide evidence about the non-redundant vital role of DYRK1A and suggest a conserved mode of action that determines normal growth and brain size in both mice and flies (Fotaki et al., 2002).

Dyrk1A transcription seems to start during mitosis of neuroepithelial progenitor cells, maintains during G1, and stops before S phase (Hämmerle et al., 2002). Dyrk1A proteins remain during S and G2 phases, but seem to be down-regulated to very low levels before the next mitosis (Hämmerle et al., 2002). The transient expression of Dyrk1A on a single cell cycle, which precedes the onset of neurogenesis, suggests that Dyrk1A may be restricted to a particular cell cycle within the lineage of neuroepithelial progenitor cells (Hämmerle et al., 2002).

At a biochemical level, DYRK1A can phosphorylate cyclin L2 protein (de Graaf et al., 2004) indicates its role in cell cycle regulation. Moreover, DYRK1A overexpression was associated with an increase in phosphorylation of the Forkhead transcription factor FKHR and with high levels of cyclin B1, suggesting a correlation in vivo between DYRK1A overexpression and cell cycle protein alteration. In addition, altered phosphorylation of transcription factors of CREB family (cyclic AMP response binding protein) was observed, supporting a role of DYRK1A overexpression in the neuronal abnormalities seen in Down syndrome and suggesting that this pathology is linked to altered levels of proteins involved in the regulation of cell cycle (Branchi et al., 2004).

These results suggest a central role of DYRK1A in the pathways of cell cycle control and its overexpression participate to neurogenesis alteration in Down syndrome patients.

Contribution of Chromosome 21 Genes in Tumors and Cancers

Genomic alteration is commonly manifested by structural or numerical chromosomal aberrations (Roschke et al., 2003). These genetic abnormalities illustrate the evolution of the cancer cell resulting in a multi-step series of genetic changes that lead to essential alterations in cell physiology such as loss of growth controls and normal apoptotic response as well as sustained angiogenesis, invasion, and metastasis (Hanahan and Weinberg, 2000).

Trisomy 21 pathogenesis could contribute to the general understanding of the oncogenic mechanisms of chromosomal aneuploidies, the most common abnormalities in cancer (Fults et al., 1990; Yamamoto et al., 1999; von Deimling et al., 2000). Thus, increased transcription of the oncogenes and tumor suppressor genes in the trisomic chromosome 21 may participate to cancer in Down syndrome patients (Satgé et al., 1997).

Some genes located on human chromosome 21 have important roles in solid tumors and cancer that are related to an alteration of the cell cycle.

Single Minded Gene (SIM2)

The human SIM2 (Chen et al., 1995; Dahmane et al., 1995), located within the DSCR of chromosome 21, shows a high homology with the Drosophila single minded (sim), a master gene of the midline development in the central nervous system that functions as transcriptional regulator in cell fate determination (Crews et al., 1988; Thomas et al., 1988; Nambu et al., 1991). SIM2 encodes helix-loop-helix transcription factor (Crews et al., 1988), belong to a family of transcriptional repressors (Chrast et al., 1997; Moffett et al., 1997), involved in brain development and neuronal differentiation (Goshu et al., 2004; Rachidi et al., 2005).

Sim2 mutant mice are lethal in the early post-natal days and show skeletal alteration due probably to cell proliferation defects (Goshu et al., 2002). Functional studies indicated that SIM2 protein control the Sonic hedgehog (Shh) expression in the brain (Epstein et al., 2000), involved in cell growth and differentiation in the brain. Moreover, SIM2 can inhibit cell cycle by inhibition of cyclin-E expression (Meng et al., 2006).

Interestingly, in addition to a role in brain and neuronal development, SIM2 represent the strongest candidate gene involved in cancer being involved in the pathogenesis of solid tumors (Deyoung et al., 2002; Deyoung et al., 2003a) and overexpression of SIM2 was associated with tumors of the colon, pancreas, and prostate (Deyoung et al., 2003a; DeYoung et al., 2003b). Moreover, SIM2 gene was expressed in the colon, prostate and pancreatic carcinomas, but not in their corresponding normal tissues, and its expression was seen in a stage-specific manner in colon and prostate tumors (Deyoung et al., 2002).

The gene expression profiling demonstrated that SIM2 is among the highly up-regulated genes in 29 prostate cancers (Halvorsen et al., 2005). SIM2 was significantly co-expressed and increased in prostate cancer. Tumor cell expression of SIM2 protein was associated with adverse clinico-pathologic factors like increased pre-operative serum prostate specific antigen, high histologic grade, invasive tumor growth with extra-prostatic extension, and increased tumor cell proliferation by Ki-67 expression. Recently, SIM2 protein expression was significantly associated with reduced patient survival and the increased expression of SIM2 protein is a novel marker of aggressive prostate cancer (Halvorsen et al., 2007).

The antisense inhibition of SIM2 expression in a colon cancer cell line resulted in inhibition of gene expression, growth inhibition and induction of apoptosis in vitro as well as inhibition of tumor growth in nude mouse tumoriginicity models (Deyoung et al., 2003a). The induction of apoptosis by the antisense SIM2 could involve a block of cell cycle, induction of differentiation or the activation of apoptotic cascades. These results suggested that SIM2 might have both diagnostic and therapeutic utility. Moreover, the inhibition by antisense technology of SIM2 expression in the CAPAN-1 pancreatic cancer cell line, causing a pronounced growth inhibition and induced cell death through apoptosis, provide a rationale for preclinical testing of the SIM2 antisense drug in pancreatic cancer models (DeYoung et al., 2003b).

The suggested stimulatory effect of SIM2 antisense on tumor cell apoptotic regulation (Aleman et al, 2005) and inhibition of cell cycle by SIM2 (Meng et al., 2006) indicate that inhibition of tumor growth by antisense blocking of the SIM2 in colon cancer may be due to an influence on cell cycle regulation.

T-Cell Lymphoma Invasion and Metastasis 1 Gene, TIAM1

The chromosome 21 gene TIAM1 (T-cell lymphoma invasion and metastasis 1) has been identified to have important roles in the progression of epithelial cancers, especially carcinomas of the breast and colon.

Tumors that do occur in Tiam1$^{(-/-)}$ mice are more likely to progress, suggesting that in skin carcinogenesis, Tiam1 is an inhibitor of tumor development (Malliri et al., 2002).

The role of TIAM1 gene in breast carcinoma progression has been investigated in several studies. The cellular adhesion molecule and receptor for hyaluronic acid binding receptor CD44 (Lesley et al., 1993; Underhill et al., 1987) binds to Tiam1 at the PHn–CC–Ex region in SP1 murine breast carcinoma cells. Furthermore, the binding of HA to CD44v3 causes an increase in Tiam1-mediated Rac activation and promotes cytoskeleton-mediated tumor cell migration (Bourguignon et al., 2000a). In addition, these findings provide more evidence for the importance of the PHn–CC–Ex domain in regulating TIAM1 localization to the plasma membrane. Similarly, in SP1 cells, Tiam1 binds to the cytoskeletal protein ankyrin, causing Tiam1-mediated Rac1 activation as well as an increase in breast tumor cell migration and invasion (Bourguignon et al., 2000b). TIAM1 is a target of heregulin (HRG) signaling. HRG-induced cellular ruffles, loss of intracellular adhesiveness, and increased cell migration all can be mimicked by overexpression of a fully functional Tiam1 construct, and overexpression of TIAM1 increases the migratory and invasive phenotype of breast tumor cell lines. Furthermore, in human breast carcinomas, a close correlation was observed between increased Tiam1 expression and increased tumor grade (Adam et al., 2001). These results suggest that increased TIAM1 expression and/or activity may promote progression of breast carcinoma.

Colon carcinoma cell lines selected for increased metastatic potential in nude mice (Morikawa et al., 1988a; 1988b; Minard et al., 2005; 2006) express more TIAM1 protein expression than their parental line. These data indicate that Tiam1 may have a role in the progression and metastasis of colon carcinomas and that TIAM1 regulates cell adhesion, migration and apoptosis in colon tumor cells.

To test the hypothesis that TIAM1 is a determinant of proliferation and metastasis in colorectal cancer, RNA interference (RNAi) study examined the effect of the inhibition of TIAM1 expression on proliferation and metastasis. It has been found that the silencing of TIAM1 resulted in the effective inhibition of in vitro cell growth and of the invasive ability of colorectal cancer cells. These results suggest that TIAM1 truly plays a causal role in the metastasis of colorectal cancer and that RNAi-mediated silencing of TIAM1 may provide an opportunity to develop a new treatment strategy for colorectal cancer (Liu et al., 2006).

Contribution of Chromosome 21 Genes in Leukemia

Down syndrome children have an approximately 20-fold increased risk of developing acute lymphoblastic leukemia and acute myeloid leukemia compared to normal children (Fong and Broder, 1987; Lange, 2000; Taub, 2001; Yang et al., 2002).

The acute myeloid leukemia in Down syndrome children displays unique characteristics including a predominance of the acute megakaryocytic leukemia phenotype (Ravindranath et al., 1992; 1996; Zipursky et al., 1994; 2001; Lange et al., 1998; Kojima et al., 2000; Gamis et al., 2003; Zeller et al., 2005; Creutzig et al., 2005) which is estimated to occur at a 600 fold greater frequency in DS children compared to non-Down syndrome children. (Zipursky et al., 1994; Hasle, 2001).

Interestingly, three copies of chromosome 21 are common in childhood acute lymphoblastic leukaemia (Forestier et al., 2003; Harrison, 2001; Ross et al., 2003), suggesting that trisomy 21 plays an important role in leukemogenesis.

Cooperation of Chromosome 21 Genes with GATA1 During Leukemogenesis

The association between Down syndrome and childhood megakaryoblastic disorders has led to an intensive search for leukemogenic genes on chromosome 21 that may cause the differentiation arrest and initiate the leukemia. The transient myeloproliferative disorder and megakaryoblastic leukemia of Down syndrome are associated with mutations in the GATA1 gene, located on the X chromosome, which encodes a zinc-finger transcription factor that is essential for normal erythroid and megakaryocytic differentiation, in combination with trisomy 21.

Somatic mutations in exon 2 of GATA1 have been detected exclusively in almost all cases in Down syndrome-related acute megakaryocytic leukemia and transient myeloproliferative disorder, a "preleukemia" that may be present in as many as 10% of newborn infants with Down syndrome, but not in non- Down syndrome-related acute myeloid leukemia or non- acute megakaryocytic leukemia Down syndrome-related leukaemia cases (Wechsler et al., 2002; Mundschau et al., 2003; Hitzler et al., 2003; Groet et al., 2003; Rainis et al., 2003; Xu et al., 2003; Ahmed et al., 2004; Crispino et al., 2005). The effect of mutations is to introduce early stop codons that result in the synthesis of a shorter GATA1 (GATA1s) protein translated from a downstream initiation site. GATA1 mutations in Down syndrome are believed to cause accumulation of poorly differentiated megakaryocytic precursors (Wechsler et al., 2002).

GATA1 mutations occur before birth and they were also found in fetal liver of aborted Down syndrome fetuses (Taub et al., 2004; Ahmed et al., 2004). The mutagenesis of the GATA1 gene is a very early event in the development of Down syndrome-related acute megakaryocytic leukemia, during the process of multistep leukemogenesis. The acquired mutations in GATA1 are probably responsible for the differentiation arrest and the initiation of clonal proliferation of immature megakaryoblasts.

Surprisingly, knock-in of the mutated GATA1s into the GATA1 locus resulted in normal adult megakaryopoieis (Li et al., 2005). However, examination of the fetal liver revealed abundant proliferation of megakaryocytic progenitors. This study proposed the existence of a fetal hematopoietic progenitor that is sensitive to a dominant pro-proliferative effect of GATA1s. The presence of trisomy 21 enhances the survival and proliferation of these fetal cells resulting in a congenital pre-leukemia syndrome. The gene expression profiling of Down syndrome megakaryocytic leukemias provides further support for a dominant pro-proliferative role for the GATA1s mutation (Bourquin et al., 2006). A family with a germ line GATA1s mutation has been reported in which the individuals do not develop leukaemia (Hollanda et al., 2006). This indicate that these mutations are necessary but insufficient for the development of the full blown acute megakaryocytic leukemia that affects some of these patients during early childhood and may not be the first event because the GATA1 gene is located on the X chromosome.

Some genes on chromosome 21 have been demonstrated to cooperate with the GATA1 mutant during leukemogenesis and seem to be involved in the mechanisms of a lineage-specific malignant conversion. These genes on chromosome 21 promote proliferation and provide a survival advantage to cells that acquire mutations in GATA1 indicating the key role of trisomy 21 as a developmental genetic model.

Acute Myeloid Leukaemia Gene, AML1

The oncogene AML1, also known as RUNX1 or CBFA2 (Levanon et al., 2001; Speck et al., 2002), is an established regulator of hematopoiesis and megakaryopoiesis (Okuda et al., 1996; Song et al., 1999; Michaud et al., 2003). AML1 haploinsufficiency in humans impairs some aspect of megakaryopoiesis, as is also seen in mouse models (Okuda et al., 1996; Song et al., 1999).

AML1 is commonly mutated and involved in various translocations in both myeloid and lymphoid leukemias and most of these abnormalities cause loss of function.

In humans, hereditary loss-of-function mutations of AML1 cause the autosomal dominant familial platelet disorder with a predisposition to develop acute myeloid leukemia (Song et al., 1999). In addition, inherited mutations in AML1 causing haplo-insufficiency with a low level of expression in hematopoietic stem cells lead to a syndrome of familial thrombocytopenia and increased susceptibility to leukaemia (Escher et al., 2007).

Most recently, it was reported that AML1 transcripts, are downregulated in Down syndrome megakaryoblasts compared to non-Down syndrome megakaryoblasts, suggesting that AML1 is linked to the megakaryocytic lineage (Bourquin et al., 2006). In addition to its gene dosage effects, AML1 coexpresses and cooperates with GATA1 in megakaryocytic differentiation (Elagib et al., 2003).

The rare human syndromes determined by AML1 mutations and other functional studies (Elagib et al., 2003; Ichikawa et al., 2004) suggest that AML1 seem to play a major role in megakaryopoiesis and megakaryocytic differentiation in a dose-dependent manner.

ETS Transcription Factors ETS2 and ERG

ETS2 and ERG, both ETS transcription factor family members, are over-expressed in adult myeloid leukaemia cases with complex acquired karyotypes involving chromosome 21

(Baldus et al., 2004). This suggests that these two genes may also play important roles in leukemogenesis in individuals with Down syndrome, since both genes are located on chromosome 21q22 within the minimal Down syndrome region (Rahmani et al., 1989).

In hematopoietic cells, the proto-oncogene ETS2 is abundantly expressed in monocytes and macrophages of the myeloid lineage but not in granulocytes (Boulukos et al., 1990) and has been implicated in regulating megakaryocytic gene expression (Lemarchandel et al., 1993), macrophage differentiation (Aperlo et al., 1996), and T-cell development (Zaldumbide et al., 2002). Constitutive ETS2 expression transforms cultured cells and makes them tumorigenic (Seth et al., 1989). In prostate cancer, ETS2 is over-expressed (Liu et al., 1997) and its function is required to maintain the transformed state of human prostate cancer cells (Sementchenko et al., 1998).

Transgenic mice over-expressing ETS2 developed a smaller thymus and lymphocyte abnormalities, similar to features observed in Down syndrome. In all circumstances of ETS2 overexpression, the increased apoptosis correlated with increased p53 and alterations in downstream factors in the p53 pathway. In the human HeLa cancer cell line, transfection with functional p53 enables ETS2 overexpression to induce apoptosis. Furthermore, crossing the ETS2 transgenic mice with $p53^{(-/-)}$ mice genetically rescued the thymic apoptosis phenotype (Wolvetang et al., 2003). Therefore, we conclude that overexpression of human chromosome 21-encoded ETS2 induces apoptosis that is dependent on p53 (Wolvetang et al., 2003).

The proto-oncogene ERG, an ETS transcription factor on chromosome 21q, involved in megakaryopoiesis and megakaryocytic leukemias is induced upon megakaryocytic differentiation of erythroleukemia cells (Rainis et al., 2005; Marcucci et al., 2005). Overexpression of ERG in an erythroleukemia cell line causes a phenotypic shift from the erythroid into the megakaryocytic lineage (Rainis et al., 2005). This support that over-expressed ERG from trisomy 21 promotes fetal megakaryopoiesis for a differentiation arresting mutations. This promegakaryocytic pressure may, in turn, enhance the selection and proliferation of hematopoietic progenitor cells with GATA1 mutations, leading to an arrest in differentiation (Rainis et al., 2005).

Basic Leucine Zipper Transcription Factor 1, BACH1

BACH1 gene (BTB and CNC homolog 1) transcription activator or repressor is localized to chromosome 21q22.1 (Blouin et al., 1998; Ohira et al., 1998) and have been identified to be involved in leukemogenesis in trisomic patients.

Transgenic expression of BACH1 in megakaryocytes in vivo results in thrombocytopenia. This is most likely due to maturation arrest of megakaryocytic cells and resulting dysfunctions in platelet formation (Toki et al., 2005). Interestingly, the phenotype of BACH1 transgenic mice resembles that of the p45 (nuclear factor erythroid 2, NF-E2) null mice, with similarities in the megakaryocytic phenotype (Shivdasani et al., 1995; Lecine et al., 1998), suggesting the presence of common or overlapping mechanisms underlying the impaired megakaryopoiesis observed in these mutant mice.

The BACH1 transgenic mice show that terminal differentiation of the megakaryocytes, as well as platelet production, were markedly impaired by the overexpression of BACH1, indicating that the chromosome 21 gene BACH1 control the terminal megakaryocytic differentiation (Toki et al., 2005). These results support that BACH1 is one of the candidates for genes that cooperate with the mutated GATA1 at a very early stage of leukemogenesis in Down syndrome. Moreover, it has been demonstrated that the mutation in GATA1, in addition to down-regulate the expression of ALM1, also increases the expression of BACH1 (Bourquin et al., 2006) leading to a further block of megakaryocytic differentiation.

Conclusions

Trisomy 21 could provide a general model for the role of aneuploidy in leukemogenesis, tumorigenesis, cancerigenesis and neurogenesis abnormalities, and may be considered as an interesting developmental genetic model to study and decipher the molecular and cellular mechanisms involved in these different abnormalities.

Trisomy 21 is a good model of collaborating pro-proliferation and differentiation arresting mutations in leukaemia, tumors and cancer. It is also a good model of the parallel amplification of multiple genes, which may act cooperatively in the same leukemogenic, tumorigenic, cancerigenic or neurogenic pathway. Finally, trisomy 21 is a good model for deciphering the targets genes and regulatory factors working in the cell cycle events in body and brain. A better knowledge of the function of genes mapping to chromosome 21 may contribute to the understanding of the several mechanisms of diseases, acting in patients with DS, such as leukaemia, tumors, cancer, and neurogenesis alterations related to cell cycle control alterations.

Some members of the phosphorylation pathways that control the cell cycle are encoded by specific chromosome 21 genes, particularly the minibrain gene Dyrk1A, suggesting some alteration in cell cycle control in Down syndrome individuals. The involvement of Dyrk1A gene in the regulation of neural proliferation on the basis of alterations in the proliferation centres of brain of Drosophila minibrain (MNB) mutants and the altered levels of cell cycle regulators in transgenic mice over-expressing Dyrk1A as well as the phosphorylation of cell cycle regulators by DYRK1A support the evolutionary conservation of DYRK1A role as regulator of proliferation in neural progenitor and its role in cell cycle regulation.

Further studies of the three oncogenes AML1, ERG and ETS2, highly related to the genesis of leukemias, could be of the most interest to contribute to the general understanding of the oncogenic mechanisms of chromosomal aneuploidies, the most common abnormalities in cancer.

The direct involvement of the human chromosome 21 Single minded 2 gene (SIM2) in the inhibition of gene expression, growth inhibition, inhibition of tumor growth as well as induction of apoptosis that could involve a block of cell cycle, establish SIM2 gene as a molecular target for cancer therapeutics and have additional important implications for the future diagnosis and treatment of specific solid tumors as well as for further understanding the cancer risk in Down syndrome patients.

The human chromosome 21 genes, AML1, ERG, ETS2, BACH-1, TIAM-1, SIM2 and DYRK1A play important roles in cell cycle and cell growth and could with the recent progress, especially in developmental genetics, in genomics and proteomics promise significant advancements and elucidation of the molecular and cellular mechanisms involved in the Down syndrome associated diseases related to cell cycle alterations, such as leukaemia, tumors and cancer, in the perspective of the development of new drugs and treatments.

Acknowledgments

We are grateful to our colleagues of Department of Molecular Biology-Jacques Monod at the Pasteur Insitute (Paris) for their advices and support. We also thank J.M. Delabar (University Paris 7) and L. Peltzer (University of French Polynesia) for their continuous support.

References

Adam, L., Vadlamudi, R. K., McCrea, P. & Kumar, R. (2001) Tiam1 overexpression potentiates heregulin-induced lymphoid enhancer factor-1/beta-catenin nuclear signaling in breast cancer cells by modulating the intercellular stability. *J. Biol. Chem., 276,* 28443–28450.

Ahmed, M., Sternberg, A., Hall, G., Thomas, A., Smith, O., O'Marcaigh, A., Wynn, R., Stevens, R., Addison, M., King, D., Stewart, B., Gibson, B., Roberts, I. & Vyas, P. (2004) Natural history of GATA1 mutations in Down Syndrome. *Blood. 103,* 2480–2489.

Aleman, M. J., DeYoung, M. P., Tress, M., Keating, P., Perry, G. W. & Narayanan, R. (2005) Inhibition of Single Minded 2 gene expression mediates tumor-selective apoptosis and differentiation in human colon cancer cells. *Proc. Natl. Acad. Sci. 102,* 12765-12770.

Aperlo, C., Pognonec, P., Stanley, E. R. & Boulukos, K. E. (1996) Constitutive c-ets2 expression in M1D+ myeloblast leukemic cells induces their differentiation to macrophages. *Mol. Cell. Biol., 16,* 6851–6858.

Aylward, E. H., Habbak, R., Warren, A. C., Pulsifer, M. B., Barta, P. E., Jerram, M. & Pearlson, G. D. (1997) Cerebellar volume in adults with Down syndrome. *Arch. Neurol., 54,* 209-212.

Bahn, S., Mimmack, M., Ryan, M., Caldwell, M. A., Jauniaux, E., Starkey, M., Svendsen, C. N. & Emson, P. (2002) Neuronal target genes of the neuron-restrictive silencer factor in neurospheres derived from fetuses with Down's syndrome: a gene expression study. *Lancet. 359,* 310–315.

Baldus, C. D., Liyanarachchi, S., Mrozek, K., Auer, H., Tanner, S. M., Guimond, M., Ruppert, A. S., Mohamed, N., Davuluri, R. V., Caligiuri, M. A., Bloomfield, C. D. & de la Chapelle, A. (2004) Acute myeloid leukemia with complex karyotypes and abnormal chromosome 21: amplification discloses overexpression of APP, ETS2, and ERG genes. *Proc. Natl. Acad. Sci., 101,* 3915–3920.

Baxter, L. L., Moran, T. H., Richtsmeier, J. T., Troncoso, J. & Reeves, R. H. (2000) Discovery and genetic localization of Down syndrome cerebellar phenotypes using the Ts65Dn mouse. *Hum. Mol. Genet., 9,* 195–202.

Becker, L. E., Armstrong, D. L. & Chan, F. (1986) Dendritic atrophy in children with Down's syndrome. *Ann. Neurol., 20,* 520–532.

Becker, L., Mito, T., Takashima, S. & Onodera, K. (1991) Growth and development of the brain in Down syndrome. In C. H. Epstein (Ed.), *The Morphogenesis of Down syndrome* (pp. 133–152). New York, Wiley-Liss.

Becker, W. & Joost, H. G. (1999) Structural and functional characteristics of Dyrk, a novel subfamily of protein kinases with dual specificity. *Prog. Nucleic Acid Res. Mol. Biol., 62,* 1-17.

Blouin, J. L., Duriaux Sail, G., Guipponi, M., Rossier, C., Pappasavas, M. P. & Antonarakis, S. E. (1998) Isolation of the human BACH1 transcription regulator gene, which maps to chromosome 21q22.1. *Hum. Genet., 102,* 282-288.

Boulukos, K. E., Pognonec, P., Sariban, E., Bailly, M., Lagrou, C. & Ghysdael, J. (1990) Rapid and transient expression of Ets2 in mature macrophages following stimulation with cMGF, LPS, and PKC activators. *Genes Dev., 4,* 401–409.

Bourguignon, L. Y., Zhu, H., Shao, L. & Chen, Y. W. (2000a) CD44 interaction with tiam1 promotes Rac1 signaling and hyaluronic acid-mediated breast tumor cell migration. *J. Biol. Chem., 275,* 1829–1838.

Bourguignon, L. Y., Zhu, H., Shao, L. & Chen, Y. W. (2000b) Ankyrin–Tiam1 interaction promotes Rac1 signaling and metastatic breast tumor cell invasion and migration. *J. Cell. Biol., 150,* 177–191.

Bourquin, J. P., Subramanian, A., Langebrake, C., Reinhardt, D., Bernard, O., Ballerini, P., Baruchel, A., Cavé, H., Dastugue, N., Hasle, H., Kaspers, G. L., Lessard, M., Michaux, L., Vyas, P., van Wering, E., Zwaan, C. M., Golub, T. R. & Orkin, S. H. (2006) Identification of distinct molecular phenotypes in acute megakaryoblastic leukemia by gene expression profiling. *Proc. Natl. Acad. Sci., 103,* 3339–3344.

Branchi, I., Bichler, Z., Minghetti, L., Delabar, J. M., Malchiodi-Albedi, F., Gonzalez, M. C., Chettouh, Z., Nicolini, A., Chabert, C., Smith, D. J., Rubin, E. M., Migliore-Samour, D. & Alleva, E. (2004) Transgenic mouse in vivo library of human Down syndorme Critical Region 1: association between *DYRK1A* overexpession, brain development abnormalities, and cell cycle protein alteration. *J. Neuropathol. Exp. Neurol., 63,* 429-440.

Chen, H., Chrast, R., Rossier, V., Gos, A., Antonarakis, S. E., Kudoh, J., Yamaki, A., Shindoh, N., Maeda, H., Minoshima, S., et al. (1995) Single-minded and Down syndrome? *Nat. Genet., 10,* 9-10.

Chrast, R., Scott, H. S., Chen, H., Kudoh, J., Rossier, C., Minoshima, S., Wang, Y., Shimizu, N. & Antonarakis, S. E. (1997) Cloning of two human homologs of the Drosophila single-minded gene SIM1on chromosome 6q and SIM2 on 21qwithin the Down syndrome chromosomal region. *Genome Res., 7,* 615-624.

Clark, S., Schwalbe, J., Stasko, M. R., Yarowsky, P. J. & Costa, A. C. (2006) Fluoxetine rescues deficient neurogenesis in hippocampus of the Ts65Dn mouse model for Down syndrome. *Exp. Neurol., 200,* 256–261.

Cohen, W. I. (1999) Health care guidelines for individuals with Down syndrome: 1999 revision. *Down Syndrome Quarterly, 4,* 1-16.

Connor, J. M., & Ferguson-Smith, M. A. (1997) *Essential medical genetics* (5th edn). Oxford, Blackwell Science.

Contestabile, A., Fila, T., Ceccarelli, C., Bonasoni, P., Bonapace, L., Santini, D., Bartesaghi, R. & Ciani, E. (2007) Cell cycle alteration and decreased cell proliferation in the hippocampal dentate gyrus and in the neocortical germinal matrix of fetuses with Down syndrome and in Ts65Dn mice. *Hippocampus, 17,* 665-678.

Cooper, J. D., Salehi, A., Delcroix, J. D., Howe, C. L., Belichenko, P. V., Chua-Couzens, J., Kilbridge, J. F., Carlson, E. J., Epstein, C. J., & Mobley, W. C. (2001) Failed retrograde transport of NGF in a mouse model of Down's syndrome: reversal of cholinergic neurodegenerative phenotypes following NGF infusion. *Proc. Natl. Acad. Sci., 98,* 10439-10444.

Coyle, J. T., Oster-Granite, M. L. & Gearhart, J. D. (1986) The neurobiologic consequences of Down syndrome. *Brain Res. Bull., 16,* 773–787.

Creutzig, U., Reinhardt, D., Diekamp, S., Dworzak, M., Stary, J. & Zimmermann, M. (2005) AML patients with Down syndrome have a high cure rate with AML-BFM therapy with reduced dose intensity. *Leukemia, 19,* 1355–1360.

Crews, S. T., Thomas, J. B. & Goodman, C. S. (1988) The Drosophila singleminded gene encodes a nuclear protein with sequence similarity to the per gene product, *Cell, 52,* 143–151.

Crispino, J. D. (2005) GATA1 in normal and malignant hematopoiesis. *Semin. Cell Dev. Biol., 16:* 137-147.

Cunningham, J. J. & Roussel, M. F. (2001) Cyclin-dependent kinase inhibitors in the development of the central nervous system. *Cell Growth Differ., 12,* 387–396.

Dahmane, N., Charron, G., Lopes, C., Yaspo, M. L., Maunoury, C., Decorte, L., Sinet, P. M., Bloch, B. & Delabar, J. M. (1995) Down syndrome critical region contains a gene homologous to Drosophila sim expressed during rat and human central nervous system development, *Proc. Natl. Acad. Sci. USA, 92,* 9191–9195.

Dalton, A. J. & Crapper-McLachlan, D. R. (1986) Clinical expression of Alzheimer's disease in Down syndrome. *Psychiatr. Clin. North. Am., 9,* 959–70.

Davisson, M. T., Schmidt, C. & Akeson, E. C. (1990) Segmental trisomy of murine chromosome 16: a new model system for studying Down syndrome. *Prog. Clin. Biol. Res., 360,* 263-280.

von Deimling, A., Fimmers, R., Schmidt, M. C., Bender, B., Fassbender, F., Nagel, J., Jahnke, R., Kaskel, P., Duerr, E. M., Koopmann, J., Maintz, D., Steinbeck, S., Wick, W., Platten, M., Müller, D. J., Przkora, R., Waha, A., Blümcke, B., Wellenreuther, R., Meyer-Puttlitz, B., Schmidt, O., Mollenhauer, J., Poustka, A., Stangl, A. P., Lenartz, D. & von Ammon, K. (2000) Comprehensive allelotype and genetic analysis of 466 human nervous system tumors. *J. Neuropathol. Exp. Neurol., 59,* 544–558.

Delabar, J. M., Theophile, D., Rahmani, Z., Chettouh, Z., Blouin, J. L., Prieur, M., Noel, B. & Sinet, P. M. (1993) Molecular mapping of twenty-four features of Down syndrome on chromosome 21. *Eur. J. Hum. Genet., 1,* 114–124.

DeYoung, M. P., Scheurle, D., Damania, H., Zylberberg, C. & Narayanan, R. (2002) Down's syndrome-associated single minded gene as a novel tumor marker. *Anticancer Res., 22,* 3149-3157.

DeYoung, M. P., Tress, M. & Narayanan, R. (2003a) Identification of Down's syndrome critical locus gene SIM2-s as a drug therapy target for solid tumors. *Proc. Natl. Acad. Sci., 100,* 4760-4765.

DeYoung, M. P., Tress, M. & Narayanan, R. (2003b) Down's syndrome- associated SingleMinded 2 gene as a pancreatic cancer drug therapy target. *Cancer Lett., 200,* 25-31.

Elagib, K. E., Racke, F. K., Mogass, M., Khetawat, R., Delehanty, L. L. & Goldfarb, A. N. (2003) RUNX1 and GATA-1 coexpression and cooperation in megakaryocytic differentiation. *Blood. 101,* 4333–4341.

Epstein, C. J. (1995) Down syndrome (trisomy 21). In C. R. Scriver, A. L., Beaudet, W. S., Sly, & D. Valle (Eds.), *The Metabolic and Molecular Bases of Inherited Disease* (pp.749–794) New York, McGraw-Hill.

Epstein, D. J., Matinu, L., Michaud, J. L., Losos, K. M., Fan, C. M. & Joyner, A. L. (2000) Members of the bHLH-PAS family regulate Shh transcription in forebrain regions of the mouse CNS. *Development. 127,* 1701-1709.

Escher, R., Wilson, P., Carmichael, C., Suppiah, R., Liu, M., Kavallaris, M., Cannon, P., Michaud, J. & Scott, H. S. (2007) *Blood Cells Mol. Dis., 39,* 107-114.

Ferencz, C., Neill, C. A., Boughman, J. A., Rubin, J. D., Brenner, J. I. & Perry, L. W. (1989) Congenital cardiovascular malformations associated with chromosome abnormalities: An epidemiologic study. *J. Pediatr., 114,* 79-86.

Fong, C. T. & Brodeur, G. M. (1987) Down syndrome and leukemia: epidemiology, genetics, cytogenetics and mechanisms of leukemogenesis. *Cancer Genet. Cytogenet., 28,* 55-76.

Forestier, E., Heim, S., Blennow, E., Borgstrom, G., Holmgren, G., Heinonen, K., Johannsson, J., Kerndrup, G., Andersen, M. K., Lundin, C., Nordgren, A., Rosenquist, R., Swolin, B. & Johansson, B. (2003) Cytogenetic abnormalities in childhood acute myeloid leukaemia: a Nordic series comprising all children enrolled in the NOPHO-93-AML trial between 1993 and 2001. *Br. J. Haematol., 121,* 566–577.

Fotaki, V., Dierssen, M., Alcantara, S., Martinez, S., Marti, E., Casas, C., Visa, J., Soriano, E., Estivill, X. & Arbones, M. L. (2002) Dyrk1A haploinsufficiency affects viability and causes developmental delay and abnormal brain morphology in mice. *Mol. Cell. Biol., 22,* 6636-6647.

Fotaki, V., Martinez De Lagran, M., Estivill, X., Arbones, M. & Dierssen, M. (2004) Haploinsufficiency of Dyrk1A in mice leads to specific alterations in the development and regulation of motor activity. *Behav. Neurosci., 118,* 815–821.

Fults, D., Pedone, C. A., Thomas, G. A. & White, R. (1990) Allelotype of human malignant astrocytoma. *Cancer Res., 50,* 5784–5789.

Gamis, A. S., Woods, W. G., Alonzo, T. A., Buxton, A., Lange, B., Barnard, D. R., Gold, S., Smith, F. O., Children's Cancer Group Study 2891. (2003) Increased age at diagnosis has a significantly negative effect on outcome in children with Down syndrome and acute myeloid leukemia: a report from the Children's Cancer Group Study 2891. *J. Clin. Oncol., 21,* 3415–3422.

Goshu, E., Jin, H., Fasnacht, R., Sepenski, M., Michaud, J. L. & Fan, C. M. (2002) Sim2 mutant have developmental defects not overlapping with those of the Sim1 mutants. *Mol. Cell Biol., 22,* 4147-4157.

Goshu, E., Jin, H., Lovejoy, J., Marion, J. F., Michaud, J. L. & Fan, C. M. (2004) Sim2 contributes to neuroendocrine hormone gene expression in the anterior hypothalamus. *Mol. Endocrinol., 18,* 1251-1262.

de Graaf, K., Hekerman, P., Spelten, O., Herrmann, A., Packman, L. C., Bussow, K., Muller-Newen, G. & Becker, W. (2004) Characterization of cyclin L2, a novel cyclin with an arginine/serine-rich domain: phosphorylation by DYRK1A and colocalization with splicing factors. *J. Biol. Chem., 279,* 4612–4624.

Groet, J., Ives, J. H., Jones, T. A., Danton, M., Flomen, R. H., Sheer, D., Hrascan, R., Pavelic, K. & Nizetic, D. (2000) Narrowing of the region of allelic loss in 21q11-21 in squamous non-small cell lung carcinoma and cloning of a novel ubiquitin-specific protease gene from the deleted segment. *Genes Chromosomes Cancer, 27,* 153-161.

Groet, J., McElwaine, S., Spinelli, M., Rinaldi, A., Burtscher, I., Mulligan, C., Mensah, A., Cavani, S., Dagna-Bricarelli, F., Basso, G., Cotter, F. E. & Nizetic, D. (2003) Acquired mutations in GATA1 in neonates with Down's syndrome with transient myeloid disorder. *Lancet. 361,* 1617-1620.

Guimera, J., Casas, C., Pucharcos, C., Solans, A., Domench, A., Planas, A. M., Ashley, J., Lovett, M., Estivill, X., & Pritchard, M. A. (1996) A human homologue of Drosophila minibrain (MNB) is expressed in the neuronal regions affected in Down syndrome and maps to the critical region. *Hum. Mol. Genet., 5,* 1305–1310.

Halvorsen, O. J., Oyan, A. M., Bo, T. H., Olsen, S., Rostad, K., Haukaas, S. A., Bakke, A. M., Marzolf, B., Dimitrov, K., Stordrange, L., Lin, B., Jonassen, I., Hood, L., Akslen, L. A. & Kalland, K. H. (2005) Gene expression profiles in prostate cancer: association with patient subgroups and tumour differentiation. *Int. J. Oncol., 26,* 329-336.

Halvorsen, O. J., Rostad, K., Oyan, A. M., Puntervoll, H., Bo, T. H., Stordrange, L., Olsen, S., Haukaas, S. A., Hood, L., Jonassen, I., Kalland, K. H. & Akslen, L. A. (2007) Increased expression of SIM2-s protein is a novel marker of aggressive prostate cancer. *Clin. Cancer Res., 13,* 892-897.

Hämmerle, B., Vera-Samper, E., Speicher, S., Arencibia, R., Martinez, S. & Tejedor, F.J. (2002) Mnb/Dyrk1A is transiently expressed and asymmetrically segregated in neural progenito cells at the transition to neurogenic divisions. *Dev. Biol., 246,* 259-273.

Hämmerle, B., Elizalde, C. & Tejedor, F. J. (2008) The spatio-temporal and subcellular expression of the candidate Down syndrome gene Mnb/Dyrk1A in the developing mouse brain suggests distinct sequential roles in neuronal development. *Eur. J. Neurosci., 27,* 1061–1074.

Hanahan, D. & Weinberg, R. A. (2000) The hallmarks of cancer. *Cell, 100,* 57-70.

Harrison, C. J. (2001) The detection and significance of chromosomal abnormalities in childhood acute lymphoblastic leukaemia. *Blood Rev., 15,* 49–59.

Hasle, H. (2001) Pattern of malignant disorders in individuals with Down's syndrome. *Lancet Oncol., 2,* 429-436.

Haydar, T. F., Nowakowski, R. S., Yarowsky, P. J. & Krueger, B. K. (2000) Role of founder cell deficit and delayed neuronogenesis in microencephaly of the trisomy 16 mouse. *J. Neurosci., 20,* 4156-4164.

Hitzler, J. K., Cheung, J., Li, Y., Scherer, S. W. & Zipursky, A. (2003) GATA1 mutations in transient leukemia and acute megakaryoblastic leukemia of Down syndrome. *Blood. 101,* 4301-4304.

Hollanda, L. M., Lima, C. S., Cunha, A. F., Albuquerque, D. M., Vassallo, J., Ozelo, M. C., Joazeiro, P. P., Saad, S. T. & Costa, F. F. (2006) An inherited mutation leading to production of only the short isoform of GATA-1 is associated with impaired erythropoiesis. *Nature Genet., 38,* 807-812.

Ichikawa, M., Asai, T., Saito, T., Yamamoto, G., Seo, S., Yamazaki, I., Yamagata, T., Mitani, K., Chiba, S., Ogawa, S., Kurokawa, M. & Hirai, H. (2004) AML-1 is required for megakaryocytic maturation and lymphocytic differentiation, but not for maintenance of hematopoietic stem cells in adult hematopoiesis. *Nature Med., 10,* 299-304.

Jernigan, T. L., Bellugi, U., Sowell, E., Doherty, S. & Hesselink, J. R. (1993) Cerebral morphologic distinctions between Williams and Down syndromes. *Arch. Neurol., 50,* 186–191.

Kannan, N. & Neuwald, A. F. (2004) Evolutionary constraints associated with functional specificity of the CMGC protein kinases MAPK, CDK, GSK, SRPK, DYRK, and CK2alpha. *Protein Sci. 13,* 2059–2077.

Kojima, S., Sako, M., Kato, K., Hosoi, G., Sato, T., Ohara, A., Koike, K., Okimoto, Y., Nishimura, S., Akiyama, Y., Yoshikawa, T., Ishii, E., Okamura, J., Yazaki, M., Hayashi, Y., Egushi, M., Tsukimoto, I. & Ueda, K. (2000) An effective chemotherapeutic regimen for acute myeloid leukaemia and myelodysplastic syndrome in children with Down's syndrome. *Leukemia. 14,* 786–791.

Korenberg, J. R., Chen, X. N., Schipper, R., Sun, Z., Gonsky, R., Gerwehr, S., Carpenter, N., Daumer; C., Dignan, P., Disteche, C., Graham, Jr. J. M., Hugdins, L., McGillivray, B., Miyazaki, K., Ogasawara, N., Park, J. P., Pagon, R., Pueschel, S., Sack, G., Say, B., Schuffenhauer, S., Soukup, S. & Yamanaka, T. (1994) Down syndrome phenotypes: The consequences of chromosomal imbalance. *Proc. Natl. Acad. Sci. USA., 91,* 4997-5001.

Lange B. (2000) The management of neoplastic disorders of haematopoiesis in children with Down's syndrome. *Br. J. Haematol., 110:* 512-524.

Lange, B. J., Kobrinsky, N., Barnard, D. R., Arthur, D. C., Buckley, J. D., Howells, W. B., et al. (1998) Distinctive demography, biology, and outcome of acute myeloid leukemia and myelodysplastic syndrome in children with Down syndrome: Children's Cancer Group Studies 2861 and 2891. *Blood. 91:* 608–615.

Lecine, P., Villeval, J. L., Vyas, P., Swencki, B., Xu, Y. & Shivdasani, R. A. (1998) Mice lacking transcription factor NF-E2 provide in vivo validation of the proplatelet model of thrombocytopoiesis and show a platelet production defect that is intrinsic to megakaryocytes. *Blood. 92,* 1608-1616.

Lejeune, J. (1990) Pathogenesis of mental deficiency in trisomy 21. *Am. J. Med. Genet. Suppl., 7,* 20-30.

LeJeune, J., Gautier, M., & Turpin, R. (1959) Etude des chromosomes somatiques de neufs enfants Mongoliens, *C. R. Acad. Sci. Paris. 248,* 1721-1722.

Lemarchandel, V., Ghysdael, J., Mignotte, V., Rahuel, C. & Romeo, P. H. (1993) GATA and Ets cis-acting sequences mediate megakaryocytespecific expression. *Mol. Cell. Biol., 13,* 668–676.

Lesley, J., Hyman, R. & Kincade, P. W. (1993) CD44 and its interaction with extracellular matrix. *Adv. Immunol., 54,* 271–335.

Levanon, D., Glusman, G., Bangsow, T., Ben-Asher, E., Male, D. A., Avidan, N., Bangsow, C., Hattori, M., Taylor, T. D., Taudien, S., Blechschmidt, K., Shimizu, N., Rosenthal, A., Sakaki, Y., Lancet, D. & Groner, Y. (2001) Architecture and anatomy of the genomic locus encoding the human leukemia-associated transcription factor RUNX1/AML1. *Gene, 262,* 23-33.

Li, Z., Godinho, F. J., Klusmann, J. H., Garriga-Canut, M., Yu, C. & Orkin, S. H. (2005) Developmental stage-selective effect of somatically mutated leukemogenic transcription factor GATA1. *Nature Genet., 37,* 613-619.

Liu, A. Y., Corey, E., Vessella, R. L., Lange, P. H., True, L. D., Huang, G. M., Nelson, G. M. & Hood, L. (1997) Identification of differentially expressed prostate genes: increased expression of transcription factor ETS-2 in prostate cancer. *Prostate. 30,* 145–153.

Liu, L., Zhang, Q., Zhang, Y., Wang, S. & Ding, Y. (2006) Lentivirus-mediated silencing of Tiam1 gene influences multiple functions of a human colorectal cancer cell line. *Neoplasia. 8,* 917-924.

Lochhead, P. A., Sibbet, G., Morrice, N. & Cleghon, V. (2005) Activation-loop autophosphorylation is mediated by a novel transitional intermediate form of DYRKs. *Cell. 121,* 925–936.

Malliri, A., van der Kammen, R. A., Clark, K., Van, D. V., Michiels, F. & Collard, J. G. (2002) Mice deficient in the Rac activator Tiam1 are resistant to Ras-induced skin tumours. *Nature. 417,* 867–871.

Marcucci, G., Baldus, C. D., Ruppert, A. S., Radmacher, M. D., Mrozek, K., Whitman, S. P., Kolitz, J. E., Edwards, C G., Vardiman, J. W., Powell, B. L., Baer, M. R., Moore, J. O., Perrotti, D., Caligiuri, M. A., Carroll, A. J., Larson, R. A., de la Chapelle, A. & Bloomfield, C. D. (2005) Overexpression of the ETS-related gene, ERG, predicts a worse outcome in acute myeloid leukaemia with normal karyotype: a Cancer and Leukemia Group B study. *J. Clin. Oncol., 23,* 9234-42.

Marin-Padilla, M. (1976) Pyramidal cell abnormalities in the motor cortex of a child with Down's syndrome. *J. Comp. Neurol., 167,* 63–75.

Meng, X., Shi, J., Peng, B., Zou, X. & Zhang, C. (2006) Effect of mouse Sim2 gene on the cell cycle of PC12 cells. *Cell Biol. Int., 30,* 349-353.

Michaud, J., Scott, H. S. & Escher, R. (2003) AML1 interconnected pathways of leukemogenesis. *Cancer Invest., 21,* 105–136.

Minard, M. E., Herynk, M. H., Collard, J. G. & Gallick, G. E. (2005) The guanine nucleotide exchange factor Tiam1 increases colon carcinoma growth at metastatic sites in an orthotopic nude mouse model. *Oncogene. 24,* 2568-2573.

Minard, M. E., Ellis, L. M. & Gallick, G. E. (2006) Tiam1 regulates cell adhesion, migration and apoptosis in colon tumor cells. *Clin. Exp. Metastasis. 23,* 301-313.

Moffett, P., Reece, M. & Pelletier, J. (1997) The murine Sim-2 gene product inhibits transcription by active repression and functional interference. *Mol. Cell. Biol., 17,* 4933-4947.

Moran, T. H., Capone, G. T., Knipp, S., Davisson, M. T., Reeves, R. H. & Gearhart, J. D. (2002) The effects of piracetam on cognitive performance in a mouse model of Down's syndrome. *Physiol. Behav., 77,* 403-409.

Morikawa, K., Walker, S. M., Nakajima, M., Pathak, S., Jessup, J. M. & Fidler, I. J. (1988a) Influence of organ environment on the growth, selection, and metastasis of human colon carcinoma cells in nude mice. *Cancer Res., 48,* 6863–6871.

Morikawa, K., Walker, S. M., Jessup, J. M. & Fidler, I. J. (1988b) In vivo selection of highly metastatic cells from surgical specimens of different primary human colon carcinomas implanted into nude mice. *Cancer Res., 48,* 1943–1948.

Mundschau, G., Gurbuxani, S., Gamis, A. S., Greene, M. E., Arceci, R. J. & Crispino, J. D. (2003) Mutagenesis of GATA1 is an initiating event in Down syndrome leukemogenesis. *Blood. 101,* 4298-4300.

Murdoch, J. C., Rodger, J. C., Rao, S. S., Fletcher, C. D. & Dunnigan, M.G. (1977) Down's syndrome: An atheroma-free model. *Br. Med. J., 2,* 226-228.

Nambu, J. R., Lewis, J. O., Wharton, K. A. & Crews, S. T. (1991) The Drosophila single-minded gene encodes a helix-loop-helix protein that acts as a master regulator of CNS midline development. *Cell. 67,* 1157-1167.

Ohira, M., Seki, N., Nagase, T., Ishikawa, K., Nomura, N. & Ohara, O. (1998) Characterization of a human homolog (BACH1) of the mouse Bach1 gene encoding a BTB-basic leucine zipper transcription factor and its mapping to chromosome 21q22.1. *Genomics. 47,* 300-306.

Okuda, T., van Deursen, J., Hiebert, S. W., Grosveld, G. & Downing, J. R. (1996) AML1, the target of multiple chromosomal translocations in human leukemia, is essential for normal fetal liver hematopoiesis. *Cell. 84,* 321–330.

Oppenheim, R. W., Cole, T. & Prevette, D. (1989) Early regional variations in motoneuron numbers arise by differential proliferation in the chick embryo spinal cord. *Dev. Biol. 133,* 468–474.

Pennington, P. F., Moon, J., Edgin, J., Stedron, J. & Nadel, L. (2003) The neuropsychology of Down syndrome: evidence for hippocampal dysfunction. *Child Dev., 74,* 75-93.

Puffenberger, E. G., Kauffman, E. R., Bolk, S., Matise, T. C., Washington, S. S., Angrist, M., Weissenbach, J., Garver, K. L., Mascari, M. & Ladda, R. (1994) Identity-by-descent and association mapping of a recessive gene for Hirschsprung disease on human chromosome 13q22, *Hum. Mol. Genet., 3,* 1217–1225.

Pulsifer, M B. (1996) The neuropsychology of mental retardation. *J. Int. Neuropsychol. Soc., 2,* 159-176.

Rachidi, M., Lopes, C., Charron, G., Delezoide, A. L., Paly, E., Bloch, B. & Delabar, J. M. (2005) Spatial and temporal localization during embryonic and fetal human development of the transcription factor SIM2 in brain regions altered in Down syndrome. *Int. J. Dev. Neurosci., 23,* 475-484.

Rachidi, M. & Lopes, C. (2007) Molecular Mechanisms of Mental Retardation in Down Syndrome. In *Focus in Mental Retardation Research.* (pp. 77-134). Nova Science Publishers, Inc.

Rahmani, Z., Blouin, J. L., Creau-Goldberg, N., Watkins, P. C., Mattei, J. F., Poissonnier, M., Prieur, M., Chettouh, Z., Nicole, A., Aurias, A., Sinet, P. M. & Delabar, J. M. (1989) Critical role of the D21S55 region on chromosome 21 in the pathogenesis of Down syndrome. *Proc. Natl. Acad. Sci. USA, 86,* 5958–5962.

Rainis, L., Bercovich, D., Strehl, S., Teigler-Schlegel, A., Stark, B., Trka, J., Amariglio, N., Biondi, A., Muler, I., Rechavi, G., Kempski, H., Haas, O. A. & Izraeli, S. (2003) Mutations in exon 2 of GATA1 are early events in megakaryocytic malignancies associated with trisomy 21. *Blood. 102,* 981-986.

Rainis, L., Toki, T., Pimanda, J. E., Rosenthal, E., Machol, K., Strehl, S., Göttgens, B., Ito, E. & Izraeli, S. (2005) The proto-oncogene ERG in megakaryoblastic leukemias. *Cancer Res., 65,* 7596–7602.

Ravindranath, Y., Abella, E., Krischer, J. P., Wiley, J., Inoue, S., Harris, M., Chauvenet, A., Alvarado, C. S., Dubowy, R., Ritchey, A. K., Land, V., Steuber, C. P. & Weinstein, H. (1992) Acute myeloid leukemia (AML) in Down's syndrome is highly responsive to chemotherapy: experience on Pediatric Oncology Group AML study 8498. *Blood. 80,* 2210–2214.

Ravindranath, Y., Yeager, A. M., Chang, M. N., Steuber, C. P., Krischer, J., Graham-Pole, J., Carroll, A., Inoue, S., Camitta, B. & Weinstein, H. J. (1996) Acute myeloid leukemia in children: a randomized comparative study of purged autologous bone marrow transplantation versus intensive multiagent consolidation chemotherapy in first remission. Pediatric Oncology Group Study-POG 8821. *N. Engl. J. Med., 334,* 1428–1434.

Raz, N., Torres, I. J., Briggs, S. D., Spencer, W. D., Thornton, A. E., Loken, W. J., Gunning, F. M., McQuain, J. D., Driesen, N. R. & Acker, J.D. (1995) Selective neuroanatomic abnormalities in Down_s syndrome and their cognitive correlates: evidence from MRI morphometry. *Neurology. 45,* 356–366.

Risser, D., Lubec, G., Cairns, N. & Herrera-Marschitz, M. (1997) Excitatory amino acids and monoamines in parahippocampal gyrus and frontal cortical pole of adults with Down syndrome. *Life Sci*., *60,* 1231–1237.

Roper, R. J. & Reeves, R.H. (2006) Understanding the basis for Down syndrome phenotypes. *PloS Genet., 2,* e50.

Roschke, A. V., Tonon, G., Gehlhaus, K. S., McTyre, N., Bussey, K. J., Lababidi, S., Scudiero, D. A., Weinstein, J. N. & Kirsch, I. R. (2003) Karyotypic complexity of the NCI-60 drug-screening panel. *Cancer Res., 63,* 8634–8647.

Ross, M. E., Zhou, X., Song, G., Shurtleff, S. A., Girtman, K., Williams, W. K., Liu, H. C., Mahfouz, R., Raimondi, S. C., Lenny, N., Patel, A. & Downing, J. R. (2003) Classification of pediatric acute lymphoblastic leukemia by gene expression profiling, *Blood. 102,* 2951–2959.

Rueda, N., Mostany, R., Pazos, A., Florez, J. & Martinez-Cué, C. (2005) Cell proliferation is reduced in the dentate gyrus of aged but not young Ts65Dn mice, a model of Down syndrome. *Neurosci. Lett., 380,* 197-201.

Satgé, D., Sasco, A. J., Cure, H., Leduc, B., Sommelet, D. & Vekemans, M. J. (1997) *Cancer.* 80, 929–935.

Schmidt-Sidor, B., Wisniewski, K. E., Shepard, T. H. & Sersen, E. A. (1990) Brain growth in Down syndrome subjects 1–22 weeks of gestation age and birth to 60 months. *Clin. Neuropathol., 4,* 181–190.

Schneider, C., Risser, D., Kirchner, L., Kitzmuller, E., Cairns, N., Prast, H., Singewald, N. & Lubec, G. (1997) Similar deficits of central histaminergic system in patients with Down syndrome and Alzheimer disease, *Neurosci. Lett*., *222,* 183–186.

Sementchenko, V. I., Schweinfest, C. W., Papas, T. S. & Watson, D. K. (1998) ETS2 function is required to maintain the transformed state of human prostate cancer cells. *Oncogene. 17,* 2883–2888.

Seth, A., Watson, D. K., Blair, D. G., Papas, T. S. (1989) c-ets-2 protooncogene has mitogenic and oncogenic activity. *Proc. Natl. Acad. Sci., 86,* 7833–7837.

Shivdasani, R. A., Rosenblatt, M. F., Zucker-Franklin, D., Jackson, C. W., Hunt, P., Saris, C. J. & Orkin, S. H. (1995) Transcription factor NF-E2 is required for platelet formation independent of the actions of thrombopoietin/MGDF in megakaryocyte development. *Cell. 81,* 695-704.

Song, W. J., Sullivan, M. G., Legare, R. D., Hutchings, S., Tan, X., Kufrin, D., Ratajczak, J., Resende, I. C., Haworth, C., Hock, R., Loh, M., Felix, C., Roy, D. C., Busque, L., Kurnit, D., Willman, C., Gewirtz, A. M., Speck, N. A., Bushweller, J. H., Li, F. P., Gardiner, K., Poncz, M., Maris, J. M. & Gilliland, D. G. (1999) Haploinsufficiency of CBFA2 causes familial thrombocytopenia with propensity to develop acute myelogenous leukaemia. *Nature Genet. 23,* 166–175.

Speck, N. A. & Gilliland, D. G. (2002) Core-binding factors in haematopoiesis and leukaemia. *Nature Rev. Cancer. 2,* 502-513.

Stasko, M. R. & Costa, A. C. (2004). Experimental parameters affecting the Morris water maze performance of a mouse model of Down syndrome. *Behav. Brain Res., 154,* 1–17.

Stoll, C., Alembik Y., Dott, B. & Roth, M. P. (1990) Epidemiology of Don syndrome in 118,265 consecutive births. *Am. J. Med. Genet., 7(Suppl.),* 79-83.

Takashima, S., Becker, L. E., Armstrong, D. L. & Chan, F. (1981) Abnormal neuronal development in the visual cortex of the human fetus and infant with Down's syndrome. A quantitative and qualitative Golgi study. *Brain Res., 225,* 1–21.

Takashima, S., Iida, K., Mito, T. & Arima, M. (1994) Dendritic and histochemical development and aging in patients with Down's syndrome. *J. Intellect. Disabil. Res., 38(Pt3),* 265–273.

Taub, J. W. (2001) Relationship of chromosome 21 and acute leukemia in children with Down syndrome. *J. Pediatr. Hematol. Oncol., 23,* 175–178.

Taub, J. W., Mundschau, G., Ge, Y., Poulik, J. M., Qureshi, F., Jensen, T., James, S. J., Matherly, L. H., Wechsler, J. & Crispino, J. D. (2004) Prenatal origin of GATA1 mutations may be an initiating step in the development of megakaryocytic leukemia in Down syndrome. *Blood. 104,* 1588-1589.

Teipel, S. J. & Hampel, H. (2006) Neuroanaotmy of Down syndrome in vivo: a moldel of preclinical Alzheier's disease. *Behav. Genet., 36,* 405-415.

Teipel, S. J., Schapiro, M. B., Alexander, G. E., Krasuski, J. S., Horwitz, B., Hoehne, C., Moller, H. J., Rapoport, S. I. & Hampel, H. (2003) Relation of corpus callosum and hippocampal size to age in non-demented adults with Down's syndrome. *Am. J. Psychiatry. 160,* 1870-1878.

Tejedor, F., Zhu, X. M., Kaltenbach, E., Ackermann, A. Baumann, A., Canal, I., Heisenberg, M., Fischbach, K. F. & Pongs, O. (1995) Minibrain: a new protein kinase family involved in postembryonic neurogenesis in Drosophila. *Neuron. 14,* 287–301.

Thomas, J. B., Crews, S. T. & Goodman, C. S. (1988) Molecular genetics of the single-minded locus: a gene involved in the development of the Drosophila nervous system. *Cell. 52,* 133-141.

Toki, T, Katsuoka, F., Kanezaki, R., Xu, G., Kurotaki, H., Sun, J., Kamio, T., Watanabe, S., Tandai, S., Terui, K., Yagihashi, S., Komatsu, N., Igarashi, K., Yamamoto, M. & Ito, E. (2005) Transgenic expression of BACH1 transcription factor results in megakaryocytic impairment. *Blood. 105,* 3100-3108.

Underhill, C. B., Green, S. J., Comoglio, P. M. & Tarone, G. (1987) The hyaluronate receptor is identical to a glycoprotein of Mr 85,000 (gp85) as shown by a monoclonal antibody that interferes with binding activity. *J. Biol. Chem., 262,* 13142–13146.

Wang, P. P, Doherty, S., Hesselink, J. R. & Bellugi, U. (1992) Callosal morphology concurs with neurobehavioral and neuropathological findings in two neurodevelopmental disorders. *Arch. Neurol., 49,* 407–411.

Wechsler, J., Greene, M., McDevitt, M. A., Anastasi, J., Karp, J. E., Le Beau, M. M. & Crispino, J. D. (2002) Acquired mutations in GATA1 in the megakaryoblastic leukemia of Down syndrome. *Nature Genet., 32,* 148-152.

Weis, S., Weber, G., Neuhold, A. & Rett, A. (1991) Down syndrome: MR quantification of brain structures and comparison with normal control subjects. *Am. J. Neuroradiol., 12,* 1207–1211.

Wisniewski, K. E. (1990) Down syndrome children often have brain with maturation delay, retardation of growth, and cortical dysgenesis. *Am. J. Med. Genet. 7,* 274–281.

Wisniewski, K. E. & Bobinski, M. (1991) Hypothalamic abnormalities in Down syndrome, *Prog. Clin. Biol. Res., 373,* 153–167.

Wisniewski, K. E. & Schmidt-Sidor, B. (1989) Postnatal delay of myelin formation in brains from Down syndrome infants and children. *Clin. Neuropathol., 8,* 55–62.

Wolvetang, E. J., Wilson, T. J., Sanij, E., Busciglio, J., Hatzistavrou, T., Seth, A., Hertzog, P. J. & Kola, I. (2003) ETS2 overexpression in transgenic models and in Down syndrome predisposes to apoptosis via the p53 pathway. *Hum. Mol. Genet., 12,* 247-255.

Xu, G., Nagano, M., Kanezaki, R., Toki, T., Hayashi, Y., Taketani, T., Taki, T., Mitui, T., Koike, K., Kato, K., Imaizumi, M., Sekine, I., Ikeda, Y., Hanada, R., Sako, M., Kudo, K., Kojima, S., Ohneda, O., Yamamoto, M. & Ito, E. (2003) Frequent mutations in the GATA-1 gene in the transient myeloproliferative disorder of Down's syndrome. *Blood. 102,* 2960-2968.

Yamamoto, N., Uzawa, K., Miya, T., Watanabe, T., Yokoe, H., Shibahara, T., Noma, H. & Tanzawa, H. (1999) Frequent allelic loss/imbalance on the long arm of chromosome 21 in oral cancer: evidence for three discrete tumor suppressor gene loci. *Oncol. Rep., 6,* 1223–1227.

Yang, E. J., Ahn, Y. S. & Chung, K. C. (2000) Protein kinase Dyrk1 activates cAMP response element-binding protein during neuronal differentiation in hippocampal progenitor cells. *J. Biol. Chem., 276,* 39819–39824.

Yang, Q., Rasmussen, S. A. & Friedman, J. M. (2002) Mortality associated with Down's syndrome in the USA from 1983 to 1997: A population-bases study. *Lancet. 359,* 1019-1025.

Yla-Herttuala, S., Luoma, J., Nikkari, T. & Kivimäki, T. (1989) Down's syndrome and atherosclerosis. *Atherosclerosis. 76,* 269-272.

Zaldumbide, A., Carlotti, F., Pognonec, P. & Boulukos, K. E. (2002) The role of the Ets2 transcription factor in the proliferation, maturation, and survival of mouse thymocytes. *J. Immunol., 169,* 4873–4881.

Zeller, B., Gustafsson, G., Forestier, E., Abrahamsson, J., Clausen, N., Heldrup, J., Hovi, L., Jonmundsson, G., Lie, S. O., Glomstein, A. & Hasle, H.; Nordic Society of Paediatric Haematology and Oncology (NOPHO). (2005) Acute leukaemia in children with Down syndrome: a population-based Nordic study. *Br. J. Haematol., 128,* 797–804.

Zipursky, A., Poon, A. & Doyle, J. (2001) Leukemia in Down syndrome. A review. *Pediatr. Hematol. Oncol., 9,* 139-149.

Zipursky, A., Thorner, P., De Harven, E., Christensen, H. & Doyle, J. (1994) Myelodysplasia and acute megakaryoblastic leukemia in Down's syndrome. *Leuk. Res., 18,* 163–171.

In: Advances in Medicine and Biology. Volume 20
Editor: Leon V. Berhardt
ISBN 978-1-61209-135-8

Chapter 5

ALTERATIONS OF AUTONOMIC MODULATION OF HEART RATE VARIABILITY IN EATING DISORDERS

P. Ameri*[*1], A. De Negri[1], M. Casu[1], A. Brugnolo[2], P. F. Cerro[3], V. Patrone[1], E. Pino[1], A. M. Ferro[3], G. Rodriguez[2] and G. Murialdo[1]

[1]Department of Endocrine and Medical Sciences, Internal Medicine Unit, University of Genoa, Italy

[2]Department of Endocrine and Medical Sciences, Neuropathophysiology Unit, University of Genoa, Italy

[3]Centre for Eating Disorders and Psychological Disturbances of Adolescence, Department of Mental Health, Azienda Sanitaria Locale 2 Savonese, Savona, Italy

ABSTRACT

Eating disorders, which include anorexia nervosa (AN) and bulimia nervosa (BN), are characterized by significant alterations of many systems maintaining body homeostasis. With respect to the cardiovascular system, a series of studies have been performed to assess whether heart rate variability (HRV), which is an indicator of cardiac autonomic regulation, changes in patients affected by eating disorders in comparison to healthy subjects. Even if some discrepancies exist, possibly due to different methods of investigation and/or criteria for patient selection, most results show the existence of a

* Correspondence: Pietro Ameri, M.D. Dipartimento di Scienze Endocrinologiche e Metaboliche, Università di Genova, Viale Benedetto XV, 6, I-16132 Genova (Italy). Phone: +39.010.3537975. Fax: +39.010.3538977. E-mail: pietroameri@unige.it

predominance of vagal over sympathetic activity in the modulation of HRV in both AN and BN. This cardiac sympathovagal imbalance may be the pathophysiological basis for the higher risk of sudden cardiac death observed in eating disorders. The pathological modifications of HRV and cardiac autonomic modulation might be secondary to the endocrine abnormalities associated with AN and, to a lesser extent, BN. Indeed, alterations of HRV have been described in classical endocrine diseases, like diabetes mellitus or acromegaly. In particular, hormones controlling caloric intake and body fat composition may play a primary role. In the second part of this chapter, after a comprehensive review of the available data about HRV in eating disorders, the possible relationships between HRV and hormonal changes are discussed.

Introduction

Eating disorders, including anorexia nervosa (AN), bulimia nervosa (BN), binge eating disorder (BED) and their variants, are characterized by a disturbance in eating behaviors associated with distress and excessive concerns about body weight and image [1]. They have a complex etiology, with participation of genetic, neurochemical, psychodevelopmental and sociocultural factors, and affect the person on the whole, involving both the psychic dimension and most biological functions [2,3].

In the two major clinical forms, i.e., AN and BN, self-starvation and/or eliminative practices can be so severe that they cause substantial health problems [1-3], as confirmed by the several medical findings that emerge during clinical examination [4]. Body adaptation to restriction, eliminative behaviors and overexercising may also lead to pathological changes in itself, especially when starvation becomes extreme [3,5]. In this regard, an example is given by the need of sufficient leptin levels for the recovery from amenorrhoea in AN [6].

Table 1. Main symptoms, signs and medical complications concerning the cardiovascular system in anorexia nervosa and bulimia nervosa

Postural and non-postural hypotension
Sinus bradycardia
Atrial and ventricular arrhythmias
Electrocardiographic abnormalities:
Low QRS voltage
Prolonged QT interval
Prominent U wave
Left ventricular changes:
Decreased left ventricular mass
Decreased cavity size
Modified diastolic dynamics
Mitral valve prolapse
Cardiomyopathy
Acrocyanosis

Among the alterations occurring in eating disorders, those concerning the cardiovascular system particularly impact on the natural history of the disease, because of their significant contribution to morbidity and mortality [2,3,7]. The main clinical manifestations secondary to cardiovascular dysfunction in AN and BN are listed in Table 1.

In AN patients, blood pressure regulation as well as heart structure and activity have been shown to be impaired [8-13]. Blood pressure values are diminished, this finding being primarily referred to hypovolemia due to starvation, purging and/or vomiting. Abnormalities in the mitral valve motion, slowed transmitral flow velocity in late diastole, and reduced left ventricular mass and filling have been documented [9-11]. Moreover, decreased potassium concentrations and variations in other electrolyte levels may lead to systolic dysfunction, cardiac output reduction, QT prolongation, and arrhythmias [8,12].

An overall mortality rate of 0.56 percent per year has been reported in AN, which is 12 times higher than that among young women in the general population [14]. The modifications in the cardiovascular system, especially those affecting the generation and propagation of the electrical impulses within the heart, contribute to the increased mortality in AN and in BN [7,8,15,16]. Although suicide widely accounts for the higher standardized mortality ratio among people with AN [17,18], a significant amount of deaths result from starvation and purging-related cardiac arrhythmias [19]. Indeed, an increased incidence of sudden death associated with arrhythmias and acute hemodynamic failure has been described, in particular when alcohol and other substance abuse is combined with binging/purging behavior [15].

Therefore, studies of cardiac pathophysiology in eating disorders may be useful to improve basic knowledge and, possibly, open new diagnostic and therapeutic perspectives.

In this paper, we focused the attention on changes in heart rate variability (HRV) and eventually in the underlying autonomic control of heart function. HRV can be investigated by means of simple and non-invasive techniques, which are being increasingly used for the study of many other diseases, too.

HRV Analysis: Methodological and General Aspects

Conventionally, HRV defines the physiological oscillations that the intervals between successive heart beats and thus consecutive instantaneous heart rates (HR) present over time [20,21].

HRV can be documented and analyzed by different non-invasive and easily reproducible techniques, all based on electrocardiogram (ECG), which identifies QRS complexes, reflecting ventricular depolarization and contraction in response to sinus node pacing. In ECG recordings the instantaneous HR as well as the duration of the intervals between successive QRS complexes (R-R intervals) can be calculated. Once determined, HR and R-R intervals can be analyzed as a temporal series (so-called "time-domain analysis"), applying simple mathematical operations, such as the calculation of mean HR or R-R interval, or more complex statistical and geometric measurements [20,21]. Alternatively, the oscillations characterizing HR and R-R intervals can be analyzed as a function of the frequency ("frequency-domain analysis"), using both parametric and non-parametric methods. At present, power spectral analysis (PSA) is the technique currently used for frequency-domain analysis. Frequency-domain analysis allows us to obtain a spectrum of the various frequencies that participate in creating HRV and to quantify each of them by means of computerized elaboration of ECG data, applying the Fourier's or other mathematical transformations [20,21].

Both 24-hour Holter and short-term ECG are suitable for PSA [20]. The first enables a more detailed evaluation, including the assessment of HRV variations in ambulatory subjects throughout a prolonged time period, including waking and sleeping, usually 24 hours [20]. Nevertheless, PSA of short-term ECG can represent a less expensive and easily administrable tool in the clinical setting [20].

Short-term ECG are recorded for 5 to 15 minutes in supine resting subjects, keeping breathing rate synchronized with a metronome set at a proper frequency [22,23]. PSA identifies three different peaks within the total spectrum of HRV, corresponding to the very low frequency (VLF), the low frequency (LF), and the high frequency (HF) components—or powers—of HRV (Figure 1).

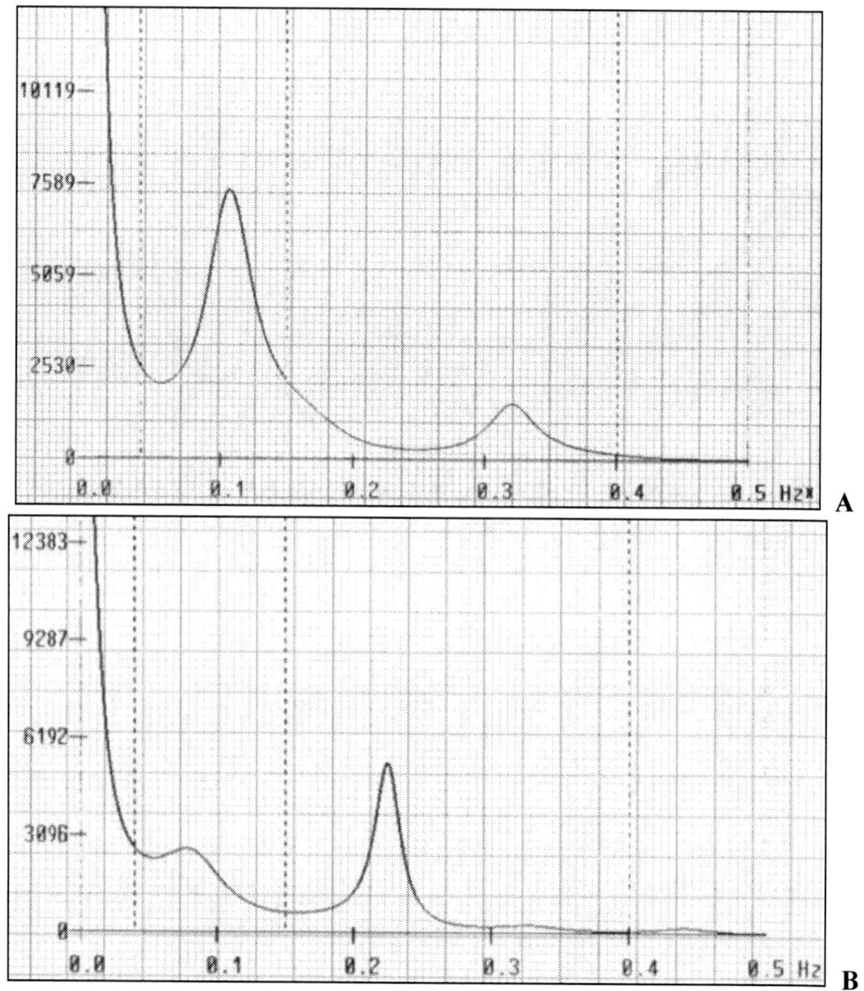

Figure 1. Heart rate variability spectrum obtained in supine resting condition by power spectral analysis of a 10 minute electrocardiogram in a healthy subject (A) and in a woman with anorexia nervosa (B). Note the increased high frequency power, indicative of vagal hyperactivity, in the patient affected by anorexia nervosa. HF: high frequency power; LF: low frequency power; VLF: very low frequency power.

The LF component of the spectrum is thought to be related to both sympathetic and parasympathetic tone [24-26], while the HF one seems predominantly associated with the vagal activity connected with respiration [24,27] (Table 2).

Table 2. Principal components of heart rate variability on power spectral analysis of short-term electrocardiograms and respective main modulating factors. Hz: Hertz

Spectral power	Frequency	Modulation
High frequency- HF	0.15-0.50 Hz	vagal activity connected with respiration
Low frequency- LF	0.04-0.15 Hz	vagal and sympathetic activity
Very low frequency- VLF	0.00-0.04 Hz	vagal activity, renin-angiotensin system, leptin, thermoregulation, others?

Consequently, the study of HRV by PSA gives information about the regulation exerted by the autonomic nervous system on cardiac frequency. With this respect, the sympathovagal balance can be assessed by calculating the LF/HF and LF/(LF+HF) ratios, which reflect the reciprocal relationships and variations between the two main powers of HRV [20,24,28]. However, further factors likely modulate HRV, as it is in the case of the VLF component, which has been related to various mechanisms such as vagal activity, renin-angiotensin system, plasma leptin concentrations, and thermoregulation [20,29,30]. In short-term PSA, however, the reliability of the calculation of the VLF power is dubious, and most experts recommend to exclude this component from the analysis [20].

Short-term ECG recording may be integrated by the dynamic lying-to-standing challenge, usually performed in a controlled and standardized manner using a tilting table (tilt-table test) [22,23]. This test is based on the rapid passive achievement of the upright position and allows us to evaluate whether the harmonic change in sympathetic and vagal nerve activity required from lying-to-standing effectively happens. In fact, in healthy individuals lying-to-standing implies a redistribution of blood volume, which elicits a real drop in cardiac output and blood pressure, compensated by reflex increased sympathetic and decreased vagal outflow [31]. This autonomic response translates into modifications of PSA: the LF power markedly increases, while the HF component drops (Figure 2A), with a consequent increase of the orthostatic LF/HF and LF/(LF+HF) ratios [20,28].

The analysis of HRV, most often by means of PSA, has progressively gained diffusion for basic investigations as well as in clinical practice. Indeed, it has been demonstrated that HRV is altered in many diseases, with diagnostic and prognostic implications.

For example, in diabetes mellitus reduced HRV is considered as an earliest predictor of cardiovascular autonomic neuropathy [32]. Besides clinical manifestations of diabetes, such as orthostatic hypotension or silent myocardial ischemia, autonomic neuropathy-related modifications of HRV have been suggested to be strictly associated with increased risk for mortality and probably also for major cardiovascular events and sudden death [33].

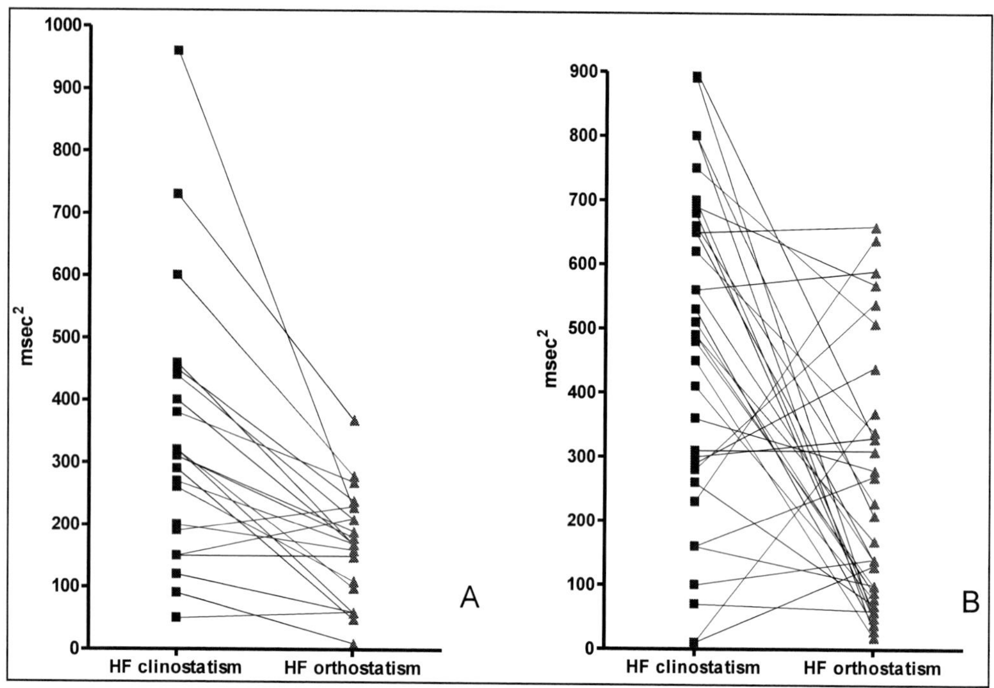

Figure 2. Changes in the absolute values of the high frequency component of heart rate variability from lying to standing in healthy subjects (A) and in patients with anorexia nervosa (B). Note the absence of the physiological decrease of the high frequency power or even its paradoxical increase in a subgroup of patients affected by anorexia nervosa. HF: high frequency power.

Heart Rate Variability in Eating Disorders

Eating disorders represent another field in which HRV has been extensively investigated. Alterations in both time-domain and frequency-domain parameters, indicative of cardiovascular autonomic dysfunction, have been documented in AN and in a less manner in BN [8,11,12,16,34-49]. Up to date, a single report regarded patients with BED [50]. The main studies of HRV in eating disorders and the corresponding references are summarized in Table 3. The total of studied patients was 370 for AN and 90 for BN. As predictable on the basis of the much higher incidence of eating disorders in women [51], almost all the subjects evaluated were female. Thus, the data presented and the conclusions drawn should be confirmed in a larger population and validated for the minority of male patients.

In AN, HRV has been studied by examining both short-term and 24 hour Holter ECG. Since a significant correlation between 10 minute and 24 hour measurements was demonstrated by direct comparison in AN patients, the determination of resting HRV could be used as a preliminary screening for the more comprehensive 24 hour Holter study [42].

An increase in HRV emerged from the analysis of resting short-term ECG in AN. The amplification of HRV appeared to be associated with vagal hyperactivity and sympathetic deficiency, revealed by higher HF power, lower LF power and LF/HF or LF/(LF+HF) ratios of PSA as compared with healthy controls [8,34,37,38,41,42,45,48,49]. An example of PSA abnormalities at rest in one paradigmatic case of AN is given in Figure 1B.

Table 3. Characteristics and main results of the principal studies investigating power spectral analysis of heart rate variability in anorexia nervosa (AN) and bulimia nervosa (BN). All the investigations included healthy subjects for comparison, with the exception of n. 45. Ref: reference; Pts: number of patients evaluated; ECG: electrocardiogram; HF: high frequency power; LF: low frequency power

Authors	Ref	Pts	Study characteristics	Main findings in patients as compared with controls
Rissanen et al. 1998	[36]	25 BN	short-term ECG, lying baseline and after 8 we of fluoxetine + behavioral therapy	↑ basal HF ↓ HF after therapy
Rechlin et al. 1998	[37]	48 AN	short-term ECG, lying and standing	↓ LF during lying failure of LF to increase with standing
Casu et al. 2002	[38]	13 AN	short-term ECG, lying and standing	failure of HF to decrease with standing
Galetta et al. 2003	[11]	25 AN	short-term ECG, lying and standing 24-h Holter ECG	↑ HF, ↓ LF/HF sinus bradycardia
Mont et al. 2003	[39]	31 AN	24-h Holter ECG baseline and after 3-18 mo of therapy	↑ basal HF normalization after refeeding sinus bradycardia
Melanson et al. 2004	[42]	6 AN	short-term ECG, lying 24-h Holter ECG	↓ LF and HF correlation between short-term and 24-h measurements of HRV
Cong et al. 2004	[43]	6 AN 8 BN	24-h Holter ECG	↑ HF, ↓ LF/HF
Facchini et al. 2006	[8]	29 AN	short-term ECG	↓ LF/HF sinus bradycardia hypokaliemia-related QT prolongation
Yoshida et al. 2006	[45]	9 AN	24-h Holter ECG baseline and after 3-5 days of therapy	↓ R-R interval, ↓ HF and ↑ LF/HF after treatment
Platisa et al. 2006	[46]	17 AN	24-h Holter ECG	"acute" AN: ↑ HF and ↓ LF "chronic" AN: ↓ HF and further ↓ LF
Tanaka et al. 2006	[47]	22 BN	short-term ECG, lying basal and after 16 we of paroxetine + behavioral therapy	↑ basal HF, ↓ basal LF/HF↓ HF after treatment
Friederich et al. 2006		38	short-term ECG, during resting and performing mental tasks	↓ HF during mental stress
Murialdo et al. 2007	[49]	34 AN 16 BN	short-term ECG, lying and standing	↓ LF/(LF+HF) during lying ↓ LF and LF/(LF+HF) during standing

The dysfunction in the autonomic control of HR was evident in supine subjects, but it was chiefly disclosed and emphasized during lying-to-standing (Figure 2B). In AN women, the HF power failed to decrease with standing differently to what physiologically verifies in healthy subjects [34,38]. Moreover, a sympathetic failure occurred, as revealed by an inadequate or even absent rise of the LF component after the postural change [38,49].

The findings on PSA of 24 hour ECG were similar to those obtained by PSA of short-term ECG and confirmed the increase of HRV and HF power, as well as the decrease of LF power. The imbalanced sympathovagal modulation of HR was proved by the reduction of LF/HF and LF/(LF+HF) ratios also in this instance [11,35,39,40,42,43,45,46].

It has been recently proposed that these modifications in HRV are peculiar of an early phase of the disease, while opposite changes develop along with the progression of the disorder [46]. In a direct comparative study, eight women with AN lasting for a mean of 12±5 months, defined as acute, had increased HRV and HF power on PSA, while the same measures were decreased in nine women with disease lasting for 36±9 months, considered as chronically ill. Moreover, in chronic AN the LF component was further decreased [46]. However, it is still unclear whether duration of AN may affect HRV, considering also that other discordant data have been published with regard to a possible relationship between illness course and HRV parameters [41,49].

In subjects with AN, PSA showed that HR can be slowed up to sinus bradycardia [8,11,37,39,45]. Therefore, in AN the finding of bradycardia, which is observed in approximately 41% of outpatients on physical examination [4], may be considered as an easily identifiable clinical sign of the predominance of vagal over sympathetic influences in HR regulation. Interestingly, in one study of cardiac abnormalities in AN the percentage of subjects presenting sinus bradycardia rose from 35% to 60% during sleep [39]. However, it is possible that slowing of HR is secondary to other factors which intervene in determining the sinus node pacing rate and are altered in AN, like thyroid hormones [8,45].

A number of studies have shown a trend toward reversion of HRV abnormalities during and after successful treatment of AN, as attested by refeeding and restoration of normal body weight [34,37,39,43]. Again, simply monitoring HR during the 24 hours might be useful: in fact HR could represent a reliable sign of the trend to normalization of cardiac autonomic influences, because it rises within a few days after beginning of refeeding [45]. In six patients with AN, who were in various stages of refeeding and body weight recovery, reduced HRV, with significantly lower time-domain and frequency-domain measures, was even observed in comparison to healthy control subjects [42].

Although the available data concerning HRV in AN sufficiently agree on the whole, some discrepancies exist. Among possible explanations, the criteria of selection of patients might be chiefly relevant. For example, age of the patients, subtype and severity of the disease, hospitalization and previous therapies to correct medical complications, as low potassium concentration, are all factors that can count in determining the present characteristics of AN and thus give rise to different HRV measures. Consistently with this hypothesis, a correlation between HRV data and body weight or body mass index was reported [37,41,46]. However, we were unable to confirm such correlation, as well as that between PSA indices and age [49].

As aforementioned, fewer investigations on HRV have been performed in BN than in AN. Among them, two studies compared HRV parameters in subjects with BN versus healthy individuals [36,47], whereas three included also AN patients in the comparison [43,48,49].

When compared with age-matched controls, PSA of short-term ECG of women with BN displayed a significant increase in the HF power, suggestive of an enhanced cardiac vagal tone [36,47]. Moreover, treatments with selective serotonin reuptake inhibitors, like fluoxetine or paroxetine, in combination with cognitive-behavioral therapy reduced cardiac vagal tone [36,47], but vagal outflow remained unmodified after the administration of placebo plus behavioral therapy [36].

HRV was comparatively assessed in AN and BN patients by PSA of both short-term [48,49] and 24 hour ECG [43]. The two disorders seem to share analogous modifications of HRV, despite the differences in clinical manifestations and body weight and composition.

In basal condition, a reduction of the LF/(HF) and LF/(LF+HF) ratios on short-term and 24 hour PSA, together with an augmented HF power over the 24 hour period, was found in both AN and BN patients, despite the first group had significantly lowered body weight and body mass index than the second did [43,49]. Lying-to-standing disclosed further modifications in HRV modulation which were similar in the two eating disorders. In fact, differently from what we observed in healthy controls, in both AN and BN the LF component failed to increase after standing or even paradoxically diminished, with a consequent decreased LF/(LF+HF) ratio, as depicted in figure 3 [49]. Nevertheless, a cardiac autonomic dysfunction which was more severe in AN than in BN has been recently described [48], in keeping with older preliminary conclusions based on the simple analysis of HR and respiratory sinus arrhythmia on standard ECG [52].

With respect to such discord between different studies, it is crucial to remind that AN and BN are parts of a continuous spectrum of disease, although they should be well defined by currently adopted diagnostic criteria, and that such patients may modify their behaviors and shift from one condition into the other, or even to other clinical types of eating disorders [1-3]. Thus, the drastic division of subjects between AN or BN groups may be arbitrary and sometimes even fallacious, generating confounding data.

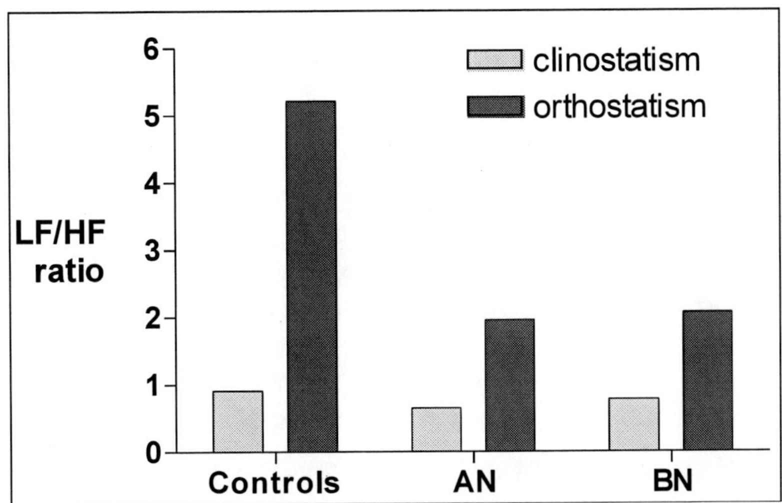

Figure 3. Ratio between the low frequency and the high frequency powers in clinostatism and orthostatism in healthy controls and in patients affected by anorexia nervosa or bulimia nervosa. HF: high frequency power; LF: low frequency power; AN: anorexia nervosa; BN: bulimia nervosa.

Within eating disorders, BED has been characterized later than AN and BN. It typically occurs in obese individuals which suffer from anxiety and/or depression, and present with episodes of overeating without subsequent compensatory behaviors [53]. A single study, performed in 38 obese women with BED, showed a greater reduction in the HF power during mental stress as compared with healthy controls matched for sex, age and body mass index [50]. Thus further investigations on cardiovascular sympathovagal function in subjects with BED are mandatory.

The alterations of the HRV autonomic control in eating disorders may produce some clinical manifestations, like bradycardia and orthostatic hypotension, as previously discussed. Moreover, in 25 subjects with AN time-domain HRV indices, consistent with vagal hyperactivity, appeared to be also correlated with the echocardiographic evidence of peculiar ventricular diastolic dynamics, such as reduced left ventricular mass and lower peak velocity transmitral flow during late diastole, but not in early diastole [11].

Finally, chronotropic incompetence, defined as the inability to achieve the 80% of HR reserve during exercise, was documented in a subgroup of AN patients with a significant reduction in the LF power and the LF/HF ratio [40].

HRV impairment might also manifest itself with the most dramatic complication of eating disorders, i.e. the occurrence of ventricular arrhythmias resulting in sudden cardiac death, especially frequent in AN patients with severe starvation.

Most studies of HRV in AN agree with an increase in HRV and parasympathetic activity, which could favor cardiac early after-depolarization, possibly resulting in fatal arrhythmias, mainly when associated to hypokaliemia.

Hypokaliemia plays an important role in inducing changes in heart repolarization and probably in triggering ventricular dysfunction and arrhythmias. Facchini et al. [8] studied the relationships between serum potassium level, QT interval, the occurrence of ventricular arrhythmias and HRV in the frequency domain in 29 patients with AN compared with 14 healthy females. Decreased LF/HF ratio indicating a vagal hyperactivity was documented in AN, but QT duration was found to be prolonged only in the setting of severe hypokaliemia (<2 mEq/l), exposing these patients to an increased risk of sudden cardiac events.

Refeeding in a small group of women affected by AN led to a decrease in HRV [42]. Depressed HRV is a well established risk factor for sudden cardiac death. As a consequence, the condition of recovering AN might be considered equal to that of individuals affected by heart failure or acute myocardial infarction [16,42].

It is conceivable that the variations in HRV in eating disorders are only a side of an overall loss of the sympathovagal balance, since extravascular functions under the autonomic regulation, such as pupillary light reflex, can also be compromised [44].

Endocrine Changes and HRV in Eating Disorders: A Perspective for Future Investigations

Although it has been clearly shown that HRV is abnormal in eating disorders, possibly as a consequence of a dysfunction of the autonomic nervous system *in toto*, the causes and mechanisms leading to the above discussed pathological modifications are still unknown. Malnutrition, independently from the presence of a concomitant eating disorder, can alter cardiac electrical activity and autonomic regulation. Indeed, in malnourished animal models significant alterations of sympathetic activity, HR and HRV take place [54,55]. The deficiency of specific nutrients may be important: one example in this regard is given by data on humans, showing that vitamin B12 deficient subjects present a decrease of the sympathetic and, to a lesser extent, parasympathetic indices of HRV [56].

Mirroring the physiologic influences of many hormones on HRV, various endocrine diseases can be characterized by a cardiac sympathovagal imbalance, such as diabetes mellitus [33], thyroid cancer on chronic l-thyroxine therapy to suppress thyroid stimulating hormone secretion [22], and acromegaly [23]. Thus, abnormalities of HRV in eating disorders might be related to endocrine dysregulations.

Levels of estrogens are very low in AN [2,57] and female sex hormones can participate in the regulation of HRV [58]. Nevertheless, the contribution of changing sex steroid concentrations to the HRV pattern in eating disorders has not been investigated yet.

The reduction of the LF power in AN may be view as one aspect of a general sympathetic hypoactivity, finalized to prevent excessive caloric consumption in a context of negative energy balance. In other terms, the decrease in sympathetic nerve outflow would act as a physiological adaptation response to starvation, along with the reduction of the secretion and/or metabolic activation rate of some hormones, like triiodothyronine (fT3), whose values are in fact lowered in AN [57], or catecholamines. Consistently, the LF component of HRV was reported to be positively correlated with the plasmatic concentration of metanephrine and normetanephrine in AN patients [40]. Yoshida et al. [45] confirmed the existence of parallel variations of HR, LF power, norepinephrine and fT3 values in AN. However, HRV measurements did not significantly correlate with fT3 levels in another study [8].

In hypertensive subjects, impairment of the renin-angiotensin-aldosterone system was described to translate in alterations of HRV [59]. The peculiar behaviors of patients with eating disorders, such as strict dieting, strenuous exercise, repeated vomiting and abuse of laxatives and/or diuretics, may impinge upon the activity of the renin-angiotensin-aldosterone axis [60,61] and eventually HRV.

In parallel with the deepening of the knowledge about "classical" endocrine dysfunctions [57], growing evidence suggests that alterations of the release and/or action of hormones controlling food intake and energy homeostasis also occur in eating disorders [5].

Leptin is one of these hormones, whose production by adipocytes is directly related to the amount of body fat mass [62,63]. Consistently with its crucial participation in the processes aimed at preserving energy balance, its levels vary in eating disorders. Serum leptin concentration is markedly low in AN [64,65], whereas either normal or decreased values have been found in BN [66,67]. It was extensively reported that leptin may exert a dose-dependent modulation on the autonomic nervous system by stimulating the sympathetic outflow, possibly justifying the correlation between body weight and increased blood pressure in obese individuals [68-72]. Furthermore, an effect of leptin on HRV was also described [27]. Therefore, we attempted to correlate the PSA parameters with serum leptin concentrations in AN and BN inpatients [49]. However, no significant relationship was demonstrated, despite the existence of a direct correlation between the hormone levels and blood pressure values [49].

Changes in the concentrations of adiponectin, an anorexigenic protein secreted by the adipose tissue, have been described in AN, more in the restrictive form than in the purging one, as well as in BN [73,74], but no information about HRV and/or cardiovascular autonomic activity in relationship with adiponectin values is available. Thus, addressed studies would be invaluable.

Similarly, ghrelin could be the object of new investigations. Ghrelin is a peptide, initially isolated from the stomach and afterwards detected in the hypothalamus and in the pituitary gland, which stimulates growth hormone and other pituitary hormone release [75,76]. Since it is also involved in energy homeostasis maintenance, mainly by acting on the hypothalamic neurons which promote food intake [76], it is conceivable that it takes part in eating disorder pathophysiology. Elevated ghrelin levels were reported in AN [77,78] and BN [79], even if the latter finding was questioned [80].

Tanaka et al. [47] noticed that ghrelin significantly decreased in BN patients after combined treatment with the selective serotonin reuptake inhibitor paroxetine and cognitive-behavioral therapy sessions, similarly to the HF power, as aforementioned. Therefore, it would be worth investigating whether ghrelin and cardiac vagal modulation are somehow reciprocally related, given their concomitant changes.

Modifications in other peptides participating to the regulation of energy balance and cardiovascular functions might exert a role in eating disorders [5]. In particular, elevated corticotropin releasing hormone (CRH) concentrations were found in the cerebrospinal fluid of patients with AN [81,82], that is consistent with the hyperactivity of the hypothalamus-pituitary-adrenal axis which characterizes the disease. However, the increased levels of corticotropin releasing hormone may be also related to other features of AN, as variations in appetite, anxiety threshold, locomotor activity as well as in function of the cardiovascular and autonomic systems [83,84]. Toghether with corticotrophin releasing hormone, neuropeptide Y might be another important effector of the alterations of HRV in eating disorders [85].

CONCLUSIONS

Several investigators analyzed HRV in the two major eating disorders, AN and BN, mostly by means of PSA. As a whole, the results disclosed that HRV is similarly altered in AN and BN, because of an imbalance between cardiac sympathetic and vagal activities, with predominance of the latter. These changes are demonstrable by PSA of short-term ECG, recorded for some minutes in resting conditions, and can be confirmed on 24-hour Holter ECG, obtained throughout the entire day. Of note, PSA of short-time ECG appears as reliable as the study of 24-hour Holter recordings, so that the first might be used for a first-step screening evaluation. In addition, short-term PSA can be enriched with the evaluation of the modifications of HRV indices during lying-to-standing, as assessed by the tilt-table test.

The cardiac sympathovagal imbalance may become clinically apparent with other manifestations besides HRV abnormalities, like bradycardia, orthostatic hypotension, and eventually chronotropic incompetence or ventricular arrhythmias.

A challenging field of research is given by the study of the relationship between changes of HRV and of hormones involved in energy balance, whose contribution to the pathophysiology of eating disorders is becoming clearer.

Researchers should focus their attention on the other entities of the spectrum of eating disorders, such as BED, and on the comparison between HRV in AN and in malnutrition due to other diseases. Moreover, the effects of refeeding on HRV and cardiac autonomic regulation need to be accurately investigated, in view of the dramatic consequences of rapid nutritional and caloric restoration, up to life-threatening cardiac events [86].

REFERENCES

[1] American Psychiatric Association. *Diagnostic and statistical manual of mental disorders,* 4th ed (DSM-IV). Washington: American Psychiatric Association, 1994.

[2] Becker, AE; Grinspoon, SK; Klibanski, A; Herzog, DB. Eating disorders. *N Engl J Med*, 1999,340,1092-8.

[3] Fairburn, C.G. and Harrison, P.J. Eating disorders. *Lancet*,2003,361,407-16.

[4] Miller, KK; Grinspoon, SK; Ciampa, J; Hier, J; Herzog, D; Klibanski, A. Medical findings in outpatients with anorexia nervosa. *Arch. Intern. Med*, 2005,165,561-6.

[5] Torsello, A; Brambilla, F; Tamiazzo, L; Bulgarelli, I; Rapetti, D; Bresciani, E; Locatelli, V. Central dysregulations in the control of energy homeostasis and endocrine alterations in anorexia and bulimia nervosa. *J. Endocrinol. Invest*, 2007,30,962-76.

[6] Brambilla, F; Monteleone, P; Bortolotti, F, Dalle Grave, R; Todisco, P; Favaro, A; Santonastaso, P; Ramacciotti, C; Paoli, R; Maj, M. Persistent amenorrhoea in weight-recovered anorexics: psychological and biological aspects. *Psychiatry Res*, 2003,118,249-57.

[7] Emborg, C. Mortality and causes of death in eating disorders in Denmark 1970-1993: a case register study. *Int. J. Eat Disord*, 1999,25,243-51.

[8] Facchini, M; Sala, L; Malfatto, G; Bragato, R; Redaelli, G; Invitti, C. Low-K^+ dependent QT prolongation and risk for ventricular arrhythmia in anorexia nervosa. *Int J. Cardiology*, 2006,106,170-6.

[9] St. John-Sutton, MG; Plappert, T; Crosby, L; Douglas, P; Mullen, J; Reichek N. Effects of reduced left ventricular mass on chambers architecture, load, and function: a study of anorexia nervosa. *Circulation*, 1985,72,991-1000.

[10] De Simone, G; Scalfi, L; Galderisi, M; Celentano, A; Di Biase, G; Tammaro, P; Garofalo, M; Mureddu, GF; de Divitiis, O; Contaldo, F. Cardiac abnormalities in young women with anorexia nervosa. *Br. Heart J.*, 1994,71,287-92.

[11] Galetta, F; Franzoni, F; Prattichizzo, F; Rolla, M; Santoro, G; Pentimone, F. Heart rate variability and left ventricular diastolic function in anorexia nervosa. *J. Adolesc. Health*, 2003,32,416-21.

[12] Roche, F; Estour, B; Kadem, M; Millot, L; Pichot, V; Duverney, D; Gaspoz, JM; Barthélémy, JC. Alteration of the QT rate dependence in anorexia nervosa. *Pacing Clin. Electrophysiol*, 2004,27,1099-104.

[13] McCallum, K; Bermudez, O; Ohlemeyer, C; Tyson, E; Portilla, M; Ferdan, B. How should the clinician evaluate and manage the cardiovascular complications of anorexia nervosa? *Eat. Disord.*, 2006,14,73-80.

[14] Sullivan PF. Mortality in anorexia nervosa. *Am. J. Psychiatry*, 1995,152,1073-4.

[15] Neumarker KJ. Mortality and sudden death in anorexia nervosa. *Int. J. Eat. Disord.*, 1997,21,205-12.

[16] Krantz, MJ and Mehler, PS. Heart rate variability in women. *Arch. Intern. Med*, 2006,166,247.

[17] Harris, EC and Barraclough, B. Excess mortality of mental disorder. *Br. J. Psychiatry*, 1998,173,11-53.

[18] Keel, PK; Dorer, DJ; Eddy, KT; Franko, D; Charatan, DL; Herzog, DB. Predictors of mortality in eating disorders. *Arch. Gen. Psychiatry*, 2003,60,179-83.

[19] Crisp, AH; Callender, JS; Halek, C; Hsu, LK. Long-term mortality in anorexia nervosa. A 20-year follow-up of the St George's and Aberdeen cohorts. *Br. J. Psychiatry*, 1992,161,104-7.

[20] Task Force of The European Society of Cardiology and The North American Society of Pacing and Electrophysiology. Guidelines. Heart rate variability. Standards of measurement, physiological interpretation, and clinical use. *Eur. Heart J.*, 1996,17,354-81.

[21] Seely, AJ and Macklem, PT. Complex systems and the technology of variability analysis. *Crit. Care*, 2004,8,R367-84.

[22] Casu, M; Cappi, C; Patrone, V; Repetto, E; Giusti, M; Minuto, F; Murialdo, G. Sympatho-vagal control of heart rate variability in patients treated with suppressive doses of L-thyroxine for thyroid cancer. *Eur. J. Endocrinol*, 2005,152,819-24.

[23] Resmini, E; Casu, M; Patrone, V; Murialdo, G; Bianchi, F; Giusti, M; Ferone, D; Minuto, F. Sympathovagal imbalance in acromegalic patients. *J. Clin. Endocrinol. Metab.*, 2006,91,115-20.

[24] Lipsitz, LA; Mietus, J; Moody, GB; Goldberger, AL. Spectral characteristics of heart rate variability before and during postural tilt. *Circulation*, 1990,81,1803-10.

[25] Bernardi, L; Leuzzi, S; Radaelli, A; Passino, C; Johnston, JA; Sleight, P. Low-frequency spontaneous fluctuations of R-R interval and blood pressure in conscious humans: a baroreceptor or central phenomenon? *Clin. Sci.*, 1994,87,649-54.

[26] Cooke, WH; Hoag, JB; Crossman, AA; Kuusela, TA; Tahvanainen, KU; Eckberg, DL. Human responses to upright tilt: a window on central autonomic integration. *J. Physiol.*, 1999,517,617-28.

[27] Pomeranz, B; Macaulay, RJ; Caudill, MA; Kutz, I; Adam, D; Gordon, D; Kilborn, KM; Barger, AC; Shannon, DC; Cohen, RJ; et al. Assessment of autonomic function in humans by heart rate spectral analysis. *Am. J. Physiol.*, 1985,248,H151-3.

[28] Malliani, A; Pagani, M; Lombardi, F; Cerutti, S. Cardiovascular neural regulation explored in the frequency domain. *Circulation*, 1991,84,482-91.

[29] Takabatake, N; Nakamura, H; Minamihaba, O; Inage, M; Inoue, S; Kagaya, S; Yamaki, M; Tomoike, H. A novel pathophysiologic phenomenon in cachexic patients with obstructive pulmonary disease: relationship between the circadian rhythm of circulating leptin and the very low-frequency component of heart rate variability. *Am. J. Resp. Crit. Care Med.*, 2001,63,1314-9.

[30] Bonaduce, D; Marciano, F; Petretta, M; Migaux, ML; Morgano, G; Bianchi, V; Salemme, L; Valva, G; Condorelli, M.. Effects of converting enzyme inhibition on heart period variability in patients with acute myocardial infarction. *Circulation*, 1994,90,108-13.

[31] Freeman R. Neurogenic orthostatic hypotension. *N Engl J Med*, 2008,358,615-24.

[32] Ziegler D. Diabetic cardiovascular autonomic neuropathy: prognosis, diagnosis and treatment. *Diabetes Metab. Rev.*, 1994,10,339-83.

[33] Vinik, AI; Maser, RE; Mitchell, BD; Freeman, R. Diabetic autonomic neuropathy. *Diabetes Care*, 2003,26,1553-79.

[34] Kreipe, RE; Goldstein, B; DeKing, DE; Tipton, R; Kempski, MH. Heart rate power spectrum analysis of autonomic dysfunction in adolescents with anorexia nervosa. *Int. J. Eat Disord.*, 1994,16,159-65.

[35] Petretta, M; Bonaduce, D; Scalfi, L; de Filippo, E; Marciano, F; Migaux, ML; Themistoclakis, S; Ianniciello, A; Contaldo, F. Heart rate variability as a measure of autonomic nervous system function in anorexia nervosa. *Clin. Cardiol.*, 1997,20,219-24.

[36] Rissanen, A; Naukkarinen, H; Virkkunen, M; Rawlings, RR; Linnoila, M. Fluoxetine normalizes increased cardiac vagal tone in bulimia nervosa. *J. Clin. Psychopharmacol*, 1998,18,26-32.

[37] Rechlin, T; Weis, M; Ott, C; Bleichner, F; Joraschky, P. Alterations of autonomic cardiac control in anorexia nervosa. *Biol. Psychiatry*, 1998,43,358-63.

[38] Casu, M; Patrone, V; Gianelli, MV; Marchegiani, A; Ragni, G; Murialdo, G; Polleri A. Spectral analysis of R-R interval variability by short-term recording in anorexia nervosa. *Eat. Weight Disord.*, 2002,7,239-43.

[39] Mont, L; Castro, J; Herreros, B; Paré, C; Azqueta, M; Magriña, J; Puig, J; Toro, J; Brugada, J. Reversibility of cardiac abnormalities in adolescents with anorexia nervosa after weight recovery. *J. Am. Acad. Child Adolesc. Psychiatry*, 2003,42,808-13.

[40] Roche, F; Barthélémy, JC; Garet, M; Costes, F; Pichot, V; Duverney, D; Kadem, M; Millot, L; Estour, B. Chronotropic incompetence to exercise separates low body weight from established anorexia nervosa. *Clin. Physiol. Funct. Imaging*, 2004,24, 270-5.

[41] Wu, Y; Nozaki, T; Inamitsu, T; Kubo, C. Physical and psychological factors influencing heart rate variability in anorexia nervosa. *Eat Weight Disord*, 2004,9,296-9.

[42] Melanson, EL; Donahoo, WT; Krantz, MJ; Poirier, P; Mehler, PS. Resting and ambulatory heart rate variability in chronic anorexia nervosa. *Am. J. Cardiol.*, 2004,94,1217-20.

[43] Cong, ND; Saikawa, R; Ogawa, R; Hara, M; Takahashi, N; Sakata, T. Reduced 24 hour ambulatory blood pressure and abnormal heart rate variability in patients with dysorexia nervosa. *Heart*, 2004,90,563-4.

[44] Bär, KJ; Boettger, S; Wagner, G; Wilsdorf, C; Gerhard, UJ; Boettger, MK; Blanz, B; Sauer, H. Changes of pain perception, autonomic function, and endocrine parameters during treatment of anorectic adolescents. *J. Am. Acad. Child Adolesc. Psychiatry*, 2006,45,1068-76.

[45] Yoshida, NM; Yoshiuchi, K; Kumano, H; Sasaki, T; Kuboki, T. Changes in heart rate with refeeding in anorexia nervosa: a pilot study. *J. Psychosom Res.*, 2006,61,571-5.

[46] Platisa, MM; Neswtorovic, Z; Damjanovic, S; Gal, V. Linear and non-linear heart rate variability measures in chronic and acute phase of anorexia nervosa. *Clin. Physiol. Funct. Imaging*, 2006,26,54-60.

[47] Tanaka, M; Nakahara, T; Muranaga, T; Kojima, S; Yasuhara, D; Ueno, H; Nakazato, M; Inui, A. Ghrelin concentrations and cardiac vagal tone are decreased after pharmacologic and cognitive-behavioral treatment in patients with bulimia nervosa. *Horm. Behav.*, 2006,50,261-5.

[48] Vigo, DE; Castro, MN; Dörpinghaus, A; Weidema, H; Cardinali, DP; Siri, LN; Rovira, B; Fahrer, RD; Nogués, M; Leiguarda, RC; Guinjoan, SM. Nonlinear analysis of heart rate variability in patients with eating disorders. *World J. Biol. Psychiatry*, 2007,11,1-7 [Epub ahead of print].

[49] Murialdo, G; Casu, M; Falchero, M; Brugnolo, A; Patrone, V; Cerro, PF; Ameri, P; Andraghetti, G; Briatore, L; Copello, F; Cordera, R; Rodriguez, G; Ferro, AM. Alterations in the autonomic control of heart rate variability in patients with anorexia or bulimia nervosa: correlations between sympathovagal activity, clinical features, and leptin levels. *J. Endocrinol. Invest*, 2007,30,356-62.

[50] Friederich, HC; Schild, S; Schellberg, D; Quenter, A; Bode, C; Herzog, W; Zipfel, S. Cardiac parasympathetic regulation in obese women with binge eating disorder. *Int. J. Obes* 2006,30,534-42.

[51] Hoek, HW and van Hoeken, D. Review of the prevalence and incidence of eating disorders. *Int. J. Eat. Disord.,* 2003,34,383-96.

[52] Kennedy, SH and Heslegrave, RJ. Cardiac regulation in bulimia nervosa. *J. Psychiatr. Res.* 1989,23,267-73.

[53] Wilfley, DE; Wilson, GT; Agras, WS. The clinical significance of binge eating disorder. *Int. J. Eat Disord.,* 2003,34 Suppl,S96-106.

[54] Sakaguchi, T; Arase, K; Fisler, JS; Bray, GA. Effect of starvation and food intake on sympathetic activity. *Am. J. Physiol.*, 1988,255(2 Pt 2),R284-8.

[55] Oliveira, EL; Cardoso, LM; Pedrosa, ML; Silva, ME; Dun, NJ; Colombari, E; Moraes, MF; Chianca, DA Jr. A low protein diet causes an increase in the basal levels and variability of mean arterial pressure and heart rate in Fisher rats. *Nutr. Neurosci.*, 2004,7,201-5.

[56] Aytemir, K; Aksöyek, S; Büyükasik, Y; Haznedaroğlu, I; Atalar, E; Ozer, N; Ovünç, K; Ozmen, F; Oto, A. Assessment of autonomic nervous system functions in patients with vitamin B12 deficiency by power spectral analysis of heart rate variability. *Pacing Clin. Electrophysiol.,* 2000,23,975-8.

[57] Muñoz, MT and Argente, J. New concepts in anorexia nervosa. *J. Pedriatic. Endocrinol. Metab.,* 2004,17,473-80.

[58] Princi, T; Parco, S; Accardo, A; Radillo, O; De Seta, F; Guaschino, S. Parametric evaluation of heart rate variability during the menstrual cycle in young women. *Biomed. Sci. Instrum.*, 2005,41,340-5.

[59] Virtanen, R; Jula, A; Kuusela, T; Helenius, H; Voipio-Pulkki, LM. Reduced heart rate variability in hypertension: associations with lifestyle factors and plasma renin activity. *J. Hum. Hypertens,* 2003,17,171-9.

[60] Mizuno, O; Tamai, H; Fujita, M; Kobayashi, N; Komaki, G; Matsubayashi, S; Nakagawa, T. Aldosterone responses to angiotensin II in anorexia nervosa. *Acta Psychiatr. Scand.,* 1992,86,450-4.

[61] Mizuno, O; Tamai, H; Fujita, M; Kobayashi, N; Komaki, G, Matsubayashi, S; Nakagawa, T. Vascular responses to angiotensin II in anorexia nervosa. *Biol. Psychiatry,* 1993,34,401-6.

[62] Ahima, RS and Fliers, JS. Leptin. *Ann Rev Physiol,* 2000,62,413-37.

[63] Chan, JL; Heist, K; De Paoli, AM; Veldhuis, JD; Mantzoros, CS. The role of falling leptin levels in the neuroendocrine and metabolic adaptation to short-term starvation in healthy men. *J. Clin. Invest,* 2003,111,1409-21.

[64] Grinspoon, S; Gulick, T; Askari, H; Landt, M; Lee, K; Anderson, E; Ma, Z; Vignati, L; Bowsher, R; Herzog, D; Klibanski, A. Serum leptin levels in women with anorexia nervosa. *J. Clin. Endocrinol. Metab.,* 1996,81,3861-3.

[65] Chan, JI and Mantzoros, CS. Role of leptin in energy-deprivation states: normal human physiology and clinical implications for hypothalamic amenorrhoea and anorexia nervosa. *Lancet,* 2005,366,74-85.

[66] Calandra, C; Musso, F; Musso, R. The role of leptin in the etiopathogenesis of anorexia nervosa and bulimia. *Eat. Weight Disord.,* 2003,8,130-7.

[67] Monteleone, P; Martiadis, V; Colurcio, B; Maj, M. Leptin secretion is related to chronicity and severity of the illness in bulimia nervosa. *Psychosom. Med.* 2002,64,874-9.

[68] Haynes, WG; Morgan, DA; Walsh, SA; Mark, AL; Sivitz, WL. Receptor mediated regional sympathetic nerve activation by leptin. *J. Clin. Invest*, 1997,100,270-8.

[69] Paolisso, G; Manzella, D; Montano, N; Gambardella, A; Varricchio, M. Plasma leptin concentrations and cardiac autonomic nervous system in healthy subjects with different body weights. *J. Clin. Endocrinol. Metab.*, 2000,85,1810-4.

[70] Correia, ML; Morgan, DA; Sivitz, WI; Mark, AL; Haynes, WG. Leptin acts in the central nervous system to produce dose dependent changes in arterial pressure. *Hypertension*, 2001,37,936-42.

[71] Tank, J; Jordan J; Diedrich A; Schroeder, C; Furlan, R; Sharma, AM; Luft, FC; Brabant, G. Bound leptin and sympathetic outflow in nonobese men. *J. Clin. Endocrinol. Metab*, 2003,88,4955-9.

[72] Eikelis, N; Schlaich, M; Aggarwal, A; Kaye, D; Esler, M. Interactions between leptin and the human sympathetic nervous system. *Hypertension*, 2003,41,1072-9.

[73] Tagami, T; Satoh, N; Usui, T; Yamada, K; Shimatsu, A; Kuzuya, H. Adiponectin in anorexia nervosa and bulimia nervosa. *J. Clin. Endocrinol. Metab.*, 2004,89,1833-7.

[74] Housova, J; Anderlova, K; Krizová, J; Haluzikova, D; Kremen, J; Kumstyrová, T; Papezová, H; Haluzik, M. Serum adiponectin and resistin concentrations in patients with restrictive and binge/purge form of anorexia nervosa and bulimia nervosa. *J. Clin. Endocrinol. Metab.*, 2005,90,1366-70.

[75] Kojima, M; Hosoda, H; Date, Y; Nakazato, M; Matsuo, H; Kangawa, K. Ghrelin is a growth-hormone-releasing acylated peptide from stomach. *Nature*, 1999,402,656-60.

[76] Hosoda, H; Kojima, M; Kangawa, K. Biological, physiological, and pharmacological aspects of ghrelin. *J. Pharmacol. Sci.*, 2006,100,398-410.

[77] Otto, B; Cuntz, U; Fruehauf, E; Wawarta, R; Folwaczny, C; Riepl, RL; Heiman, ML; Lehnert, P; Fichter, M; Tschöp, M. Weight gain decreases elevated plasma ghrelin concentrations of patients with anorexia nervosa. *Eur. J. Endocrinol.*, 2001,145,669-73.

[78] Misra, M; Miller, KK; Kuo, K; Griffin, K; Stewart, V; Hunter, E; Herzog, DB; Klibanski, A. Secretory dynamics of ghrelin in adolescent girls with anorexia nervosa and healthy adolescents. *Am. J. Physiol. Endocrinol. Metab*, 2005,289,E347-56.

[79] Tanaka, M; Naruo, T; Muranaga, T; Yasuhara, D; Shiiya, T; Nakazato, M; Matsukura, S; Nozoe, S. Increased fasting plasma ghrelin levels in patients with bulimia nervosa. *Eur. J. Endocrinol.*, 2002,146,R1-3.

[80] Nakazato, M; Hashimoto, K; Shiina, A; Koizumi, H; Mitsumoti, M; Imai, M; Shimizu, E; Iyo, M. No changes in serum ghrelin levels in female patients with bulimia nervosa. *Prog. Neuropsychopharmacol Biol. Psychiatry*, 2004,28,1181-4.

[81] Hotta, M; Shibasaki, T; Masuda, A; Imaki, T; Demura, H; Ling, N; Shizume, K. The responses of plasma adrenocorticotropin and cortisol to corticotropin-releasing

hormone (CRH) and cerebrospinal fluid immunoreactive CRH in anorexia nervosa patients. *J Clin Endocrinol Metab*, 1986,62,319-24.

[82] Kaye, WH; Gwirtsman, HE; George, DT; Ebert, MH; Jimerson, DC; Tomai, TP; Chrousos, GP; Gold, PW. Elevated cerebrospinal fluid levels of immunoreactive corticotropin-releasing hormone in anorexia nervosa: relation to state of nutrition, adrenal function, and intensity of depression. *J. Clin. Endocrinol Metab.*,1987,64,203-8.

[83] Krahn, DD and Gosnell, BA. Corticotropin-releasing hormone: possible role in eating disorders. *Psychiatr. Med.*, 1989,7,235-45.

[84] Kooh, GF and Heinrichs, SC. A role for corticotropin releasing factor and urocortin in behavioral responses to stressors. *Brain Res.*, 1999,848,141-52.

[85] Oświecimska, J; Ziora, K; Geisler, G; Broll-Waśka, K. Prospective evaluation of leptin and neuropeptide Y (NPY) serum levels in girls with anorexia nervosa. *Neuro. Endocrinol. Lett.*, 2005,26,301-4.

[86] Mehler, PS and Crews, CK. Refeeding the patient with anorexia nervosa. *Eat. Disord.*, 2001,9,167-71.

In: Advances in Medicine and Biology. Volume 20
Editor: Leon V. Berhardt

ISBN 978-1-61209-135-8

Chapter 6

Antimicrobial-Resistant Gram-Positive Cocci in the Third Millennium: Novel Pharmaceutical Weapons and Their Therapeutic Perspectives

*Roberto Manfredi**
Department of Internal Medicine,
Aging, and Nephrologic Diseases, Division of Infectious Diseases,
"Alma Mater Studiorum" University of Bologna,
S. Orsola Hospital, Bologna, Italy

Abstract

Introduction. Methicillin- and also vancomycin (glycopeptide)-resistant Gram-positive organisms have emerged as an increasingly problematic cause of hospital-acquired infections, and are also spreading into the community. Vancomycin (glycopeptide) resistance has emerged primarily among Enterococci, but the MIC values of vancomycin are also increasing for the entire *Staphylococcus* species, worldwide.

Materials and Methods. The aim of our review is to evaluate the efficacy and tolerability of newer antibiotics with activity against methicillin-resistant and glycopeptide-resistant Gram-positive cocci, on the grounds of our experience at a tertiary care metropolitan Hospital, and the most recent literature evidence in this field.

Results. Quinupristin-dalfopristin, linezolid, daptomycin, and tigecycline show an excellent *in vitro* activity, comparable to that of vancomycin and teicoplanin for methicillin-resistant staphylococci, and superior to that of vancomycin for vancomycin-

* Correspondence: Roberto Manfredi, MD. Associate Professor of Infectious Diseases, University of Bologna. c/o Infectious Diseases, S. Orsola Hospital. Via Massarenti 11. I-40138 Bologna, Italy. Telephone: +39-051-6363355. Telefax: +39-051-343500. E-mail: Roberto.manfredi@unibo.it

resistant isolates. Dalbavancin, televancin and oritavancin are new glycopeptide agents with excellent activity against Gram-positive cocci, and have superior pharmacodynamics properties compared to vancomycin. We review the bacterial spectrum, clinical indications and practical use, pharmacologic properties, and expected adverse events and contraindications associated with each of these novel antimicrobial agents, compared with the present standard of care.

Discussion. Quinupristin-dalfopristin is the drug of choice for vancomycin-resistant *Enterococcus faecium* infections, but it does not retain effective activity against *Enterococcus faecalis*. Linezolid activity is substantially comparable to that of vancomycin in patients with methicillin-resistant *Staphylococcus aureus* (MRSA) pneumonia, although its penetration into the respiratory tract is exceptionally elevated. Tigecycline has activity against both *Enterococus* species and MRSA; it is also active against a broad spectrum of Enterobacteriaceae and anaerobes, which allows its use for intra-abdominal, diabetic foot, and surgical infections. Daptomycin has a rapid bactericidal activity for *Staphylococcus aureus* and it is approved in severe complications, like bacteremia and right-sided endocarditis. It cannot be used to treat pneumonia and respiratory diseases, due to its inactivation in the presence of pulmonary surfactant.

Keywords: Resistant Gram-positive cocci, Staphylococci, Enterococci, Pneumococci, Streptococci, epidemiology, clinical issues, novel antimicrobial compounds, characteristics, literature evidence.

Gram-positive cocci have re-emerged as predominant pathogens of human hosts within the past 10–15 years. After the introduction of penicillin over 60 years ago, infections by *Staphylococcus aureus*, *Streptococcus pyogenes* and *Streptococcus pneumoniae* finally became treatable. Within a short period of time, however, *S. aureus* developed resistance to penicillin. As a conseqience, penicillinase-resistant penicillins were successfully introduced in the early 1960s. Concomitantly, resistance emerged for the penicillinase-resistant penicillins, and finally methicillin (oxacillin)-resistant *S. aureus* (MRSA) became a major hospital-acquired pathogen. Vancomycin (belonging to the class of glycopeptides) remained an active agent against MRSA and coagulase-negative Staphylococci, so that it was increasingly used during the subsequent years, until now. From the 1990s to the present, however, the emergence of resistance to vancomycin also occurred in a significant proportion [1-3]. First among these organisms were *Enterococcus faecium* and *Enterococcus faecalis* [4]. Subsequently, vancomycin (glycopeptide)-resistant enterococci (VRE) became major hospital-acquired pathogena. In the past several years, MRSA were also spreading clonally into the community (the so-called “CA-MRSA”), leading to increased use of vancomycin-teicoplanin therapy [5] . In the late 1990’s, glycopeptide resistance was reported for coagulase-negative Staphylococci [6] and then, *S. aureus* (the so-called vancomycin-intermediate *S. aureus*, or VISA, and the so-called glycopeptide-intermediate *S aureus*, or GISA). The first reported isolation of VISA occurred in Japan in 1997 [7] and more than one hundred VISA isolates have been reported in the subsequent years [8]. In the year 2002, three vancomycin-resistant *S. aureus* (VRSA) strains isolated from clinical specimens of American patients were found to have high-level resistance to vancomycin (MIC >32 ug/ml) [9].

Although a number of cases of VRSA have since been described [10], these isolates fortunately have not yet become widespread.

The re-emergence of Gram-positive cocci have been well established in the setting of hospital-acquired infections, but community-acquired infections due to MRSA have become increasingly problematic during the last years [11-13]. Foreign body infections and bacteremia caused by coagulase-negative Staphylococci have also increased during time [14]. As a result, vancomycin-teicoplanin usage has increased in both inpatients and outpatiens. Although the majority of *S. aureus* strains remain susceptible *in vitro* to vancomycin, its efficacy against methicillin-sensitive *S. aureus* (MSSA) is inferior to that of penicillinase-resistant penicillins, and beta-lactam derivatives as a whole [15, 16].

Actually, MRSA is born as a multi-drug resistant pathogen. Resistance to the macrolides, lincosamides, aminoglycosides and all beta-lactam agents as a group, as well as fluoroquinolones, are also seen when MRSA is of clinical concern. Rifampin should not be used as a single agent due to rapid emergence of resistance in these microorganisms, while doxycycline and trimethoprim-sulfamethoxazole (cotrimoxazole), are bacteriostatic rather than bactericidal in their mechanisms of action [17].

S aureus is well known to be a virulent and invasive pathogen. It produces a variety of pyrogenic toxins and superantigens which contribute to its overall virulence [18]. The presence of the Panton-Valentine leukocidin may predispose to invasive skin and soft tissue infections, and also necrotizing pneumonias and other necrotizing infectious localizations. MRSA infection often has its origin from a localized skin infection, with subsequent contiguous or hematogenous spread to lungs, heart (endocarditis), central nervous system (CNS), and sometimes bones and joints and other organs and sites [19]. The prolonged duration of treatment with vancomycin or teicoplanin for severe infections, like endocarditis and osteomyelitis, may lead to more frequent and severe adverse effects (especially nephropathy, serum electrolyte imbalance, and myelotixicity). While VISA/GISA and VRSA infections have only rarely been reported, clinical hetero-resistant populations of VISA (with strains shosing MIC values > 4-16 mcg/ml) have been isolated following prolonged administration of glycopeptides. Moreover, pharmacodynamics of vancomycin may have led to unappreciated under-dosing of vancomycin, thereofore predisposing to microbial resistance [20]. Coagulase-negative Staphylococci have the capability to produce a glycocalyx enabling them to attach to prosthetic materials [21]. Biofilm formation on the surfaces of medical devices (i.e., prosthetic devices, central vascular catheters), provides a protected environment for coagulase-negative Staphylococci; this biofilm formation impedes antibiotic penetration and reduces target site formation [21, 22]. As expected, catheter-related blood stream infections, CNS ventricular shunt infections, prosthetic joint infections and prosthetic valve endocarditis are commonly caused by coagulase-negative Staphylococci [23].

The large majority of these microorganisms usually are or become resistant to methicillin. Intermediate resistance to vancomycin was first reported among coagulase-negative Staphylococci, several years before it occurred among *S. aureus* strains. Unlike *S. aureus*, infections by coagulase-negative Staphylococci on prosthetic hardware tend to be insidious and more chronic. Therapy often requires a combined medical- surgical approach with removal of the infected device, and prolonged duration (usually exceeding four weeks) of antibiotic therapy thereafter.

Vancomycin-resistant Enterococci (VRE) are primarily associated with healthcare institutional acquisition in patients with co-morbid conditions. Since their peak incidence around the year 2000, several new antibiotics with excellent activity against VRE have been introduced into clinical practice in the meantime.

On the other hand, *S. pneumoniae* is the most frequent cause of community-acquired pneumonia (CAP). It accounts for at least one third of patients with CAP. The crude incidence of this common pathogen rises to greater than 50%, if respiratory culture with Gram stains, and urinary antigen for *S .pneumoniae* are systematically performed. Associated bacteremia occurs in 20% of pneumococcal pneumonias and mortality is notably higher than for other respiratory pathogens. *In vitro* resistance of *S. pneumoniae* to penicillin as currently defined by Clinical Laboratory Standards Institute (CLSI) criteria, does not necessarily correlate with clinical failure. Specifically, penicillins have been favorably efficacious for pneumonia caused by penicillin-resistant pneumococci [24, 25]. These resistant isolates are often also resistant to macrolides, and *in vitro* resistance to macrolide does appear to correlate with clinical outcome [26, 27].

In adults patients, *S. pneumoniae* also represents the most common cause of meningitis. Empiric therapy for meningitis with ceftriaxone and vancomycin pending antibiotic susceptibility testing is often employed. Data from a large scale observational study of pneumococcal meningitis suggests that combination therapy may be superior to monotherapy [28].

Groups A Streptococci (whose leading organism is *Streptococcus pyogenes*), as well as other beta-hemolytic Streptococci, are often associated with life-threatening infections, especially involviong the skin and soft tissues. Group B, C, F, and G beta-hemolytic Streptococci can also cause invasive infection and becteremia. *Streptococcus agalactiae* (belonging to group B Streptococci), is a common cause of neonatal sepsis. Fortunately, susceptibility to penicillin remains stable for the majority of the above-mentioned Streptococci.

Epidemiological Experience at a Major Tertiary Care Hospital in Northern Italy

A prospective, microbiological surveillance study has been going on for a decade at our academic Hospital (S. Orsola-Malpighi Hospital, Bologna, Italy), in order to check the epidemiological-clinical evolution of bacterial and fungal infections occurring among inpatients. All microorganisms isolated and identified from sterile sites (i.e., blood cultures, protected bronchoalveolar lavage, urine culture, and so on), are systematically tested for *in vitro* susceptibility against a consistent panel of antimicrobial compounds, and data are reported quarterly (A. Nanetti and S. Ambretti, unpublished data). Each bacterial isolate cultured from a single patient within one month, is counted only once.

Some data regarding *in vitro* sensitivity rates of *Staphylococcus aureus* and Enterococci from the year 2005 to the first nine months of the year 2008, are reported in Table 1.

With regard to *S. aureus*, we underline a progressive reduction of methicillin resistance rate (from 56.8% of strains in the year 2005, up to 44.7% of overall tested strains of the year 2008), while we confirm a maintained 100% susceptibility to all available glycopeptides.

Table 1. Microbiological figures from patients hospitalized at our tertiary-care Hospital (S. Orsola-Malpighi Hospital, Bologna, Italy), year 2005–2008

***In vitro* antimicrobial susceptibility of *Staphylococcus aureus* strains isolated from inpatients (years 2005-2008)**

Antibiotic susceptibility	Year 2005 (190 strains)	Year 2006 (167 strains)	Year 2007 (131 strains)	Year 2008 (Jan. to Sep.) (103 strains)
Penicillin	7.9	7.8	9.9	7.8
Amoxicillin-clavulanate	43.2	48.5	48.9	55.3
Cefotaxime/Ceftriaxone	42.9	48.5	48.9	55.3
Methicillin/Oxacillin	43.2	48.5	49.6	55.3
Erithromycin	34.2	48.5	51.9	55.3
Clindamycin	34.9	48.5	51.9	54.4
Chloramphenicol	84.7	88.6	87.8	83.5
Rifampicin	60.2	64.0	62.1	80.4
Cotrimoxazole	87.8	98.2	93.1	95.1
Gentamicin	31.1	41.0	43.5	49.5
Vancomycin/Teicoplanin	100.0	100.0	100.0	100.0

***In vitro* antimicrobial susceptibility of Enterococci isolated from inpatients (years 2005–2008)**

Antibiotic susceptibility	Year 2005 (206 strains)	Year 2006 (151 strains)	Year 2007 (155 strains)	Year 2008 (Jan. to Sep.) (108 strains)
Penicillin	50.0	49.0	51.6	50.9
Ampicillin	50.0	49.0	52.3	51.9
Tetracyclin	46.6	45.0	38.7	38.0
Vancomycin/Teicoplanin	98.1	95.4	94.8	96.3
Linezolid	100,0	100,0	100,0	100,0
Daptomycin	N/A	N/A	N/A	100.0
Quinupristin/Dalfopristin	100,0	100,0	100,0	100,0

Of major interest, the introduction of novel guidelines for a correct antimicrobial use in medicine and surgery, led to a progressive reduction of the frequency of antibiotic-resistant strains, as demostrated by an increased mean *in vitro* susceptibility rate against almost all tested compounds demostrated in the year 2008, *versus* years 2005-2007, together with an apparent, progressive trend toward reduced resistances during the examined temporal span (year 2005 to year 2008). Furthermore, elevated sensitivity rates are also found for a number of "older" molecules, like cotrimoxazole (87.9% to 98.2%), chloramphenicol (83.5% to 88.6%), followed by rifampicin (60.2% to 80.4%), clindamycin (34.9% to 54.4%), and erythromycin (34.2% to 55.3%).

These last compounds, which are largely available and do not imply increased costs of administration, are expected to be clinically effective, alone and/or in combination with anti-Gram-positive agents (Table 1).

When examining the temporal trend of isolation of Enterococci (as a group) in the same time period (year 2005, up to September 2008), we notice a full sensitivity to novel compounds (i.e. linezolid, daptomycin, and quinupristin/dalfopristin), and a maintained *in vitro* effectiveness of glycopeptides (94.8% to 98.1%). Vancomycin-resistant Enterococci (also called VRE) were isolated infrequently: only 9 cases in the year 2005, 6 in the year 2006, 8 in the year 2007, and only three cases in the first 9 months of the year 2009 (A. Nanetti and S. Ambretti, unpublished data). Also in the case of Enterococci, a number of "older" compounds still retain an effective activity against Enterococci, as demostrated by susceptibility rates of penicillin (49.0% to 51.6%), ampicillin (49.0% to 52.3%), and tetracyclines (38.0% to 46.6%). Also in this case, these compounds may act favorably or may be a part of a combination regimen, after *in vitro* sensitivity assays (Table 1).

Novel Antibacterial Agents with Enhanced Activity against Resistant Gram Positive Cocci

The following antibacterial agents have been approved during the last five years: quinupristin /dalfopristin [29-32], linezolid [33, 34], daptomycin [35-37], and tigecycline [38, 39]. Novel glycopeptide agents under study include dalbavancin [40], telavancin, and oritavancin [41]. Even novel cephalosporins (i.e. cefbiprole), and fluroquinolones (i.e. garenofloxacin) with enhanced activity against MRSA are in the pipeline. Some features of these novel antimicrobial molecule are summarized in Table 2 (microbiological spectrum, clinical indication, adverse events), and in Table 3 (selected pharmacological features).

Quinupristin/Dalfopristin

The so-called streptogramin antibiotic, quinupristin/dalfopristin, is a combination of two semisynthetic pristinamycin derivatives, which are represented by quinupristin and dalfopristin, in a 30:70 ratio. Resistance to quinupristin/dalfopristin can occur by several mechanisms increasing enzymatic modification, active transport of specific efflux pumps mediated by an adenosine triphosphate-binding proteinm and alteration of the target site. Resistance is rare for Streptococci and *Enterococcus faecium* [42].

Table 2. Novel antimicrobial agents for the management of resistant Gram-positive infections. Microbiological, clinical, and therapeutic features as of end of year 2008

Drug	Class	Microbiologically effective on				Adverse events	Clinical indications for at end-2008 (infections)						
		MRSA	MRSE	PRSP	VRE		Blood stream	Skin-soft tissue	Hospital-acquired pneumonia	Endo-carditis	Bone And joint	CNS	Intra abdominal
Vancomycin	Glycopeptide	+	+	+	Not VISA-VRSA	Nephro- and ototoxicity Red man syndrome	+	+	+	+	+	+	+
Teicoplanin	Glycopeptide	+	+	+	Not Van-A	N/D	+	+	-	-	-	-	-
Quinupristin/ Dalfopristin	Streptogramin	+	+	+	*E. faecium*	Hepatic, phlebitis Artho-myalgias	+	+	-	-	-	-	-
Linezolid	Oxazolidinone	+	+	+	+	Neuropathy, myelotixicity, serotonin syndrome	+	+	+	-	+	-	-
Daptomycin	Lipopeptide	+	+	+	+	Arthro-myalgias, CPK rise	+	+	-	+	-	-	-
Tigecycline	Glycopeptide	+	+	+	+	Nausea, diarrhea		+	-	-	-	-	+
Dalbavancin	Glycopeptide	+	+	+	Not VAN-A	Gastrointestinal, hypokaleamia	+	+	-	-	-	-	-
Oritavancin	Glycopeptide	+	+	+	+	N/D	-	+	-	-	-	-	-
Telavancin	Lipoglycopeptide	+	+	+	+	Altered taste, CNS, phlebitis7	-	+	-	-	-	-	-

Table 3. Novel antimicrobial agents for the management of resistant Gram-positive infections. Selected pharmacogical features, updated at the end of year 2008

Drug	Class	Pharmacodynamics	Protein binding (%)	Elimination route	Dosage adjustement	
					Renal	Hepatic
Vancomycin	Glycopeptide	AUC/MIC	10-55	Renal	+	-
Teicoplanin	Glycopeptide	AUC/MIC	90	Renal	+	-
Quinupristin/Dalfopristin	Streptogramin	AUC/MIC	N/A	Hepatic/feces	N/A	+
Linezolid	Oxazolidinone	AUC/MIC	31	Hepatic	N/A	N/A
Daptomycin	Lipopeptide	AUC/MIC	92	Renal	+	N/A
Tigecycline	Glycopeptide	Time above MIC	68	Biliary	N/A	+
Dalbavancin	Glycopeptide	AUC/MIC	>95	Renal	+	N/A
Oritavancin	Glycopeptide	AUC/MIC	N/A	Renal	Renal	N/A
Telavancin	Lipoglycopeptide	AUC/MIC	N/A	Renal	+	-

This streptogramin combination acts synergistically to inhibit bacterial protein synthesis at the ribosome level. Quinupristin/dalfopristin is therefore active against *Staphylococcus aureus* (including MRSA strains), *Streptococcus pneumoniae*, and Gram-positive anaerobes such as *Clostridium* spp., *Peptococcus* spp., and *Peptostreptococcus* spp. It is effective against vancomycin-sensitive as well as vancomycin-resistant *Enterococcus faecium* (VREF), but has little *in vitro* activity against *Enterococcus faecalis*, so that it cannot be recommended until final speciation of Enterococcal organisms is concluded. Dalfopristin/quinipristin association inhibits cytochrome P450 3A4, and can inhibit agents metabolized through this pathway.

Dosage adjustments may be needed in patients with hepatic dysfunction. Renal function has minimal impact on the agent's pharmacokinetics. A post-antibiotic effect is observed at 4-5 hours at 4X MIC versus Staphylococci, 7-9 hours for Streptococci, and only 4 hours for Enterococci [43].

The registered clinical indications for quinupristin/dalfopristin use include intraabdominal infections, bacteremia, urinary tract infection and skin and soft tissue infections in which Enterococci may play a relevant pathogenic role. Overall clinical success rate for patients with vancomycin-resistant *E. faecium* (VREF) proved to be 74%, while overall clinical and bacteriological success rate was 66% [44]. Patients with bacteremia, those on mechanical ventilation, and those undergoing surgery had a worse outcome as might be expected [44]. The most common and notable adverse events were arthralgias and myalgias. In a comparative clinical trial of therapy for Gram-positive skin and soft tissue infections, *S. aureus* was the most frequent pathogen isolated [45]. The clinical success rate of quinupristin/dalfopristin was comparable (68%) to the comparator agents (71%) (cefazolin, oxacillin or vancomycin). A higher incidence of drug-related adverse events occurred with quinopristin/dalfopristin as compared to other agents [46]. For those patients receiving comparator agents, the most common reason for discontinuation was treatment failure (12%) [46]. Furthermore, quinupristin/dalfopristin was compared to vancomycin in patients with hospital-acquired pneumonia [47]. Successful outcomes were similar at 56% for quinupristin/dalfopristin and 58% for vancomycin. The bacteriologic success rate was identical for both antibiotic groups, at around 54% of treated cases. Quinupristin/dalfopristin has been also used to treat patients infected by *S. aureus* intolerant of or failing standard therapies [48].

Ninety patients were treated an average of 28 days with a 71% clinical outcome of cure or improvement and bacteriologic outcome of eradication or presumed eradication. Infections included bone and joint, skin and soft tissue, bacteremia, endocarditis and respiratory tract involvement. Adverse events included arthralgias mainly (11%), myalgias (9%) and nausea (9%). However, in patients with hepatic dysfunction or liver transplantation and concurrent receipt of immunosuppressive chemotherapy, the incidence of arthralgias approached 50% of treated subjects [49, 50].

Linezolid

Linezolid is an oxazolidinone antibiotic with activity against Gram-positive pathogens including VRE, MRSA, and VISA. The unique mechanism of action of linezolid involves the inhibition of bacterial protein synthesis through binding to the domain V regions of the 23 Sr

RNA gene 46. Resistance to linezolid requires mutations of multiple gene copies, and seems an infrequent phenomenon. Linezolid is 100% bioavailable when given by either oral or intravenous route. Maximal plasma levels are achieved within 1-2 hours after oral dosing. Protein binding is only around 30% with free distribution to well-perfused tissues. The drug does not require dosage alteration in the presence of renal failure, and no interaction exists for cytochrome P450 enzymes (and drugs metabolized through this last pathway). Linezolid and its two metabolites are decreased with hemodialysis, so dosing should occur postdialysis [51].

Linezolid is currently approved for skin and soft tissue infections and pneumonia due to susceptible pathogens [52]. In two controlled trials of hospital-acquired pneumonia, a trend was seen for linezolid superiority over vancomycin [53, 54]. There is little data (mainly based on observational studies), on the utility of linezolid for either bacteremia [55], or osteomyelitis [56, 57].

Based on a rabbit model, linezolid does not have sufficient CSF penetration, and should not be recommended for pneumococcal meningitis [33]. However, CNS penetration appears adequate to treat CSF shunt infections [58], and brain abscesses, too.

The myelotoxicity (especially the thrombocytopenia), is the most common serious adverse event caused by linezolid [59]; it can be ameliorated or prevented by co-administration of pyridoxine (Vitamin B6) [60-63]. Both peripheral and optic neuropathy have been reported with prolonged use of greater than four consecutive weeks [64, 65]. Lactic acidosis has also been reported and is not associated with duration of administration [65, 66]. Interaction exists between linezolid and serotonin-reuptake inhibitors (antidepressants drugs). In these last cases, a minority of patients might develop the so-called serotonin syndrome (fever, agitation with mental status changes and tremors). Due to its weak activity as a monoamine oxidase- inhibitor, linezolid should not be used concomitantly with agents such as tramadol, pethidne, duloxetine, venlafaxine, milnacipran, sibutramine, chlorpeniramine, brompheniramine, cyproheptadine, citalopram, and paroxetine [65, 67]. Drug metabolites may accumulate in the event of severe renal failure.

Daptomycin

Daptomycin is the first in a new class of antimicrobial agents: a lipopeptide antibiotic with activity against *S. aureus* (including methicillin-resistant strains), beta-hemolytic Groups A, B C and G Streptococci, and Enterococci, including ampicillin-and vancomycin- resistant strains. The mechanism of action of daptomycin is unique as le molecule causes a calcium ion dependent disruption of bacterial cell membrane potential resulting in an efflux of potassium which inhibits RNA, DNA and protein synthesis. Rare instances of resistance have occurred in clinical trials, although the mechanism of resistance has not yet been clerarly identified to date. Daptomycin was shown to have a rapidly bactericidal effect *in vitro* against Gram-positive drug-resistant pathogens. Its activity is concentration-dependent and once daily dosing is associated with significant post-antibiotic effect.

Daptomycin is highly protein bound (around 92%), with a terminal half-life of 8 hours, which allows for once daily dosing. Post-antibiotic effect proves dose dependent, and is reduced in the presence of albumin (i.e. essudates). The drug volume of distribution is low (0.1L/kg) and the Cmax (54.6mcg/ml) is unchanged at steady state is achieved by day three of administration in humans. Cmax concentrations occur at the end of a 30 minute infusion.

Dosage needs to be reduced and dosing interval extended to every 48 hours in patients with reduced creatinine clearance <30mL/min; the same occurs for patients on either hemodialysis or peritoneal dialysis; the daptomycin dose is in these patients becomes 4 mg/kg every 48 hours. Daptomycin should be administered after hemodialysis as approximately 15% is cleared per 4 hour hemodialysis session. On the other hand, no dose adjustments for hepatic dysfunction are required.

In early clinical trails conducted in the years 1980's-1990's, daptomycin was given in divided daily doses of 2 mg/kg every 12 hours for skin and soft tissue infection and 3 mg/kg every 12 hours for bacteremia, achieving good clinical and bacteriological outcomes. However, rise in serum creatine phosphokinase (CPK), with myalgias and muscle weakness led to initial abandonment of this promising antibiotic. However, myopathy was reversible upon cessation of the drug. With the advent of MRSA infections, daptomycin has been re-examined and resurrected, and its dosage has been increased to 4 mg/kg daily for skin and soft-tissue infection [68], and up to 6 mg/kg daily for bacteremia and endocarditis [69]. Both indications have been approved by the United States FDA. Otherwise, daptomycin is not approved for the treatment of bacterial pneumonia; its efficacy is significantly compromised by its interaction with pulmonary surfactant [70]. Significant drug-drug interaction occurs with the statins, and patients receiving HMG-CoA reductase inhibitors; these drugs should be suspended and avoided while the patient is undergoing a daptomycin course.

Tigecycline

Tigecycline is a novel glycylcycline molecule, which is a derivative of the tetracycline minocycline. Resistance to the tetracycline class is classically mediated by ribosomal protection mechanisms or by active efflux. Tigecycline has more potent activity against tetracycline-resistant organisms, and maintains a broad antibacterial spectrum against Gran-positive and also Gram-negative pathogens. Tigecycline binds more avidly to the ribosome and either does not induce efflux proteins or is not readily exported by efflux proteins [38]. Resistant clinical isolates were associated with up-regulation of chromosomially mediated efflux pumps. Unlike original tetracyclines, tigecycline has a large volume of distribution (above 10 L/kg), the protein binding is approximately 68%, the terminal half-life of elimination is 36 hours, and less than 15% of the native drug is excreted unchanged in the urine. Clinical trials have been conducted in patients with complicated skin and soft tissue infections and intraabdominal infections for which the drug gained its United States FDA approval.

Based on *in vitro* susceptibility data, tigecycline has a broad spectrum of activity against both Gram-positive cocci (including methicillin-resistant Staphylococci or MRSA, penicillin-resistant *Streptococcus pneumoniae*, beta-hemolytic group A and group B Streptococci, Enterococci (vancomycin-susceptible ones), and *Listeria monocytogenes.* Unlike other new agents for Gram-positive cocci, tigecycline also has extensive activity against Gram-negative pathogens including *Haemophilus influenzae, Neisseria spp* [11], Enterobacteriaceae, and non-lactose fermenters other than *Pseudomonas aeruginosa.* The MIC_{90} values for *Proteus* spp., *Providentia* spp., and *Burkholderia cepacia* is 8mcg/mL, which limits its utility in infections caused by these mentioned pathogens.

Tigecycline needs no reduction in renal impairment and it is not dialyzable. Patients with severe hepatic dysfunction (Child-Pugh stage C liver disease), should receive a lower dosage. Tigecycline activity is dependent on the time above the MIC, and the drug concentrations should be above the MIC values for at least 50% of the dosing interval.

The expected adverse effects of tigeciclyne are primarily gastrointestinal in origin, with nausea, vomiting, diarrhea and heartburn as the most frequent. As with all tetracyclines, tigecycline is contraindicated for pregnant females and for children less than 8 years of age [71]. Drug interactions of tigecycline with either digoxin or warfarin do not alter the effect of either drug. Tigecycline does not inhibit metabolism mediated by the cytochrome P450 isoforms IA2, 2C8, 2C9, 2C19, 2D6 and 3A4, so that no drug-drug interaction are expected with drugs metabolized by these cytochrome isoforms.

Dalbavancin

Dalbavancin is a true second-generation glycopeptide. Its unique pharmacokinetic profile allows once weekly dosing. It is not active against VRE, but has an excellent activity against MRSA, *S. pyogenes* and *S. pneumoniae* as well as vancomycin-susceptible Enterococci. It is bactericidal and synergistic with ampicillin against Van-A type Enterococci. The mechanism of action is the inhibition of the cell wall peptidoglycan cross-linking.

The daily dosage is 1000 mg i.v. once, followed by 500 mg i.m. 7 days later; the terminal half-life of dalbavancin is 9-12 days in humans due to protein binding of greater than 95%. Animal models of infection show excellent activity against MRSA or GISA endocarditis, penicillin-resistant *Streptococcus pneumoniae* [12] pneumonia or MRSA pouch infection, and septicemia due to Staphylococci, Streptococci or Enterococci. This antibiotic has been evaluated for catheter-related bacteremia [72] and skin and soft tissue infections [73]. Dalbavancin was effective and well tolerated in adult patients with catheter-related bacteremia caused by coagulase-negative Staphylococci, MSSA and MRSA in a comparative trial with vancomycin. In skin and soft tissue infections, a 92-94% microbiological and clinical response respectively was found in an open label phase 2 comparative dosing trial [73]. Clinical success at follow-up visits for the two dose dalbavancin group was 80% for MRSA *versus* 50% for comparator therapy (which included beta-lactams, clindamycin, vancomycin and linezolid, respectively).

Oritavancin

Oritavancin is another derivative of vancomycin: it's a chloroeremomycin with the substitution of vancosamine by epi-vancosamine. It has a similar spectrum of activity to vancomycin but with consistently lower MIC values (<1mg/L). No resistance to oritavancin has been noted among *S. aureus* strains including VISA strains, but VAN-A and VAN-B strains of Enterococci with reduced susceptibility to oritavancin have been obtained *in vitro*. The known mechanisms of resistance of oritavancin are: 1) complete elimination of D-Ala-entry precursors; 2) mutations in the VAN Sb sensor of the VAN B cluster; or 3) the expression of Van Z (the precise function of which is still unknown).

Oritavancin shows rapid, concentration dependent bactericidal activity with a concentration-dependent post-antibiotic effect exerted against both VRE and MRSA. Oritavancin activity is negatively affected by large inoculum and its activity *versus* VRE was slightly reduced in stationary phase or in acidic foci of infection. In animal models, its efficacy has been demonstrated for experimental MRSA endocarditis and *S. pneumoniae* meningitis [74, 75]. In a reliable endocarditis model, the addition of gentamicin proved to be synergistic, and able to prevent the emergence of resistant mutants. With regard to skin and soft tissue infections, oritavancin proved to be at leat equivalent to vancomycin, for both clinical and bacteriological cure (about 78% of cure rate) [76].

Telavancin

Televancin is a rapidly bactericidal lipoglycopeptide analog of vancomycin. The mechanism of action of this moleculr is by inhibition of peptidoglycan chain formation through blockage of both the transpeptidation and transglycosylation steps; and by a direct effect on the bacterial membrane dissipating membrane potential and effecting changes in cellular permeability. The *in vitro* activity of telavancin demonstrates enhanced activity against MRSA, penicillin-resistant *S. pneumoniae,* GISA and Van-A type Enterococci. Telavancin achieves a higher volume of distribution into tissues and a prolonged half-life [77]. A high level of protein binding (93%) occurs in human plasma and repetitive dosing does not lead to accumulation. The terminal half-life is 7-9 hours at doses above 5 mg/kg [78]. Telavancin exhibit time-dependent killing activity [79]. Telavancin and its comparators of vancomycin or beta-lactam agent have been compared in a phase 2 clinical trial for skin and skin-structure infections. Clinical cure rates were similar at 92% in for telavancin *versus* 96% for comparator agents. Microbiologic rates of cure were noted to be 93% in the telavancin group and 95% among the comparator group [80]. For complicated skin and soft tissue infections, clinical cure rates were at 96% for telavancin and 90% for comparator agents. Microbiologic eradication was better with telavancin (92%) *versus* comparator agents (78%, p=0.07) [80]. Telavancin is currently under assessment in phase 3 trials of hospital-acquired pneumonia. Adverse events associated with telavancin among evaluated patients included vomiting, paresthesias and dyspnea. Laboratory abnormalities included microalbuminemia and a decreased platelet count [81].

Clinical Indications of Novel Antibiotics with Expanded Spectrum against Resistant Gram-Positive Cocci

Skin and Soft-Tissue Infections

Skin and soft-tissue infections caused by Gram-positive cocci range from a simple cellulitis to life-threatening necrotizing fasciitis. All of the newer agents have been studied for such infections and have been found to be efficacious (Table 2). Most of the patients in these studies had less severe infections than necrotizing fasciitis as that infection requires a surgical approach as well as antibiotic therapy. All five FDA-approved agents, i.e. quinupristin/dalfopristin, linezolid, daptomycin, tigecycline and vancomycin are appropriate choices for an effective treatment of Gram-positive pathogens. Only tigecycline has activity against Gram-negative bacilli pathogens. So, tigecycline may have a major role for diabetic foot infections and infected decubitus ulcers which may be co-infected by anaerobic bacteria and aerobic Gram-negative bacilli, in addition to Gram-positive cocci.

Bone and Joint Infections

With regard to osteomyelitis and joint infections, Gram-positive cocci largely predominate over other microbial patogens. *S. aureus*, both MSSA and MRSA, as well as coagulase-negative Staphylococci account for greater than 50% of recovered pathogens. Unfortunately, only few studies have prospectively investigated the above-mentioned newer antibiotics in these infections [56, 57].

Aneziokoro *et al.* evaluated 20 patients who received linezolid for osteomyelitis for 6 weeks or more in a retrospective non-comparative study [82]. Fifty-five percent of cases (11 patients) achieved a cure with follow-up periods ranging from 6 to 49 months (median of 36 months). Prospective comparative studies of efficacy in bone and joint infections have not been reported to date. In two retrospective studies, 22 patients with osteomyelitis and three subjects with septic joint infections were treated with daptomycin [83, 84]. MRSA was the predominant pathogen in over 75% of patients. Daptomycin was used as salvage therapy, and its usual dose was 6 mg/kg per day. Clinical success rate was about 90%; follow up periods were one year or less.

Limited data has been published with respect to bone and joint infections for dalbavancin, tigecycline or quinupristin/dalfopristin in humans. In a rabbit model of MRSA osteomyelitis, the combination of rifampin and tigecycline was compared to vancomycin with or without rifampicin, tigecycline alone, and vancomycin alone [85].

All regimens were effective (in about 90% of episodes). Untreated rabbits had spontaneous cure in 26% of cases (4/15). Tigecycline concentrations are higher in infected bone than in non-infected bone. A rabbit model of quinupristin/dalfopristin prosthetic joinT infection with MRSA was compared to vancomycin with or without rifampicin, showing an equivalent outcome [86].

Pneumonia and Lower Respiratory Tract Infections

Pneumonia due to Gram-positive cocci is common. In the community, infection is usually due to *S. pneumoniae* and occasionally *S. aureus.* Hospital-acquired pneumonia (HAP) is often caused by MRSA organisms. Linezolid was comparable to vancomycin in the therapy of MRSA-associated VAP, although a trend was seen for linezolid superiority [53, 54].

Daptomycin is not indicated for pneumonia due to its interaction with surfactant [70], while tigecycline is undergoing clinical evaluation. Quinupristin/dalfopristin has been compared to vancomycin for hospital-acquired pneumonia [47].

One hundred and 71 patients had similar clinical response rates of about 57% respectively. Drug discontinuation adverse events occurred more frequently in the quinupristin/dalfopristin group (15%), as compared to vancomycin. Only two isolated of the 87 overall strains were shown to have decreased susceptibility to quinupristin/dalfopristin during and after treatment.

Intra-abdominal Infections

Of the newer antibiotics, only tigecycline has been approved for intra-abdominal infections. As mentioned, tigecycline's broader spectrum of activity includes Gram-negative bacilli and anaerobic bacilli. Linezolid, daptomycin, and quinupristin/dalfopristin can be used in combination with antibiotics with Gram-negative spectrum of activity such as aztreonam, and especially carbapenems, fluoroquinolones, and aminoglycosides. Of concern, quinupristin/dalfopristin has no activity against *E. faecalis*.

Bacteremia and Endocarditis

Daptomycin and quinipristin/dalfopristin have been approved by the United States FDA organisms for the treatment of Gram-positive bacteremia. In addition, daptomycin has been approved for use in *S. aureus* right-sided endocarditis [87]. Dalbavancin, linezolid, tigecycline and oritavancin have not yet been approved for bacteremia due to gram-positive cocci. Linezolid has been evaluated for Gram-positive bacteria [55, 88, 89]. Among 108 bacteremic patients receiving linezolid, eradication was seen in 91% and clinical cure was seen in 94% of episodes [55].

On the other hand, linezolid is still not approved for catheter-related becteremia and endocarditis. A randomized study of 726 patients with catheter-related bacteremia received linezolid or vancomycin; an excess number of deaths were seen for patients receiving linezolid due mainly to Gram-negative rods implicated in these infections [90].

Based on 23 case reports and three case series, a total of 63% (21/33) of patients with endocarditis were successfully cured after linezolid administration [91]. MRSA and vancomycin-intermediate *S.aureus* were the most commonly isolated cocci (24.2% and 30.3% of cases, respectively). Five cases were succesfully treated with linezolid monotherapy.

Potential Synergistic Interactions of Newer Antibiotics: *In Vitro* Studies

In vitro interaction between the new antistaphylococcal antibiotics were virtually always indifferent, therefore leading to a possible additive effect, although a few experiments showed possible (Table 4) [89, 92-95, 98-107]. For instance, a synergistic interaction was found for quinupristin/dalfopristin and vancomycin in two independent studies [92, 93].

On the other hand, antagonistic interactions were demonstrated for the combination of linezolid plus vancomycin [94], and linezolid plus gentamicin [95].

It should be emphasized that *in vitro* interaction may not translate into clinical efficacy. Quinupristin/daflopristin in combination with vancomycin appeared to be favorable for the management of MRSA infections responding poorly to vancomycin [96]. However, we have to specify that the MRSA isolates were of a specific genotype, accessory gene regulator (agr), which has been linked to vancomycin treatment failure [96]. Nevertheless, such information may be useful if innovative combination therapy needs to be administered to severely ill patients with invasive *S. aureus* infection unresponsive to monotherapy.

Controlled clinical trials using combinations including these new agents are indicated for patients with severe, life-threatening infections caused by gram-positive cocci, and randomized trials are strongly warranted in this somewhat unexplored field.

Table 4. Some experimental studies conducted *in vitro* or on animal models, regarding possible interactions between the different antimicrobial agents effective on Gram-positive cocci [89, 92-95, 98-107]

Reference quotation	Combination	Pathogens	Interaction
[97]	Daptomycin+vancomycin	GISA	Additive
[97]	Daptomycin+gentamicin	GISA	Additive
[93]	Daptomycin+gentamicin	MSSA/MRSA	Enhanced time-kill
[98]	Daptomycin+gentamicin	MSSA/MRSA	Increased killing
[99]	Daptomycin+rifampicin	MRSA	Additive
[100]	Daptomycin+gentamicin+rifampicin	MRSA	Additive
[97]	Linezolid+vancomycin	GISA	Additive
[94]	Linezolid+vancomycin	MSSA/MRSA	Antagonistic
[89, 95]	Linezolid+vancomycin	MRSA	Indifferent
[101]	Linezolid+vancomycin	MSSA/MRSA/MRSE	Increased killing
[94, 98]	Linezolid+gentamicin	MSSA/MRSA	Indifferent
[95]	Linezolid+gentamicin	MRSA	Antagonistic
[94]	Linezolid+rifampicin	MSSA/MRSA	Indifferent
[95]	Linezolid+rifampicin	MSSA/MRSA	Synergistic
[102]	Linezolid+rifampicin	MSSA	Indifferent
[101]	Linezolid+Quinupristin/dalfopristin	MRSA	Increased killing
[97]	Quinupristin/dalfopristin+vancomycin	GISA	Additive-Synergistic
[103]	Quinupristin/dalfopristin+vancomycin	MSSA/MRSA	Additive
[101]	Quinupristin/dalfopristin+vancomycin	MRSA	Increased killing
[92]	Quinupristin/dalfopristin+vancomycin	MSSA/MRSA	Synergistic
[97]	Quinupristin/dalfopristin+gentamicin	GISA	Indifferent
[104]	Quinupristin/dalfopristin+rifampicin	MSSA	Increased killing
[105]	Quinupristin/dalfopristin+rifampicin	MRSA	Synergistic
[106]	Tigecycline+vancomycin	MRSA	Indifferent
[107]	Tigecycline+gentamicin	MRSA/GISA	Increased killing
[106]	Tigecycline+rifampicin	MRSA	Indifferent

REFERENCES

[1] Foster JK, Lentino JR, Strodtman R, DiVincenzo C. Comparison of *in vitro* activity of quinolone antibiotics and vancomycin against gentamicin- and methicillin-resistant *Staphylococcus aureus* by time-kill kinetic studies. *Antimicrob Agents Chemother* 1986; 30:823-7.

[2] Tallent SM, Bischoff T, Climo M, Ostrowsky B, Wenzel RP, Edmond MB. Vancomycin susceptibility of oxacillin-resistant *Staphylococcus aureus* isolates causing nosocomial bloodstream infections. *J. Clin. Microbiol* 2002; 40:2249-50.

[3] Sieradzki K, Leski T, Dick J, Borio L, Tomasz A. Evolution of a vancomycin-intermediate *Staphylococcus aureus* strain in vivo: multiple changes in the antibiotic resistance phenotypes of a single lineage of methicillin-resistant *S. aureus* under the impact of antibiotics administered for chemotherapy. *J. Clin. Microbiol* 2003; 41:1687-93.

[4] Murray BE. Vancomycin-resistant enterococci. *Am. J. Med.* 1997; 102:284-93.

[5] Okuma K, Iwakawa K, Turnidge JD, et al. Dissemination of new methicillin-resistant Staphylococcus aureus clones in the community. *J. Clin. Microbiol.* 2002; 40:4289-94.

[6] Schwalbe RS, Stapleton JT, Gilligan PH. Emergence of vancomycin resistance in coagulase-negative staphylococci. *N. Engl. J. Med.*1987; 316:927-31.

[7] Hiramatsu K, Hanaki H, Ino T, Yabuta K, Oguri T, Tenover FC. Methicillin-resistant *Staphylococcus aureus* clinical strain with reduced vancomycin susceptibility. *J. Antimicrob Chemother 1997*; 40:135-6.

[8] Appelbaum PC. MRSA-the tip of the iceberg. *Clin. Microbiol Infect* 2006; 12 *Suppl* 2:3-10.

[9] MMWR. *Staphylococcus aureus* resistant to vancomycin-United States 2002. Vol. 51 (26): 565.567RE, 2002.

[10] Chang S, Sievert DM, Hageman JC, et al. Infection with vancomycin-resistant *Staphylococcus aureus* containing the vanA resistance gene. *N. Engl J. Med.* 2003; 348:1342-7.

[11] King MD, Humphrey BJ, Wang YF, Kourbatova EV, Ray SM, Blumberg HM. Emergence of community-acquired methicillin-resistant *Staphylococcus aureus* USA 300 clone as the predominant cause of skin and soft-tissue infections. *Ann. Intern. Med.* 2006; 144:309-17.

[12] Moellering RC, Jr. The growing menace of community-acquired methicillin-resistant *Staphylococcus aureus*. Ann. Intern Med. 2006; 144:368-70.

[13] Noskin GA, Rubin RJ, Schentag JJ, et al. The burden of *Staphylococcus aureus* infections on hospitals in the United States: an analysis of the 2000 and 2001 Nationwide Inpatient Sample Database. *Arch. Intern Med.* 2005; 165:1756-61.

[14] Rupp ME, Archer GL. Coagulase-negative Staphylococci: pathogens associated with medical progress. *Clin. Infect Dis*. 1994; 19:231-43; quiz 244-5.

[15] Schaaff F, Reipert A, Bierbaum G. An elevated mutation frequency favors development of vancomycin resistance in *Staphylococcus aureus*. *Antimicrob Agents Chemother* 2002; 46:3540-8.

[16] Chang FY, Peacock JE, Jr., Musher DM, et al. *Staphylococcus aureus* bacteremia: recurrence and the impact of antibiotic treatment in a prospective multicenter study. *Medicine* (Baltimore) 2003; 82:333-9.
[17] Markowitz N, Quinn EL, Saravolatz LD. Trimethoprim-sulfamethoxazole compared with vancomycin for the treatment of *Staphylococcus aureus* infection. *Ann. Intern. Med.* 1992; 117:390-8.
[18] Becker K, Friedrich AW, Lubritz G, Weilert M, Peters G, Von Eiff C. Prevalence of genes encoding pyrogenic toxin superantigens and exfoliative toxins among strains of *Staphylococcus aureus* isolated from blood and nasal specimens. *J. Clin. Microbiol.* 2003; 41:1434-9.
[19] Lowy FD. *Staphylococcus aureus* infections. *N. Engl. J. Med.* 1998; 339:520-32.
[20] Sakoulas G, Moellering RC, Jr., Eliopoulos GM. Adaptation of methicillin-resistant *Staphylococcus aureus* in the face of vancomycin therapy. *Clin. Infect Dis*. 2006; 42 Suppl 1:S40-50.
[21] Donlan RM, Costerton JW. Biofilms: survival mechanisms of clinically relevant microorganisms. *Clin. Microbiol. Rev*. 2002; 15:167-93.
[22] Caiazza NC, O'Toole GA. Alpha-toxin is required for biofilm formation by *Staphylococcus aureus*. *J. Bacteriol.* 2003; 185:3214-7.
[23] von Eiff C, Peters G, Heilmann C. Pathogenesis of infections due to coagulase-negative Staphylococci. *Lancet Infect Dis* 2002; 2:677-85.
[24] Yu VL, Chiou CC, Feldman C, et al. An international prospective study of pneumococcal bacteremia: correlation with *in vitro* resistance, antibiotics administered, and clinical outcome. *Clin. Infect Dis*. 2003; 37:230-7.
[25] Peterson LR. Penicillins for treatment of pneumococcal pneumonia: does in vitro resistance really matter? *Clin. Infect Dis*. 2006; 42:224-33.
[26] Lonks JR, Garau J, Gomez L, et al. Failure of macrolide antibiotic treatment in patients with bacteremia due to erythromycin-resistant *Streptococcus pneumoniae*. *Clin. Infect. Dis*. 2002; 35:556-64.
[27] Schentag JJ, Klugman K.P., Yu V.L., et al. *Streptococcus pneumoniae* bacteremias: pharmacodynamic correlations with outcome and macrolide resistance: a controlled study. In*t. J. Antimicrob Agents* 2007; 30:264-9.
[28] Greenberg D DR, Klugman K, Madhi SA, Feldman C, Roberts S, Morris A, Chedid MBF, Chiou CC, Yu VL. *Streptococcus pneumoniae* serotypes causing meningitis in children and adults. *Proceedings of the 14th ICAAC Conference*. Washington, DC., 2004.
[29] Carpenter CF, Chambers HF. Daptomycin: another novel agent for treating infections due to drug-resistant gram-positive pathogens. *Clin. Infect Dis*. 2004; 38:994-1000.
[30] Fenton C, Keating GM, Curran MP. Daptomycin. *Drugs* 2004; 64:445-55.
[31] Steenbergen JN, Alder J, Thorne GM, Tally FP. Daptomycin: a lipopeptide antibiotic for the treatment of serious Gram-positive infections. *J. Antimicrob. Chemother* 2005; 55:283-8.
[32] Schriever CA, Fernandez C, Rodvold KA, Danziger LH. Daptomycin: a novel cyclic lipopeptide antimicrobial. *Am. J. Health Syst. Pharm.* 2005; 62:1145-58.
[33] Moellering RC. Linezolid: the first oxazolidinone antimicrobial. *Ann. Intern Med.* 2003; 38:135-42.

[34] Birmingham MC, Rayner CR, Meagher AK, Flavin SM, Batts DH, Schentag JJ. Linezolid for the treatment of multidrug-resistant, gram-positive infections: experience from a compassionate-use program. *Clin. Infect. Dis*. 2003; 36:159-68.

[35] LaPlante KL, Rybak MJ. Daptomycin - a novel antibiotic against Gram-positive pathogens. *Expert Opin. Pharmacother* 2004; 5:2321-31.

[36] Jeu L, Fung HB. Daptomycin: a cyclic lipopeptide antimicrobial agent. *Clin. Ther* 2004; 26:1728-57.

[37] Alder JD. Daptomycin: a new drug class for the treatment of Gram-positive infections. *Drugs Today* (Barc) 2005; 41:81-90.

[38] Livermore DM. Tigecycline: what is it, and where should it be used? *J. Antimicrob. Chemother* 2005; 56:611-4.

[39] Pankey GA. Tigecycline. *J.Antimicrob. Chemother* 2005; 56:470-80.

[40] Van Bambeke F, Van Laethem Y, Courvalin P, Tulkens PM. Glycopeptide antibiotics: from conventional molecules to new derivatives. *Drugs* 2004; 64:913-36.

[41] Virginlar N. MA. Glycopeptides (Dalbavancin, Oritavancin, Teicoplanin, Vancomycin). In: Yu VL, Eds. *Antimicrobial Therapy and Vaccines*. Vol. II: Antimicrobial Agents: www.antimicrobe.org, 2004.

[42] Hershberger E, Donabedian S, Konstantinou K, Zervos MJ. Quinupristin-dalfopristin resistance in gram-positive bacteria: mechanism of resistance and epidemiology. *Clin. Infect. Dis*. 2004; 38:92-8.

[43] Speciale A, La Ferla K, Caccamo F, Nicoletti G. Antimicrobial activity of quinupristin/dalfopristin, a new injectable streptogramin with a wide Gram-positive spectrum. *Int. J. Antimicrob. Agents* 1999; 13:21-8.

[44] Moellering RC, Linden PK, Reinhardt J, Blumberg EA, Bompart F, Talbot GH. The efficacy and safety of quinupristin/dalfopristin for the treatment of infections caused by vancomycin-resistant *Enterococcus faecium*. Synercid Emergency-Use Study Group. *J. Antimicrob Chemother* 1999; 44:251-61.

[45] Nichols RL, Graham DR, Barriere SL, et al. Treatment of hospitalized patients with complicated gram-positive skin and skin structure infections: two randomized, multicentre studies of quinupristin/dalfopristin versus cefazolin, oxacillin or vancomycin. Synercid Skin and Skin Structure Infection Group. *J. Antimicrob Chemother* 1999; 44:263-73.

[46] Meka VG, Pillai SK, Sakoulas G, et al. Linezolid resistance in sequential *Staphylococcus aureus* isolates associated with a T2500A mutation in the 23S rRNA gene and loss of a single copy of rRNA. *J. Infect. Dis*. 2004; 190:311-7.

[47] Fagon J, Patrick H, Haas DW, et al. Treatment of gram-positive nosocomial pneumonia. Prospective randomized comparison of quinupristin/dalfopristin versus vancomycin. Nosocomial Pneumonia Group. *Am. J. Respir Crit. Care Med*. 2000; 161:753-62.

[48] Drew RH, Perfect JR, Srinath L, Kurkimilis E, Dowzicky M, Talbot GH. Treatment of methicillin-resistant *Staphylococcus aureus* infections with quinupristin-dalfopristin in patients intolerant of or failing prior therapy. For the Synercid Emergency-Use Study Group. *J. Antimicrob Chemother* 2000; 46:775-84.

[49] Carver PL, Whang E, VandenBussche HL, Kauffman CA, Malani PN. Risk factors for arthralgias or myalgias associated with quinupristin-dalfopristin therapy. *Pharmacotherapy* 2003; 23:159-64.

[50] Raad I, Hachem R, Hanna H. Relationship between myalgias/arthralgias occurring in patients receiving quinupristin/dalfopristin and biliary dysfunction. *J. Antimicrob. Chemother* 2004; 53:1105-8.

[51] Stalker DJ, Jungbluth GL. Clinical pharmacokinetics of linezolid, a novel oxazolidinone antibacterial. *Clin. Pharmacokinet* 2003; 42:1129-40.

[52] Weigelt J, Itani K, Stevens D, Lau W, Dryden M, Knirsch C. Linezolid versus vancomycin in treatment of complicated skin and soft tissue infections. *Antimicrob Agents Chemother* 2005; 49:2260-6.

[53] Rubinstein E, Cammarata S, Oliphant T, Wunderink R. Linezolid (PNU-100766) versus vancomycin in the treatment of hospitalized patients with nosocomial pneumonia: a randomized, double-blind, multicenter study. *Clin. Infect. Dis.* 2001; 32:402-12.

[54] Wunderink RG, Rello J, Cammarata SK, Croos-Dabrera RV, Kollef MH. Linezolid vs vancomycin: analysis of two double-blind studies of patients with methicillin-resistant *Staphylococcus aureus* nosocomial pneumonia. *Chest* 2003; 124:1789-97.

[55] Rayner CR, Forrest A, Meagher AK, Birmingham MC, Schentag JJ. Clinical pharmacodynamics of linezolid in seriously ill patients treated in a compassionate use programme. *Clin. Pharmacokinet* 2003; 42:1411-23.

[56] Rayner CR, Baddour LM, Birmingham MC, Norden C, Meagher AK, Schentag JJ. Linezolid in the treatment of osteomyelitis: results of compassionate use experience. *Infection* 2004; 32:8-14.

[57] Razonable RR, Osmon DR, Steckelberg JM. Linezolid therapy for orthopedic infections. *Mayo Clin. Proc.* 2004; 79:1137-44.

[58] Cook AM, Ramsey CN, Martin CA, Pittman T. Linezolid for the treatment of a heteroresistant *Staphylococcus aureus* shunt infection. *Pediatr. Neurosurg.* 2005; 41:102-4.

[59] Rho JP, Sia IG, Crum BA, Dekutoski MB, Trousdale RT. Linezolid-associated peripheral neuropathy. *Mayo Clin. Proc.* 2004; 79:927-30.

[60] Spellberg B, Yoo T, Bayer AS. Reversal of linezolid-associated cytopenias, but not peripheral neuropathy, by administration of vitamin B6. *J. Antimicrob Chemother* 2004; 54:832-5.

[61] Young LS. Hematologic effects of linezolid *versus* vancomycin. *Clin. Infect Dis.* 2004; 38:1065-6.

[62] Rao N, Ziran BH, Wagener MM, Santa ER, Yu VL. Similar hematologic effects of longterm linezolid and vancomycin therapy in a prospective observational study of patients with orthopedic infections. *Clin. Infect. Dis.* 2004; 38:1058-64.

[63] Nasraway SA, Shorr AF, Kuter DJ, O'Grady N, Le VH, Cammarata SK. Linezolid does not increase the risk of thrombocytopenia in patients with nosocomial pneumonia: comparative analysis of linezolid and vancomycin use. *Clin. Infect Dis.* 2003; 37:1609-16.

[64] Kulkarni K, Del Priore LV. Linezolid induced toxic optic neuropathy. *Br. J. Ophthalmol.* 2005; 89:1664-5.

[65] Narita MT, B Yu VL. Linezolid-associated peripheral and optic neuropathy, lactic acidosid and serotonin syndrome: a review. *Pharmacotherapy* 2007; 27:1189-97.

[66] Soriano A, Miro O, Mensa J. Mitochondrial toxicity associated with linezolid. *N. Engl J. Med.* 2005; 353:2305-6.

[67] Bernard L, Stern R, Lew D, Hoffmeyer P. Serotonin syndrome after concomitant treatment with linezolid and citalopram. *Clin. Infect Dis*. 2003; 36:1197.

[68] Arbeit RD, Maki D, Tally FP, Campanaro E, Eisenstein BI. The safety and efficacy of daptomycin for the treatment of complicated skin and skin-structure infections. *Clin. Infect. Dis*. 2004; 38:1673-81.

[69] Fowler VG, Cosgrove S. Abrutyn E, et al. Daptomycin *vs*. Standard Therapy for Staphylococcus aureus Bacteremia (SAB) and Infective Endocarditis (SAIE). *45th Annual Interscience Congress on Antimicrobial Agents and Chemotherapy*. Washington DC, 2005.

[70] LaPlante KL, Rybak, M.J. *Daptomycin. Antimicrobial Therapy and Vaccines*. Vol. II: Antimicrobial Agents: www.antimicrobe.org.

[71] Zhanel GG, Homenuik K, Nichol K, et al. The glycylcyclines: a comparative review with the tetracyclines. *Drugs* 2004; 64:63-88.

[72] Raad I, Darouiche R, Vazquez J, et al. Efficacy and safety of weekly dalbavancin therapy for catheter-related bloodstream infection caused by gram-positive pathogens. *Clin. Infect Dis*. 2005; 40:374-80.

[73] Seltzer E, Dorr MB, Goldstein BP, Perry M, Dowell JA, Henkel T. Once-weekly dalbavancin *versus* standard-of-care antimicrobial regimens for treatment of skin and soft-tissue infections. *Clin. Infect Dis*. 2003; 37:1298-303.

[74] Kaatz GW, Seo SM, Aeschlimann JR, Houlihan HH, Mercier RC, Rybak MJ. Efficacy of LY333328 against experimental methicillin-resistant Staphylococcus aureus endocarditis. *Antimicrob Agents Chemother* 1998; 42:981-3.

[75] Gerber J, Smirnov A, Wellmer A, et al. Activity of LY333328 in experimental meningitis caused by a Streptococcus pneumoniae strain susceptible to penicillin. *Antimicrob Agents Chemother* 2001; 45:2169-72.

[76] Giamarellou H ORW, Harris H, Owen S, Porter S, Loutit J. Phase 3 trial comparing 3-7 days of oritavancin vs. 10-14 days of vancomycin/cephalexin in the treatment of patients with complicated skin and skin structure infections (CSSI). In: *Program and abstracts of the 43rd ICAAC Conference*, Chicago, IL, September 14-17,2003, 2003. American Society of Microbiology.

[77] Barrett JF. Recent developments in glycopeptide antibacterials. *Curr Opin Investig Drugs* 2005; 6:781-90.

[78] Shaw JP, Seroogy J, Kaniga K, Higgins DL, Kitt M, Barriere S. Pharmacokinetics, serum inhibitory and bactericidal activity, and safety of telavancin in healthy subjects. *Antimicrob Agents Chemother* 2005; 49:195-201.

[79] Hegde SS, Reyes N, Wiens T, et al. Pharmacodynamics of telavancin (TD-6424), a novel bactericidal agent, against gram-positive bacteria. *Antimicrob Agents Chemother* 2004; 48:3043-50.

[80] Stryjewski ME, Chu VH, O'Riordan WD, et al. Telavancin versus standard therapy for treatment of complicated skin and skin structure infections caused by gram-positive bacteria: FAST 2 study. *Antimicrob Agents Chemother* 2006; 50:862-7.

[81] Stryjewski ME, O'Riordan WD, Lau WK, et al. Telavancin versus standard therapy for treatment of complicated skin and soft-tissue infections due to gram-positive bacteria. *Clin. Infect. Dis*. 2005; 40:1601-7.

[82] Aneziokoro CO, Cannon JP, Pachucki CT, Lentino JR. The effectiveness and safety of oral linezolid for the primary and secondary treatment of osteomyelitis. *J. Chemother* 2005; 17:643-50.

[83] Finney MS, Crank CW, Segreti J. Use of daptomycin to treat drug-resistant Gram-positive bone and joint infections. *Curr. Med. Res. Opin*. 2005; 21:1923-6.

[84] Anthony S.J. HM, Angelos E, Stratton CW. Clinical Experience with daptomycin in Patients with Orthopedic-Related Infections. *43rd IDSA Annual Meeting 2005*. San Francisco, CA, 2005.

[85] Yin LY, Lazzarini L, Li F, Stevens CM, Calhoun JH. Comparative evaluation of tigecycline and vancomycin, with and without rifampicin, in the treatment of methicillin-resistant *Staphylococcus aureus* experimental osteomyelitis in a rabbit model. *J. Antimicrob. Chemother* 2005; 55:995-1002.

[86] Howden BP, Ward PB, Charles PG, et al. Treatment outcomes for serious infections caused by methicillin-resistant *Staphylococcus aureus* with reduced vancomycin susceptibility. *Clin. Infect Dis*. 2004; 38:521-8.

[87] Fowler VG, Jr., Boucher HW, Corey GR, et al. Daptomycin versus standard therapy for bacteremia and endocarditis caused by *Staphylococcus aureus*. *N. Engl. J. Med*. 2006; 355:653-65.

[88] Woods CW, Cheng AC, Fowler VG, Jr., et al. Endocarditis caused by *Staphylococcus aureus* with reduced susceptibility to vancomycin. *Clin. Infect Dis*. 2004; 38:1188-91.

[89] Chiang FY, Climo M. Efficacy of linezolid alone or in combination with vancomycin for treatment of experimental endocarditis due to methicillin-resistant *Staphylococcus aureus*. *Antimicrob Agents Chemother* 2003; 47:3002-4.

[90] FDA. Information for Healthcare Professionals: Linezolid (marketed as Zyvox) March 16, 2007, 2007.

[91] Falagas ME, Manta KG, Ntziora F, Vardakas KZ. Linezolid for the treatment of patients with endocarditis: a systematic review of the published evidence. *J. Antimicrob. Chemother* 2006; 58:273-80.

[92] Kang SL, Rybak MJ. In-vitro bactericidal activity of quinupristin/dalfopristin alone and in combination against resistant strains of *Enterococcus* species and *Staphylococcus aureus*. J. *Antimicrob. Chemother* 1997; 39 Suppl A:33-9.

[93] Tsuji BT, Rybak MJ. Short-course gentamicin in combination with daptomycin or vancomycin against *Staphylococcus aureus* in an *in vitro* pharmacodynamic model with simulated endocardial vegetations. *Antimicrob Agents Chemother* 2005; 49:2735-45.

[94] Grohs P, Kitzis MD, Gutmann L. *In vitro* bactericidal activities of linezolid in combination with vancomycin, gentamicin, ciprofloxacin, fusidic acid, and rifampin against *Staphylococcus aureus*. *Antimicrob Agents Chemother* 2003; 47:418-20.

[95] Jacqueline C, Caillon J, Le Mabecque V, et al. *In vitro* activity of linezolid alone and in combination with gentamicin, vancomycin or rifampicin against methicillin-resistant *Staphylococcus aureus* by time-kill curve methods. *J. Antimicrob. Chemother* 2003; 51:857-64.

[96] Moise-Broder PA, Sakoulas G, Eliopoulos GM, Schentag JJ, Forrest A, Moellering RC, Jr. Accessory gene regulator group II polymorphism in methicillin-resistant *Staphylococcus aureus* is predictive of failure of vancomycin therapy. *Clin. Infect. Dis*. 2004; 38:1700-5.

[97] Tsuji BT, Rybak MJ. Etest synergy testing of clinical isolates of *Staphylococcus aureus* demonstrating heterogeneous resistance to vancomycin. *Diagn. Microbiol. Infect Dis.* 2006; 54:73-7.

[98] LaPlante KL, Rybak MJ. Impact of high-inoculum *Staphylococcus aureus* on the activities of nafcillin, vancomycin, linezolid, and daptomycin, alone and in combination with gentamicin, in an in vitro pharmacodynamic model. *Antimicrob Agents Chemother* 2004; 48:4665-72.

[99] Sakoulas G, Eliopoulos GM, Alder J, Eliopoulos CT. Efficacy of daptomycin in experimental endocarditis due to methicillin-resistant *Staphylococcus aureus*. *Antimicrob Agents Chemother* 2003; 47:1714-8.

[100] Baltch A RW, Bopp L. et al. Killing of methicillin-resistant *Staphylococcus aureus* by daptomycin, gentamicin, and rifampin, singly and in combination, in broth and in human monocyte-derived macrophages, with and without GM-CSF and Interferon Activation. *Proceedings of the 15th ICAAC Conference, 2005, Abstract E-1741, American Society of Microbiology*, Washington, 2005.

[101] Allen GP, Cha R, Rybak MJ. *In vitro* activities of quinupristin-dalfopristin and cefepime, alone and in combination with various antimicrobials, against multidrug-resistant Staphylococci and Enterococci in an *in vitro* pharmacodynamic model. *Antimicrob Agents Chemother* 2002; 46:2606-12.

[102] Dailey CF, Pagano PJ, Buchanan LV, Paquette JA, Haas JV, Gibson JK. Efficacy of linezolid plus rifampin in an experimental model of methicillin-susceptible *Staphylococcus aureus* endocarditis. *Antimicrob Agents Chemother* 2003; 47:2655-8.

[103] Kang SL, Rybak MJ, McGrath BJ, Kaatz GW, Seo SM. Pharmacodynamics of levofloxacin, ofloxacin, and ciprofloxacin, alone and in combination with rifampin, against methicillin-susceptible and -resistant *Staphylococcus aureus* in an *in vitro* infection model. *Antimicrob Agents Chemother* 1994; 38:2702-9.

[104] Zarrouk V, Bozdogan B, Leclercq R, et al. Activities of the combination of quinupristin-dalfopristin with rifampin *in vitro* and in experimental endocarditis due to *Staphylococcus aureus* strains with various phenotypes of resistance to macrolide-lincosamide-streptogramin antibiotics. *Antimicrob Agents Chemother* 2001; 45:1244-8.

[105] Sambatakou H, Giamarellos-Bourboulis EJ, Grecka P, Chryssouli Z, Giamarellou H. *In vitro* activity and killing effect of quinupristin/dalfopristin (RP59500) on nosocomial *Staphylococcus aureus* and interactions with rifampicin and ciprofloxacin against methicillin-resistant isolates. *J. Antimicrob. Chemother* 1998; 41:349-55.

[106] Petersen PJ, Labthavikul P, Jones CH, Bradford PA. *In vitro* antibacterial activities of tigecycline in combination with other antimicrobial agents determined by chequerboard and time-kill kinetic analysis. *J. Antimicrob. Chemother* 2006; 57:573-6.

[107] Mercier RC, Kennedy C, Meadows C. Antimicrobial activity of tigecycline (GAR-936) against *Enterococcus faecium* and *Staphylococcus aureus* used alone and in combination. *Pharmacotherapy* 2002; 22:1517-23.

In: Advances in Medicine and Biology. Volume 20
Editor: Leon V. Berhardt

ISBN 978-1-61209-135-8

Chapter 7

PROMISING APPLICATIONS FOR TLR AGONISTS AS IMMUNE ADJUVANTS

Bara Sarraj[1] and Yanal Murad*[2]

[1]Department of Lab Medicine, Children's Hospital Boston, Boston, MA, USA
Divisions of Nephrology and Organ Transplantation, Feinberg School of Medicine, Northwestern University, Chicago, IL, USA
[2]Department of Experimental Surgery, Duke University, Durham, NC, USA
National Research Council of Canada, Institute for Biological Sciences, Ottawa, ON

ABSTRACT

Toll like receptors (TLRs) are part of the innate immune system, and they belong to the pattern recognition receptors (PRR) family, which is designed to recognize and bind certain molecules that are restricted to pathogens, like LPS and CpG. Different TLR signals converge through few common adapter proteins to relay their signals, a process that results in the activation of several genes essential for mounting an immune response. The wide distribution of TLRs on hematopoietic and non-hematopoietic cells, and their high potential for activating the host immune system makes TLR ligands great adjuvant candidates. They will elicit their function on a wide variety of cells, and stimulate both the innate and adaptive immune systems. This review will describe some of the TLR agonists that are considered as major targets for the development of agonists that could serve as adjuvants and agents of immunotherapy. Several of these TLR agonists, including TLR2, TLR4, TLR7 and TLR9 agonists are being developed and tested in clinical trials.

* E-mail : yanal10@hotmail.com

INTRODUCTION

Adjuvants are immune response or drug action enhancers that, along with antigens, comprise the primary components of effective vaccines. Major components of adjuvants are bacterial or viral extracts, mineral oils and aluminum hydroxide metal [1]. Adjuvants mostly prime antigen-presenting myeloid dendritic cells (mDCs) and interferon-producing plasmacytoid DCs (pDCs). Known examples of adjuvants are Freund's and Bacille Calmette-Guérin (BCG) [1]. The immune system is divided into innate and adaptive, connected by antigen-presenting cells. The adaptive immunity, which is highly sophisticated, and present only in higher organisms, detects non-self antigens and pathogens through recognition of peptide antigens using antigen receptors expressed on the surface of B and T cells. In contrast, the innate immune system is conserved among multicellular organisms, and is designed to recognize pathogen-associated molecular patterns (PAMPs), which are restricted to pathogens and absent in vertebrates, through pattern recognition receptors (PRR).

PRRs are germ line encoded molecules, present on different immune cells and are divided into intracellular and extracellular innate receptors [2, 3]. Characterized intracellular innate receptors (TLR-independent) include anti-bacterial nucleotide-domain oligomerization (NOD)-like receptors (NLR) and anti-viral retinoic-acid-inducible gene I (RIG)-like receptors (RLR) [4-6]. Among the best known and studied PRRs are Toll Like Receptors (TLRs). TLRs were first discovered in vertebrates on the basis of their homology with Drosophila Toll protein, a molecule that stimulates the production of antimicrobial proteins in *Drosophila melanogaster* [7, 8]. TLRs belong to the Toll superfamily, which also include the IL-1 receptor (IL-1R) and IL-18R. There are at least 12 members of the TLR family in vertebrates discovered to date (9 in humans) (Table 1).

Table 1. TLRs, agonists and adjuvants

TLR	Ligands	Expression/location	Adjuvant
TLR1	Lipoproteins	Cell surface	Pam3Cys*
TLR2	Lipoteichoic acid	intracellular Cell surface	Pam3Cys/MALP2/OspA*
TLR3	dsRNA	intracellular	PolyI:C
TLR4	LPS	intracellular Cell surface	MPL
TLR5	Flagellin	intracellular Cell surface	Flagellin
TLR6	Lipoproteins	Cell surface	MALP-2*
TLR7	Viral ssRNA	intracellular	Imidazoquinolins/polyU**
TLR8	Viral ssRNA	intracellular	Imidazoquinolins/polyU**
TLR9	CpG DNA	intracellular	CpG ODNs (table 3)

* TLR2 in dimerization with TLR1/TLR6/unknown, respectively.

** TLR7/8 heterodimer.

Monophosphoryl lipid A, MPL; *Mycoplasma* macrophage-activating lipopeptide 2, MALP-2; Outer surface lipoprotein, OspA.

The wide distribution of TLRs (from DCs, macrophages, and NK cells to B and T cells and even endothelial and epithelial cells), and their high potential for activating the host immune system makes the TLR ligands great adjuvant candidates, since they will elicit their function on a wide variety of cells, and stimulate both the innate and adaptive immune system (Table 2).

In fact, TLRs and other PRRs are widely distributed on haematopoietic and non-haematopoietic cells, with each cell type expressing a typical pattern of PRRs. This pattern of expression is further modulated or modified by the activation, maturation or differentiation of the cells [3].

Table 2. TLR cell distribution [1, 2, 9]

TLR	Cells
TLR1	Leukocytes
TLR2	Gr, Mf, NK, DC, T, B, Fb, Ep.
TLR3	Human and murine mDCs
TLR4	Gr, Mf, NK, DC, T, B, Fb, Ep.
TLR5	Ep, DCs
TLR6	DCs
TLR7	Immune cells, human pDCs, murine pDCs
TLR8	Human monocytes and myeloid DCs
TLR9	Immune cells, human pDCs and B cells, murine mDCs and pDCs

Granulocyte, Gr; macrophage, Mf; natural killer cell, NK; dendritic cell, DC; fibroblast, Fb; epithelial cell, Ep; myeloid DC, mDC; plasmacytoid DC, pDC.

STRUCTURE

TLRs are single-pass transmembrane proteins with an N-terminal Leucine-rich repeats (LRRs), a transmembrane domain, and a C-terminal cytoplasmic domain.

The extracellular domain of TLRs, which can recognize specific pathogen components, contains 19–25 tandem copies of the LRR motif, each of which is 24–29 amino acids in length [10], but despite the conservation among LRR domains, different TLRs can recognize several structurally unrelated ligands [11-13]. TLR structures show high similarity to that of the interleukin-1 receptor (IL-1R) family, and share IL-1R intracellular domain, a conserved region of ~200 amino acids in their cytoplasmic tails, which is known as the Toll/IL-1R (TIR) domain [14, 15].

The history of adjuvants that target TLRs date back to William Coley, who used bacterial extracts for the treatment of cancer [16]. More recently, many established and experimental vaccines incorporate TLR ligands in prophylactic vaccines against infectious diseases, as well as in therapeutic immunization for diseases like cancer. The engagement of TLR signaling pathways maybe used to trigger an adaptive immune response that could boost vaccine responses [9]. It has been reported that in most cases, signaling through TLRs will favor a Th1-type immune response, which is needed for the protection against most pathogens [17]

Signaling: TLR Activation

The activation of TLRs begins with the ligand binding to extracellular LRRs and, either through receptor oligomerization and/or induction of a conformational change across the plasma membrane; this will induce the recruitment/activation of adapter proteins through the Toll/IL-1 Receptor (TIR) domain.

TLRs recruit a specific combination of TIR-domain-containing adaptors, including myeloid differentiation primary response gene 88 (MyD88), TIR-containing adaptor protein/MyD88-adaptor-like (TIRAP/MAL), TIR-domain-containing adaptor inducing interferon-β (IFN-β)/TIR-domain-containing adaptor molecule 1 (TRIF/TICAM1) and TRIF-related adaptor molecule 2/TIR-domain-containing adaptor molecule (TRAM/TICAM2) [14, 18, 19] (Figure 1).

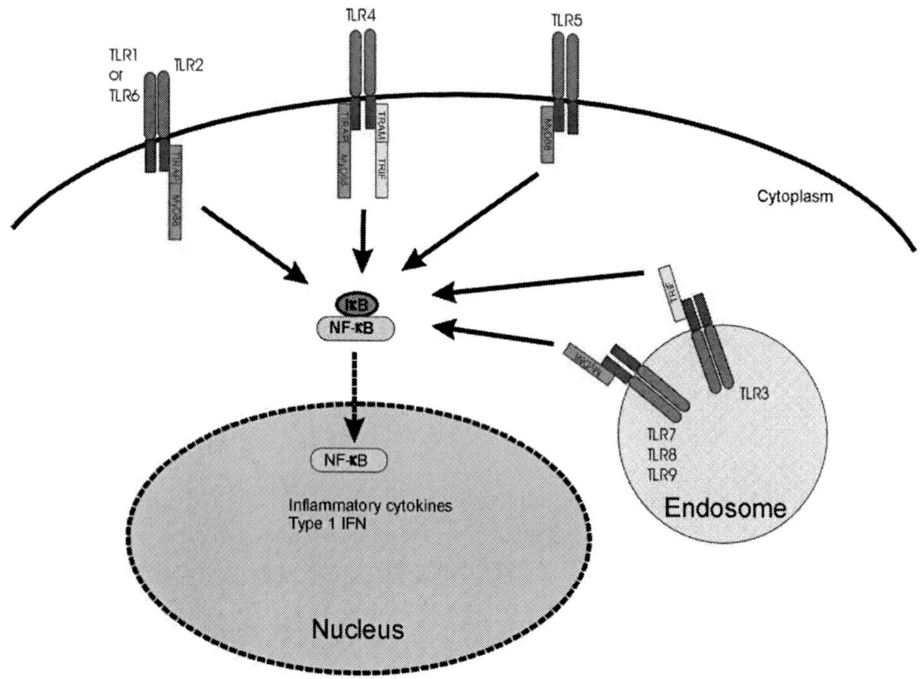

Figure 1. Overview of different TLR signaling pathways, which leads to the activation of NF-κB, through either MyD88 or TRIF mediated pathways. All TLRs signal through MyD88, except for TLR3 and TLR 2 subfamily (TLR 1, TLR2 and TLR6). TLR4 can signal through either pathway. Different TLR ligands are summarized in table 1.

All TLR molecules use MyD88 for signaling, except TLR3, which uses TRIF instead [20]. MyD88 serves as the sole signal relay in TLR5, TLR7, and TLR9. That was indicated when it was shown that NF-κB activation and inflammatory cytokine induction is defective in MyD88-deficient mice [21, 22]. However, the response is normal in mice deficient for TIRAP, TRIF or TRAM, indicating that MyD88 is used as the sole adaptor by these TLRs. TLR2 requires TIRAP as an adaptor to bridge between MyD88 and TLR2 [23, 24], while TLR4 uses all of these adaptors (MyD88, TRIF, TRAM, and TIRAP) for signaling [23, 25, 26]. These adapters lead to the activation of canonical IKKβ-dependent complexes, degradation of IκBα and IκBβ, and liberation of, primarily, RelA and c-Rel nuclear factor-κB (NF-κB) components (Figure 1).

MyD88-Dependent Pathway

The MyD88-dependent pathway signals via MyD88, IRAK, and TRAF6 and leads to NF-κB activation. MyD88 is composed of two domains: a TIR domain and a death domain. Upon TLR activation, through its death domain, MyD88 interacts with the death domains of members of the IRAK (IL-1 receptor-associated kinase) family of protein kinases, including IRAK1, IRAK2, IRAK4 and IRAK-M [14, 18]. It is believed that all MyD88- utilizing TLRs directly recruit MyD88, except for TLR2 and TLR4, which use a TIR domain containing adaptor (TIRAP) to bridge MyD88 to TLR. Once phosphorylated, IRAKs dissociate from MyD88 and interact with TRAF6, a member of the TRAF family. TRAF6, an E3 ligase, forms a complex with Ubc13 and Uev1A to promote the synthesis of lysine 63-linked polyubiquitin chains, which in turn activate TAK1, a MAPKKK [27]. This activation of TAK1 will lead to the activation of IKK complex, and eventually NF-κB, through yet undefined mediators [28, 29].

MyD88-Independent (TRIF-Dependent) Pathway

This pathway, also called TRIF-dependent pathway, activates NF-κB through the RIP1/TRAF6–TAK1–IKKα/β pathway. This pathway was discovered when MyD88$^{-/-}$ cells displayed partial NF-κB activation when exposed to LPS [30]. Also, when cells are stimulated through TLR3 and TLR4, TRIF will mediate the activation of NF-κB in the absence of MyD88 [31]. For TLR3, all downstream signaling appears to be TRIF-mediated, while TLR4 can signal through either MyD88 or TRIF pathways.

TLR Agonists as Adjuvants

Prevention of infectious diseases through immunization was discovered and mentioned in the late 18th century, and is considered as one of the greatest achievements of modern medicine. Nonetheless, scientists are still facing considerable challenges for improving the efficacy of vaccines, whether it is for the development of new prophylactic vaccines for infectious agents, or for therapeutic immunization for noninfectious diseases such as cancer [9]. Regardless of the vaccine target, successful immunization results in activation of adaptive immunity, which might be accomplished, in part, through stimulation of the TLRs. The involvement of TLR pathways in activating adaptive immunity and vaccine responses is well established, and is explained by the control exerted by TLRs over the activation of adaptive immunity [32]. This control is mediated through several mechanisms, including TLR-mediated activation of dendritic cells (DCs) by upregulating chemokine, MHC and costimulatory receptors, producing cytokines and chemokines, suppression of T regulatory cells (T_{regs}) and reversal of tolerance [1, 9].

How Do TLR Agonists Stimulate Adaptive Immunity?

We summarize here some of the mechanisms TLR agonists are thought to work to stimulate adaptive immunity

Cross-Priming of CD8 T Cells

DCs can display exogenously internalized antigens on MHC class I molecules to CD8 T cells, a process termed cross-priming. Cross priming is very important in terms of vaccination against intracellular antigens that CD8 T cells will efficiently eliminate [9]. Murine DCs had been shown to cross present dsRNA or CpG to CTLs in vivo [33, 34].

IL-12 and Th1 Responses

The stimulation of TLRs, in most instances, will favor the development a Th1 type of response, which is required for the protection against most pathogens [9, 35]. Also, it seems that IL-12 plays an important role in this process and in the stimulation of adaptive immunity [17, 36]. The synergistic effects of certain TLR ligands are marked by the production of IL-12 p70 (the active form of IL12), while stimulation of DCs with a single TLR ligand will normally induce negligible levels of biologically active IL12 [3, 37].

Reversal of Tolerance

Several studies have shown that the engagement of TLRs can overcome peripheral tolerance leading to specific CTL responses in therapeutic tumor vaccines, thus enhancing the vaccine efficiency [38-40]. Tolerance through T_{regs} (CD^+ $CD25^+$) can be overcome by stimulating TLRs on DCs, which was found to be dependent on IL-6 production, since $IL6^{-/-}$ mice were not able to prime pathogen-specific T cells in the presence of T_{reg} cells [9, 38]. Moreover, Stimulation of TLR8 on T_{reg} cells was found to reverse their suppression function [40].

Upregulation of Co-Stimulatory Molecules

Activation of TLRs leads to the upregualtion of several costimulatory molecules, like CD40, CD80, CD86, and CD70, on antigen presenting cells [32], which has been shown to improve vaccine efficacy [41-43]. Moreover, TLR activation will lead to DC maturation, and accumulation of MHC class II and antigen complexes on the cell surface [44, 45].

TLR AGONISTS WITH POTENTIAL CLINICAL USE

TLR2

TLR2 recognizes a variety of microbial components, including lipoproteins from pathogens including Gram-negative bacteria, Mycoplasma and spirochets [46-48], peptidoglycan and lipoteichoic acid from Gram-positive bacteria [49-52], Lipoarabinomannan from mycobacteria [51-53] and several other microbial components from other microorganisms [54]. Probably the best known TLR2 agonist is the outer-surface lipoprotein (OspA) of *Borrelia burgdorferi,* which has been used as a vaccine for Lyme disease [55]. It has been hypothesized that OspA signals through TLR1/TLR2 heterodimer, since neither TLR1$^{-/-}$ nor TLR2$^{-/-}$ mice were able to mount a protective immune response after OspA vaccination; also, low responders to the vaccine were shown to have low surface expression of TLR1, but unchanged expression of TLR2 [56].

Another aspect of TLR2 ligand recognition involves interaction with other TLRs, including TLR1 and TLR6. Several reports suggested cooperation between TLR2 and TLR6, including co-immunoprecipitation of the two proteins [57], or the recognition of macrophage activating lipopeptide (MALP-2) only in the presence of both TLR2 and TLR6 [57]. MALP-2 is a synthetic lipopeptide with two long chain fatty acid ester residues that signals through TLR2 (associated with TLR6), activates nuclear transcription factor NF-κB, induces the synthesis of a number of cytokines and chemokines, depending on its target cell, and induces maturation of dendritic cells [58].

TLR4

One of the best studied interactions between a microbial pathogen-associated molecular pattern and a TLR is the recognition of lipopolysaccharide (LPS), the major component of the outer membrane of Gram-negative bacteria, by TLR4. It is known that the adjuvant activity of bacterial cell walls is responsible for their ability to activate the innate immune system through a cognate receptor. LPS, the TLR4 ligand, has been experimentally shown to be a potent adjuvant for vaccines, although its extreme toxicity prevents its use in humans. The adjuvant effect of LPS is solely dependent on TLR4-mediated, MyD88-dependent signaling. The dependence of LPS signaling on a single protein was revealed by the identification, in 1965, of a remarkable phenotype in mice marked by a very specific and profound insensitivity to LPS. The phenotype was first observed in mice of the C3H/HeJ strain, and neither the lethal effect of LPS nor any of the cellular effects of LPS occurred in these mice. This included the well-known adjuvant effect of LPS [59].

LPS is one of the most potent inducers of the immune system, and it was revealed later how it is recognized by a complex cascade of extracellular "pattern recognition receptors", which chaperone the LPS from the bacterial membrane to the transmembrane receptor TLR4. Two additional proteins have been identified that interact with TLR4 and are implicated in LPS recognition, MD-2 and CD14. MD-2 is a secreted molecule that physically interacts with TLR4. CD14 is a glycosylphosphatidylinositol-linked membrane protein (devoid of a cytoplasmic signaling domain) with LPS binding activity. Collectively TLR4, MD-2, and CD14 form a trimolecular LPS receptor complex [60, 61]. It is interesting to notice that TLR4 is the only known TLR which can utilize all five TIR- domain-containing adaptor proteins: MyD88, TIRAP, TRIF, TRAM, and SARM [19].

The lipid A portion of LPS was shown to be responsible for both the adjuvant and the toxic properties of LPS [62, 63]. Monophosphoryl lipid A (MPL) was developed by Ribi and co-workers as an attenuated derivative of LPS that lacks many of the endotoxic properties of the parent molecule and yet retains potent adjuvant and immunostimulating activities [64]. It was found that the removal of the phosphate group from the reducing end sugar of the lipid A disaccharide decreased the toxicity of the molecule 100 to 1000 folds without affecting the immunostimulating activity. The resulting derivative, which had only one phosphate group, was called monophosphoryl lipid A [64-66].

Subsequently, it was determined that removal of an ester-linked fatty acid group from the 3-position further reduced the pyrogenic properties without substantially affecting the adjuvant properties. The resulting 3-*O*-deacylated monophosphoryl lipid A (MPL), which is isolated and structurally derivatized from LPS of *Salmonella minnesota* R595, has proven to be a safe and effective vaccine adjuvant [67, 68]. In 1981, Takayama *et al.* isolated a nontoxic lipid A fraction, inducing tumor and metastases regression in animal models: ONO-4007 (sodium 2-deoxy-2-[3S-(9-phenylnonanoyloxy) tetradecanoyl] amino-3-O-(9-phenylnonanoyl)-Dglucopyranose 4-sulphate), developed by Ono Pharmaceutical Co (Osaka, Japan) [69]. Several animal studies also showed that ONO-4007 had remarkable and selective efficacy against TNF-α-sensitive tumors [70, 71]. Other partial agonists for TLR4, like DT-5461, SDZ MLR 953 and GLA-60, have been described, and although their development as anticancer agents was discontinued, their pharmacological profile deserves to be considered [72-74].

TLR5

TLR5 recognizes the bacterial protein flagellin, a 55-kDa monomeric protein constituent of flagella, which is the motility apparatus used by many microbial pathogens. The major protein constituent of bacteria flagella is also a potent activator of the innate immune responses [75]. The purification of culture supernatants of *Listeria monocytogenes* containing TLR5-stimulating activity led to the identification flagellin as the active component [75]. TLR5 is expressed by epithelial cells, monocytes, and immature DCs.

Biochemical analysis of *Salmonella* flagellin revealed that both conserved domains within both N- and C-termini of the protein are important in inducing proinflammatory responses in cultured intestinal epithelial cells [76, 77]. Recent studies also suggested that the hypervariable domain is not involved in proinflammatory activation. A recombinantly expressed flagellin mutant, in which the central hypervariable domain was deleted, was shown not to detract from its ability to induce NF-κB signaling, suggesting that the highly conserved N and C termini are sufficient for TLR5 activation [77]. A more recent study mapped the precise sequences responsible for the activity of flagellin on both the C- and N-termini of the protein [78].

CBLB502, a TLR5 agonist peptide, was found to protect against both major acute radiation toxicities of the hematopoietic system (HP) and gastrointestinal tract (GI). It was also found to reduce radiation toxicity without diminishing the therapeutic antitumor effect of radiation and without promoting radiation-induced carcinogenicity [79]. The results from this paper suggest that TLR5 agonists may be valuable as both adjuvants for cancer radiotherapy and protectants or mitigators for radiation emergencies [79]. In another study, a recombinant protein vaccine that was developed for West Nile virus, elicited a strong WNV-E-specific immunoglobulin G antibody response that neutralized viral infectivity and conferred protection against a lethal WNV challenge in C3H/HeN mice. The vaccine was designed by fusing a modified version of bacterial flagellin (STF2 Delta) to the EIII domain of the WNV envelope protein [80-82].

TLR7/8

TLR 7 and 8 were originally found to be expressed intracellularly and both interact with imidazoquinoline as a ligand in vitro. It has been suggested that TLR 7 and 8 differ in cell targets and cytokine production [83], but both induce the immune response to ssRNA human parechovirus 1 [84] and group B coxsackieviruses [85]. Both receptors are expressed on human bone marrow progenitor cells and ligation results in myeloid lineage differentiation [86]. TLR7 is also expressed on the surface of chronic lymphocytic leukemia cells [87]. Imiquimod, a synthetic TLR7 (porcine TLR7/8 [88]) agonist, is being used topically to treat a number of human pathological conditions such as genital warts and basal cell carcinoma. Imiquimod enhanced anti-melanoma vaccine effects in mice [89] and probably empowered DCs effector functions in basal cell carcinoma patients [90]. Further, a cryosurgery/imiquimod combined treatment of murine tumor had better curative outcome than cryosurgery alone [91]. Isatoribine, a selective TLR7 agonist, reduced plasma hepatitis C viral load in humans [92]. TLR7 had been suggested to be involved in murine autoimmunity and DC dysregulation [93, 94]. 852A, a specific TLR 7 agonist, is in phase I and II trial for advanced cancer patients [95, 96]. Treatment with S-28463, an imiquimod family member reduced allergic asthma in mice and associated symptoms in rats [97, 98] and TLR7/8 dual agonist 3M-011 protected rats against H3N2 influenza virus [99].

TLR9

Among the different TLRs, probably the greatest interest now revolves around TLR9, which recognizes bacterial and viral DNA with unmethylated CpG motifs in double stranded DNA (CpG DNA). TLR9 (in common with other members of the TLR family, including TLR3 and TLR7) is localized to the endosomal and lysosomal compartments within the cells, and is expressed primarily in B cells and pDCs [100-102].

The activation of TLR9 on pDCs induces the secretion of IFN-α, which drives the migration of pDCs to the marginal zone of lymph nodes, where they stimulate T cells [103] . Stimulation of TLR9 on human B cells induces proliferation, enhanced antigen specific antibody production, and IFN-α production. TLR9 activation, along with B cell antigen receptor, induces naïve B cell differentiation into plasma cells. Activation of TLR9 alone will drive the differentiation of memory B cells into plasma cells [104].

Three classes of CpG oligonucleotide ligands for TLR9 have been described, and can be distinguished by different sequence motifs and different abilities to stimulate IFN- α secretion and maturation of pDCs [105]. These structurally distinct classes of CpG ODN are class A, B and C. Class A, also known as type D, has a backbone made of phosphodiester bonds. This class has poly-G motifs flanking a central palindromic sequence. This class strongly induces IFN-α secretion from pDCs, induces pDC maturation, but is a weak stimulator of B cell proliferation [106, 107]. Class B CpG ODN, also known as type K, includes the most studied TLR9 agonist, CpG-7909. This class has a phosphorothioate backbone and is a weak inducer of IFN-α secretion but a strong stimulator of B cell proliferation and pDC maturation. The third class, known as class C, has intermediate immune properties between classes A and B, where it induces both IFN-α secretion and B cell stimulation. This class also has a phosphorothioate backbone, and its sequence combines structural elements of both CpG

ODN-A and CpG ODN-B. The most potent sequence M362, contains a 5′ end ‘TCGTCG’ motif and a ‘GTCGTT’ motif; both are present in CpG ODN-B (CPG-7909) and a palindromic sequence characteristic for CpG ODN-A (ODN 2216) [108, 109].

Short synthetic ODNs containing the immune stimulatory CpG motifs (CpG ODNs) have been used as a vaccine adjuvant in the treatment of cancer, asthma, allergies and infections [105, 110]. Several companies, such as Pfizer, Coley, Dynavax, and Idera, have produced CpG ODNs for clinical use (Table 3). A number of CpG ODNs have been or are being currently tested in multiple phase II and III trials as adjuvants to cancer vaccines and in combination with other conventional cancer therapies [110, 111]

TLR9 agonist CpG ODN treatment shifts the Th1/Th2 cytokine balance towards dominance of Th1 cytokines with increased synthesis of IgG2a, IFN, IL-6 and IL-12. This shift in cytokine balance makes CpG ODN useful for use in the treatment of several conditions where this balance is skewed towards a Th2 type reaction, in addition to its use as an adjuvant in vaccination protocols. Overproduction of Th2 cytokines and overexpression of IgE is the hallmark of allergic reactions[112], so the Th1-biased immune effect of CpG ODN has been applied in the development of allergy vaccines. A conjugate of CpG ODN to ragweed allergen has been evaluated as an allergy vaccine in allergic patients [113]. The immune response was successfully redirected towards a nonallergic and noninflammatory Th1 type, with a significant clinical benefit and reduced allergic symptoms.

CpG ODN was also used as a mono-therapy for different viral and bacterial infections. CpG 10101 (class C ODN) was tested in a Phase Ib trial for treating hepatitis C virus positive patients, where doses of up to 0.75 mg/kg were given weekly. The use of CPG 10101 caused a dose-dependent decrease in blood viral RNA levels, which was associated with biomarkers of TLR9 activation, like NK cell activation and serum IFN-α.

Although activation of the innate immune system could be demonstrated by cytokine and chemokine release, further studies are needed to demonstrate an adaptive immune response, including a virus-specific T cell response. CpG ODNs have been used as adjuvants for hepatitis B vaccination in at least two studies [114, 115]. In a randomized, double-blind Phase I dose escalation study, healthy volunteers were vaccinated at 0, 4 and 24 weeks by intramuscular injection with Engerix-B (GlaxoSmithKline) mixed with saline (control) or with CPG-7909.

Table 3. TLR9 agonists in clinical use

CpG ODN	Class	Company	Uses
PF 3512676	B	Pfizer	Cancer monotherapy or in combination with other cancer therapies, vaccines
			Hepatitis C virus
ODN CpG	C	Coley	Allergies, combination with monoclonal
10101	B	Dynavax	antibodies in cancer therapy, vaccines
ODN 1018 ISS			Cancer monotherapy
			Combination with chemotherapy
IMO-2055		Idera	

Anti-HBs appeared significantly sooner and were significantly higher in CPG-7909 recipients compared to control subjects, and most CPG-7909 vaccinated subjects developed protective levels of anti-HB IgG within just two weeks of the priming vaccine dose. A trend towards higher rates of positive cytotoxic T-cell lymphocyte responses was noted in the two higher dose groups of CPG-7909 compared to controls [115].

CONCLUSION

Our ability to dissect the immune response against pathogens, accompanied with advances in the field of molecular biology, have led to the introduction of more specific treatments that could substitute the old and crude way of using live attenuated pathogens, whole inactivated organisms, and inactivated toxins, for vaccines, which resulted in many undesirable side effects. Instead, recombinant proteins and synthetic peptides were used in newer vaccines, but they yielded poor immune responses due to their poor immunogenicity, unless combined with immunostimulatory adjuvants to induce a potent immune response.

Advances in our understanding of TLR functions enabled scientists to develop and use TLR agonists as vaccine adjuvants. In fact, many promising experimental vaccine strategies for both infectious diseases and for malignancies include the use of TLR ligands as adjuvants. Also, TLR agonists could be a very important asset in the development of non-antibiotic agents for the treatment of infection with multi-drug resistant pathogens. Moreover, our understanding of the downstream signaling of TLRs, and how they exert their functions could allow us to use reagents to enhance the signaling pathways, or modulate the TLR system to better fight infection or cancer.

REFERENCES

[1] Seya T, Akazawa T, Tsujita T, Matsumoto M. Role of Toll-like receptors in adjuvant-augmented immune therapies. *Evid Based Complement Alternat Med.* 2006 Mar;3(1):31-8; discussion 133-7.

[2] Ishii KJ, Akira S. Toll or toll-free adjuvant path toward the optimal vaccine development. *Journal of clinical immunology*. 2007 Jul;27(4):363-71.

[3] Trinchieri G, Sher A. Cooperation of Toll-like receptor signals in innate immune defence. *Nature reviews*. 2007 Mar;7(3):179-90.

[4] Creagh EM, O'Neill LA. TLRs, NLRs and RLRs: a trinity of pathogen sensors that co-operate in innate immunity. *Trends in immunology*. 2006 Aug;27(8):352-7.

[5] Meylan E, Tschopp J, Karin M. Intracellular pattern recognition receptors in the host response. *Nature*. 2006 Jul 6;442(7098):39-44.

[6] Takeuchi O, Akira S. Signaling pathways activated by microorganisms. *Current opinion in cell biology*. 2007 Apr;19(2):185-91.

[7] Lemaitre B. The road to Toll. *Nature reviews*. 2004 Jul;4(7):521-7.

[8] Medzhitov R, Preston-Hurlburt P, Janeway CA, Jr. A human homologue of the Drosophila Toll protein signals activation of adaptive immunity. *Nature*. 1997 Jul 24;388(6640):394-7.

[9] van Duin D, Medzhitov R, Shaw AC. Triggering TLR signaling in vaccination. *Trends in immunology*. 2006 Jan;27(1):49-55.

[10] Bell JK, Mullen GE, Leifer CA, Mazzoni A, Davies DR, Segal DM. Leucine-rich repeats and pathogen recognition in Toll-like receptors. *Trends in immunology*. 2003 Oct;24(10):528-33.

[11] Janeway CA, Jr., Medzhitov R. Innate immune recognition. *Annual review of immunology*. 2002;20:197-216.

[12] Medzhitov R. Toll-like receptors and innate immunity. *Nature reviews*. 2001 Nov;1(2):135-45.

[13] Akira S. Toll-like receptors and innate immunity. *Advances in immunology*. 2001;78:1-56.

[14] Akira S, Takeda K. Toll-like receptor signalling. *Nature reviews*. 2004 Jul;4(7):499-511.

[15] Slack JL, Schooley K, Bonnert TP, Mitcham JL, Qwarnstrom EE, Sims JE, et al. Identification of two major sites in the type I interleukin-1 receptor cytoplasmic region responsible for coupling to pro-inflammatory signaling pathways. *The Journal of biological chemistry*. 2000 Feb 18;275(7):4670-8.

[16] Wiemann B, Starnes CO. Coley's toxins, tumor necrosis factor and cancer research: a historical perspective. *Pharmacology and therapeutics*. 1994;64(3):529-64.

[17] Brightbill HD, Libraty DH, Krutzik SR, Yang RB, Belisle JT, Bleharski JR, et al. Host defense mechanisms triggered by microbial lipoproteins through toll-like receptors. *Science (*New York, NY. 1999 Jul 30;285(5428):732-6.

[18] West AP, Koblansky AA, Ghosh S. Recognition and signaling by toll-like receptors. *Annual review of cell and developmental biology*. 2006;22:409-37.

[19] O'Neill LA, Bowie AG. The family of five: TIR-domain-containing adaptors in Toll-like receptor signalling. *Nature reviews*. 2007 May;7(5):353-64.

[20] Yamamoto M, Sato S, Hemmi H, Hoshino K, Kaisho T, Sanjo H, et al. Role of adaptor TRIF in the MyD88-independent toll-like receptor signaling pathway. *Science* (New York, NY. 2003 Aug 1;301(5633):640-3.

[21] Hemmi H, Kaisho T, Takeuchi O, Sato S, Sanjo H, Hoshino K, et al. Small anti-viral compounds activate immune cells via the TLR7 MyD88-dependent signaling pathway. *Nature immunology*. 2002 Feb;3(2):196-200.

[22] Hoshino K, Kaisho T, Iwabe T, Takeuchi O, Akira S. Differential involvement of IFN-beta in Toll-like receptor-stimulated dendritic cell activation. *International immunology*. 2002 Oct;14(10):1225-31.

[23] Kawai T, Takeuchi O, Fujita T, Inoue J, Muhlradt PF, Sato S, et al. Lipopolysaccharide stimulates the MyD88-independent pathway and results in activation of IFN-regulatory factor 3 and the expression of a subset of lipopolysaccharide-inducible genes. *J Immunol.* 2001 Nov 15;167(10):5887-94.

[24] Horng T, Barton GM, Flavell RA, Medzhitov R. The adaptor molecule TIRAP provides signalling specificity for Toll-like receptors. *Nature*. 2002 Nov 21;420(6913):329-33.

[25] Yamamoto M, Sato S, Hemmi H, Sanjo H, Uematsu S, Kaisho T, et al. Essential role for TIRAP in activation of the signalling cascade shared by TLR2 and TLR4. *Nature.* 2002 Nov 21;420(6913):324-9.

[26] Yamamoto M, Sato S, Hemmi H, Uematsu S, Hoshino K, Kaisho T, et al. TRAM is specifically involved in the Toll-like receptor 4-mediated MyD88-independent signaling pathway. *Nature immunology*. 2003 Nov;4(11):1144-50.

[27] Chen ZJ. Ubiquitin signalling in the NF-kappaB pathway. *Nature cell biology*. 2005 Aug;7(8):758-65.

[28] Sato S, Sanjo H, Takeda K, Ninomiya-Tsuji J, Yamamoto M, Kawai T, et al. Essential function for the kinase TAK1 in innate and adaptive immune responses. *Nature immunology*. 2005 Nov;6(11):1087-95.

[29] Kawai T, Akira S. TLR signaling. *Seminars in immunology*. 2007 Feb;19(1):24-32.

[30] Kawai T, Adachi O, Ogawa T, Takeda K, Akira S. Unresponsiveness of MyD88-deficient mice to endotoxin. *Immunity*. 1999 Jul;11(1):115-22.

[31] Oshiumi H, Matsumoto M, Funami K, Akazawa T, Seya T. TICAM-1, an adaptor molecule that participates in Toll-like receptor 3-mediated interferon-beta induction. *Nature immunology*. 2003 Feb;4(2):161-7.

[32] Iwasaki A, Medzhitov R. Toll-like receptor control of the adaptive immune responses. *Nature immunology*. 2004 Oct;5(10):987-95.

[33] Schulz O, Diebold SS, Chen M, Naslund TI, Nolte MA, Alexopoulou L, et al. Toll-like receptor 3 promotes cross-priming to virus-infected cells. *Nature*. 2005 Feb 24;433(7028):887-92.

[34] Heit A, Schmitz F, O'Keeffe M, Staib C, Busch DH, Wagner H, et al. Protective CD8 T cell immunity triggered by CpG-protein conjugates competes with the efficacy of live vaccines. *J. Immunol*. 2005 Apr 1;174(7):4373-80.

[35] Spellberg B, Edwards JE, Jr. Type 1/Type 2 immunity in infectious diseases. *Clin. Infect. Dis*. 2001 Jan;32(1):76-102.

[36] Trinchieri G. Interleukin-12 and the regulation of innate resistance and adaptive immunity. *Nature reviews*. 2003 Feb;3(2):133-46.

[37] Gautier G, Humbert M, Deauvieau F, Scuiller M, Hiscott J, Bates EE, et al. A type I interferon autocrine-paracrine loop is involved in Toll-like receptor-induced interleukin-12p70 secretion by dendritic cells. *The Journal of experimental medicine*. 2005 May 2;201(9):1435-46.

[38] Pasare C, Medzhitov R. Toll pathway-dependent blockade of CD4+CD25+ T cell-mediated suppression by dendritic cells. *Science* (New York, NY. 2003 Feb 14;299(5609):1033-6.

[39] Yang Y, Huang CT, Huang X, Pardoll DM. Persistent Toll-like receptor signals are required for reversal of regulatory T cell-mediated CD8 tolerance. *Nature immunology*. 2004 May;5(5):508-15.

[40] Peng G, Guo Z, Kiniwa Y, Voo KS, Peng W, Fu T, et al. Toll-like receptor 8-mediated reversal of CD4+ regulatory T cell function. *Science* (New York, NY. 2005 Aug 26;309(5739):1380-4.

[41] Borst J, Hendriks J, Xiao Y. CD27 and CD70 in T cell and B cell activation. *Current opinion in immunology*. 2005 Jun;17(3):275-81.

[42] Diehl L, den Boer AT, Schoenberger SP, van der Voort EI, Schumacher TN, Melief CJ, et al. CD40 activation in vivo overcomes peptide-induced peripheral cytotoxic T-lymphocyte tolerance and augments anti-tumor vaccine efficacy. *Nature medicine*. 1999 Jul;5(7):774-9.

[43] Chen Z, Dehm S, Bonham K, Kamencic H, Juurlink B, Zhang X, et al. DNA array and biological characterization of the impact of the maturation status of mouse dendritic cells on their phenotype and antitumor vaccination efficacy. *Cellular immunology*. 2001 Nov 25;214(1):60-71.

[44] Hertz CJ, Kiertscher SM, Godowski PJ, Bouis DA, Norgard MV, Roth MD, et al. Microbial lipopeptides stimulate dendritic cell maturation via Toll-like receptor 2. *J. Immunol.* 2001 Feb 15;166(4):2444-50.

[45] Cella M, Engering A, Pinet V, Pieters J, Lanzavecchia A. Inflammatory stimuli induce accumulation of MHC class II complexes on dendritic cells. *Nature*. 1997 Aug 21;388(6644):782-7.

[46] Aliprantis AO, Yang RB, Mark MR, Suggett S, Devaux B, Radolf JD, et al. Cell activation and apoptosis by bacterial lipoproteins through toll-like receptor-2. *Science* (New York, NY. 1999 Jul 30;285(5428):736-9.

[47] Aliprantis AO, Yang RB, Weiss DS, Godowski P, Zychlinsky A. The apoptotic signaling pathway activated by Toll-like receptor-2. *The EMBO journal*. 2000 Jul 3;19(13):3325-36.

[48] Lien E, Sellati TJ, Yoshimura A, Flo TH, Rawadi G, Finberg RW, et al. Toll-like receptor 2 functions as a pattern recognition receptor for diverse bacterial products. *The Journal of biological chemistry*. 1999 Nov 19;274(47):33419-25.

[49] Schwandner R, Dziarski R, Wesche H, Rothe M, Kirschning CJ. Peptidoglycan- and lipoteichoic acid-induced cell activation is mediated by toll-like receptor 2. *The Journal of biological chemistry*. 1999 Jun 18;274(25):17406-9.

[50] Yoshimura A, Lien E, Ingalls RR, Tuomanen E, Dziarski R, Golenbock D. Cutting edge: recognition of Gram-positive bacterial cell wall components by the innate immune system occurs via Toll-like receptor 2. *J. Immunol.* 1999 Jul 1;163(1):1-5.

[51] Underhill DM, Ozinsky A, Smith KD, Aderem A. Toll-like receptor-2 mediates mycobacteria-induced proinflammatory signaling in macrophages. *Proceedings of the National Academy of Sciences of the United States of America*. 1999 Dec 7;96(25):14459-63.

[52] Lehner MD, Morath S, Michelsen KS, Schumann RR, Hartung T. Induction of cross-tolerance by lipopolysaccharide and highly purified lipoteichoic acid via different Toll-like receptors independent of paracrine mediators. *J. Immunol.* 2001 Apr 15;166(8):5161-7.

[53] Means TK, Wang S, Lien E, Yoshimura A, Golenbock DT, Fenton MJ. Human toll-like receptors mediate cellular activation by Mycobacterium tuberculosis. *J. Immunol.* 1999 Oct 1;163(7):3920-7.

[54] Takeda K, Kaisho T, Akira S. Toll-like receptors. *Annual review of immunology*. 2003;21:335-76.

[55] Steere AC, Sikand VK, Meurice F, Parenti DL, Fikrig E, Schoen RT, et al. Vaccination against Lyme disease with recombinant Borrelia burgdorferi outer-surface lipoprotein A with adjuvant. Lyme Disease Vaccine Study Group. *The New England journal of medicine*. 1998 Jul 23;339(4):209-15.

[56] Alexopoulou L, Thomas V, Schnare M, Lobet Y, Anguita J, Schoen RT, et al. Hyporesponsiveness to vaccination with Borrelia burgdorferi OspA in humans and in TLR1- and TLR2-deficient mice. *Nature medicine*. 2002 Aug;8(8):878-84.

[57] Ozinsky A, Underhill DM, Fontenot JD, Hajjar AM, Smith KD, Wilson CB, et al. The repertoire for pattern recognition of pathogens by the innate immune system is defined by cooperation between toll-like receptors. *Proceedings of the National Academy of Sciences of the United States of America*. 2000 Dec 5;97(25):13766-71.

[58] Schneider C, Schmidt T, Ziske C, Tiemann K, Lee KM, Uhlinsky V, et al. Tumour suppression induced by the macrophage activating lipopeptide MALP-2 in an ultrasound guided pancreatic carcinoma mouse model. *Gut*. 2004 Mar;53(3):355-61.

[59] Beutler B, Jiang Z, Georgel P, Crozat K, Croker B, Rutschmann S, et al. Genetic analysis of host resistance: Toll-like receptor signaling and immunity at large. *Annual review of immunology*. 2006;24:353-89.

[60] da Silva Correia J, Soldau K, Christen U, Tobias PS, Ulevitch RJ. Lipopolysaccharide is in close proximity to each of the proteins in its membrane receptor complex. transfer from CD14 to TLR4 and MD-2. *The Journal of biological chemistry*. 2001 Jun 15;276(24):21129-35.

[61] da Silva Correia J, Ulevitch RJ. MD-2 and TLR4 N-linked glycosylations are important for a functional lipopolysaccharide receptor. *The Journal of biological chemistry*. 2002 Jan 18;277(3):1845-54.

[62] Galanos C, Luderitz O, Rietschel ET, Westphal O, Brade H, Brade L, et al. Synthetic and natural Escherichia coli free lipid A express identical endotoxic activities. *European journal of biochemistry / FEBS*. 1985 Apr 1;148(1):1-5.

[63] Takada H, Kotani S. Structural requirements of lipid A for endotoxicity and other biological activities. *Critical reviews in microbiology*. 1989;16(6):477-523.

[64] Takayama K, Qureshi N, Ribi E, Cantrell JL. Separation and characterization of toxic and nontoxic forms of lipid *A. Reviews of infectious diseases*. 1984 Jul-Aug;6(4):439-43.

[65] Qureshi N, Mascagni P, Ribi E, Takayama K. Monophosphoryl lipid A obtained from lipopolysaccharides of Salmonella minnesota R595. Purification of the dimethyl derivative by high performance liquid chromatography and complete structural determination. *The Journal of biological chemistry*. 1985 May 10;260(9):5271-8.

[66] Madonna GS, Peterson JE, Ribi EE, Vogel SN. Early-phase endotoxin tolerance: induction by a detoxified lipid A derivative, monophosphoryl lipid A. *Infection and immunity*. 1986 Apr;52(1):6-11.

[67] Thoelen S, Van Damme P, Mathei C, Leroux-Roels G, Desombere I, Safary A, et al. Safety and immunogenicity of a hepatitis B vaccine formulated with a novel adjuvant system. *Vaccine*. 1998 Apr;16(7):708-14.

[68] Stoute JA, Slaoui M, Heppner DG, Momin P, Kester KE, Desmons P, et al. A preliminary evaluation of a recombinant circumsporozoite protein vaccine against Plasmodium falciparum malaria. RTS,S Malaria Vaccine Evaluation Group. The New *England journal of medicine*. 1997 Jan 9;336(2):86-91.

[69] Takayama K, Ribi E, Cantrell JL. Isolation of a nontoxic lipid A fraction containing tumor regression activity. *Cancer research*. 1981 Jul;41(7):2654-7.

[70] Yang D, Satoh M, Ueda H, Tsukagoshi S, Yamazaki M. Activation of tumor-infiltrating macrophages by a synthetic lipid A analog (ONO-4007) and its implication in antitumor effects. *Cancer Immunol Immunother*. 1994 May;38(5):287-93.

[71] Kuramitsu Y, Nishibe M, Ohiro Y, Matsushita K, Yuan L, Obara M, et al. A new synthetic lipid A analog, ONO-4007, stimulates the production of tumor necrosis

factor-alpha in tumor tissues, resulting in the rejection of transplanted rat hepatoma cells. *Anti-cancer drugs*. 1997 Jun;8(5):500-8.

[72] Garay RP, Viens P, Bauer J, Normier G, Bardou M, Jeannin JF, et al. Cancer relapse under chemotherapy: why TLR2/4 receptor agonists can help. *European journal of pharmacology*. 2007 Jun 1;563(1-3):1-17.

[73] Sato K, Yoo YC, Mochizuki M, Saiki I, Takahashi TA, Azuma I. Inhibition of tumor-induced angiogenesis by a synthetic lipid A analogue with low endotoxicity, DT-5461. *Jpn J Cancer Res*. 1995 Apr;86(4):374-82.

[74] Kiani A, Tschiersch A, Gaboriau E, Otto F, Seiz A, Knopf HP, et al. Downregulation of the proinflammatory cytokine response to endotoxin by pretreatment with the nontoxic lipid A analog SDZ MRL 953 in cancer patients. *Blood*. 1997 Aug 15;90(4):1673-83.

[75] Hayashi F, Smith KD, Ozinsky A, Hawn TR, Yi EC, Goodlett DR, et al. The innate immune response to bacterial flagellin is mediated by Toll-like receptor 5. *Nature*. 2001 Apr 26;410(6832):1099-103.

[76] Eaves-Pyles T, Murthy K, Liaudet L, Virag L, Ross G, Soriano FG, et al. Flagellin, a novel mediator of Salmonella-induced epithelial activation and systemic inflammation: I kappa B alpha degradation, induction of nitric oxide synthase, induction of proinflammatory mediators, and cardiovascular dysfunction. *J. Immunol*. 2001 Jan 15;166(2):1248-60.

[77] Eaves-Pyles TD, Wong HR, Odoms K, Pyles RB. Salmonella flagellin-dependent proinflammatory responses are localized to the conserved amino and carboxyl regions of the protein. *J. Immunol*. 2001 Dec 15;167(12):7009-16.

[78] Murthy KG, Deb A, Goonesekera S, Szabo C, Salzman AL. Identification of conserved domains in Salmonella muenchen flagellin that are essential for its ability to activate TLR5 and to induce an inflammatory response in vitro. *The Journal of biological chemistry*. 2004 Feb 13;279(7):5667-75.

[79] Burdelya LG, Krivokrysenko VI, Tallant TC, Strom E, Gleiberman AS, Gupta D, et al. An agonist of toll-like receptor 5 has radioprotective activity in mouse and primate models. *Science* (New York, NY. 2008 Apr 11;320(5873):226-30.

[80] McDonald WF, Huleatt JW, Foellmer HG, Hewitt D, Tang J, Desai P, et al. A West Nile virus recombinant protein vaccine that coactivates innate and adaptive immunity. *The Journal of infectious diseases*. 2007 Jun 1;195(11):1607-17.

[81] Huleatt JW, Jacobs AR, Tang J, Desai P, Kopp EB, Huang Y, et al. Vaccination with recombinant fusion proteins incorporating Toll-like receptor ligands induces rapid cellular and humoral immunity. *Vaccine*. 2007 Jan 8;25(4):763-75.

[82] Huleatt JW, Nakaar V, Desai P, Huang Y, Hewitt D, Jacobs A, et al. Potent immunogenicity and efficacy of a universal influenza vaccine candidate comprising a recombinant fusion protein linking influenza M2e to the TLR5 ligand flagellin. *Vaccine*. 2008 Jan 10;26(2):201-14.

[83] Gorden KB, Gorski KS, Gibson SJ, Kedl RM, Kieper WC, Qiu X, et al. Synthetic TLR agonists reveal functional differences between human TLR7 and TLR8. *J. Immunol*. 2005 Feb 1;174(3):1259-68.

[84] Triantafilou K, Vakakis E, Orthopoulos G, Ahmed MA, Schumann C, Lepper PM, et al. TLR8 and TLR7 are involved in the host's immune response to human parechovirus 1. *European journal of immunology*. 2005 Aug;35(8):2416-23.

[85] Triantafilou K, Orthopoulos G, Vakakis E, Ahmed MA, Golenbock DT, Lepper PM, et al. Human cardiac inflammatory responses triggered by Coxsackie B viruses are mainly Toll-like receptor (TLR) 8-dependent. *Cell Microbiol.* 2005 Aug;7(8):1117-26.

[86] Sioud M, Floisand Y, Forfang L, Lund-Johansen F. Signaling through toll-like receptor 7/8 induces the differentiation of human bone marrow CD34+ progenitor cells along the myeloid lineage. *Journal of molecular biology.* 2006 Dec 15;364(5):945-54.

[87] Garantziotis S, Hollingsworth JW, Zaas AK, Schwartz DA. The effect of toll-like receptors and toll-like receptor genetics in human disease. *Annual review of medicine.* 2008;59:343-59.

[88] Zhu J, Lai K, Brownile R, Babiuk LA, Mutwiri GK. Porcine TLR8 and TLR7 are both activated by a selective TLR7 ligand, imiquimod. *Molecular immunology.* 2008 Jun;45(11):3238-43.

[89] Craft N, Bruhn KW, Nguyen BD, Prins R, Lin JW, Liau LM, et al. The TLR7 agonist imiquimod enhances the anti-melanoma effects of a recombinant Listeria monocytogenes vaccine. *J. Immunol.* 2005 Aug 1;175(3):1983-90.

[90] Stary G, Bangert C, Tauber M, Strohal R, Kopp T, Stingl G. Tumoricidal activity of TLR7/8-activated inflammatory dendritic cells. *The Journal of experimental medicine.* 2007 Jun 11;204(6):1441-51.

[91] Redondo P, del Olmo J, Lopez-Diaz de Cerio A, Inoges S, Marquina M, Melero I, et al. Imiquimod enhances the systemic immunity attained by local cryosurgery destruction of melanoma lesions. *The Journal of investigative dermatology.* 2007 Jul;127(7):1673-80.

[92] Horsmans Y, Berg T, Desager JP, Mueller T, Schott E, Fletcher SP, et al. Isatoribine, an agonist of TLR7, reduces plasma virus concentration in chronic hepatitis C infection. *Hepatology (*Baltimore, Md. 2005 Sep;42(3):724-31.

[93] Deane JA, Pisitkun P, Barrett RS, Feigenbaum L, Town T, Ward JM, et al. Control of toll-like receptor 7 expression is essential to restrict autoimmunity and dendritic cell proliferation. *Immunity.* 2007 Nov;27(5):801-10.

[94] Barrat FJ, Meeker T, Chan JH, Guiducci C, Coffman RL. Treatment of lupus-prone mice with a dual inhibitor of TLR7 and TLR9 leads to reduction of autoantibody production and amelioration of disease symptoms. *European journal of immunology.* 2007 Dec;37(12):3582-6.

[95] Dudek AZ, Yunis C, Harrison LI, Kumar S, Hawkinson R, Cooley S, et al. First in human phase I trial of 852A, a novel systemic toll-like receptor 7 agonist, to activate innate immune responses in patients with advanced cancer. *Clin. Cancer Res.* 2007 Dec 1;13(23):7119-25.

[96] Dummer R, Hauschild A, Becker JC, Grob JJ, Schadendorf D, Tebbs V, et al. An exploratory study of systemic administration of the toll-like receptor-7 agonist 852A in patients with refractory metastatic melanoma. *Clin. Cancer Res.* 2008 Feb 1;14(3):856-64.

[97] Moisan J, Camateros P, Thuraisingam T, Marion D, Koohsari H, Martin P, et al. TLR7 ligand prevents allergen-induced airway hyperresponsiveness and eosinophilia in allergic asthma by a MYD88-dependent and MK2-independent pathway. *American journal of physiology.* 2006 May;290(5):L987-95.

[98] Camateros P, Tamaoka M, Hassan M, Marino R, Moisan J, Marion D, et al. Chronic asthma-induced airway remodeling is prevented by toll-like receptor-7/8 ligand

S28463. *American journal of respiratory and critical care medicine*. 2007 Jun 15;175(12):1241-9.

[99] Hammerbeck DM, Burleson GR, Schuller CJ, Vasilakos JP, Tomai M, Egging E, et al. Administration of a dual toll-like receptor 7 and toll-like receptor 8 agonist protects against influenza in rats. *Antiviral research*. 2007 Jan;73(1):1-11.

[100] Leifer CA, Kennedy MN, Mazzoni A, Lee C, Kruhlak MJ, Segal DM. TLR9 is localized in the endoplasmic reticulum prior to stimulation. *J. Immunol.* 2004 Jul 15;173(2):1179-83.

[101] Latz E, Schoenemeyer A, Visintin A, Fitzgerald KA, Monks BG, Knetter CF, et al. TLR9 signals after translocating from the ER to CpG DNA in the lysosome. *Nature* immunology. 2004 Feb;5(2):190-8.

[102] Ahmad-Nejad P, Hacker H, Rutz M, Bauer S, Vabulas RM, Wagner H. Bacterial CpG-DNA and lipopolysaccharides activate Toll-like receptors at distinct cellular compartments. *European journal of immunology*. 2002 Jul;32(7):1958-68.

[103] Faith A, Peek E, McDonald J, Urry Z, Richards DF, Tan C, et al. Plasmacytoid dendritic cells from human lung cancer draining lymph nodes induce Tc1 responses. *Am. J. Respir. Cell Mol. Biol.* 2007 Mar;36(3):360-7.

[104] Bernasconi NL, Onai N, Lanzavecchia A. A role for Toll-like receptors in acquired immunity: up-regulation of TLR9 by BCR triggering in naive B cells and constitutive expression in memory B cells. *Blood.* 2003 Jun 1;101(11):4500-4.

[105] Krieg AM. Therapeutic potential of Toll-like receptor 9 activation. *Nature reviews*. 2006 Jun;5(6):471-84.

[106] Krug A, Rothenfusser S, Selinger S, Bock C, Kerkmann M, Battiany J, et al. CpG-A oligonucleotides induce a monocyte-derived dendritic cell-like phenotype that preferentially activates CD8 T cells. *J. Immunol.* 2003 Apr 1;170(7):3468-77.

[107] Kerkmann M, Costa LT, Richter C, Rothenfusser S, Battiany J, Hornung V, et al. Spontaneous formation of nucleic acid-based nanoparticles is responsible for high interferon-alpha induction by CpG-A in plasmacytoid dendritic cells. *The Journal of biological chemistry*. 2005 Mar 4;280(9):8086-93.

[108] Hartmann G, Battiany J, Poeck H, Wagner M, Kerkmann M, Lubenow N, et al. Rational design of new CpG oligonucleotides that combine B cell activation with high IFN-alpha induction in plasmacytoid dendritic cells. *European journal of immunology*. 2003 Jun;33(6):1633-41.

[109] Krieg AM. CpG motifs in bacterial DNA and their immune effects. *Annual review of immunology*. 2002;20:709-60.

[110] Murad YM, Clay TM, Lyerly HK, Morse MA. CPG-7909 (PF-3512676, ProMune): toll-like receptor-9 agonist in cancer therapy. *Expert opinion on biological therapy*. 2007 Aug;7(8):1257-66.

[111] Krieg AM. Development of TLR9 agonists for cancer therapy. *The Journal of clinical investigation*. 2007 May;117(5):1184-94.

[112] Hayashi T, Raz E. TLR9-based immunotherapy for allergic disease. *The American journal of medicine*. 2006 Oct;119(10):897 e1-6.

[113] Creticos PS, Schroeder JT, Hamilton RG, Balcer-Whaley SL, Khattignavong AP, Lindblad R, et al. Immunotherapy with a ragweed-toll-like receptor 9 agonist vaccine for allergic rhinitis. *The New England journal of medicine*. 2006 Oct 5;355(14):1445-55.

[114] Halperin SA, Van Nest G, Smith B, Abtahi S, Whiley H, Eiden JJ. A phase I study of the safety and immunogenicity of recombinant hepatitis B surface antigen co-administered with an immunostimulatory phosphorothioate oligonucleotide adjuvant. *Vaccine.* 2003 Jun 2;21(19-20):2461-7.

[115] Cooper CL, Davis HL, Morris ML, Efler SM, Adhami MA, Krieg AM, et al. CPG 7909, an immunostimulatory TLR9 agonist oligodeoxynucleotide, as adjuvant to Engerix-B HBV vaccine in healthy adults: a double-blind phase I/II study. *Journal of*

In: Advances in Medicine and Biology. Volume 20
Editor: Leon V. Berhardt

ISBN 978-1-61209-135-8

Chapter 8

CURRENT TRENDS IN GLAUCOMA: WHAT'S ABOUT NEUROPROTECTION?

Sergio Pinar[1,2] and Elena Vecino[1*]

[1] Cell Biology Department (University of the Basque Country UPV/EHU)
[2]Ophthalmology at Cruces Hospital, Bilbao, Spain

ABSTRACT

Neuroprotection was initially investigated for disorders of the central nervous system such as amyotrophic lateral sclerosis, Alzheimer's disease, Parkinson and head trauma, but only few therapies have been approved (nemantine for Azheimer's dementia).

Neuroprotection is a process that attempts to preserve the remaining cells that are still vulnerable to damage. The main aim of neuroprotective therapy is to apply pharmacologic or other means to attenuate the hostility of the environment surrounding the degenerating cells, or to supply the cells with the tools to deal with this aggression, providing resilience to the insult. By definition, glaucoma neuroprotection must be considered independent of intraocular pressure lowering, and the target neurons should be in the central visual pathway, including retinal ganglion cells.

Several agents have been reported neuroprotective in glaucoma, both in clinical studies, such as Ca2+ channel blockers, and in experimental studies, such as betaxolol, brimonidine, NMDA antagonists, Nitric Oxide Synthase inhibitors, neurotrophins and ginkgo biloba extract. However, to establish neuroprotective drugs for glaucoma, well-designed clinical trials are required, specially randomized clinical trials comparing neuroprotective treatments to placebo, and although there is laboratory evidence for glaucoma neuroprotection by several drugs, we still evidence from randomized clinical trials.

* Corresponding Author: Elena Vecino, Departamento de Biología Celular e Histología, Grupo de Oftalmo-Biología Experimental (GOBE), Facultad de Medicina, Universidad del País Vasco, 48940 Leioa, Vizcaya, SPAIN, e.mail: elena.vecino@ehu.es

The ideal anti-glaucoma drug would be one that when applied topically, reduces IOP, but also probes to reach the retina in appropriate amounts, and activates specific receptors in the retina to attenuate retinal ganglion cell death.

INTRODUCTION

1. Epidemiology And Risk Factors In Glaucoma

Glaucoma is an optic neuropathy that affects nearly 60 million people, being expected to reach 79,6 million by year 2020 [Quigley et al., 2006]. Nowadays it is considered the second leading cause of blindness worldwide. Glaucoma is associated with selective death of retinal ganglion cells (RGC) [Quigley, 1995; Mittag et al., 2000; Cordeiro et al., 2004; Libby et al., 2005; García-Valenzuela et al., 2005; Urcola et al., 2006; Hernández et al., 2008] and structural changes in optic nerve head. This loss of RGC and previous axonal dysfunction are thought to be the cause of visual field changes in those affected by glaucoma [Buckingham et al., 2008].

Elevated intraocular pressure (IOP) has been established as the main risk factor for developing glaucoma. It is known that ocular hypertension is due to impairment in the removal of aqueous humour via the trabecular meshwork of multiple causes (Figures 1 and 2). That's why a frequent and repetitive examination of IOP has great value when following patients diagnosed of glaucoma. However, single IOP measurements do not necessarily provide adequate information about pressure, as the normal diurnal variation in IOP can be on the order of 6 mm Hg, and even larger in eyes with glaucoma, reaching 30 mm Hg [Newell and Krill, 1964; Kitazawa and Horie, 1975]. Furthermore, the fact that measurement of IOP alone is of limited value is reinforced by the existence of patients with NTG, where we can find evidence of optic nerve damage, and IOP measurements within the normal statistical range. This concept emphasizes the important role played by other pressure-indepentent risk factors in the development and progression of glaucomatous neuropathy. In addition, progression of glaucomatous damage continues in many patients (up to one-sixth of patients with glaucoma) despite attenuation of the initial injury (control of initially raised IOP) [Brubaker, 1996; Lisegang, 1996].

Electroretinography in glaucoma eyes can also show deterioration. ERG demonstrates an attenuation of both a-waves and b-waves after the induction of ocular hypertension using cauterization of episcleral vein experimental glaucoma model in animals [Mittag et al., 2000; Bayer et al., 2001], and also a deterioration of inner retinal function, measured by an attenuation of positive and negative scotopic threshold response (STR) [Li et al., 2006].

Primary open-glaucoma (POAG) is considered to be the most common subtype of glaucoma, and its pathogenesis remains a fascinating area for new research studies. It is almost certain that POAG develops in a multifactorial manner, and some of the risk factors may determine the appearance of the damaged glaucomatous optic nerve head [Broadway et al., 1999]. Numerous potential secondary risk factors for the development of glaucoma have been studied. First of all, high ocular tension (above the normal population level, 21 mmHg) is established as the primary risk factor [Hart et al., 1979; Leibowitz et al., 1980; Armaly et al., 1980; Drance et al., 1981; Sommer, 1989; Quigley et al., 1994]. However, the existence of patients with normal-tension glaucoma (NTG) has suggested that secondary factors play an important role in the development of glaucomatous optic neuropathy in these patients whose IOPs lie within the normal statistical range [Broadway et al., 1999].

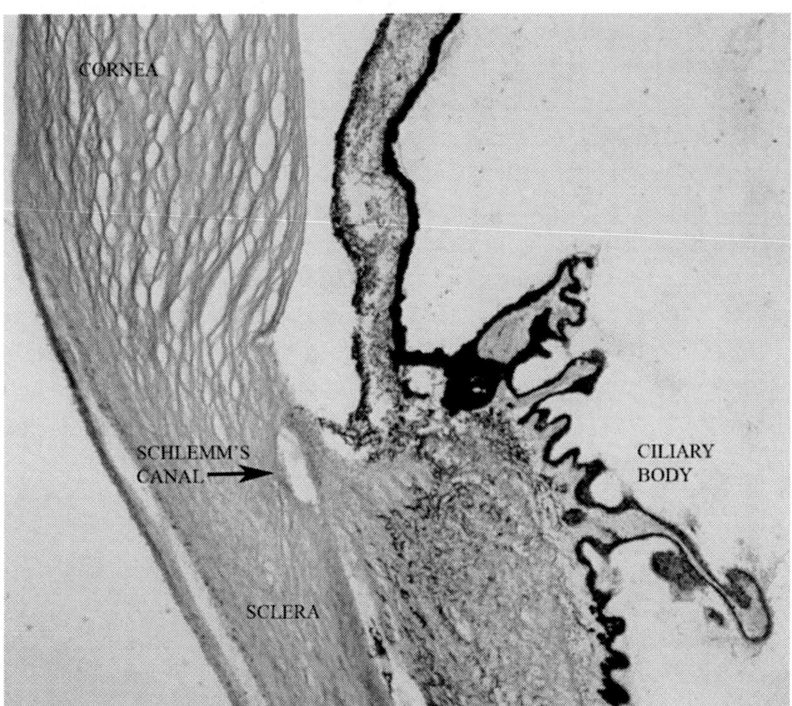

Figure 1 ↑ Trabeculum and Schlemm's channel

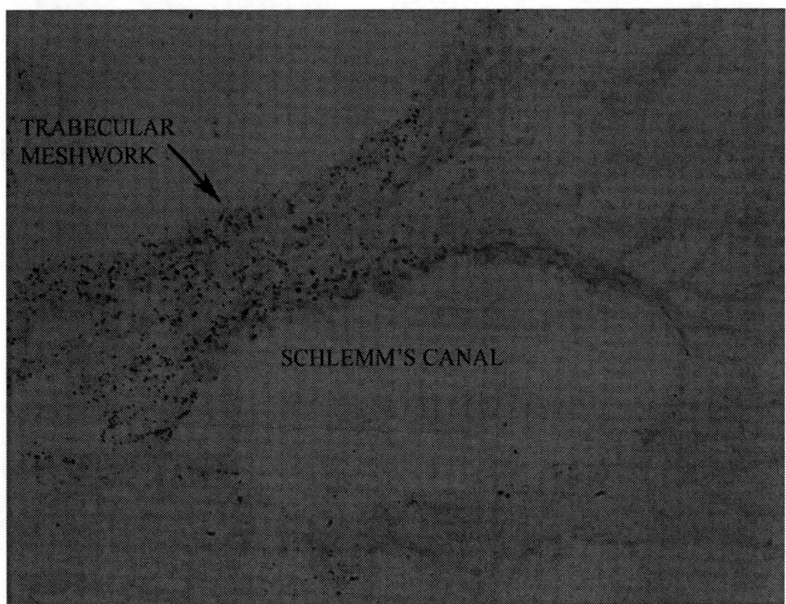

Figure 2 ↑ Pigmented trabeculum and Schlemm's channel

Cardiovascular disease has also been implicated as potential risk factors [Spaeth et al., 1975; Drance et al., 1978; Goldberg et al., 1981; Demailly et al., 1984; Freyler et al., 1988; Schulzer et al., 1990; Carter et al., 1990]. Other specific cardiovascular risk factors have also been proposed, such as systemic hypertension [Wilson et al., 1987; Tielsch et al., 1995;], hypotension, diabetes [Gramer et al., 1985; Hayreh et al., 1994; Graham et al., 1995; Tielsch et al., 1995], increased blood viscosity or platelet aggregation [Klaver et al., 1985; Trope et al., 1987; Hoyng et al., 1992], vasospasm [Flammer et al., 1987; Gasser and Flammer, 1987; Drance et al., 1988; Gasser et al., 1990; Gasser and Flammer, 1991] and migraine [Corbett et al., 1985; Phelps and Corbett, 1985]. Increasing age, family history of glaucoma and black race are also identified risk factors for developing glaucoma [Drance et al., 1978; Hart et al., 1979; Sommer et al., 1991; Quigley et al., 1994; Hollows and Graham., 1996; Shin et al., 1977]. Moreover, myopia has been identified as a risk factor for developing glaucoma [Daubs and Crick, 1981]. Probably, thin eye wall and large globe increases the influence of intraocular pressure over the optic disk. Pseudoexfoliation syndrome has also been probed an important risk factor to develop glaucoma (Figure 3)

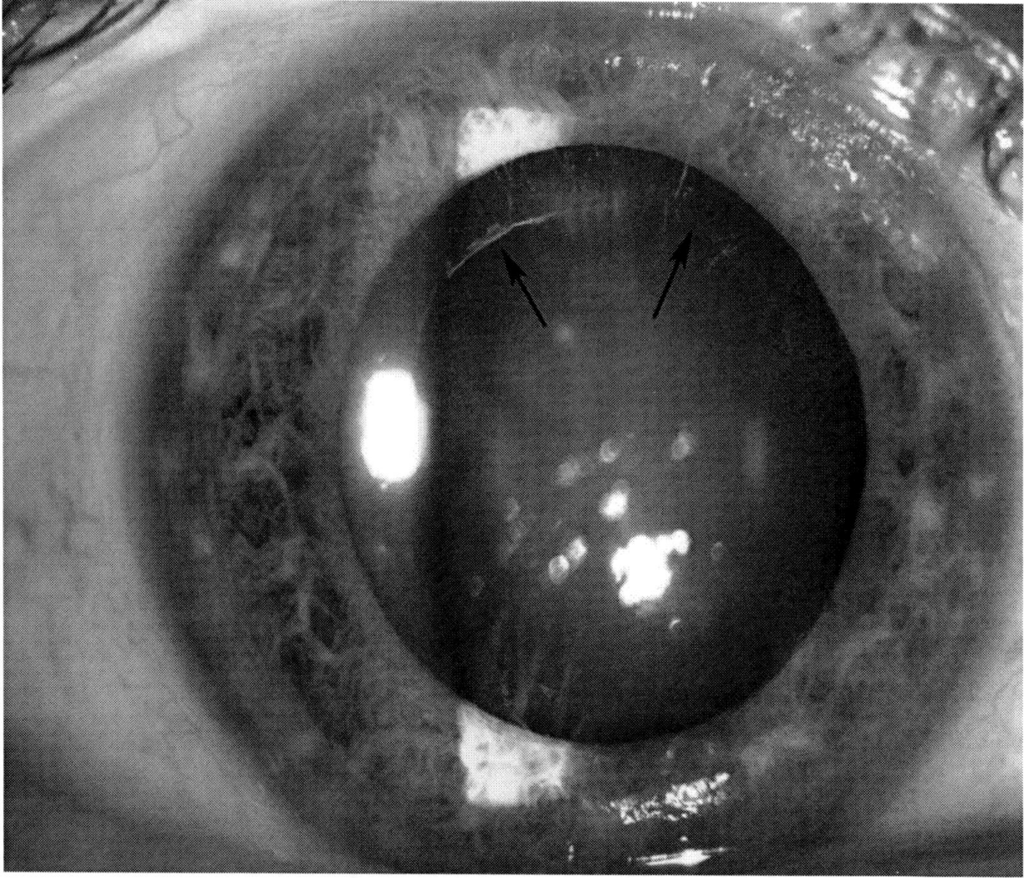

Figure 3 ↑ Patient with pseudoexfoliation syndrome

2. Biphasic Theory For Glaucoma Damage

2.1. Mechanical and vascular factors trigger glaucoma pathogenesis

Most authors blame that the initial site of damage in glaucoma is the lamina cribosa [Quigley and Addicks, 1980, 1981; Quigley et al., 1981]. A combined effect of vascular risk factors susceptibility and the triggering effect of increased intraocular pressure would lead to an initial damage of the axons at the level of lamina cribosa. However, another possibility will be the direct damage of the pressure apply to the retinal ganglion cell bodies at the level of the retina that will induce changes in the molecular structure of these cells and thus inducing the apoptosis of the RGCs.

Elevated IOP could induce the optic disc cupping causing optic nerve axonal compression at the lamina cribosa, and blockage of axoplasmic retrograde flow of the RGC axons. Glial cells present in the optic nerve head and lamina cribosa that will have modified their homeostasis at this level and influence in the RGC axons [Ransom and Behar, 2003]. These glial cells support neurons, regulate extracellular K+ levels, remove glutamate and GABA neurotransmitters (specially in synapses), renew precursor in the synthesis of glutamate, regulate extracellular pH levels and osmolarity and supply to the neuron the energy needed by the provision of lactate and glucose from glycogen stores [Ransom et al., 2003]. This primary insult may alter the gene expression of glial cells and their behaviour. In addition, both retrograde and anterograde axonal transport is blocked at the lamina cribosa in glaucomatous eyes [Minckler et al., 1977; Radius and Anderson, 1981; Dandona et al., 1991].

In the other hand, the vascular factors can also play an important role in the initial insult to the optic nerve head. Although the density of capillaries is similar in normal eyes and those affected by glaucoma [Quigley et al., 1984], increased IOP may lead to reduced blood flow in the capillaries of the optic nerve head, probably because of a faulty autorregulation of blood flow in the optic disk [Grunwald et al., 1984]. Astrocytes may also induce vasoconstriction of regional small capillaries by the release of vasoactive peptides during stress, associated to increased intracellular Ca2+ [Mulligan and MacVicar, 2004]. This isquemic condition and the deplection of energy stores can affect axonal Na+/K+ ATPase, that would increase intracellular Na+, leading to an overload of intracellular Ca2+ due to a greater Na+/Ca2+ exchanger activity [Osborne and Schwartz, 1994; Stys et al., 2004].

Both mechanical and vascular initial damage could estimulate axonal degeneration due to the lack of energy supply, neurotrophin withdrawal because of axoplasmic transport block, or the local involvement of the glial cells residing in the optic nerve head and lamina cribosa. (Figure 4)

2.2. Secondary degeneration

After an initial insult, that distorts the normal function of axons and astroglia in the optic nerve head and lamina cribosa, several molecular and cellular changes occur, leading to the death of the injured neurons and the neighbouring intact neurons through secondary degeneration.

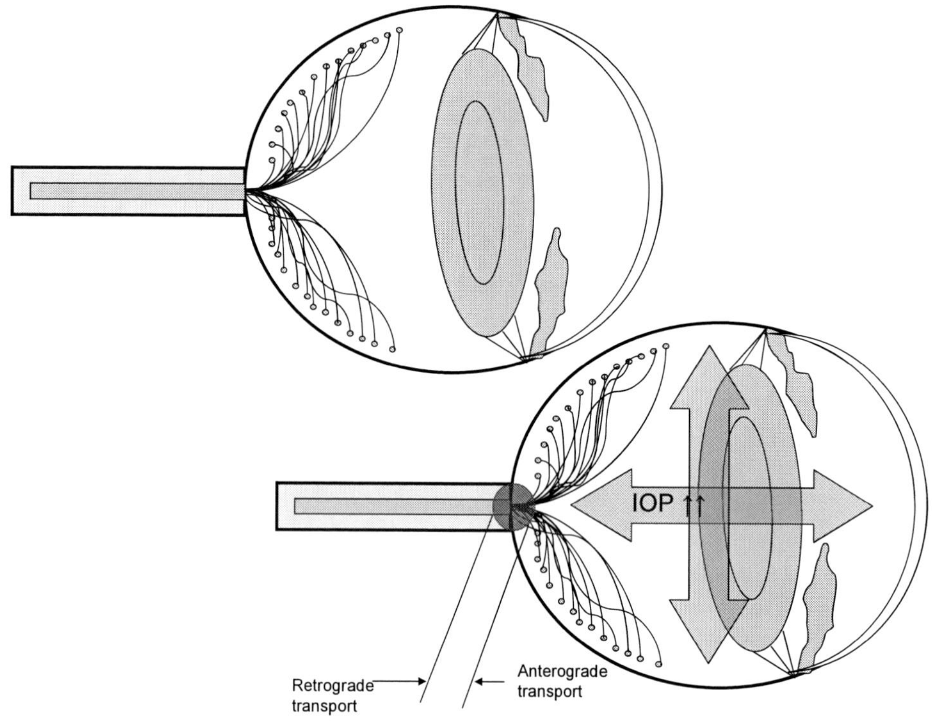

Figure 4 ↑ Effect of raised intraocular pressure on retinal ganglion cells

a) Molecular changes

Neurotrophin deprivation

An alteration of axoplasmic transport of soluble growth factors called neurotrophins from target neurons in the superior colliculus and lateral geniculate nucleus in the brain. RGCs as all other neurons in CNS require support of neurotrophins, as these small peptides regulate cellular metabolism by the activation of neuronal target-cell receptors. [Minckler et al., 1977; Radius and Anderson, 1981; Dandona et al., 1991; McKinnon, 1997]. It could also be an alteration in the endogenous retinal cell secretion of growth factors, since we have demonstrated that Müller glia sents factors that can neuroprotect RGCs from death, at least in vitro (García et al., 2002, 2003). Moreover, an increase in the expression of BDNF from RGCs have been also demonstrated to be a possible mechanism of survival of the retina after an ischaemic insult (Vecino et al., 98, 99).

Activation of the caspase enzyme family

RGC injury in ONT and experimental glaucoma involve activation of caspases 3 and 8, as well as the mitogen-activated protein kinase pathway [Isenmann and Bahr., 1997; Kikuchi et al., 2000; Levkovitch-Verbin et al., 2005].

Glutamate induced excitotoxicity

An excess of glutamate in the retina could damage the neurons hyperstimulating the ionotropic N-methyl-D-aspartate (NMDA) receptors, and thus generating a toxic influx of extracellular Ca2+. Abnormally high Ca2+ concentration leads to inappropriate activation of nucleases, proteases and lipases, that directly leads to the generation of free radicals and activation of the nitric oxide pathway [Dreyer and Lipton, 1993; Dreyer et al., 1996; Napper et al., 1999; Naskar and Dreyer, 2001; Martin et al., 2002].

ATP releases

There is evidence that acute glaucoma leads to an elevation in extracellular ATP levels that damages RGCs [Bao et al., 2004; Zhang et al., 2005; Dahl and Locovei, 2006; Spray et al., 2006; Zhang et al., 2006; Resta et al., 2007].

TNF-α

Growing evidence supports that this proinflammatory cytokine may lead to retinal ganglion cell death during glaucoma [Gottschall and Yu, 1995; Goureau et al., 1997; Downen et al., 1999; Tezel and Wax, 2000; Liefner et al., 2000; Yan et al., 2000; Tezel et al., 2001; Tezel and Yang, 2004; Ahmed et al., 2004; Desai et al., 2004; Tezel et al., 2004; Guo et al., 2005]

Oxidative stress and free radicals

ROS get accumulated in cells that undergo oxidative stress, and they react with nitric oxide producing free radicals. As a final consecuence, this oxidative chain leads to mitochondrial dysfuncticon, DNA degradation, and cell failure. Free radicals cause extensive damage to the RGCs and their axons [Oku et al., 1997; Levkovitch-Verbin et al., 2000; Tezel, 2006].

Increased accumulation of self-proteins

This phenomenon is common for many neurodegenerative diseases. As in Alzheimer's disease, it has been reported in glaucoma the accumulation of amyloid precursor protein [McKinnon et al. 2002].

Increased expression of pro-apoptotic genes

During glaucoma there is an unbalance between pro-apoptotic and anti-apoptotic proteins. In fact, there is an increased expression of pro-apoptotic genes, such as bax [Oltvai and Korsmeyer, 1994] and a downregulation of anti-apoptotic genes, such as bcl-xL [Levin et al., 1997].

b) Cellular changes

"Activation" response of glial cells

After the primary injury to the optic nerve head, glia surrounding damaged axons alter their gene expression profile.

This "activated" microglia can directly cause damage to axons and neighbouring cells as it has been suggested to synthesize and release nitric oxide [Neufeld et al., 1999; Liu and Neufeld, 2000]. Nitric oxide neurotoxicity occurs through a reaction with superoxide anion, forming peroxynitrite and free radical species. Peroxinitrite S-nitrosilates both proteins and nucleic acids in neighbouring cells and axons [Farkas and Grosskreutz, 2001].

Muller cell gliosis

Müller cells are important for normal retinal function. Reactive Müller cells protect the tissue in most forms of retinal injury and disease from further damage and preserve tissue function by the release of antioxidants and neurotrophic factors, and may contribute to retinal regeneration. However, Müller cell gliosis can also contribute to neurodegeneration and impedes regenerative processes in the retinal tissue by the formation of glial scars [Osborne et al., 2004; Brigmann et al., 2009].

Astrocyte induced vasoconstriction

Activated glia may induce vasoconstriction of regional small capillaries in the optic nerve head. During time of stress (under conditions of increased intracellular Ca^{2+}), astrocytic end-feet, that surround small capillaries would release vasoactive peptides [Mulligan and MacVicar, 2004]. This secondary vasoconstriction associated to the initial vascular primary injury caused by mechanical compression of the optic nerve head and lamina cribosa may lead to reduced blood flow in the capillaries of the optic nerve head.

3.Glaucoma And Neuroprotection

Neuroprotection in glaucoma consists in preventing the death of marginally damaged neurons and the secondary denegaration of those undergoing the hostile environment created by the initial insult. In other words, neuroprotection attempts to provide protection to such retinal ganglion cells that continue to remain at risk [Chew and Ritch, 1997].

A neuroprotective drug is expected to prevent death of RGC in presence of chronic stress by attenuating the hostility of the environment or supplying the cells with the tools to deal with those chances [Kaufman et al., 1999].

The pharmacological profile of neuroprotective drugs should fulfil these four criteria [Wheeler et al., 2001]:

To have a specific target (receptors) in the retina

To exhibit neuroprotective activity and have a measurable effect on RGC survival

To reach the retina/vitreous in neuroprotective concentrations after clinical dosing

(First three criteria must be obtained in animal models)

Neuroprotective activity must be demonstrated in randomized, controlled, clinical trials in humans.

Glaucoma neuroprotection offers potential as a complementary therapy to IOP-lowering for patients with previous severe damage and for those in whom pressure-lowering agents are ineffective to stop progression.

However, efficacy for neuroprotective agents has not yet been proven in glaucoma, and although there is laboratory evidence for glaucoma neuroprotection by several drugs, but still evidence lacks from randomized clinical trials [Weinreb, 2007].

In order to design clinical trials for neuroprotection, primary outcome measures can be:

-Functional visual acuity (The Ischemic Optic Neuropathy Decompression Trial Research Group, 2000),

-Visual field loss, although it may require several years before meaningful changes appear. Measurements of visual field has become standard way to assess the functional progression of glaucoma.

-Some new methods for assessing visual function, like:

-Electrophysiological testing [Berson et al., 1993]

-Frecuency Doubling Technology (FDT),

-Short Wave Length Automated Perimetry (SWAP),

-Contrast sensitivity,

-The Heidelberg Retinal Tomograph (HRT)

-The GDx nerve fiber analyzer

-The optical Coherence Tomograph (OCT).

3.1. Neuroprotective Drugs In Glaucoma

3.1.1.Ocular hypotensive drugs

a). Antiglaucomatous agents

α-2 agonists (Brimonidine)

Treatment with intraperitoneal brimonidine leads to a total rescue of RGCs in glaucomatous eyes; however, the intraperitoneal application of brimonidine in animals has not been found to reduce IOP [Hernández et al., 2008]. The neuroprotective effect of brimonidine has been studied in several experimental models of optic neuropathy, including optic nerve injury [Yoles et al., 1999; Ruíz et al., 2000], transitory ischemia [Donello et al., 2001; Lafuente et al., 2001], and experimental glaucoma [Ahmed et al. , 2001; Wheeler and Woldemussie, 2001; Wheeler et al., 2003; Hernández et al., 2008].

Brimonidine treatment may exert its neuroprotective effect through inhibition of the apoptotic cascade, reduction of glutamate toxicity, or enhancing the expression of BDNF. Thus, brimonidine could activate an anti-apoptotic pathway in RGCs by inducing the expression of anti-apoptotic genes such as bcl2 and bcl-xl [Lai et al., 1999]. Activation of RGC alpha-2 adrenergic receptors may inhibit pro-apoptotic mitochondrial signaling [Wheeler et al., 2001; Wheeler et al., 2001]. Also, brimonidine-mediated neuroprotection could also be due to the inhibition of glutamate-mediated excitiotoxicity [Vorwerk et al., 1996]. Brimonidine could also exert its beneficial effects via neurotrophins. Thus, intravitreal injection of brimonidine has been shown to upregulate brain-derived neurotrophin factor (BDNF) expression in rat RGCs [Gao et al., 2002].

ß blockers (betaxolol, metipranolol and timolol):

The ability to confer neuroprotection to retinal neurones is a common feature of three ophthalmic beta adrenoceptor antagonists (betaxolol, metipranolol and timolol).

Betaxolol, is able to protect retinal neurones in vitro and ganglion cells in vivo from the detrimental effects of either ischemia reperfusion or from excitotoxicity, after topical application. The neuroprotective effect of betaxolol is thought not to be elicited through an interaction with beta-adrenoceptors, but by its ability to reduce influx of sodium and calcium through voltage-sensitive calcium and sodium channels. Betaxolol has also been shown to increase blood flow velocity in the optic nerve head tissue [Tamaki et al., 1999].

Improvements have been made in measuring ocular blood flow velocity, but measurement difficulties still persist [Harris et al., 1999].

Metipranolol and timolol when topically applied behave like betaxolol. They all attenuate the detrimental effect of ischemia-reperfusion [Wood et al., 2003].

Muscarinic receptor agonist (pilocarpine):

Pretreatment of pilocarpine prevents glutamate-induced neuron death through M1 muscarinic receptor. The antiapoptotic effect of pilocarpine has been associated with maintaining calcium homeostasis, recovering mitochondrial membrane potential, and regulating the expression of Bcl-2 and Caspase-3 [Zhou et al., 2008].

b). Cannabinoids

Hepler and Frank first demonstrated that smoking marijuana cigarettes could lower IOP by up to 45% (Hepler and Frank, 1971). Later works showed that delta-9-tetrahydrocannabinol (Δ9-THC) lowered IOP intravenously [Purnell and Gregg, 1975], orally [Merritt et al., 1980] or by inhalation [Merritt et al., 1980]. Other results suggested that topical application of a synthetic cannabinoid receptor agonist (WIN55212-2), lowers IOP in human glaucoma, and that the IOP-lowering effects of WIN55212-2 are, probably, mediated through CB1 cannabinoid receptors [Porcella et al., 2001].

Nowadays, there is important progress in the role of cannabinoids in affording neuroprotection in traumatic, ischemic, inflammatory and neurotoxic damage in neurons [Van der Stelt and Di Marzo, 2005; de Lago and Fernández-Ruíz, 2007; Mechoulam and Shohami, 2007].

Other studies have shown that CB1 agonists (THC and cannabidiol) protect RGCs from glutamate-induced excitotoxicity [El Remessy et al., 2003; Opere et al., 2006], and ischemia secondary to an experimental model of glaucoma in rat [Crandall et al., 2007].

HU-211 is a nonpsychotropic cannabinoid that behaves as a noncompetitive NMDA antagonist [Feigenbaum et al., 1989], and systemic administration provides neuroprotection for RGCs after nerve axotomy [Yoles et al., 1996].

3.2. Other drugs

a). Tetracyclines

Minocycline

Minocycline hydrochloride is a second-generation tetracycline, commomnluy used in humans because of its beneficial antimicrobial and anti-inflammatory actions. Minocycline effectively crosses the blood-brain barrier [Aronson, 1980]. This drug also has remarkable neuroprotective qualities in animal models of cerebral ischemia, traumatic brain injury, Huntington disease, and Parkinson disease [Yrjanheikki et al., 1998; Yrjanheikki et al. 1999; Tikka and Koistinaho 2001; Du et al., 2001; Arvin et al., 2002].

Levkovitch-Verbin et al. in 2006 demonstrated that systemic administration of minocycline significantly delays apoptosis of RGCs after severe injury of optic nerve transaction (ONT) [Levkovitch-Verbin et al., 2006]. Its effect is associated with inhibition of inducible nitric oxide synthase [Du et al. 2001], caspase 1 and caspase 3 expression [Chen et al., 2000; Sanchez et al., 2001]. It also inhibits p38 mitogen-activated protein kinase [Yrjanheikki et al., 1998; Yrjanheikki et al., 1999] and cytochrome c release [Zhu et al., 2002]. It also depresses oxygen radical release and matrix metalloproteinase activity [Gabler and Creamer, 1991; Golub et al., 1991; Sadowski and Steinmeyer, 2001; Ryan et al., 2001; Brundula et al., 2002; Power et al., 2003]. It has also been found neuroprotective in models of photoreceptor death [Baptiste et al., 2004; Hughes et al., 2004; Zhang et al., 2004].

Doxycycline

Doxycycline is another semisynthetic second-generation tetracycline with well-known effects of neuroprotection [Gabler et al., 1992; Smith and Gabler, 1994; Smith and Gabler, 1995]. However, it hasn't shown as much neuroprotective effect as minocycline in most studies.

b). Ca 2+ channel blockers

In patients with low-tension glaucoma, there was a significant difference in the progression of visual field defects, with only two of 18 eyes (11%) of patients taking calcium channel blockers, compared to ten of 18 eyes (56%) of controls showing new visual field defects. Similarly, low-tension glaucoma patients taking calcium channel blocker therapy demonstrated no evidence of progressive optic nerve damage, compared to eight of 18 control eyes (44%). However, such differences didn't appear in patients with open-angle glaucoma, suggesting that calcium channel blockers may be useful in the management of low-tension glaucoma [Netland et al., 1993].

Nifedipine

There is clinical evidence of nifedipine's neuroprotective effects, as in a significant number of patients with NTG who took nifedipine, visual field got improved after 6 months follow-up [Kittazawa et al., 1989].

Flunarizine

Flunarizine has been demonstrated to enhance RGC survival after optic nerve transection in mice [Eschweiler and Bahr, 1993].

c). Antioxidants

Several natural compounds possess potential antioxidative value. Polyphenolic compounds, such as tea, red wine, dark chocolate, or coffee have antioxidative properties. Coffee contains 3-methyl-1,2-cyclopentanedione (MCP), capable of scavenging peroxynitirite. Red wine-polyphenols (e.g., resveratrol), exert vasoprotective effects by inhibiting the synthesis of endothelin-1. Dark chocolate decreases blood pressure and improves endothelium-dependant vasorelaxation. Omega-3-fatty acids and magnesium can also improve blood flow regulation Anthocyanosides (bilberries) owe their antioxidant effects to their particular chemical structure. Other antioxidants include ubiquinone and melatonin, and heat shock proteins can be induced naturally by the use of sauna baths [Mozaffarieh and Flammer, 2007; Mozaffarieh et al., 2008]. (Figure 5)

Ginkgo biloba extract (GBE) exerts protective effects against free radical damage and lipid peroxidation, preserves mitochondrial metabolism and ATP production and also behaves as a scavenger of superoxide radicals and nitric oxide [Janssens et al., 1995].

GBE on glaucoma patients improves ocular blood flow. A Phase I cross-over trial of GBE with placebo control in 11 healthy volunteers was performed. Color Doppler imaging was used to measure ocular blood flow before and after treatment. Ginkgo biloba extract did not alter arterial blood pressure, heart rate, or IOP, but significantly increased diastolic velocity in the ophthalmic artery in the ophthalmic artery in the ophthalmic artery [Chung et al., 1999].

Natural compounds with antioxidative value
Polyphenolic Flavonoids
Tea
Coffee
Wine
Dark Chocolate
Ginkgo Bilboa
Anthocyanosides
Vaccinium Myrtillus (Bilberry)
Vitamins
Alpha Lipoic Acid (ALA)
Thiamin (Vitamin B1)
Cobalamin (Vitamin B12)
Miscellaneous
Ubiquinone (Coenzyme Q10)
Melatonin

Figure 5 ↑ Natural compounds with antioxidative value

d) NMDA antagonists, p38 MAPK, and caspase inhibitors

Open-channel blocker of the NMDA receptors, memantine, blocks only excessive NMDA receptor activity while leaving normal function relatively intact.

This NMDA antagonist is neuroprotectant by preventing excessive calcium influx, inhibition of over-stimulation of the NMDA receptor, and inhibiting OPA1 release from mitochondria by blockage of NMDA receptor. OPA1 effect is accompanied by increased Bcl-2 expression, decreased Bax expression and apoptosis blockade [Ju et al., 2008].

Memantine effectively blocks the excitotoxic response of retinal ganglion cells both in culture and in vivo [Vorwerk et al., 1996], and has been proben neuroprotectant after systemic administration [Lagreze et al., 1998]. Although memantine has demonstrated exciting results for neuroprotection in laboratories, the phase III clinical trial of Memantine failed to prove such activity [Ge, 2008].

NMDA receptors and downstream signaling pathways, triggered by p38 mitogen activated protein kinase (MAPK) and caspases, are also potential targets of intervention for glaucoma, and improved memantine derivatives, p38 MAPK, and caspase inhibitors seem plausible therapeutics to prevent RGC death [Seki and Lipton, 2008].

e). 17beta-estradiol

17beta-estradiol (E2) is a steroid hormone, which has been shown to increase the viability, survival, and differentiation of primary neuronal cultures from different brain areas including amygdala, hypothalamus, and neocortex.

17 beta-estradiol (E2), has shown cytoprotective activities in animal models of stroke, myocardial infarct and neurodegenerative diseases.

Systemic administration of E2 significantly reduces RGC loss induced by acute increase of IOP in rat. In addition, pretreatment with E2, 30 min before ischemia, minimizes the elevation of glutamate observed during the reperfusion period. These effects seem to be in part mediated by the activation of the estrogen receptor, since a pre-treatment with ICI 182-780, a specific estrogen receptor antagonist, partially counteracts the neuroprotection afforded by the estrogen [Russo et al., 2008].

Later studies suggested the possible involvement of estrogen receptor/Akt/CREB/ thioredoxin-1, and estrogen receptor/MAPK/NF-kappaB, in estrogen-mediated retinal ganglion cell protection [Zhou et al., 2007].

However, three synthetic estrogen analogues (ZYC-1, ZYC-3, ZYC-10) showed efficacy of neuroprotection against glutamate-induced RGC cell death. These compounds seem to affect neuroprotection via pathways independent of the classical estrogen receptors, as inclusion of an estrogen receptor antagonist (ICI 182,780) did not reverse their neuroprotective properties against glutamate insult. These results support the hypothesis that estrogen analogues may be useful in neuroprotection of retinal ganglion cells in ocular pathologies such as glaucoma [Kumar et al., 2005]. This receptor –independent effect may be explained by the fact that estrogens and novel analogs prevent cell death in large measure by maintaining functionally intact mitochondria [Simpkins et al., 2005].

f). Nitric oxide synthase (NOS-2) inhibitors (Aminoguanidine)

Nitric-oxide synthase (NOS-2) is elevated in the optic nerve heads from human glaucomatous eyes and from rat eyes with chronic, moderately elevated IOP.

Aminoguanidine, a relatively specific inhibitor of NOS-2, has been shown to protect RGCs in a rat model of glaucoma. Pharmacological neuroprotection by inhibition of NOS-2 may prove useful for the treatment of patients with glaucoma [Neufeld et al., 1999].

g). Erythropoietin

Erythropoietin (EPO) plays an important role in the brain's response to neuronal injury [*Siren et al., 2001]*. EPO is also involved in the wound healing process [*Haroon et al., 2003]*. In the DBA/2J mouse model of glaucoma EPO promoted RGC survival without affecting IOP. These results suggest that EPO may be a potential therapeutic neuroprotectant in glaucoma [Zhong et al., 2007].

4. Future Neuroprotective Therapies For Glaucoma

Targeted approaches are being made to investigate new therapies to afford neuroprotection in glaucoma.

4.1. Neurotrophins

BDNF, ciliary neurotrophin factor and basic fibroblastic growth factor showed to protect human RGC from death in culture and in vivo [Rabacchi et al., 1994]. Glial cell line–derived neurotrophic factor also showed neuroprotection of RGCs in mice and it's still under study [Ward et al., 2007]. Müller secreted factors have been demonstrated to neuroprotect RGCs from death in primary cell cultures (García et al, 2003).

Neurotrophin administration in glaucoma is being evaluated, however the main problem may be in delivering these substances so that they could reach the retina.

4.2. Controlling Infamation: Anti-TNF-α-Strategies

Tumor necrosis factor-alpha (TNF-α) is a proinflammatory cytokine implicated in the immune response. TNF-α is produced by macrophages, lymphoid cells, endothelial cells, fibroblastas and also by glial cells, and, growing evidence supports that it may lead to neuronal cell death [Downen et al., 1999]. TNF-α is increased in the retina and optic nerve head in glaucomatous eyes. TNF-α is secreted by stressed glial cells and it can induce RGC death through TNF receptor-1-mediated caspase cascade, mitochondrial dysfunction, and oxidative damage (accumulation of reactive oxygen species (ROS)) [Tezel and Wax, 2000; Tezel and Yang, 2004]. In adittion, immunohistochemical analysis of human eyes revealed an increased immunolabeling for TNF-α and TNF-R1 in the optic nerve head [Yan et al., 2000] and retina [Tezel et al., 2001] of glaucomatous eyes relative to controls, showing a predominant localization in RGCs and their axons, which are the sensitive targets for cytotoxic effects of TNF-α. Findings have also demonstrated an upregulation in gene expression for TNF-α and TNF-R1 in ocular hypertensive eyes [Ahmed et al., 2004]. Other findings provide evidence that TNF death receptor signaling is involved in the secondary degeneration of RGCs associated with JNK signaling [Tezel et al., 2004]. TNF-α in glaucoma may also be associated with another neurodestructive role by its ability to induce a highly potent secondary oxidant, nitric oxide [Goureau et al., 1997].

TNF-α also activates matrix metalloproteinases [Gottschall and Yu, 1995], associated with neurotoxicity [Guo et al., 2005]. TNF-α is also a potent stimulator of endothelin-1 (ET-1) synthesis and secretion in optic nerver head astrocytes, a potent vasoactive peptide associated with glaucomatous neurodegeneration [Desai et al., 2004].

TNF-α appears to play an important role in the immune response, ranging effects from inflammation to apoptosis, through Wallerian degeneration [Liefner et al., 2000].

Glial cells, both activated during glaucoma and by TNF-alpha, and secreting TNF -alpha, could serve as antigen-presenting cells and thus constitute a new way to neuroprotection [Baudouin and Liang, 2006].

Targeting TNF-α signaling for RGC rescue should block the caspase cascade and improve the ability of these neurons to survive. However, ability of anti- TNF-α strategies against axonal injury also needs to be clarified.

4.3 Inmunomodulatory Compounds: Glatiramer Acetate(GA)

Prevention of ganglion cells loss was also observed by prior immunization of animals using a synthetic polymer close to myelin (COP1), capable of stimulating a specific lymphocyte reaction of neuronal impairment without inducing uveitis [Baudouin and Liang, 2006].

Afterwards, in a rat model of experimental chronic glaucoma, vaccination slowed down the progression of glaucoma, by making milieu of the nerve and retina less hostile to RGC survival [Schwartz et al., 2007].

4.4 Gene Therapy

Several gene families have been identified to play roles in RGCs apoptosis or survival. Caspases stimulate apoptosis and carry out disassembly of the cells [Hetts, 1998]. Tumor suppression protein, p53 can upregulate the expression of the pro-apoptotic gene bax and down-regulate the expression of the anti-apoptotic gene bcl-2 [Nickells, 1999].

In the other hand, Bcl-2 and related proteins are important inhibitors of apoptosis. They inhibit intermediate proteins that activate caspases [Adams and Cory, 1998].

Transgenic mice allowed expression of the apoptosis-inhibiting gene Bcl-2 in rat neurons, with a 50% increase in retinal ganglion cell number. A gene can be delivered to the damaged tissue via viruses, artificial liposomes, and direct transfer, and even induce the cell to express some genes by its own regulatory pathways.

Recently, rat RGCs were able to be transfected by recombinant adeno-associated virus – BDNF (rAAV-BDNF) in vitro. The transfected cells could express BDNF gene at the level of both mRNA and protein. Apoptosis rate was lower in the transfected cells. This study indicated that rAAV-BDNF transfection can be used for the potential gene therapy in glaucomaneuroprotection [Li et al., 2008].

4.5 siRNA

Since its discovery in plants in the early 1990s, RNA interferente (RNAi) has begun to emerge as a promising technology to apply to therapeutics. This phenomenon consist of a specific and selective inhibition of gene expression in an efficient manner. RNAi is mediated by small interfering RNA (siRNA), consisting of 19-23 nucleotide double-stranded RNA, that promote specific nucleolytic cleavage of mRNA targets through an RNA-induced silencing complex. In this way, siRNAs offers a powerful tool that can be used to determine functional significance of newly identified molecules as treatment targets. This technology can also serve along with other genomic or pharmacologic treatments to provide neuroprotection in glaucoma.

Besides, siRNA can be locally and topically administrated for glaucoma in controlled doses, avoiding to treat the whole body.

4.6 Stem Cells

The eye has several advantages that make it a good candidate for gene therapy. It is a small organ, relatively isolated, and most targeted cells are not undergoing cell division, so there is less risk of oncogenesis.

Numerous stem cell sources exist, with embryonic and fetal stem cells liable to be superseded by adult-derived cells. Stem cell transplantation is currently being explored as a therapy for many neurodegenerative diseases including glaucoma. Cellular therapies have the potential to provide chronic neuroprotection after a single treatment, and possible neuroprotective mechanisms offered by stem cell transplantation include the supply of neurotrophic factors and the modulation of matrix metalloproteinases and other components of the CNS environment to facilitate endogenous repair [Bull et al., 2008].

Acknowledgments

We are grateful for support from the American Glaucoma Foundation (USA), Spanish Ministry of Science and Technology (SAF2007-62060), Grupos Consolidados del Gobierno Vasco, Gangoiti Foundation, Red Patología Ocular RD07/0062, Fundaluce, ONCE and BIOEF08/ER/006.

References

Adams, JM; Cory, S. The Bcl-2 functions in an antioxidant pathway to prevent apoptosis. Science, 1998; 281: 1322-1326.

Ahmed, F; Brown, KM; Stephan, DA; Morrison, JC; Johnson, EC; Tomarev, SI. Microarray analysis of changes in mRNA levels in the rat retina after experimental elevation of intraocular pressure. Invest Ophthalmol Vis Sci, 2004; 45: 1247-1258.

Ahmed, F; Hegazi, K; Chaundhray, P; Sharma, SC. Neuroprotective effect of α2 agonist (brimonidine) on adult rat retinal ganglion cells after increased intraocular pressure. Brain Res, 2001; 913: 133-139.

Armaly, MF; Krueger, DE; Maunder, L; Becker, B ; Hetherington, JJr; Kolker, AE; Levene, RZ; Maumenee, AE; Pollack, IP; Shaffer, RN. Biostatistical analysis of the collaborative glaucoma study. I. Summary report of the risk factors for glaucomatous visual-field defects. Arch Ophthalmol, 1980; 98: 2163-2171.

Aronson, AL. Pharmacotherapeutics of the newer tetracyclines. J Am Vet Med Assoc, 1980; 176: 1061-1068

Arvin, KL; Han, BH; Du, Y; Lin, SZ; Paul, SM; Holtzman, DM. Minocycline markedly protects the neonatal brain against hypoxic-ischemic injury. Ann Neurol, 2002; 52: 54-61.

Bao, L; Locovei, S; Dahl, G. Pannexin membrane channels are mechanosensitive conduits for ATP. FEBS Lett 2004, 572: 65–68.

Baptiste, DC; Hartwick, AT; Jollimore, CA; Baldridge, WH; Seigel, GM; Kelly, ME. An investigation of the neuroprotective effects of tetracycline derivatives in experimental models of retinal cell death. Mol Pharmacol, 2004; 66: 1113-1122.

Baudouin, C; Liang, H. Vaccine for glaucoma, myth or reality? J Fr Ophtalmol, 2006; 29 (2):9-12

Bayer, AU; Danias, J; Brodie, S; Maag, KP; Chen, B; Shen, F; Podos, SM; Mittag TW. Electrorretinographic abnormalities in a rat glaucoma model with chronic elevated intraocular pressure. Exp Eye Res, 2001; 72: 667-77.

Berson, EC; Rosner, B; Sandberg, MA; Hayes, KC; Nicholson, BW; Weisel-Di Franco, C; Willett W. A randomized trial of vitamin A and vitamin E supplementation for retinitis pigmentosa. Arch Ophthalmol 1993; 111(6): 761-772.

Brigmann A, Iaudiev I, Pannicke T, Nurm A, Hollborn M, Wiedemann P, Osborne NN, Reichenbach A. Cellular signalling and factor involved in Müller cell gliosis: neuroprotective and detrimental effects. Prog Retin Eye Res 2009; 28(6): 423-451.

Brundula, V; Rewcastle, NB; Metz, LM; Bernard, CC; Yong, VW. Targeting leukocyte MMPs and transmigration: minocycline as a potential therapy for multiple sclerosis. Brain, 2002; 125: 1297-1308.

Bull, ND; Johnson, TV; Martin, KR. Stem cells for neuroprotection in glaucoma. Prog Brain Res, 2008; 173: 511-519.

Carter, CJ; Brooks, DE; Doyle, DL; Drance, SM. Investigations into a vascular etiology of low-tension glaucoma. Ophthalmology, 1990; 97: 49-55.

Chen, M; Ona, VO; Li, M ; Ferrante, RJ; Fink, KB; Zhu, S; Bian, J; Guo, L; Farrell, LA; Hersch, SM; Hobbs, W; Vonsattel, JP; Cha, JH; Friedlander, RM. Minocycline inhibits caspase-1 and caspase-3 expression and delays mortality in a transgenic mouse model of Huntington disease. Nat Med, 2000; 6: 797-801.

Chew, SJ; Ritch, R. Neuroprotection: the next break through in glaucoma? Proceedings of the third Annual Optic Nerve Rescue and Restoration Think Tank. J Glaucoma, 1997; 6: 263-266.

Chung, HS; Harris, A; Kristinsson, JK; Ciulla, TA; Kagemann, C; Ritch, R. Ginkgo biloba extract increases ocular blood flow velocity. J Ocul Pharmacol Ther, 1999; 15(3):233-240.

Corbett, JJ; Phelps, CD; Eslinger, P; Montague, PR. The neurologic evaluation of patients with low-tension glaucoma. Invest Ophthalmol Vis Sci, 1985; 26: 1101-1104.

Cordeiro, MF; Guo, L; Luong, V; Harding, G; Wang, W; Jones, HE; Moss, SE; Sillito, AM; Fitzke, FW. Real-time imaging of single nerve cell apoptosis in retinal neurodegeneration. Proc Natl Acad Sci USA, 2004; 101: 13352–13356.

Crandall, J; Matragoon, S; Khalifa, YM; Borlongan, C; Tsai, NT; Caldwell, RB; Liou, GI. Neuroprotective and intraocular pressure-lowering effects of Delta9-tetrahydrocannabinol in a rat model of glaucoma. Ophthalmic Res, 2007; 39: 69-75.

Dahl, G; Locovei, S. Pannexin: To gap or not to gap, is that a question? IUBMB Life, 2006; 58: 409–419.

Dandona, L; Hendrickson, A; Quigley, HA. Selective effects of experimental glaucoma on axonal transport by retinal ganglion cells to the dorsal lateral geniculate nucleus. Invest Ophthalmol Vis Sci, 1991; 32: 484–491.

Daubs, JG; Crick, RP. Effect of refractive error on the risk of ocular hypertension and open angle glaucoma. Trans Ophthalmol Soc UK, 1981; 101: 121-126.

de Lago, E; Fernández-Ruíz, J. Cannabinoids and neuroprotection in motor-related disorders. CNS Neurol Disord Drug Targets, 2007; 6: 337-387.

Demailly, P; Cambien, P; Plouin, PD; Bason, P; Chevallier, B. Do patients with low-tension glaucoma have particular cardiovascular characterisitics? Ophthalmologica 1984; 188: 65-75.

Desai, D; He, S; Yorio, T; Krishnamoorthy, RR; Prassana, G. Hypoxia augments TNF-alpha-mediated endothelin-1 release and cell proliferation in human optic nerve head astrocytes. Biochem Biophys Res Commun, 2004; 318: 642-648.

Donello, JE; Padillo, EU; Webster, ML; Wheeler, LA; Gil, DW. Alpha (2)-adrenoceptor agonists inhibit vitreal glutamate and aspartate accumulation and preserve retinal function after transient ischemia. J Pharmacol Exp Ther, 2001; 1: 216-223.

Downen, M; Amaral, TD; Hua, LL; Zhao, ML; Lee, SC. Neuronal death in cytokine-activated primary human brain cell culture: role of tumor necrosis factor-alpha. Glia, 1999; 28: 114-127.

Drance, SM; Douglas, GD; Wijsman, K. et al. Response of blood flow to warm and cold in normal and low-tension glaucoma patients. Am J Ophthalmol, 1988; 105: 35-39.

Drance, SM; Schulzer, M; Douglas, GR; Sweeney, VP. Use of discriminant analysis. II. Identification of persons with glaucomatous visual field defects. Arch Ophthalmol, 1978; 96: 1571-1573.

Drance, SM; Schulzer, M; Douglas, GR; Sweeney, VP. Use of discriminant analysis. II. Identification of persons with glaucomatous visual field defects. Arch Ophthalmol, 1978; 96: 1571-1573.

Drance, SM; Schulzer, M; Thomas, B; Douglas, GR. Multivariate analysis in glaucoma: use of discriminant analysis in predicting glaucomatous visual field damage. Arch Ophthalmol, 1981; 99: 1019-1022.

Dreyer, EB; Lipton, SA. A proposed role of excitatory amino acids in glaucoma visual loss. IOVS, 1993; 34: 1504.

Dreyer, EB; Zurakowski, D; Schumer, RA; Podos, SM; Lipton, SA. Elevated glutamate levels in the vitreous body of humans and monkeys with glaucoma. Arch Ophthalmol, 1996; 114: 299-305.

Du, Y; Ma, Z; Lin, S; Dodel, RC; Gao, F; Bales, KR; Triarhou, LC, Chernet, E; Perry, KW ; Nelson, DL; Luecke, S; Phebus, LA; Bymaster, FP; Paul, SM. Minocycline prevents nigrostriatal dopaminergic neurodegeneration in the MPTP model of Parkinson's disease. Proc Natl Acad Sci U S A, 2001; 98:14669-14674.

El Remessy, AB; Khalil, IE; Matragoon, S; Abou-Mohamed, G; Tsai, NJ; Roon, P; Caldwell, RB; Caldwell, RW; Green, K; Liou, GI. Neuroprotective effect of delta(9)-tetrahydrocannabinol and cannabidiol in N-mehyl-D-aspartate-induced retinal neurotoxicity: involvement of peroxynitrite. Am J Pathol, 2003; 163: 1997-2008.

Eschweiler, GW; Bahr, M. Flunarizine enhances rat retinal ganglion cells survival after axotomy. J Neurol Sci, 1993; 116: 34-40.

Farkas, RH; Grosskreutz, CL. Apoptosis, neuroprotection and retinal ganglion cell death: An overview. Int Ophthalmol Clin, 2001; 41:111-130.

Feigenbaum, JJ; Bergmann, F; Richmond, SA; Mechoulam, R; Nadler, V; Kloog, Y; Sokolovsky, M. Nonpsychotropic cannabinoid acts as a functional N-methyl-D-aspartate receptor blocker. Proc Natl Acad Sci USA, 1989; 86: 9584-9587.

Flammer, J; Guthauser, U; Mahler, M. Do ocular vasospasms help cause low-tension glaucoma? Doc Ophthalmol Proc Ser, 1987; 49: 397-399.

Freyler, H; Menapace, R: Ist die Erblindung an Glauckom vermeidbar? Spektrum Augenheilkd 2; 121, 1988;

Gabler WL, Smith J, Tsukuda N. comparison of doxycycline and a chemically modified tetracycline inhibiton of leukocyte functions. Res Commun Chem Pathol Pharmacol. 1992; 78: 151-160;

Gabler, WL; Creamer, HR. Suppression of human neutrophil functions by tetracyclines. J Periodontal Res, 1991; 26: 52-58.

Gao, H; Qiao, X; Cantor, LB; WuDun, D. Up-regulation of brain-derived neurotrophic factor expression by brimonidine in rat retinal ganglion cells. Arch Ophthalmol, 2002; 797-803.

García M, Forster V, Hicks D, Vecino E. Effects of müller glia on cell survival and neuritogenesis in adult porcine retina in vitro. Invest Ophthalmol Vis Sci, 2002; 43(12): 3735-43.

García M, Forster V,. Hicks D, Vecino E. In vivo expression of neurotrophins and neurotrophin receptors is conserved in adult porcine retina in vitro. Invest Ophthalmol Vis Sci. 2003; 44(10): 4532-4541.

Gasser, P; Flammer, J. Blood-cell velocity in the nailfold capillaries of patients with normal-tension and high-tension glaucoma. Am J Ophthalmol, 1991; 111: 585-588.

Gasser, P; Flammer, J. Influence of vasospasm on visual function. Doc Ophthalmol, 1987; 66: 3-18.

Gasser, P; Flammer, J; Guthauser, U; Mahler, F. Do vasospasms provoke ocular diseases? Angiology, 1990; 41: 213-220.

Ge, J. Glaucoma neuroprotection. How far is it from a dream to reality. Zhonghua Yan Ke Za Zhi, 2008; 44(5): 385-387 (abstract).

Goldberg, I; Hollows, FC; Kass, MA; Becker, B: Systemic factors in patients with low-tension glaucoma. Br J Ophthalmol, 1981; 65: 56.

Golub, LM;, Ramamurthy, NS; McNamara, TF; Greenwald, RA; Rifkin, BR. Tetracyclines inhibit connective tissue breakdown: new therapeutic implications for an old family of drugs. Crit Rev Oral Biol Med, 1991; 2: 297-321.

Gottschall, PE; Yu, X. Cytokines regulate gelatinase A and B (matrix metalloproteinase 2 and 9) activity in cultured rat astrocytes. J Neurochem. 1995; 64: 1513-1520.

Goureau, O; Amiot, F; Dautry, F and Courtois, Y. Control of nitric oxide production by endogenous TNF-α in mouse retinal pigemented epithelial and Muller glial cells. Biochem Biophys Res Commun, 1997; 240: 132-135.

Graham, SL; Drance, SM; Wijsman, K, Douglas, GR; Mikelberg, FS. Ambulatory blood pressure monitoring in glaucoma. The nocturnal dip. Ophthalmology, 1995; 102: 61-69.

Gramer E, Leydhecker W. Glaukom ohne Hochdruck. Klin Monatsbl Augenheilkd 1985; 262-267.

Grunwald, JE; Riva, CE; Stone, RA; Keates, EU; Petrig, BL. Retinal autoregualtion in open-angle glaucoma. Ophthalmology, 1984; 91: 1690-1694.

Haroon, ZA; Amin, K; Jiang, X; Arcasoy, MO. "A novel role for erythropoietin during fibrin-induced wound-healing response". Am J Pathol, 2003; 163: 993–1000.

Hart, WJr; Jablonski, M; Kass, MA; Becker, B. Multivariate analysis of the risk of glaucomatous field loss. Arch Opthalmol, 1979; 97: 1455-1458.

Hart, WMJr; Yablonski, M; Kass, MA; Becker, B. Multivariate analysis of the risk of glaucomatous visual field loss. Arch Ophthalmol, 1979; 97: 1455-1458.

Hayreh, SS; Zimmerman, MB; Podhajsky, P; Alward, WL. Nocturnal arterial hypotension and its role in optic nerve head and ocular ischemic disorders. Am J Ophthalmol, 1994; 117: 603-624.

Hepler, RS; Frank, IR. Marihuana smoking and intraocular pressure. JAMA, 1971. 217: 1392.

Hernández, M; Urcola, JH; Vecino, E. Retinal ganglion cell neuroprotection in a rat model of glaucoma following brimonidine, latanoprost or combined treatments. Exp Eye Res, 2008; 86: 798-806.

Hetts, SW. To die or not to die: an overview of apoptosis and its role in disease. JAMA, 1998; 279: 300-307.

Hollows, FC; Graham, PA. Intra-ocular pressure, glaucoma and glaucoma suspects in a defined population. Br J Ophthalmol, 1996; 50: 570-586.

Hoyng, PFJ; de Jong, H; Oosting, H ; Stilma, J. Platelet aggregation, disc haemorrhage and progressive loss of visual fields in glaucoma. Int Ophthalmol, 1992; 15: 65-73.

Hughes, EH; Schlichtenbrede, FC; Murphy, CC; Broderick, C; van Rooijenn, N; Ali RR; Dick, AD. Minocycline delays photoreceptor death in the rods mouse through a microglia-independent mechanism. Exp Eye Res, 2004; 78: 1077-1084.

Isenmann, S; Bahr, M. Expression of c-Jun protein in degenerating retinal ganglion cells after optic nerve lesion in the rat. Exp Neurol, 1997; 147: 28-36.

Janssens, D; Michiels, C; Delaive, E; Eliaers, F; Drieu, K; Remacle, J. Protection of hypoxia induced ATP decrease in endothelial cells by Ginkgo biloba extract and bilobalide. Biochem Pharmacol, 1995; 50: 991-999.

Ju, WK; Kim, KY; Angert, M; Duong-Polk, KX; Lindsey, JD; Ellisman, MH; Weinreb, RN. Memantine blocks mitochondrial OPA1 and cytochrome c release, and subsequent apoptotic cell death in glaucomatous retina. Invest Ophthalmol Vis Sci, 2008. 20. [Epub ahead of print]

Kaufman, PL; Gabelt, BT; Cynader, M. Introductory comments on neuroprotection. Surv Ophthalmol, 1999; 43: 89-90.

Kikuchi, M; Tenneti, L; Lipton, SA. Role of p38 mitogen-activated protein kinase in axotomy-induced apoptosis of rat retinal ganglion cells. J Neurosci, 2000; 20: 5037-5044.

Kitazawa, Y; Horie, T. Diurnal variation of intraocular pressure in primary open-angle glaucoma. Am J Ophthalmol, 1975; 79: 557-566.

Kittazawa, Y; Shirai, H; Go, FJ. The effect of Ca2+ antagonist on visual field in low-tension glaucoma. Graefes Arch Clin Exp Ophthalmol, 1989; 227: 408-412.

Klaver, JHJ; Greve, EL; Golsinga, H; Geijssen, HC, Heuvelmans, JH. Blood and plasma viscosity measurements in patients with glaucoma. Br J Ophthalmol, 1985; 69: 765-770.

Kumar, DM; Perez, E; Cai, ZY; Aoun, P; Brun-Zinkernagel, AM; Covey, DF; Simpkins, JW; Agarwal, N. Role of nonfeminizing estrogen analogues in neuroprotection of rat retinal ganglion cells. Free Radic Biol Med, 2005; 15; 39(12): 1676.

Lafuente, MP; Villegas-Pérez, MP; Sobrado-Calvo, P; García Aviles, A; Miralles de Imperial, J; Vidal-Sanz M. Neuroprotective effects of alpha (2)-selective adrenergic agonists against ischemia-induced retinal ganglion cell death. Invest Ophthalmol Vis Sci, 2001. 42: 2074-2084.

Lagreze, WA; Knorle, R; Back, M; Feuerstein, TJ. Memantine is neuroprotective in a rat model of pressure-induced retinal ischemia. Invest Ophthalmol Vis Sci, 1998; 39: 1063-1066.

Lai, RK; Chun, TY; Hasson, DW; Wheeler, LA. Activation of cell survival signalling pathway in the retina by selective alpha-2 adrenoceptor agonist brimonidine. Invest Ophthalmol Cis Sci, 1999; 10: 220-226.

Leibowitz, HM; Krueger, DE ; Maunder LR ; Milton, RC ; Kini, MM ; Kahn, HA; Nickerson, RJ; Pool, J; Colton, TL; Ganley, JP; Loewenstein, JI, Dawber, TR. The Framingham Eye Study monograph: an ophthalmological and epidemiological study of cataract, glaucoma, diabetic retinopathy, macular degeneration, and visual acuity in a general population of 2631 adults, 1973-1975. Surv Ophthalmol, 1980; 24: 335-610.

Levin, LA; Schlamp, CL; Spieldoch, RL; Geszvain, KM; Nickells RW. Identification of the bcl-2 family of genes in the rat retina. Invest Ophthalmol Vis Sci, 1997; 38: 2545-2553.

Levkovitch-Verbin, H; Harris-Cerruti, C; Groner, Y; Wheeler, LA; Schwartz, M; Yoles, E. RGC death in mice after optic nerve crush injury: oxidative stress and neuroprotection. Invest Ophthalmol Vis Sci, 2000; 41: 4169-4174.

Levkovitch-Verbin, H; Kalev-Landoy, M; Habot-Wilner, Z; Melamed, S. Minocycline delays death of retinal ganglion cells in experimental glaucoma and after optic nerve transaction. Arch Ophthalmol, 2006; 15: 520-526.

Levkovitch-Verbin, H; Quigley, HA; Martin, KR; Harizman, N; Valenta, DF; Pease, ME; Melamed, S. The transcription factor c-jun is activated in retinal ganglion cells in experimental rat glaucoma. Exp Eye Res, 2005; 80: 663-670.

Li, HY; Zhao, JL; Zhang, H. Transfection of brain-derived neurotrophic factor gene by recombinant adeno-associated virus vector in retinal ganglion cells in vitro. Zhonghua Yan Ke Za Zhi, 2008; 44(4): 354-60(abstract).

Li, RS; Tay, DK; Chan, HH; So, KF. Changes of retinal functions following the induction of ocular hypertension in rats using argon laser photocoagulation. Clin Exp Ophthalmol, 2006; 34: 575-583.

Libby, RT; Li, Y; Savinova, OV; Barter, J; Smith, RS; Nickells, RW; John, SW. Susceptibility to neurodegeneration in a glaucoma is modified by Bax gene dosage. PLoS Genet, 2005; 1: 17–26.

Liefner, M; Siebert, H; Sachse, T; Michel, U; Kollias, G; Bruck, W. The role of TNF-alpha during Wallerian degeneration. J Neuroimmunol, 2000; 108: 147-152.

Lisegang, TJ. Glaucoma. Changing concepts and future directions. Mayo Clin Proc, 1996; 7: 689-694.

Liu, B; Neufeld, AH. Expression of nitric oxide synthase-2 (NOS-2) in reactive astrocytes of the human glaucomatous optic nerve head. Glia, 2000; 30: 178–186.

Martin, KR; Levkovitch-Verbin, H; Valenta, D; Baumrind, L; Pease, ME; Quigley, HA. Retinal glutamate transporter changes in experimental glaucoma and after optic nerve transection in the rat. Invest Ophthalmol Vis Sci, 2002; 43: 2236-2243.

McKinnon, SJ. Glaucoma, apoptosis and neuroprotection. Curr Opin Ophthalmol, 1997; 8: 28-37.

McKinnon, SJ; Lehman, DM; Kerrigan-Baumrind, LA; Merges, CA; Pease, ME; Kerrigan, DF; Ransom, NL; Tahzib, NG; Reitsamer, HA; Levkovitch-Verbin, H; Quigley, DJ. and Zack, DJ. Caspase activation and amyloid precursor protein cleavage in rat ocular hypertension. Invest Ophthalmol, 2002; 43: 1077-1087.

Mechoulam, R; Shohami, E. Endocannabinoids and traumatic brain injury. Mol Neurobiol, 2007; 36: 68-77.

Merritt, JC; Crawford, WJ; Alexander, PC; Anduze, AL; Gelbart, SS. Effect of marijuana on intraocular and blood pressure in glaucoma. Ophthalmology, 1980; 87: 222-228.

Merritt, JC; McKinnon, S; Armstrong, JR; Hatem, G; Reid, LA. Oral Delta(9)-tetrahydrocannabinol, in heterogenous glaucoma. Ann Ophthalmol, 1980; 12: 947-950.

Minckler, DS; Bunt, AH; Johanson, GW. Orthograde and retrograde axoplasmic transport during acute ocular hyptertension in the monkey. Invest Ophthalmol Vis Sci, 1977; 16: 426–441.

Mittag, TW; Danias, J; Pohorenec, G; Yuan, HM; Burakgazi, E; Chalmers-Redman, R; Podos, SM; Tatton, WG. Retinal damage after 3 to 4 months of elevated intraocular pressure in a rat glaucoma model. Invest Ophthalmol Vis Sci, 2000; 41: 3451–3459.

Mozaffarieh, M; Flammer, J. A novel perspective on natural therapeutic approaches in glaucoma therapy. Expert Opin Emerg Drugs, 2007; 12(2): 195-198.

Mozaffarieh, M; Grieshaber, MC; Orgül, S; Flammer, J. The potential value of natural antioxidative treatment in glaucoma. Surv Ophthalmol, 2008; 53(5): 479-505.

Mulligan, SJ; MacVicar, BA. Calcium transients in astrocyte endfeet cause cerebrovascular constrictions. Nature, 2004; 431: 195–199.

Napper, GA; Pianta, MJ; Kalloniatis, M. Reduced glutamate uptake by retinal glial cells under ischemic/hypoxic conditions. Vis Neurosci, 1999; 16: 149-158.

Naskar, R; Dreyer, EB. New horizons in neuroprotection. Surv Ophthalmol, 2001; 45(3): 250-256.

Netland, PA; Chaturvedi, N; Dreyer, EB. Calcium channel blockers in the management of low tension and open angle glaucoma. Am J Ophthalmol, 1993; 115: 608-613.

Neufeld, AH; Sawada, A; Becker, B. Inhibition of nitric oxide synthase-2 by aminoguanidine provides neuroprotection of retinal ganglion cells in a rat model of chronic glaucoma. Proc Nat Acad Sci USA, 1999; 96: 9944-9948.

Newell, FW; Krill, AE. Diurnal tonography in normal and glaucomatous eyes. Trans Am Ophthalmol Soc, 1964; 62: 349-358.

Nickells, RW. Apoptosis of retinal ganglion cells in glaucoma: an update of the molecular pathways involved in cell death. Surv Ophthalmol, 1999; 43: 151-161.

Oku, H; Yamaguchi, H; Sugiyama, T; Kojima, S; Ota, M; Azuma, I. Retinal toxicity of nitric oxide released by administration of nitric oxide donor in the albino rabbit. Invest Ophthalmol Vis Sci, 1997; 38: 2540-2544.

Oltvai ZN; Korsmeyer SJ. Checkpoints of duelling dimmers foil death wishes. Cell, 1994; 79: 189-192.

Opere, CA; Zheng, WD; Zhao, M; Lee, JS; Kulkarni, KH; Ohia, SE. Inhibition of potassium- and ischemia-evoked [3H] D-aspartate release from isolated bovine retina by cannabinoids. Curr Eye Res, 2006; 31: 645-653.

Osborne NN, Casson RJ, Wood JP, Chidlow G, Graham M, Melen J. Retinal ischemia: mechanisms of damage and potential therapeutic strategies. Prog Retin Eye Res 2004; 23(1): 91-147.

Osborne, BA; Schwartz, LM. Essential genes that regulate apoptosis. Trends Cell Biol, 1994; 4: 394–399.

Phelps, CD; Corbett, JJ. Migraine and low-tension glaucoma. A case-control study. Invest Ophthalmol Vis Sci, 1985; 26; 1105-1108.

Porcella, A; Maxia, C; Gessa, GL; Pani, L. The synthetic cannabinoid WIN55212-2 decreases the intraocular pressure in human glaucoma resistant to conventional therapies. Eur J Neurosci, 2001 (13): 490-412.

Power, C; Henry, S; Del Bigio, MR; Larsen, PH; Corbett, D; Imai, Y; Yong, VW; Peeling, J. Intracerebral hemorrhage induces macrophage activation and matrix metalloproteinases. Ann Neurol, 2003; 53: 731-742.

Purnell, WD; Gregg JM. Delta(9)-tetrahydrocannabinol, euphoria and intraocular pressure in man. Ann Ophthalmol, 1975; 7: 921-923.

Quigley HA, Addicks EM. Regional differences in the structure of the lamina cribrosa and their relation to glaucomatous optic nerve damage. Arch Ophthalmol, 1981; 99: 137-144.

Quigley, HA. Ganglion cell death in glaucoma: pathology recapitulates ontogeny. Aust N Z J Ophthalmol, 1995; 23: 85–91.

Quigley, HA; Addicks, EM. Chronic experimental glaucoma in primates. II. Effect of extended intraocular pressure elevation on optic nerve head and axonal transport. Invest Ophthalmol Vis Sci, 1980; 19: 137–152.

Quigley, HA; Addicks, EM; Green, WR; Maumenee, AE. Optic nerve damage in human glaucoma: II. The site of injury and susceptibility to damage. Arch Ophthalmol, 1981; 99: 635–649.

Quigley, HA; Engerm, C; Katz, J; Sommer, A; Scottm, R; Gilbertm, D. Risk factors for the development of glaucomatous visual field loss in ocular hypertension. Arch Ophthalmolol, 1994; 112; 644-649,

Quigley, HA; Hohman, RM; Addicks, EM; Green, WR. Blood vessels of the glaucomatous optic disk in experimental primate and human eyes. Invest Ophthalmol Vis Sci, 1984; 25: 918-931.

Quigley, HS; Broman, AT. The number of people with glaucoma worldwide in 2010 and 2020. Br J Ophthalmol, 2006; 90(3): 253-254.

Rabacchi, SA; Ensini, M; Bonfanti, I; Grarina, A; Maffei, L. Nerve growth factor reduces apoptosis of axotomized retinal ganglion cells in the neonatal rat. Neuroscience, 1994; 63: 969-973.

Radius, RL; Anderson, DR. Rapid axonal transport in primate optic nerve. Arch Ophthalmol, 1981; 99: 650–654.

Ransom, B; Behar, T; Nedergaard, M. New roles for astrocytes (stars at last). TRENDS Neurosci, 2003; 26: 520–522 .

Resta, V; Novelli, E; Vozzi, G; Scarpa, C; Caleo, M; Ahluwalia, A; Solini, A; Santini, E; Parisi, V; Di Virgilio, F; Galli-Resta, L. Acute retinal ganglion cell injury caused by intraocular pressure spikes is mediated by endogenous extracellular ATP. Eur J Neurosci, 2007; 25: 2741–2754.

Ruiz, G; Wheeler, L; WoldeMussie, E; Wheeler, LA. Time course of pre or post treatment by brimonidine on neuroprotection in rat optic nerve injury model. Invest Ophthalmol Vis Sci, 2000; 41: 830.

Russo, R; Cavaliere, F; Watanabe, C; Nucci, C; Bagetta, G; Corasaniti, MT; Sakurada, S; Morrone, LA. 17Beta-estradiol prevents retinal ganglion cell loss induced by acute rise of intraocular pressure in rat. Prog Brain Res, 2008; 173: 583-590.

Ryan, ME; Usman, A; Ramamurthy, NS; Golub, LM; Greenwald, RA. Excessive matrix metalloproteinase activity in diabetes: inhibition by tetracycline analogues with zinc reactivity. Curr Med Chem, 2001; 8: 305-316.

Sadowski, T; Steinmeyer, J. Effects of tetracyclines on the production of matrix metalloproteinases and plasminogen activators as well as of their natural inhibitors, tissue inhibitor of metalloproteinases-1 and plasminogen activator inhibitor-1. Inflamm Res, 2001; 50: 175-182.

Sanchez Mejia, RO; Ona, VO; Li, M; Friedlander, RM. Minocycline reduces traumatic brain injury-mediated caspase-1 activation, tissue damage, and neurological dysfunction. Neurosurgery, 2001; 48: 1393-1401.

Schulzer, M; Drance, SM; Carter, CJ; Brooks, DE; Douglas, GR; Lau, W. Biostatistical evidence for two distinct chronic open angle glaucoma populations. Br J Ophthtalmol, 1990: 74: 196-200.

Schwartz, M. Modulating the immune system: a vaccine for glaucoma? Can J Ophthalmol, 2007; 42: 439-441.

Seki, M, Lipton, SA. Targeting excitotoxic/free radical signaling pathways for therapeutic intervention in glaucoma. Prog Brain Res, 2008; 173: 495-510.

Shin, DH; Becker, B; Kolker, AE. Family history in primary open-angle glaucoma. Arch Ophthalmol, 1977; 95: 598-600.

Simpkins, JW; Wang, J; Wang, X; Perez, E; Prokai, L; Dykens, JA. Mitochondria play a central role in estrogen-induced neuroprotection. Curr Drug Targets CNS Neurol Disord, 2005; 4(1): 69-83.

Sirén, AL; Fratelli, M; Brines, M; Goemans, C; Casagrande, S; Lewczuk, P; Keenan, S; Gleiter, C; Pasquali, C; Capobianco, A; Mennini, T; Heumann, R; Cerami, A; Ehrenreich, H; Ghezzi, P. *"Erythropoietin prevents neuronal apoptosis after cerebral ischemia and metabolic stress". Proc Natl Acad Sci USA, 2001; 98: 4044–4049.*

Smith, JR; Gabler, WL. Effects of doxycycline in two rat models of ischemia/reperfucion injury. Proc West Pharmacol So, 1994; 37: 3-4.

Smith, JR; Gabler, WL. Protective effects of doxycycline in mesenteric ischemia and reperfusion. Res Commun Mol Pathol Pharmacol, 1995; 88: 303-315.

Sommer, A ; Tielsch, JM ; Katz, J. et al. Relationship between intraocular pressure and primary open angle glaucoma among white and black Americans. The Baltimore Eye Survey. Arch Ophthalmol, 1991; 109: 1090-1095.

Sommer, A. Intraocular pressure and glaucoma. Am J Ophthalmol, 1989; 107: 186-188.

Spaeth, GL. Fluorescein angiography: its contributions towards understanding the mechanisms of visual loss in glaucoma. Trns Am Ophthalmol Soc, 1975; 73: 491-553.

Spray, DC; Ye, ZC; Ransom, BR. Functional connexion “hemichannels”: a critical appraisal. Glia, 2006; 54: 758–773.

Stys, PK. White matter injury mechanisms. Curr Mol Med, 2004; 4: 113–130.

Tamaki, Y; Araie, M; Tomita, K; Nagahara, M. Effect of topical betaxolol on tissue circulation in the human optic nerve head. *J Ocul Pharmacol Ther,* 1999; 15: 313-321.

Tezel, G. and Wax, MB. Increased production of tumor necrosis factor-alpha by glial cells exposed to simulated ischemia or elevated hydrostatic pressure induces apoptosis in cocultured retinal ganglion cells. J Neurosci, 2000; 20: 8693-8700.

Tezel, G. and Yang, X. Caspse-independent component of retinal ganglion cell death, in vitro. Invest Ophthalmol Vis Sci, 2004; 45: 4049-4059.

Tezel, G. Oxidative stress in glaucomatous neurodegeneration: mechanisms and consequences. Prog Retin Eye Res, 2006; 25: 490-513.

Tezel, G; Li, LY; Patil, RV. and Wax, MB. Tumor necrosis factor-alpha and its receptor-1 in the retina of normal and glaucomatous eyes. Invest Ophthalmol Vis Sci, 2001; 42: 1787-1794.

Tezel, G; Yang, X; Yang, J. and Wax, MB. Role of tumor necrosis factor receptor-1 in the death of retinal ganglion cells following optic nerve crush injury in mice. Brain Res, 2004; 996: 202-212.

Thanos S, Bahr M, Barde YA, Vanselow J. Survival and axonal elongation of adult rat retinal ganglion cells: in vitro effects of lesioned sciatic nerve and brain-derived neurotrophic factor (BDNF). Eur J Neurosci, 1989; 1: 19-26.

The AGIS Investigators. The Advanced Glaucoma Intervention Study (AGIS): 7. The relationship between control of intraocular pressure and visual field deterioration. Am J Ophthalmol, 2000; 130(4): 429-440.

The Ischemic Optic Neuropathy Desompression Trial Research Group. Optic nerve descompression surgery for nonarteritic anterior ischemic optic neuropathy (NAION) is not effective and may be harmful. The Ischemic Optic Neuropathy Descompression Trial Research Group. JAMA, 2000; 273(8): 625-632.

Tielsch, JM; Katz, J; Sommer, A; Quigley, HA; Javitt, JC. Hypertension, perfusion pressure and primary open-angle glaucoma. A population-based assessment. Arch Ophthalmol, 1995; 113: 216-221.

Tikka, TM; Koistinaho, JE. Minocycline provides neuroprotection against N-methyl-D-aspartate neurotoxicity by inhibiting microglia. J Immunol, 2001; 166: 7527-7533.

Trope, GE; Salinas, SA; Glynn, M. Blood viscosity in primary open-angle glaucoma. Can J Ophthalmol, 1987; 22: 202-204.

Urcola, JH; Hernández, M; Vecino, E. Three experimental glaucoma models in rats: Comparison of the effects of intraocular pressure elevation on retinal ganglion cell size and death. Exp Eye Res, 2006; 83: 429-437.

Van der Stelt, M; Di Marzo, V. Cannabinoid receptors and their role in neuroprotection. Neuromolecular Med, 2005; 7: 37-50.

Vecino E, Caminos E, Ugarte M, Martín-Zanca D, Osborne NN. Inmunohistochemical distribution of neurotrophins and their receptors in the rat retina and the effects of ischemia and reperfusion. Gen Pharmacol, 1998; 30(2): 305-314.

Vecino E, Ugarte M, Nash MS, Osborne NN. NMDA induces BDNF expression in the albino rat retina in vivo. Neuroreport, 1999; 10(5): 1103-1106.

Vorwerk, CK; Lipton, SA; Zurakowski, D; Hyman, BT; Sabel, BA; Dreyer, EB. Chronic low-dose glutamate is toxic to retinal ganglion cells. Toxicity blocked by mementine. Invest Ophthalmol Vis Sci, 1996; 37: 1618-1624.

Ward MS; Khoobehi, A; Lavik, EB; Langer, R; Young, MJ. Neuroprotection of retinal ganglion cells in DBA/2J mice with GDNF-loaded biodegradable microspheres. *J Pharm Sci*, 2007; 96: 558-568.

Weinreb, RN. Glaucoma neuroprotection: What is it? Why is it needed? Can J Ophthalmol, 2007; 42: 396-398.

Wheeler, LA; Gil, DW; WoldeMussie, E. Role of alpha-2 adrenergic receptors in neuroprotection and glaucoma. Surv Ophthalmol, 2001; 45(3): 290-296.

Wheeler, LA; Tatton, NA; Elstner, M. Alpha-2 adrenergic receptor activation by brimonidine reduces neuronal apoptosis through Akt (protein Kinase B) dependent new synthesis of BCL-2. Invest Ophthalmol Vis Sci, 2001; 42: 411.

Wheeler, LA; Woldemussie, E. Alpha-2 adrenergic receptor agonists are neuroprotective in experimental models of glaucoma. Eur J Ophthalmol, 2001; 11: 30-35.

Wheeler, LA; WoldeMussie, E; Lai, R. Role of alpha-2 agonist in neuroprotection. Surv Ophthalmol, 2003; 48: 47-51.

Wilson, MR; Hertzmark, E; Walker, AM; Childs-Shaw, K; Epstein, DL. A case-control study of risk factors in open angle glaucoma. Arch Ophthalmol, 1987; 105: 1066-1071.

Wood, JP; Schmidt, KG; Melena, J; Chidlow, G; Allmeier, H; Osborne, NN. The beta adrenoceptor antagonists metipranolol and timolol are retinal neuroprotectants: comparison with betaxolol. Exp Eye Res, 2003; 76(4): 505-516.

Yan, X; Tezel, G; Wax, MB; Edward DP. Matrix metalloproteinases and tumor necrosis factor alpha in glaucomatous optic nerve head. Arch Ophthalmol, 2000. 118: 666-673.

Yoles, E; Belkin, M; Schwartz, M. HU-211, a nonpsychotropic cannabinoid, produces short- and long-term neuroprotection after optic nerve axotomy. J Neurotrauma, 1996; 13: 49-57.

Yoles, E; Wheeler, LA; Schwartz, M. Alpha2-adrenoreceptor agonists are neuroprotective in a rat model of optic nerve degeneration.Invest Ophthalmol Vis Sci, 1999; 40: 65-73.

Yrjanheikki, J; Keinanen, R; Pellikka, M; Hokfelt, T; Koistinaho, J. Tetracyclines inhibit microglial activation and are neuroprotective in global brain ischemia. Proc Natl Acad Sci U S A, 1998; 95: 15769-15774.

Yrjanheikki, J; Tikka, T; Keinanen, R; Goldsteins, G; Chan, PH; Koistinaho, JA. A tetracycline derivative, micocycline, reduces inflammation and protects against focal cerebral ischemia with a wide therapeutic window. Proc Natl Acad Sci USA, 1999; 96: 13496-13500.

Zangwill, LM; Bowd, C; Berry CC; Williams, J; Blumenthal, EZ; Sánchez-Galeana, CA; Vasile, C; Weinreb, RN. Discriminating between normal and glaucomataous eyes using the Heidelberg retina tomography, GDx Nerve Fiber Analyzer, and Optical Coherence Tomograph. Arch Ophthalmol, 2001; 119(7): 985-993.

Zhang, C; Lei, B; Lam, TT; Yang, F ; Sinha, D; Tso, MO. Neuroprotection of photoreceptors by minocycline in light-induced retinal degeneration. Invest Ophthalmol Vis Sci, 2004; 45: 2753-2759.

Zhang, X; Reigada, D; Zhang, M; Laties, AM; Mitchell, CH. Increased ocular pressure increases vitreal levels of ATP. Association for Research in Vision and Ophthalmology (ARVO), 2006; Abstract 426.

Zhang, X; Zhang, M; Laties, AM; Mitchell, CH. Balance of purines may determine life or death as A3 adenosine receptors prevent loss of retinal ganglion cells following P2X7 receptor stimulation. J Neurochem, 2006; 98: 566-575.

Zhang, X; Zhang, M; Laties, AM; Mitchell, CH. Stimulation of P2X7 receptors elevates Ca2_ and kills retinal ganglion cells. Invest Ophthmol Vis Sci, 2005; 46: 2183–2191.

Zhong, L; Bradley, J; Schubert, W; Ahmed, E; Adamis, AP; Shima, DT; Robinson, GS; Ng, YS. Erythropoietin promotes survival of retinal ganglion cells in DBA/2J glaucoma mice. Invest Ophthalmol Vis Sci, 2007; 48(3): 1212-1218.

Zhou, W; Zhu, X; Zhu, L; Cui, YY; Wang, H; Qi, H; Ren, QS; Chen, HZ. Neuroprotection of muscarinic receptor agonist pilocarpine against glutamate-induced apoptosis in retinal neurons. Cell Mol Neurobiol. 2008; 28(2): 263-275.

Zhou, X; Li, F; Ge, J; Sarkisian, SRJr, Tomita H; Zaharia, A; Chodosh, J; Cao, W. Retinal ganglion cell protection by 17-beta-estradiol in a mouse model of inherited glaucoma. Dev Neurobiol, 2007; 67(5): 603-616.

Zhu, S; Stavrovskaya, IG; Drozda, M; Kim, BY; Ona, V; Li, M; Sarang, S; Liu, AS; Hartley, DM; Wu, DC; Gullans, S; Ferrante, RJ; Przedborski, S; Kristal, BS; Friedlander, RM. Minocycline inhibits cytochrome c release and delays progression of amyotrophic lateral sclerosis in mice. Nature, 2002; 417: 74-78.

In: Advances in Medicine and Biology. Volume 20
Editor: Leon V. Berhardt

ISBN 978-1-61209-135-8

Chapter 9

PHARMACOLOGICAL ASPECTS OF BORATES

Iqbal Ahmad[a], Sofia Ahmed[a], Muhammad Ali Sheraz[a4], Kefi Iqbal[b] and Faiyaz H. M. Vaid[c]

[a]Institute of Pharmaceutical Sciences, Baqai Medical University, Toll Plaza, Super Highway, Gadap Road, Karachi-74600, Pakistan

[b]Baqai Dental College, Baqai Medical University, Toll Plaza, Super Highway, Gadap Road, Karachi-74600, Pakistan

[c]Department of Pharmaceutical Chemistry, Faculty of Pharmacy, University of Karachi, Karachi-75270, Pakistan

ABSTRACT

Boric acid and sodium borate are pharmaceutical necessities and are used as anti-infective agents and as a buffer in pharmaceutical formulations. Boric acid is used as an antimicrobial agent for the treatment of Candida, Aspergillus and Trichomonas infections. It is employed in the treatment of psoriasis, acute eczema, dermatophytoses, prostrate cancer, deep wounds and ear infections. Borates are toxic in nature and produce symptoms of poisoning on exposure to mineral dust, oral ingestion and topical use. Boric acid induces developmental and reproductive toxicity in mice, rats and rabbits. It has been shown to produce cytotoxic, embryotoxic, genotoxic, ototoxic and phytotoxic effects. Borates are considered to be completely absorbed by the oral route. They are also absorbed through denuded or irritated skin. Boric acid causes changes in lipid metabolism and in acid-base equilibrium of the blood. Boric acid appears to modulate certain inflammatory mediators and to regulate inflammatory processes. It takes part in the reversible or irreversible inhibition of certain enzyme systems. Borates are widely used as a preservative and insecticide.

[4] Corresponding author: E-mail address: ali_sheraz80@hotmail.com. Tel: +92-21-4410293; Fax: +92-21-4410439.

1. Introduction

Boron is a nonmetal and exists only in the +3 oxidation state. It does not occur free in nature and is found in combination with oxygen as oxyacids including metaboric acid (HBO_2), orthoboric acid (H_3BO_3) and tetraboric or pyroboric acid ($H_2B_4O_7$), and borates such as sodium borate or borax ($Na_2B_4O_7.10H_2O$), sodium perborate ($NaBO_3.4H_2O$) and calcium borate ($CaB_4O_7.4H_2O$). Boric acid and sodium borate are the two pharmaceutical necessities and are cited in the British Pharmacopoeia (2009) and the United States Pharmacopeia-National Formulary (2007). Sodium borate is found in large quantities in California as a crystalline deposit. Boric acid is produced from sodium borate or other borates by reacting with hydrochloric acid or sulphuric acid. Humans consume daily about a milligram of boron, mostly from fruits and vegetables. At high doses, boron is a developmental and reproductive toxin in animals. Prolonged exposure to pharmacologically active levels of boric acid, the naturally occurring form of boron in human plasma, induces morphological changes in cells. Boric acid is a component of cosmetics, pharmaceuticals, agricultural products and consumer goods and is used in numerous industrial processes. The potential widespread human exposure to borates may cause hypotension, metabolic acidosis, oliguria, chemesthesis and other symptoms. The clinical uses, toxicity, pharmacokinetics, metabolism and related aspects of borates are presented in the following sections:

2. Clinical Uses

2.1. Antibacterial Agent

Boric acid and sodium borate are medicinally important compounds. Boric acid is weakly bacteriostatic and is used as a topical anti-infective in liquid dosage forms. It is non-irritating and its solutions are suitable for application to the cornea of the eye. Aqueous solutions of boric acid are employed as eyewash, mouthwash and for irrigation of the bladder. A 2.2% solution is isotonic with lacrimal fluid but such solutions will hemolyze red blood cells. It is also used as a dusting powder, on dilution with some inert material and can be absorbed through irritated skin. Sodium borate is bacteriostatic and has been employed as an ingredient of cold creams, eyewashes and mouthwashes. It is used as an alkalizing agent in denture adhesives. Sodium perborate is an oxidizing agent and is used as a local anti-infective. Various borate buffers are used to maintain the pH of alkaline pharmaceutical formulations (Soine and Wilson, 1967; O'Neil, 2001; Yelvigi, 2005; Reilly, 2006; Sweetman, 2007).

Combined treatment of purulent pyonecrotic lesions of lower extremities in diabetic patients has been performed by means of a permanent abacterial medium using 2% boric acid as an antiseptic (Shaposhnikov and Zorik, 2001). The minimum inhibitory concentration (MIC) of boric acid for bacteria is: Gram negative 800-12800 mg/L and Gram positive 1600-6400 mg/L (Grzybowska *et al.*, 2007).

Urine samples collected at home and in hospital have been preserved with boric acid for storage before processing (Porter and Brodie, 1969; Lum and Meers, 1989; Jewkes *et al.*, 1990; Meers and Chow, 1990; Lee and Critchley, 1998; Gillespie *et al.*, 1999; Thongboonkerd and Saetun, 2007; Thierauf, 2008). Estrogen metabolites in breast cancer risk, excreted in urine, are preserved with boric acid before assay with an ELISA kit (Falk *et al.*, 2000). A preservation medium composed of nutrient broth, 1.8% boric acid, and 1% sodium chloride at pH 7.0 has been described to maintain the stability of *Escherichia coli* cultures for up to 10 days at room temperature (Brodsky *et al.*, 1978).

Phenylmercuric borate is used as an alternative antimicrobial preservative to phenylmercuric acetate or phenylmercuric nitrate in local anesthetic preparations and eye drops. It is more soluble in water than phenylmercuric nitrate and is less irritant than either phenylmercuric acetate or phenylmercuric nitrate (Abdelaziz and El-Nakeeb, 1988; Kodym *et al.*, 2003; British Pharmacopoeia, 2009).

2.2. Antifungal Agent

Boric acid is a fungistatic agent and has been used in the treatment of chronic vulvovaginal candidiasis (Van Slyke *et al.*, 1981; Jovanovic *et al.*, 1991; Prutting and Cerveny, 1998; Guaschino *et al.*, 2001; Romano *et al.*, 2005; Ray *et al.*, 2007a,b; Das Neves *et al.*, 2008). It is effective against infections caused by various Candida species (*Candida albicans, Candida glabrata, Candida krusei, Candida parapsilosis*) (Sobel and Chaim, 1997; Otero *et al.*, 1999; Singh *et al.*, 2002; Sobel *et al.*, 2003; Van Kessel *et al.*, 2003; Romano *et al.*, 2005; Nyirjesy *et al.*, 2005; Deseta *et al.*, 2008), non-*Candida albicans* (Sood *et al.*, 2000; Otero *et al.*, 2002), *Trichomonas vaginalis* (Aggarwal and Shier, 2008), and *Aspergillus niger* (Avino-Martinez *et al.*, 2008). Boric acid has been used in the treatment of psoriasis (Limaye and Weightman, 1997), trophic ulcers (Izmailov and Izmailov, 1998), trabecular bone quality (Sheng *et al.*, 2001), dermatophytoses (Patel and Agrawal, 2002), acute eczema (Bai *et al.*, 2007) and neural morphallaxis (Martinez *et al.*, 2008). As a fungistatic agent, boric acid is used in the form of a 2.5% aqueous-alcoholic solution (30:70, v/v), vaginal suppositories (600 mg), intravaginal capsules (600 mg) and topical creams.

2.3. Chemopreventive Agent for Human Cancer

Toxicity and carcinogenicity studies of boric acid in male and female mice have indicated that this is a noncarcinogenic chemical (Dieter, 1994). Boric acid and boronated compounds are used as boron carriers for boron neutron capture therapy in the treatment of glioblastoma, melanoma and other conditions. The suitability of these compounds in the treatment of cancer has been evaluated on the basis of pharmacokinetic studies characterizing their biodistribution, tumor uptake and tumor selectivity (Laster *et al.*, 1991; Gregoire *et al.*, 1993) and the effect of electroporation on cell killing (Ono *et al.*, 1998a,b; Kinashi *et al.*, 2002). The effectiveness of boron neutron capture in killing tumor cells depends on the number of ^{10}B atoms delivered to the tumor, the subcellular distribution of ^{10}B and the thermal neutron fluence at the side of the tumor. The presence of 600 ppm ^{10}B (boric acid) in the cell medium during irradiation with d(14) + Be neutrons enhances the DNA damage by 20% compared to that of neutron irradiation alone (Poller *et al.*, 1996). Boric acid inhibits the proliferation of fibroblastoid murine L929-cells in a linear dose-dependent manner (Walmod *et al.*, 2004). Bor-tezomib, a dipeptide boronate proteasome inhibitor, shows activity in the treatment of multiple myeloma (Liu *et al.*, 2008).

Boric acid and 3-nitrophenyl boronic acid have been found to inhibit human prostate cancer cell proliferation. Studies using DU-145 prostate cancer cells showed that boric acid induces a cell death-independent proliferative inhibition, with little effect on cell cycle stage distribution and mitochondrial function. It causes a dose-dependent reduction in cyclins A-E, as well as MAPK proteins, suggesting their contribution to proliferative inhibition. Boric acid is involved in the inhibition of the enzymatic activity of prostate-specific antigen which is a well-established marker of prostate cancer (Gallardo-Williams *et al.*, 2003, 2004; Barranco and Eckhert, 2004, 2006). It also inhibits NAD^+ and $NADP^+$ as well as the release of stored Ca^{2+} in growing DU-145 prostate cancer cells and thus impairs Ca^{2+} signaling (Barranco *et al.*, 2009). Exposure to boric acid and calcium fructoborate results in the inhibition of the proliferation of breast cancer cells in a dose-dependent manner (Scorei *et al.*, 2008). Increased groundwater concentrations of boric acid correlate with reduced risk of prostate cancer incidence and mortality. It also improves the anti-proliferative effectiveness of chemopreventive agents, selenomethionine and genistein, which enhances the ionization radiation effect on cell kill (Barranco *et al*, 2007).

2.4. Wound Treatment

Boric acid has been found to play an important role in wound healing. A 3% boric acid solution has been used in the treatment of deep wounds with loss of tissue. Dramatic improvement in wound healing was observed through its action on the extracellular matrix. Boron derivatives (triethanolamine borate, N-diethyl- phosphoramidate- propylboronic acid, 2,2- dimethylhexyl-1,3- propanediol- aminopropylboronate and 1,2- propanediol-aminopropylboronate) tested in a study mimicked the effects of boric acid (Blech *et al.*, 1990; Borrelly *et al.*, 1991; Benderdour *et al.*, 2000). A mixture consisting of aqueous boric acid and calcium hypochlorite solutions is widely used in the management of open wound healing (Salphale and Shenoi, 2003).

The presence of boron has been found to decrease the synthesis of proteoglycan, collagen and total proteins in pelvic cartilage of chick embryo but increases the release of these macromolecules (Benderdour *et al.*, 1997). The amount of phosphorylated proteins is enhanced in presence of boron while endoprotease activity in cartilage is significantly increased. The *in vitro* effects of boric acid may explain its *in vivo* effects on wound healing (Bennderdour *et al.*, 1997). Boric acid modulates extracellular matrix and tumor necrosis factor-alpha (TNF-alpha) synthesis in human fibroblasts. Total mRNA levels are higher after boric acid treatment and reach a maximum after 6 h. The effects of boric acid observed in wound repair may be due to TNF-alpha synthesis and secretion (Benderdour *et al.*, 1998).

2.5. Ear Infections

Boric acid has been used effectively as a topical antiseptic for treating asteatosis (Karakus *et al.*, 2003), otomycosis (Del Palacio *et al.*, 2002; Ozcan *et al.*, 2003), otits externa (Mendelsohn *et al.*, 2005) and chronic suppurative otitis media in children (Gao *et al.*, 1999; Macfadyen *et al.*, 2005; Minja *et al.*, 2006). A 4% aqueous solution of boric acid has been found to produce ototoxic effects when applied to the middle ear of guinea pigs (Ozturkcan *et al.*, 2009).

2.6. Dental Treatment

Hydrated sodium tetraborate (borax) is a component of glass-ionomer restorative cements in dentistry for its useful properties (Wilson and Kent, 1972; Bansal *et al.*, 1995). Boric acid is also a component of dental cements, commonly incorporated into glass (as an ingredient in the melt) and occasionally added to the powder component of the glass-ionomer cement (Prentice *et al.*, 2006). Boric acid as HBO_3^{2-} or BO_3^{3-}, may act as a weak polyalkenoate cross-linker in the acid-base glass-ionomer reaction involved in the setting process (Bochek *et al.*, 2002). Sodium perborate (30-35%) has been used as a bleaching agent for intracoronal bleaching of teeth (Teixeira *et al.*, 2003, 2004; Shinohara *et al.*, 2004, 2005; Piemjai and Surakompontorn, 2006; Amaral *et al.*, 2008), root filled discolored teeth (Lim *et al.*, 2004), artificially discolored teeth (Lee *et al.*, 2004; Baumler *et al.*, 2006), artificially stained primary teeth (Campos *et al.*, 2007) and root filled teeth (Yui *et al.*, 2008). It has been found to be the best intracoronal bleaching agent for teeth (Timpawat *et al.*, 2005).

3. TOXICITY

3.1. Human Toxicity

The improper use of boric acid containing antiseptics is one of the most common causes of toxic accidents in newborns and infants. Health hazards may also arise from inadvertent absorption of insecticides and household products containing borates as well as from occupational accidents related to production and use of borates. Most of the boronated compounds with hypolipidemic, antiinflammatory or anticancer properties have been found to be highly toxic at the required therapeutic dosages in animals (Locatelli *et al.*, 1987; Craan *et al.*, 1997). The absorption of boric acid from damaged skin may reach toxic levels and cause fatal poisoning, particularly in infants, on topical application to burns, denuded areas, granulation tissues and serous cavities. It may cause serious poisoning from oral ingestion of as little as 5 g. Symptoms of poisoning include nausea, vomiting, abdominal pain, diarrhea, headache, depression, alopecia and visual disturbance. The kidney may be injured resulting in death. Boric acid accumulates principally in the brain, liver and body fat, with the greatest amount in the brain. It is the grey matter of the cerebrum and the spinal cord with the greatest content of boron following death.

The potential widespread human exposure to borates may cause hypotension, metabolic acidosis, oliguria, chemesthesis and other symptoms (Soine and Wilson, 1967; Browning, 1969; Gosselin *et al.*, 1976; Seigel and Wilson, 1986; Linden *et al.*, 1986; Litovitz *et al.*, 1988; Kiesche-Nesselrodt and Hooser, 1990; Werfel *et al.*, 1998; Teshima *et al.*, 2001; Beckett *et al.*, 2001; Pahl *et al.*, 2005; Reilly, 2006; Cain *et al.*, 2008).

Accidental boric acid poisoning by food containing boric acid as a preservative (Chao *et al.*, 1991; Tangermann *et al.*, 1992) and by ingestion of a boric acid containing pesticide by children (Hamilton and Wolf, 2007) has been reported. Several cases of acute ingestions of boric acid resulting in death have occurred (Restuccio *et al.*, 1992; Ishii *et al.*, 1993; Matsuda *et al.*, 2004). Human exposure to boron and its compounds can arise from a variety of natural sources, such as soil and water, or through its use in pesticides, preservatives, pharmaceuticals and household products (Siegel and Wilson, 1986; Woods, 1994; Moore, 1997; Richold, 1998; Coughlin, 1998; Hubbard, 1998; Cain *et al.*, 2008) and may be hazardous to health.

3.2. Embryotoxicity

It has been observed that fishes in the embryo-larval stage of development are sensitive to boron (0-500 μM) as boric acid. Chronic exposure below 9 μM boron/L impairs embryonic growth and above 10 mM boron/L causes death (Rowe *et al.*, 1998). Boric acid induces dysmorphogenesis in the skull, vertebral column and ribs in both human population and in laboratory animals (Tyl *et al.*, 2007).

Prenatal exposure of rats to elevated levels of boric acid (500 mg/kg) causes embryonic malformations of the axial skeleton involving the head, sternum, ribs and vertebrae (Norotsky *et al.*, 1998). Boric acid (500-750 mg/kg), administered to pregnant CD-1 mice once daily on gestational days 6-10 has been shown to produce ribogenesis, a rare effect in laboratory animals. It has been suggested that boric acid affects early processes such as gastrulation and presomitic mesoderm formation (Cherrington and Chernoff, 2002). Axial skeletal defects in boric acid-exposed rat embryos are associated with anterior shifts of hox gene expression domains (Wery *et al.*, 2003). The skeletal defects resulting from combined exposure to hyperthermia and boric acid are additive for segmentation defects and synergistic for the reduction in numbers of vertebrae (Harrouk *et al.*, 2005). Boric acid inhibits embryonic histone deacetylases in mouse suggesting a molecular mechanism for the induction of skeletal malformations and boric acid-related teratogenicity (Di Renzo *et al.*, 2007). The embryonic stem cell test (EST) has been developed as a ECVAM-validated assay to detect embryotoxicity (Peters *et al.*, 2008a). The automated image recording of contractile cardiomyocyte-like cells in the EST allows for an unbiased high throughput method to assess the relative embryotoxic potency of some test compounds. The values obtained for these compounds are: 6-aminonicotinamide (1) > valproic acid (0.007-0.013) > boric acid (0.002-0.005) > penicillin G (0.00001) (Peters *et al.*, 2008b).

Blood boron concentrations of 1.27 ± 0.298 and 1.53 ± 0.546 μg B/g are associated with the no-observed-adverse effect level (10 mg/B/kg/d) and lowest-observed-adverse-effect level (13 mg B/kg/d) with developmental toxicity in pregnant rats fed boric acid throughout gestation (Price *et al.*, 1997).

3.3. Developmental and Reproductive Toxicity

Boron occurs most frequently in nature as borates and boric acid and is essential for higher plants and vertebrates. Humans consume daily about 1 mg of boron, mostly from fruits and vegetables. At high doses, boron is a developmental and reproductive toxin in animals. An oral NOAEL (no-observed-adverse-effect-level) of 9.6 mg B/kg/day has been established for developmental toxicity in Sprague-Dawley rats fed boric acid (Pahl *et al.*, 2001). Boric acid and inorganic borates cause both the developmental and reproductive toxicity in mice (Harris *et al.*, 1992), rats (Chapin *et al.*, 1997, 1998; Fail *et al.*, 1998; Wang *et al.*, 2008), and rabbits (Price *et al.*, 1996a).

Several studies have been conducted on the developmental toxicity of boric acid in mice, rats and rabbit (Heindel *et al.* 1992, 1994; Price *et al.*, 1996b, 1998; Vaziri *et al.*, 2001). The developmental toxicity NOAEL in the rat is 0.075% boric acid (55 mg/kg/day) on gestational day 20 and 0.1% boric acid (74 mg/kg/day) on postnatal day 21 (Price *et al.*, 1996b). A human risk assessment study has been conducted to derive an appropriate safe exposure level of boric acid and borax in drinking water. A rat developmental toxicity study with a NOAEL of 9.6 mg B/kg/day was selected as a basis for the risk assessment since it represents the most sensitive endpoint of toxicity. The results showed that boron in US drinking water (0.031 B/L) is much below the toxic level (2.44 mg B/L) and does not pose any health risk to the public (Murray, 1995).The benchmark dose (BMD) approach has been proposed as an alternative basis for reference value calculations in the assessment of developmental toxicity (Allen *et al.*, 1996).

The potential reproductive toxicity of boric acid in Swiss mice has been evaluated using the Reproductive Assessment by Continuous Breeding (RACB) protocol. The results establish the reproductive toxicity of boric acid in mice and demonstrate that the male is the most sensitive sex (Fail *et al.*, 1991).

Acute oral exposure to boric acid (2000 mg/kg) adversely affects spermiation and sperm quality in the male rat. The no effect level is 500 mg/kg (Linder *et al.*, 1990). Development of testicular lesions in adult rats after treatment with high dose boric acid (6000-9000 ppm) has been observed. This is characterized by decreased fertility, sperm motility and inhibited spermiation followed by atrophy (Treinen and Chapin, 1991; Ku *et al.*, 1993a,b; Chapin and Ku, 1994; Yoshizaki *et al.*, 1999). A mechanism of the testicular toxicity of boric acid based on the involvement of serine proteases plasminogen activators in spermiation has been proposed (Ku and Chapin, 1994). Boric acid feed (4500-9000 ppm) in mice is a potent reproductive toxicant in males and females (Fail *et al.*, 1991).

Collaborative studies to evaluate toxicity on male reproductive system by repeated dose studies in rats have shown that at the appropriate dose (500 mg/kg), the testicular toxicity of boric acid can be detected after only two weeks of repeated daily oral treatment (Kudo *et al.*, 2000; Fukuda *et al.*, 2000). Boric acid has been evaluated for reproductive toxicity in *Xenopus laevis*. This model appears to be a useful tool in the initial assessment of potential reproductive toxicants for further testing (Fort *et al.*, 2001).

3.4. Genotoxicity

The genotoxic potential of boric acid in *Escherichia coli* PQ 37 has been assessed in the presence of aflatoxin B1 using SOS chromotest. Boric acid induces beta-galactosidase synthesis on the tester bacteria both in the presence and absence of S9 activation mixure. It may possibly act as a syngenotoxic and/or a cogenotoxic agent (Odunola, 1997).

The effect of boric acid as a food preservative on root tips of *Allium cepa* L has been studied. Boric acid (20 to 100 ppm) reduced mitotic division in *A. cepa* compared with that of the respective control. Mitotic index values were generally decreased with increasing concentrations of boric acid and the longer period of treatment (5 to 20 h) (Turkoglu, 2007).

The genetic effects of boric acid and borax (2.5 to 10 μM) on human whole blood cultures with and without the addition of titanium oxide (a carcinogenic / mutagenic agent) have been investigated. Both compounds have been found to cause increased resistance of DNA to damage induced by titanium oxide (Turkez, 2008).

3.5. Phytotoxicity

A phytotoxicity test has been applied to evaluate the toxicity of toxicants in Arctic soils. The phytotoxicity of boric acid in cryoturbated soils of polar region is much greater than that in other soils. A boric acid concentration of less than 150 μg/g soil is needed to inhibit root and shoot growth by 20%. Northern wheat grass (*Elymus lanceolatus*) is more sensitive to toxicants in Arctic soil than other plants (Anaka *et al.*, 2008).

4. Absorption of Boric Acid

Three percent boric acid incorporated in an anhydrous, water emulsifying ointment causes no increase of boron levels in blood and urine during a period of 1-9 days after a single topical application. The same amount of boric acid incorporated in a water-based jelly does cause an increase in blood and urine levels, beginning within 2-6 h after application. The decisive factor is the degree of liberation of boron depending on the type of vehicle (Stuttgen *et al.*, 1982). Boric acid taken orally by six male volunteers is absorbed to equal extents from a 3% water solution and an anhydrous, water emulsifying ointment. Virtually complete gastro-intestinal absorption and renal excretion are indicated by the 96 h urinary recovery amounting to 89.1-98.3% and 89.2-97.5%, of the dose ingested as solution and ointment, respectively. This indicates that the formulation of the ointment is an important factor in determining the release of boric acid on external application but it does not alter the absorption of boric acid on ingestion of the ointment (Jansen *et al.*, 1984a). The absorption of boric acid through damaged skin from a body ointment has been measured by an in-vitro permeation method. The maximum permeation was 14.1 μg/cm^2 of boric acid, which corresponds to 86% of the release from the ointment. The results indicate that the degree of permeation through damaged skin depends on the degree of liberation from the vehicle (Dusemund, 1987). The release of biocides (benzoic, sorbic and boric acids) incorporated into modified silica films have been investigated with respect to composite structure. The liberation rates of the embedded acids are proportional to the biocide-to-silica ratio and are changed by adding soluble polymers such as hydroxypropylcellulose (Bottcher *et al.*, 1999).

Dordas and Brown (2001) have suggested boron (in the form of undissociated boric acid) as an essential micronutrient for animals. They determined the permeability coefficient of boric acid in frog (*Xenopus oocytes*) as 1.5×10^{-6} cm/s, which is very close to the permeability across liposomes made with phosphatidylcholine and cholesterol (the major lipids in the oocyte membrane). The corneal penetration of CS-088, an ophthalmic agent, is significantly enhanced in the presence of EDTA / boric acid by approximately 1.6 fold. The permeability-enhancing effect of EDTA / boric acid is apparently synergistic and concentration dependent on both EDTA and boric acid (Kikuchi *et al.*, 2005a). In the corneal permeability of CS-088 (4-[1-hydroxy-1-methylethyl]-2-propyl-1-[4-[2-[tetrazole-5-yl] phenyl] phenyl] methylimidazole-5-carboxylic acid monohydrate), EDTA / boric acid significantly increase the membrane fluidity of liposomes (Kikuchi *et al.*, 2005b). The dermal route is important in exposure to biocidal products including boric acid. Repeated exposure of these products (specially undiluted) increases skin permeability causing harmful effects (Buist *et al.*, 2005).

5. Dissolution Kinetics

Robson *et al.* (2000) studied the influence of buffer composition on the release of cefuroxime axetil from stearic acid micropheres with particular emphasis on establishing the relationship between buffer composition and drug release. Drug dissolution and release from microspheres at pH 7.0 indicated marked differences in release profile with an approximate rank order of Sorensens modified phosphate buffer > citrate phosphate buffer > approximately boric acid buffer > mixed phosphate buffer.

The dissolution of probertite ($NaCaB_5O_9.5H_2O$) in boric acid solution has been investigated as a function of temperature. The increase in the boric acid concentration leads to an increase in the dissolution of probertite. However, the boric acid concentration above 5% at 60 and 80°C does not significantly affect the dissolution of probertite. The dissolution kinetics of probertite follows pseudo first-order reaction (Mergen and Demirhan, 2008).

6. Pharmacokinetics of Boric Acid

It has been demonstrated that water-emulsifying and hydrophobic ointments containing boric acid liberate only minute amounts (1-6 %) within 24 h compared with the nearly total liberation from a jelly. The pharmacokinetics rule out the risk of cumulative poisoning with topical preparations containing low amounts (upto 3%) of boric acid (Schon *et al.*, 1984). A pharmacokinetic study has been conducted by giving to eight young adult males a single dose of boric acid (562-611mg) by I.V. infusion. The plasma concentration curves best fitted a three-compartment open model. The 120 h urinary excretion was 98.7 ± 9.1% of dose, C_{tot} 54.6 ± 8.0 mL/min/1.73 m^2, $t_{1/2}$ beta 21.0 ± 4.9 h and distribution volumes V_1, V_2 and V_3: 0.251 ± 0.099, 0.456 ± 0.067 and 0.340 ± 0.128 L/kg (Jansen *et al.*, 1984b). Suicidal ingestion of about 21 g of boric acid by a young female showed the concentration of boric acid in serum and urine as 465 μg/mL and 3.40 mg/mL, respectively. The half-life of boric acid in serum (13.46 h) was decreased to 3.76 h and the total body clearance (0.99 L/h) increased to 3.53 L/h on hemodialysis. The additional removal of boric acid by the method was about 5 g. Thus hemodialysis has been found to be very useful in the treatment of acute boric acid poisoning (Teshima *et al.*, 1992; Naderi and Palmer, 2006).

The pharmacokinetics of sodium tetraborate in rats by administering 1 mL oral dose to several groups of rats (n=20) at several dose levels ranging from 0-0.4 mg/100 g body weight as boron has been studied. After 24 h the average urinary excretion rate for the element was 99.6 ± 7.9%. The data have been interpreted according to a one-compartment open model. The various parameters estimated are: absorption half-life, $t_{1/2}$ α 0.608 ± 0.432 h; elimination half-life, $t_{1/2}$ 4.64 ± 1.19 h; volume of distribution, V_d 142.0 ± 30.2 mL/100 g body weight; total clearance, C_{tot} 0.359 ± 0.0285 mL/min/100 g body weight. The maximum boron concentration (C_{max}) in serum after administration was 2.13 ± 0.270 mg/L, and the time needed to reach the maximum concentration (T_{max}) was 1.76 ± 0.887 h. The pharmacokinetic model is proposed as a useful tool to deal with the problem of environmental or industrial exposure to boron or in the case of accidental acute intoxication (Usuda *et al.*, 1998). The half-life of boric acid in humans is on the order of one day (Moseman, 1994). A comparative review of the pharmacokinetics of boric acid in humans and rodents shows remarkable similarity in their behavior (Murray, 1998).

Boric acid ingestion has been associated with greatly increased urinary excretion of riboflavin in patients, both children and adults. Most of the riboflavin is excreted within the first 24 h after ingestion of boric acid (Pinto *et al.*, 1978). Boric acid complexes with the polyhydroxyl ribitol side chain of riboflavin and causes enhanced excretion of the vitamin to promote riboflavinuria in animals and man (Pinto and Rivlin, 1987).

7. METABOLISM

Boric acid plays an important role in bone development. Dietry boron (in the form of boric acid) may be beneficial for optimum calcium metabolism and as a consequence, optimal bone metabolism. Serum levels of minerals as well as osteocalcin (a marker of bone resorptoin) are dependent to a greater extent on the hormonal (17 beta-estradiol) status of the animal. Boron treatment increases the hormonal-induced elevation of urinary calcium and magnesium. The bone mineral density of the L5 vertebra and proximal femur is highest after hormonal treatment (Gallardo-Williams *et al.*, 2003).

A boric acid solution (275 mg/L), administered to rats for 30 days orally in an "ad libitum" dose, has been found to cause changes in the lipid metabolism, in the water-mineral balance of the organism and in the acid-base equilibrium of the blood. The boric acid solution, suitably diluted with a nutritive fluid, acted spastically on the peristalsis of the small intestine of the rabbit (Drobnik and Latour, 2001).

The influence of dietary boron supplementation on some serum parameters and egg-yolk cholesterol has been studied in laying hens. Serum gamma-glutamyl transpeptidase (GGT) activity, albumin, glucose, total cholesterol and HDL– and LDL–cholesterol levels are decreased with all boron levels (10-400 mg/kg boric acid) (Eren and Uyanik, 2007). An investigation of the dietary effects of boric acid (156-2500 ppm) on the protein profiles in greater wax moth (*Galleria mellonella*) has been carried out. Many undetermined protein fractions (6.5-260 kDa), in addition to well-defined protein fractions such as lipophorins and storage proteins, have been detected in the tissues. A marked quantitative change in the 45kDa protein fraction of the hemolymph has been observed in the VIIth instar larvae reared on 2500 ppm boric acid (Hyrsl *et al.*, 2008). The presence of boron has been found to decrease the synthesis of proteoglycan, collagen and total proteins in the pelvic cartilage of chick embryo (Benderdour *et al.*, 1997).

8. Oxidative Stress

Oxidative stress plays an important role during inflammatory diseases. Boric acid has been implicated to modulate certain inflammatory mediators and regulate inflammatory processes. Boric acid could inhibit the lipopolysaccharide-induced formation of tumor necrosis factor-alpha and ameliorate the d,l-buthionine-S,R-sulfoximine induced glutathione depletion in TNF-1 cells (Cao *et al.*, 2008).

Peripheral blood cultures exposed to various doses (5-500 mg/L) of boron compounds have been tested for DNA damage and oxidative stress. It has been found that boron compounds at low doses are useful in supporting antioxidant enzyme activities in human blood cultures, though in increasing doses they constitute oxidative stress (Turkez *et al.*, 2007). The effects of boric acid induced oxidative stress on antioxidant enzymes and survivorship of wax moth larvae (*Galleria mellonella*) have been studied. The content of malondialdehyde (an oxidative stress indicator) was found to be significantly increased indicating that boric acid toxicity is related, in part, to oxidative stress management (Hyrsl *et al.*, 2007).

9. Enzyme Inhibition

Various enzymes have been found to be reversibly or irreversibly inhibited by boric acid (a competitive inhibitor). It prevents jack bean urease from irreversible pyrocatechol (Kot and Zaborska, 2003) and allicin inactivation (Juszkiewicz *et al.*, 2003). Urease from the seeds of pigeon pea is competitively inhibited by boric acid and boronic acids in the order of 4-bromophenylboronic acid > boric acid > butylboronic acid > phenylboronic acid. Urease inhibition by boric acid is maximal at pH 5.0 and minimal at pH 10.0. The trigonal planar $B(OH)_3$ form is a more effective inhibitor than the tetrahedral $B(OH)_4^-$ anionic form (Reddy and Kayastha, 2006).

The addition of boric acid to an *in vitro* pre-mRNA splicing reaction causes a dose-dependent reversible inhibition effect on the second step of splicing. The mechanism of action does not involve chelation of metal ions, hindrance of 3' splice-site, or binding to h Slu7 (Shomron and Ast, 2003). Boric acid inhibits adenosine diphosphate-ribosyl cyclase (ADP-ribosyl cyclase) non-competitively. ADP-ribosyl cyclase converts NAD^+ to cyclic ADP-ribose (cADPR) and nicotinamide. Boric acid binding to cADPR is characterized by an apparent binding constant of 655 ± 99 L/mol at pH 10.3 (Kim *et al.*, 2006).

The activity of penicillin G acylase from *Alcaligenes faecalis* is increased 7.5 fold when cells are permeabilized with 0.3% (w/v) ***cetyl trimethyl ammonium bromide*** (CTAB). The treated cells are then entrapped by polyvinyl alcohol crosslinked with boric acid and glutaraldehyde to increase the stability. The conversion yield of penicillin G to 6-aminopenicillanic acid is 75% by the immobilized system (Cheng *et al.*, 2006).

10. Synthesis of Bioactive Compounds

The 2-arachidonoylglycerol, an endogenous cannabinoid receptor ligand, has been synthesized from 1,3-benzylideneglycerol and arachidonic acid in the presence of N,N'-dicyclohexylcarbodiimide and 4-dimethylaminopyridine followed by treatment with boric acid and trimethyl borate. It stimulates NG108-15 cells to induce rapid transient elevation of the intracellular free Ca^{2+} concentrations through a CB1 receptor-dependent mechanism (Suhara *et al.*, 2000).

The synthesis of a polyglycerol with dendritic structure has been achieved by ring-opening polymerization of deprotonated glycidol. The polyglycerol is reacted with *o*-carboxymethylated chitosan dendrimer. The reaction of the dendrimer with boric acid results in a marked increase in the bulk viscosity indicating that boron can initiate the formation of a charge transfer complex. The complex shows significant activity against *Staphylococcus aureus* and *Pseudomonas aeruginosa* (Alencar de Queiroz *et al.*, 2006).

The action of boric acid at the molecular level has been examined using cell-free systems of transportation (isolated placenta nuclei) and translation (wheat germ extract). It has been found that 10 mM boric acid greatly increases RNA synthesis. Full length functional mRNA is produced because proteins of 14-80 kDa are translated (Dzondo-Gadet *et al.*, 2002).

11. Analysis in Biological Fluids

Various analytical methods have been used for the determination of boric acid and borates in biological fluids. The determination of borates in biological fluids and tissues has been carried out by colorimetric methods using carminic acid as reagent. More recently the technique of electrothermal atomic absorption spectrometry (ETAAS) has been applied to determine boron concentrations in blood in the cases of acute poisoning (Moffat *et al.*, 2004). Boric acid in biological materials has been determined by its reaction with protonated curcumin and measurement of absorbance at 550 nm (Yoshida *et al.*, 1989). Quantitative determination of percutaneous absorption of ^{10}B in ^{10}B-enriched boric acid and borates in biological materials has been achieved by inductively coupled plasma-mass spectrometry (ICP-MS) (Wester *et al.*, 1998a, 1998b, 1998c). ICP-MS has been found to be highly sensitive for the measurement of boron concentrations in blood and tissues in patients undergoing boron neutron capture therapy of certain cancers (Morten and Delves, 1999).

Boric acid undergoes molecular interaction or complex formation with a number of compounds (Saxena and Verma, 1983; Mendham *et al.*, 2000; Ahmad *et al.*, 2008, 2009a; 2009b). This property has been used for the spectrophotometric determination of aminophylline in serum (Li *et al.*, 2008), fluorimetric determination of rubomycin in blood, milk and urine (Alykov *et al.*, 1976), amperometric determination of glucose in blood or plasma (Macholan *et al.*, 1992), potentiometric determination of zolpidem hemitartrate in biological fluids (Kelani, 2004), gas chromatographic-ion trap tandem-mass spectrometric determination of organic acids in urine (Pacenti *et al.*, 2008), capillary zone electrophoretic determination of ascorbic acid and uric acid in plasma (Zinellu *et al.*, 2004), icariin and its metabolites in serum (Liu and Lou, 2004), uric acid in urine (Zhao *et al.*, 2008), α- and βamanitun in urine (Robinson-Fuentes *et al.*, 2008) and microchip electrophoretic determination of glucose in blood (Maeda *et al.*, 2007).

12. Preservation and Insect Control

12.1. Preservative

Boric acid and borates are used as wood preservatives against decay and rot fungi (*Postia placenta* and *Coniophora puteana*) (Williams and Amburgey, 1987; Temiz *et al.*, 2008) and as fire retardant in wood preservation industry (Baysal *et al.*, 2007). Boron-11 nuclear magnetic resonance imaging and spectroscopy have been used to characterize the nature and distribution of boron compounds after preservative treatment of radiate pine wood with trimethylborate. Trimethylborate undergoes rapid hydrolysis to form boric acid in pine wood (Meder *et al.*, 1999). Strong fungicide and insecticide effects have been observed on treating wood with composite films of boric acid (Bottcher *et al.*, 1999).

Borates have been used in mummification processes in Pharaonic Egypt more than 4000 years ago. Salt samples of borates used as embalming material for mummies contained 2-4 μM/g. Smaller borate concentrations (1.2 μM/g) were found in ancient bone samples (Kaup *et al.*, 2003).

12.2. Insecticide

Insecticidal use of borates is widespread because of their low mammalian toxicity and negligible insect resistance. Boric acid and borax have been used as insecticides in the control of American cockroaches (Lizzio, 1986), German cockroaches (*Blattella germanica*) (Strong *et al.*, 1993; Cochran, 1995; Miller and Kochler, 2000; Habes *et al.*, 2001; Zurek *et al.*, 2003; Gore and Schal, 2004; Gore *et al.*, 2004; Appel *et al.*, 2004; Zhang *et al.*, 2005; Kilani-Morakchi *et al.*, 2006; Wang and Bennett, 2009), adult mosquitoes (Diptera: Culicidae) (Xue and Barnard, 2003; Ali *et al.*, 2006; Xue *et al.*, 2006, 2008), cat fleas (Siphonaptera: Pulicidae) (Klotz *et al.*, 1994), house flies (Diptera: Muscidae) (Hogsette *et al.*, 2002), and Argentine ants (Hymenoptera: Formicidae) (Hooper-Bui and Rust, 2000; Klotz *et al.*, 2000; Ulloa-Chacon and Jaramillo, 2003; Rust *et al.*, 2004; Stanley and Robinson, 2007). In the larvae of the wax moth (*Galleria mellonella*) the boric acid toxicity is related, in part, to oxidative stress management (Hyrsl *et al.*, 2007).

References

Abdelaziz, A.A. and El-Nakeeb, M.A. (1988). Sporicidal activity of local anesthetics and their binary combinations with preservatives. *J. Clin. Pharm. Ther.*, *13*, 249-256.

Aggarwal, A. and Shier, R.M. (2008). Recalcitrant *Trichomonas vaginalis* infections successfully treated with vaginal acidification. *J. Obstet. Gynaecol. Can.*, *30*, 55-58.

Ahmad, I., Ahmed, S., Sheraz, M.A. and Vaid, F.H.M. (2008). Effect of borate buffer on the photolysis of riboflavin in aqueous solution. *J. Photochem. Photobiol. B: Biol.*, *93*, 82-87.

Ahmad, I., Ahmed, S., Sheraz, M.A. and Vaid, F.H.M. (2009a). Borate: toxicity, effect on drug stability and analytical applications. In M.P. Chung (Ed.), *Handbook on Borates: Chemistry, Production and Applications*. New York, NY: Nova Science Publishers Inc. (in press).

Ahmad, I., Ahmed, S., Sheraz, M.A. and Vaid, F.H.M. (2009b). Analytical applications of borates. *Mat. Sci. Res. J.,3,* (in press).

Alencar de Queiroz, A.A., Abraham, G.A., Pires Çamillo, M.A., Higa, O.Z., Silva, G.S., del Mar Fernández, M., San Román, J. (2006). Physicochemical and antimicrobial properties of boron-complexed polyglycerol-chitosan dendrimers. *J. Biomater. Sci. Polym. Ed.*, *17*, 689-707.

Ali, A., Xue, R.D. and Barnard, D.R. (2006). Effects of sublethal exposure to boric acid sugar bait on adult survival, host-seeking, blood feeding behavior, and reproduction of Stegomyia albopicta. *J. Am. Mosq. Control Assoc.*, *22*, 464-468.

Allen, B.C., Strong, P.L., Price, C.J., Hubbard, S.A. and Daston, G.P. (1996). Benchmark dose analysis of developmental toxicity in rats exposed to boric acid. *Fundam. Appl. Toxicol., 32*, 194-204.

Alykov, N.M., Karibiants, M.A. and Alykova, T.V. (1976). Fluorimetric determination of rubomycin with the boric acid reaction. *Antibiotiki*, *21*, 920-921.

Amaral, C., Jorge, A., Veloso, K., Erhardt, M., Arias, V. and Rodrigues, J.A. (2008). The effect of in-office in combination with intracoronal bleaching on enamel and dentin bond strength and dentin morphology. *J. Contemp. Dent. Pract.*, *9*, 17-24.

Anaka, A., Wickstrom, M. and Siciliano, S.D. (2008). Increased sensitivity and variability of phytotoxic responses in Arctic soils to a reference toxicant, boric acid. *Environ. Toxicol. Chem., 27*, 720-726.

Appel, A.G., Gehret, M.J. and Tanley, M.J. (2004). Effects of moisture on the toxicity of inorganic and organic insecticidal dust formulations to German cockroaches (Blattodea: Blattellidae). *J. Econ. Entomol., 97*, 1009-1016.

Avino-Martinez, J.A., Espana-Gregori, E., Peris-Martinez, C.P. and Blanes, M. (2008). Successful boric acid treatment of Aspergillus niger infection in an exenterated orbit. *Ophthal. Plast. Reconstr. Surg., 24*, 79-81.

Bai, Y.P., Yang, D.Q. and Wang, Y.M. (2007). Clinical study on treatment of acute eczema by Shuangfujin. *Zhongguo Zhong Xi Yi Jie He Za Zhi.*, *27*, 72-75.

Bansal, R.K., Tewari, U.S., Singh, P. and Murthy, D.V.S. (1995). Modified polyalkenoate (glass-ionomer) cement-a study. *J. Oral Rehabil.*, *22*, 533-537.

Barranco, W.T. and Eckhert, C.D. (2004). Boric acid inhibits human prostate cancer cell proliferation. *Cancer Lett., 216*, 21-29.

Barranco, W.T. and Eckhert, C.D. (2006). Cellular changes in boric acid-treated DU-145 prostate cancer cells. *Br. J. Cancer, 94*, 884-890.

Barranco, W.T., Hudak, P.F. and Eckhert, C.D. (2007). Evaluation of ecological and in vitro effects of boron on prostate cancer risk (United States). *Cancer Causes Control, 18*, 71-77.

Barranco, W.T., Kim, D.H., Stella, S.L. Jr. and Eckhert, C.D. (2009). Boric acid inhibits stored Ca^{2+} release in DU-145 prostate cancer cells. *Cell Biol. Toxicol.*, *25*, 309-320.

Bäumler, M.A., Schug, J., Schmidlin, P. and Imfeld, T. (2006). In vitro tests of internal tooth whitening agents on colored solutions do not replace tests on teeth. *Schweiz. Monatsschr. Zahnmed.*, *116*, 1000-1005.

Baysal, E., Altinok, M., Colak, M., Ozaki, S.K. and Toker, H. (2007). Fire resistance of Douglas fir (*Pseudotsuga menzieesi*) treated with borates and natural extractives. *Bioresour. Technol.*, *98*, 1101-1105.

Beckett, W.S., Oskvig, R., Gaynor, M.E. and Goldgeier, M.H. (2001). Association of reversible alopecia with occupational topical exposure to common borax-containing solutions. *J. Am. Acad. Dermatol.*, *44*, 599-602.

Benderdour, M., Hess, K., Gadet, M.D., Dousset, B., Nabet, P. and Belleville, F. (1997). Effect of boric acid solution on cartilage metabolism. *Biochem. Biophys. Res. Commun.*, *234*, 263-268.

Benderdour, M., Hess, K., Dzondo-Gadet, M., Nabet, P., Belleville, F. and Dousset, B. (1998). Boron modulates extracellular matrix and TNF alpha synthesis in human fibroblasts. *Biochem. Biophys. Res. Commun.*, *246*, 746-751.

Benderdour, M., Van Bui, T., Hess, K., Dicko, A., Belleville, F. and Dousset, B. (2000). Effects of boron derivatives on extracellular matrix formation. *J. Trace Elem. Med. Biol.*, *14*, 168-173.

Blech, M.F., Martin, C., Borrelly, J. and Hartemann, P. (1990). Treatment of deep wounds with loss of tissue. Value of a 3 percent boric acid solution. *Presse Med. 19*, 1050-1052.

Bochek, A.M., Yusupova, L.D., Zabivalova, N.M. and Petropavlovskii, G.A. (2002). Rheological properties of aqueous H-carboxymethyl cellulose solutions with various additives. *Russ. J. Appl. Chem.*, *75*, 645-648.

Borrelly, J., Blech, M.F., Grosdidier, G., Martin-Thomas, C. and Hartemann, P. (1991). Contribution of a 3% solution of boric acid in the treatment of deep wounds with loss of substance. *Ann. Chir. Plast. Esthet.*, *36*, 65-69.

Böttcher, H., Jagota, C., Trepte, J., Kallies, K.H. and Haufe, H. (1999). Sol-gel composite films with controlled release of biocides. *J. Control Release*, *60*, 57-65.

British Pharmacopoeia (2009). London: Her Majesty's Stationary Office, Electronic Version 11.0.

Brodsky, M.H., Ciebin, B.W. and Schiemann, D.A. (1978). Simple bacterial preservation medium and its application to proficiency testing in water bacteriology. *Appl. Environ. Microbiol.*, *35*, 487-491.

Browning, E. (1969). *Toxicity of Industrial Metals* (2nd edition, pp. 90-97). New York, NY: Appleton-Century-Crofts.

Buist, H.E., van de Sandt, J.J., van Burgsteden, J.A. and de Heer, C. (2005). Effects of single and repeated exposure to biocidal active substances on the barrier function of the skin in vitro. *Regul. Toxicol. Pharmacol.*, *43*, 76-84.

Cain, W.S., Jalowayski, A.A., Schmidt, R., Kleinman, M., Magruder, K., Lee, K.C. and Culver, B.D. (2008). Chemesthetic responses to airborne mineral dusts: boric acid compared to alkaline materials. *Int. Arch. Occup. Environ. Health*, *81*, 337-345.

Campos, S.F., César, I.C., Munin, E., Liporoni, P.C. and do Rego, M.A. (2007). Analysis of photoreflectance and microhardness of the enamel in primary teeth submitted to different bleaching agents. *J. Clin. Pediatr. Dent.*, *32*, 9-12.

Cao, J., Jiang, L., Zhang, X., Yao, X., Geng, C., Xue, X. and Zhong, L. (2008). Boric acid inhibits LPS-induced TNF-alpha formation through a thiol-dependent mechanism in THP-1 cells. *J. Trace Elem. Med. Biol.*, *22*, 189-195.

Chao, T.C., Maxwell, S.M. and Wong, S.Y. (1991). An outbreak of aflatoxicosis and boric acid poisoning in Malaysia: a clinicopathological study. *J. Pathol., 164*, 225-233.

Chapin, R.E. and Ku, W.W. (1994). The reproductive toxicity of boric acid. *Environ. Health Perspect., 102 (Suppl.7),* 87-91.

Chapin, R.E., Ku, W.W., Kenney, M.A. and McCoy, H. (1998). The effects of dietary boric acid on bone strength in rats. *Biol. Trace Elem. Res.*, *66*, 395-399.

Chapin, R.E., Ku, W.W., Kenney, M.A., McCoy, H., Gladen, B., Wine, R.N., Wilson, R. and Elwell, M.R. (1997). The effects of dietary boron on bone strength in rats. *Fundam. Appl. Toxicol.*, *35*, 205-215.

Cheng, S., Wei, D., Song, Q. and Zhao, X. (2006). Immobilization of permeabilized whole cell penicillin G acylase from Alcaligenes faecalis using pore matrix crosslinked with glutaraldehyde. *Biotechnol. Lett.*, *28*, 1129-1133.

Cherrington, J.W. and Chernoff, N. (2002). Periods of vertebral column sensitivity to boric acid treatment in CD-1 mice in utero. *Reprod. Toxicol.*, *16*, 237-243.

Cochran, D.G. (1995). Toxic effects of boric acid on the German cockroach. *Experientia, 51*, 561-563.

Coughlin, J.R. (1998). Sources of human exposure: overview of water supplies as sources of boron. *Biol. Trace Elem. Res., 66*, 87-100.

Craan, A.G., Myres, A.W. and Green, D.W. (1997). Hazard assessment of boric acid in toys. *Regul. Toxicol. Pharmacol.*, *26*, 277-280.

Das Neves, J., Pinto, E., Teixeira, B., Dias, G., Rocha, P., Cunha, T., Santos, B., Amaral, M.H. and Bahia, M.F. (2008). Local treatment of vulvovaginal candiosis: general and practical considerations. *Drugs, 68*, 1787-1802.

Del Palacio, A., Cuetara, M.S., Lopez-Suso, M.J., Amor, E. and Garau, M. (2002). Randomized prospective comparative study: short-term treatment with ciclopiroxolamine (cream and solution) versus boric acid in the treatment of otomycosis. *Mycoses, 45,* 317-328.

Deseta, F., Schmidt, M., Vu, B., Essmann, M. and Larsen, B. (2008). Antifungal mechanisms supporting boric acid therapy of *Candida vaginitis*. *J. Antimicrob. Chemother.*, *63*, 325-336.

Di Renzo, F., Cappelletti, G., Broccia, M.L., Giavini, E. and Menegola, E. (2007). Boric acid inhibits embryonic histone deacetylases: a suggested mechanism to explain boric acid-related teratogenicity, *Toxicol. Appl. Pharmacol., 220*, 178-185.

Dieter, M.P. (1994). Toxicity and carcinogenicity studies of boric acid in male and female B6C3F1 mice. *Environ. Health Perspect.*, *102*, 93-97.

Dordas, C. and Brown, P.H. (2001). Permeability and the mechanism of transport of boric acid across the plasma membrane of *Xenopus laevis oocytes*. *Biol. Trace Elem. Res.*, *81*, 127-139.

Drobnik, M. and Latour, T. (2001). Investigation of the pharmacodynamic properties of the solution of boric acid. *Rocz. Panstw. Zakl. Hig.*, *52*, 329-336.

Dusemund, B. (1987). Liberation and in vitro skin permeation of boric acid from an ointment. *Arzneimittelforsch.*, *37*, 1197-1201.

Dzondo-Gadet, M., Mayap-Nzietchueng, R., Hess, K., Nabet, P., Belleville, F. and Dousset, B. (2002). Action of boron at the molecular level: effects on transcription and translation in an acellular system. *Biol. Trace Elem. Res.*, *85*, 23-33.

Eren, M. and Uyanik, F. (2007). Influence of dietary boron supplementation on some serum metabolites and egg-yolk cholesterol in laying hens. *Acta Vet. Hung.*, *55*, 29-39.

Fail, P.A., Chapin, R.E., Price, C.J. and Heindel, J.J. (1998). General reproductive, developmental and endocrine toxicity of boronated compounds. *Reprod. Toxicol., 12,* 1-18.

Fail, P.A., George, J.D., Seely, J.C., Grizzle, T.B. and Heindel, J.J. (1991). Reproductive toxicity of boric acid in Swiss (CD-1) mice: assessment using the continuous breeding protocol. *Fundam. Appl. Toxicol.*, *17*, 225-239.

Falk, R.T., Rossi, S.C., Fears, T.R., Sepkovic, D.W., Migella, A., Adlercreutz, H., Donaldson, J., Bradlow, H.L. and Ziegler, R.G. (2000). A new ELISA kit for measuring urinary 2-hydroxyestrone, 16 alpha-hydroxyestrone, and their ratio; reproducibility, validity, and assay performance after freeze-thaw cycling and preservation by boric acid. *Cancer Epidemiol. Biomarkers Prev.*, *9*, 81-87.

Fort, D.J., Stover, E.L., Bantle, J.A., Dumont, J.N. and Finch, R.A. (2001). Evaluation of a reproductive toxicity assay using *Xenopus laevis*: boric acid, cadmium and ethylene glycol monomethyl ether. *J. Appl. Toxicol., 21*, 41-52.

Fukuda, R., Hiroda, M., Mori, I., Chatani, F., Morishima, H. and Mayahare, H. (2000). Collaborative work to evaluate toxicity on male reproductive organs by repeated dose studies in rats 24. Testicular toxicity of boric acid after 2- and 4- week administration periods. *J. Toxicol. Sci., 25* (*Spec. No.*), 233-239.

Gallardo-Williams, M.T., Maronpot, R.R., Wine, R.N., Brunssen, S.H. and Chapin, R.E. (2003). Inhibition of the enzymatic activity of prostate-specific antigen by boric acid and 3-nitrophenyl boronic acid. *Prostate, 54*, 44-49.

Gallardo-Williams, M.T., Chapin, R.E., King, P.E., Moser, G.J., Goldsworthy, T.L., Morrison, J.P. and Maronpot, R.R. (2004). Boron supplementation inhibits the growth and local expression of IGF-1 in human prostate adenocarcinoma (LNCaP) tumors in nude mice. *Toxicol. Pathol., 32*, 73-78.

Gao, Q., Liu, Z. and Cui, Y. (1999). Clinical and bacteriology observation on intraoperation usage of tarivid otic solution in treatment of chronic otitis media with cholesteatoma. *Lin Chuang Er Bi Yan Hon Ke Za Zhi, 13,* 219-220.

Gillespie, T., Fewster, J. and Masterton, R.G. (1999). The effect of specimen processing delay on borate urine preservation. *J. Clin. Pathol.*, *52*, 95-98.

Gore, J.C. and Schal, C. (2004). Laboratory evaluation of boric acid-sugar solutions as baits for management of German cockroach infestations. *J. Econ. Entomol., 97,* 581-587.

Gore, J.C., Zurek, L., Santangelo, R., Stringham, S.M., Watson, D.W. and Schal, C. (2004). Water solutions of boric acid and sugar for management of German cockroach populations in live stock production systems. *J. Econ. Entomol., 97*, 715-720.

Gosselin, R.E., Hodge, N.C. and Smith, R.P. (Eds.) (1976). *Clinical Toxicology of Commercial Products* (4th edition, pp. 63-66). Baltimore, MD: Williams and Wilkins.

Gregoire, V., Begg, A.C., Huiskamp, R., Veiryk, R. and Bartelink, H. (1993). Selectivity of boron carriers for boron neutron capture therapy: pharmacological studies with borocapture sodium, L-boronophenylalanine and boric acid in murine tumors. *Radiother. Oncol.*, *27*, 46-54.

Grzybowska, W., Młynarczyk, G., Młynarczyk, A., Bocian, E., Luczak, M. and Tyski, S. (2007). Estimation of activity of pharmakopeal disinfectants and antiseptics against Gram-negative and Gram-positive bacteria isolated from clinical specimens, drugs and environment. *Med. Dosw. Mikrobiol.*, *59*, 65-73.

Guaschino, S., Deseta, F., Sartore, A., Ricci, G., De Santo, D., Piccoli, M. and Alberico, S. (2001). Efficacy of maintenance therapy with topical boric acid in comparison with oral itraconazole in the treatment of recurrent vulvovaginal candidiasis. *Am. J. Obstet. Gynecol.*, *184*, 598-602.

Habes, D., Kilani-Morakchi, S., Aribi, N., Farine, J.P. and Soltani, N. (2001). Toxicity of boric acid to *Blattella germanica* (Dictyoptera: Blattellidae) and analysis of residues in several organs. *Meded Rijksuniv Gent Fak Landbouwkd Toegep Biol. Wet.*, *66*, 525-534.

Hamilton, R.A. and Wolf, B.C. (2007). Accidental boric acid poisoning following the ingestion of household pesticide. *J. Forensic Sci.*, *52*, 706-708.

Harris, M.W., Chapin, R.E., Lockhart, A.C. and Jokinen, M.D. (1992). Assessment of a short-term reproductive and developmental toxicity screen. *Fundam. Appl. Toxicol.*, *19*, 181-196.

Harrouk, W.A., Wheeler, K.E., Kimmel, G.L., Hogan, K.A. and Kimmel, C.A. (2005). Effects of hyperthermia and boric acid on skeletal development in rat embryos. *Birth Defects Res. B. Dev. Reprod. Toxicol.*, *74*, 268-276.

Heindel, J.J., Price, C.J. and Schwetz, B.A. (1994). The developmental toxicity of boric acid in mice, rats and rabbits. *Environ. Health Perspect.*, *102(Suppl.7)*, 107-112.

Heindel, J.J., Price, C.J., Field, E.A., Marr, M.C., Myers, C.B., Morrissey, R.E. and Schwetz, B.A. (1992). Developmental toxicity of boric acid in mice and rats. *Fundam. Appl. Toxicol.*, *18*, 266-277.

Hogsette, J.A., Carison, D.A. and Nejame, A.S. (2002). Development of granular boric acid sugar baits for house flies (Diptera: Muscidae). *J. Econ. Entomol.*, *95*, 1110-1112.

Hooper-Bui, L.M. and Rust, M.K. (2000). Oral toxicity of abamectin, boric acid, fipronil, and hydramethylnon to laboratory colonies of Argentine ants (Hymenoptera: Formicidae). *J. Econ. Entomol.*, *93*, 858-864.

Hubbard, S.A. (1998). Comparative toxicology of borates. *Biol. Trace Elem. Res.*, *66*, 343-357.

Hyrsl, P., Büyükgüzel, E. and Büyükgüzel, K. (2007). The effects of boric acid-induced oxidative stress on antioxidant enzymes and survivorship in *Galleria mellonella*. *Arch. Insect Biochem. Physiol.*, *66*, 23-31.

Hyrsl, P., Büyükgüzel, E. and Büyükgüzel, K. (2008). Boric acid-induced effects on protein profiles of Galleria mellonella hemolymph and fat body. *Acta Biol. Hung.*, *59*, 281-288.

Ishii, Y., Fujizuka, N., Takahashi, T., Shimizu, K., Tuchida, A., Yano, S., Naruse, T. and Chishiro, T. (1993). A fatal case of acute boric acid poisoning. *J. Toxicol. Clin. Toxicol.*, *31*, 345-352.

Izmailov, G.A. and Izmailov, S.G. (1998). Treatment of trophic ulcers of lower extremities by use of boric acid-hydrocortisone mixture. *Khirurgiia (Mosk.)*, *1*, 46-47.

Jansen, J.A., Schon, J.S. and Aggerbeck, B. (1984a). Gastro-intestinal absorption and in vitro release of boric acid from water-emulsifying ointments. *Food Chem. Toxicol.*, *22*, 44-53.

Jansen, J.A., Anderson, J. and Schon, J.S. (1984b). Boric acid single dose pharmacokinetics after intravenous administration to man. *Arch. Toxicol.*, *55*, 64-67.

Jewkes, F.E., McMaster, D.J., Napier, W.A., Houston, I.B. and Postlethwaite, R.J. (1990). Home collection of urine specimens-boric acid bottles or dipslides? *Arch. Dis. Child, 65*, 286-289.

Jovanovic, R., Congema, E. and Nguyen, H.T. (1991). Antifungal agents vs. boric acid for treating chronic mycotic vulvovaginitis. *J. Reprod. Med., 36*, 593-597.

Juszkiewicz, A., Zaborska, W., Sepioł, J., Góra, M. and Zaborska, A. (2003). Inactivation of jack bean urease by allicin. *J. Enzyme Inhib. Med. Chem., 18*, 419-424.

Karakus, M.E., Arda, H.N., Ikinciogullari, A., Gedikli, Y., Coskun, S., Balaban, N. and Akdogan, O. (2003). Microbiology of the external auditory canal in patients with asteatosis and itching. *Kulak. Burun. Bogaz. Ihtis. Derg., 11*, 33-38.

Kaup, Y., Schmid, M., Middleton, A. and Weser, U. (2003). Borate in mummification salts and bones from Pharaonic Egypt. *J. Inorg. Biochem., 94*, 214-220.

Kelani, K.M. (2004). Selective potentiometric determination of zolpidem hemitartrate in tablets and biological fluids by using polymeric membrane electrodes. *J. AOAC Int., 87*, 1309-1318.

Kiesche-Nesselrodt, A. and Hooser, A. (1990). Toxicology of selected pesticides, drugs, and chemicals. Boric acid. *Vet. Clin. North Am. Small Anim. Pract., 20,* 369-373.

Kikuchi, T., Suzuki, M., Kusai, A., Iseki, K. and Sasaki, H. (2005a). Synergistic effect of EDTA and boric acid on corneal penetration of CS-088. *Int. J. Pharm., 290*, 83-89.

Kikuchi, T., Suzuki, M., Kusai, A., Iseki, K., Sasaki, H. and Nakashima, K. (2005b). Mechanism of permeability-enhancing effect of EDTA and boric acid on the corneal penetration of 4-[1-hydroxy-1-methylethyl]-2-propyl-1-[4-[2-[tetrazole-5-yl] phenyl] phenyl] methylimidazole-5-carboxylic acid monohydrate (CS-088). *Int. J. Pharm., 299*, 107-114.

Kilani-Morakchi, S., Aribi, N., Farine, J.P., Smagghe, G. and Soltani, N. (2006). Cuticular hydrocarbon profiles in *Blattella germanica*: effects of halofenozide, boric acid and benfuracarb. *Commun. Agric. Appl. Biol. Sci., 71*, 555-562.

Kim, D.H., Hee, S.Q., Norris, A.J., Faull, K.F. and Eckhert, C.D. (2006). Boric acid inhibits adenosine diphosphate-ribosyl cyclase non-competitively. *J. Chromatogr. A., 1115*, 246-252.

Kinashi, Y., Masunaga, S. and Ono, K. (2002). Mutagenic effect of borocaptate sodium and boronophenylalanine in neutron capture therapy. *Int. J. Radiat. Oncol. Biol. Phys., 54*, 562-567.

Klotz, J.H., Greenberg, L., Amrhein, C. and Rust, M.K. (2000). Toxicity and repellency of borate-sucrose water baits to Argentine ants (Hymenoptera: Formicidae). *J. Econ. Entomol., 93*, 1256-1258.

Klotz, J.H., Moss, J.I., Zhao, R., Davis, L.R. Jr. and Patterson, R.S. (1994). Oral toxicity of boric acid and other boron compounds to immature cat fleas. *J. Econ. Entomol., 87,* 1534-1536.

Kodym, A., Marcinkowski, A. and Kukuła, H. (2003). Technology of eye drops containing aloe (*Aloe arborescens* Mill-Liliaceae) and eye drops containing both aloe and neomycin sulphate. *Acta Pol. Pharm., 60*, 31-39.

Kot, M. and Zaborska, W. (2003). Irreversible inhibition of jack bean urease by pyrocatechol. *J. Enzyme Inhib. Med. Chem., 18*, 413-417.

Ku, W.W. and Chapin, R.E. (1994). Mechanism of the testicular toxicity of boric acid in rats: in vivo and in vitro studies. *Environ. Health Perspect., 102 (Suppl.7)*, 99-105.

Ku, W.W., Shih, L.M. and Chapin, R.E. (1993a). The effects of boric acid on testicular cells in culture. *Reprod. Toxicol., 7*, 321-331.

Ku, W.W., Chapin, R.E., Wine, R.N. and Gladen, B.C. (1993b). Testicular toxicity of boric acid: relationship of dose to lesion development and recovery in the F344 rat. *Reprod. Toxicol., 7*, 305-319.

Kudo, S., Tanase, H., Yamasaki, M., Nakao, M., Miyata, Y., Tsuru, K. and Imai, S. (2000). Collaborative work to evaluate toxicity on male reproductive organs by repeated dose studies in rats: 23. A comparative 2- and 4- week repeated oral dose testicular toxicity study of boric acid in rats. *J. Toxicol. Sci., 25 (Spec. No.)*, 223-232.

Laster, B.H., Kahl, S.B., Popenoe, E.A., Pate, D.W. and Fairchild, R.G. (1991). Biological efficacy of boronated low-density lipoprotein for boron neutron capture therapy as measured in cell culture. *Cancer Res., 51*, 4588-4593.

Lee, Z.S. and Critchley, J.A. (1998). Simultaneous measurement of catecholamines and kallikrein in urine using boric acid preservative. *Clin. Chim. Acta, 276*, 89-102.

Lee, G.P., Lee, M.Y., Lum, S.O., Poh, R.S. and Lim, K.C. (2004). Extraradicular diffusion of hydrogen peroxide and pH changes associated with intracoronal bleaching of discoloured teeth using different bleaching agents. *Int. Endod. J., 37*, 500-506.

Li, J., Jiang, Y., Sun, T. and Ren, S. (2008). Fast and simple method for assay of ciclopirox olamine by micellar electrokinetic capillary chromatography. *J. Pharm. Biomed. Anal., 47*, 929-933.

Lim, M.Y., Lum, S.O., Poh, R.S., Lee, G.P. and Lim, K.C. (2004). An in vitro comparison of the bleaching efficacy of 35% carbamide peroxide with established intracoronal bleaching agents. *Int. Endod. J., 37*, 483-488.

Limaye, S. and Weightman, W. (1997). Effect of an ointment containing boric acid, zinc oxide, starch and petrolatum on psoriasis. *Australas. J. Dermatol., 38*, 185-186.

Linden, C.H., Hall, A.H., Kulig, K.W. and Rumack, B.H. (1986). Acute ingestions of boric acid. *J. Toxicol. Clin. Toxicol., 24*, 269-279.

Linder, R.E., Strader, L.F. and Rehnberg, G.L. (1990). Effect of acute exposure to boric acid on the male reproductive system of the rat. *J. Toxicol. Environ. Health, 31*, 133-146.

Litovitz, T.L., Klein-Schwartz, W., Oderda, G.M. and Schmitz, B.F. (1988). Clinical manifestations of toxicity in a series of 784 boric acid ingestions. *Am. J. Emerg. Med., 6*, 209-213.

Liu, J. and Lou, Y.J. (2004). Determination of icariin and metabolites in rat serum by capillary zone electrophoresis: rat pharmacokinetic studies after administration of icariin. *J. Pharm. Biomed. Anal., 36*, 365-370.

Liu, F.T., Agrawal, S.G., Movasaghi, Z., Wyatt, P.B., Rehman, I.U., Gribben, J.G., Newland, A.C. and Jia, L. (2008). Dietary flavonoids inhibit the anticancer effects of the proteasome inhibitor bortezomib. *Blood, 112*, 3835-3846.

Lizzio, E.F. (1986). A boric acid-rodenticide mixture used in the control of coexisting rodent-cockroach infestations. *Lab. Anim. Sci., 36*, 74-76.

Locatelli, C., Minoia, C., Tonini, M. and Manzo, L. (1987). Human toxicology of boron with special reference to boric acid poisoning. *G. Ital. Med. Lev., 9*, 141-146.

Lum, K.T. and Meers, P.D. (1989). Boric acid converts urine into an effective bacteriostatic transport medium. *J. Infect., 18*, 51-58.

Macfadyen, C., Gamble, C., Garner, P., Macharia, I., Mackenzie, I., Mugwe, P., Oburra, H., Otwombe, K., Taylor, S. and Williamson, P. (2005). Topical quinolone vs. antiseptic for treating chronic suppurative otitis media: a randomized controlled trial. *Trip. Med. Int. Health, 10*, 190-197.

Macholan, L., Skladal, P., Bohackova, I. and Krejci, J. (1992). Amperometric glucose biosensor with extended concentration range utilizing complexation effect of borate. *Biosens. Bioelectron.*, *7*, 593-598.

Maeda, E., Kataoka, M., Hino, M., Kajimoto, K., Kaji, N., Tokeshi, M., Kido, J., Shinohara, Y. and Baba, Y. (2007). Determination of human blood glucose levels using microchip electrophoresis. *Electrophoresis*, *28*, 2927-2933.

Martinez, V.G., Manson, J.M. and Zoran, M.J. (2008). Effects of nerve injury and segmental regeneration on the cellular correlates of neural morphallaxis. *J. Exp. Zoolog. B. Mol. Dev. Evol.*, *310*, 520-533.

Matsuda, K., Okamoto, M., Ashida, M., Ishimaru, T., Horiuti, I., Suzuki, K. and Yamamoto, S. (2004). Toxicological analysis over the past five years at a single institution, *Rinsho Byori (Japan), 52*, 819-823.

Meder, R., Franich, R.A. and Callaghan, P.T. (1999). ^{11}B magnetic resonance imaging and MAS spectroscopy of trimethylborate-treated radiata pine wood. *Solid State Nucl. Magn. Reson.*, *15*, 69-72.

Meers, P.D. and Chow, C.K. (1990). Bacteriostatic and bactericidal actions of boric acid against bacteria and fungi commonly found in urine. *J. Clin. Pathol.*, *43*, 484-487.

Mendelsohn, C.L., Griffin, C.E., Rosenkrantz, W.S., Brown, L.D. and Boord, M.J. (2005). Efficacy of boric-complexed zinc and acetic-complexed zinc otic preparations for canine yeast otitis externa. *J. Am. Anim. Hosp. Assoc., 41*, 12-21.

Mendham, J., Denney, R.C., Barnes, J.D. and Thomas, M.J.K. (2000). *Vogel's Textbook of Quantitative Chemical Analysis* (6th edition, pp. 384-385). Delhi: Dorling Kindersley.

Mergen, A. and Demirhan, M.H. (2009). Dissolution kinetics of probertite in boric acid solution. *Int. J. Miner. Process.*, *90*, 16-20.

Minja, B.M., Moshi, N.H., Ingvarsson, L., Bastos, I. and Grenner, J. (2006). Chronic suppurative otitis media in Tanzanian school children and its effects on hearing. *East Afr. Med. J., 83*, 322-325.

Moffat, A.C., Osselton, M.D. and Widdop, B., Eds. (2004). *Clarke's Analysis of Drugs and Poisons* (3rd edition, p. 269). London: Pharmaceutical Press.

Moore, J.A. (1997). An assessment of boric acid and borax using the IEHR evaluative process for assessing human developmental and reproductive toxicity of agents. Expert Scientific Committee. *Reprod. Toxicol., 11*, 123-160.

Morten, J.A. and Delves, H.T. (1999). Measurement of total boron and ^{10}B concentration and the detection and measurement of elevated ^{10}B levels in biological samples by inductively coupled plasma-mass spectrometry using the determination of ^{10}B: ^{11}B ratios. *J. Anal. Atomic Spectrom.*, *14*, 1545-1556.

Moseman, R.F. (1994). Chemical disposition of boron in animals and humans. *Environ. Health Perspect.*, *102*, 113-117.

Murray, F.J. (1995). A human health risk assessment of boron (boric acid and borax) in drinking water. *Regul. Toxicol. Pharmacol., 22*, 221-230.

Murray, F.J. (1998). A comparative review of the pharmacokinetics of boric acid in rodents and humans. *Biol. Trace Elem. Res.*, *66*, 331-341.

Naderi, A.S. and Palmer, B.F. (2006). Successful treatment of a rare case of boric acid overdose with hemodialysis. *Am. J. Kidney dis.*, *48*, e95-e97.

Narotsky, M.G., Schmid, J.E., Andrews, J.E. and Kavlock, R.J. (1998). Effects of boric acid on axial skeletal development in rats. *Biol. Trace Elem. Res.*, *66*, 373-394.

Nyirjesy, P., Alexander, A.B. and Weitz, M.V. (2005). Vaginal *Candida parapsilosis*: pathogen or bystander? *Infect. Dis. Obstet. Gynecol.*, *13*, 37-41.

Odunola, O.A. (1997). Individual and combined genotoxic response of boric acid and aflatoxin B1 in *Escherichia coli* PQ 37. *East Afr. Med. J.*, *74*, 499-502.

O'Neil, M.J. (Ed.) (2001). *The Merck Index* (13th edition). Rahway, NJ: Merck and Co., Electronic Version.

Ono, K., Kinashi, Y., Masunaga, S., Suzuki, M. and Takagaki, M. (1998a). Effect of electroporation on cell killing by boron neutron capture therapy using borocaptate sodium (^{10}B-BSH). *Jpn. J. Cancer Res.*, *89*, 1352-1357.

Ono, K., Kinashi, Y., Masunaga, S., Suzuki, M. and Takagaki, M. (1998b). Electroporation increases the effect of borocaptate (^{10}B-BSH) in neutron capture therapy. *Int. J. Radiat. Oncol. Biol. Phys.*, *42*, 823-826.

Otero, L., Palacio, V., Mendez, F.J. and Vazquez, F. (2002). Boric acid susceptibility testing of non-C. albicans Candida and *Saccharomyces cerevisiae*: comparison of three methods. *Med. Mycol.*, *40*, 319-322.

Otero, L., Fleites, A., Méndez, F.J., Palacio, V. and Vázquez, F. (1999). Susceptibility of Candida species isolated from female prostitutes with vulvovaginitis to antifungal agents and boric acid. *Eur. J. Clin. Microbiol. Infect. Dis.*, *18*, 59-61.

Ozcan, K.M., Ozcan, M., Karaarslan, A. and Karaarslan, F. (2003). Otomycosis in Turkey: predisposing factors, aetiology and therapy. *J. Laryngol. Otol.*, *117*, 39-42.

Ozturkcan, S., Dundar, R., Katilmis, H., Ilknur, A.E., Aktas, S. and Haciomeroglu, S. (2009). The ototoxic effect of boric acid solutions into the middle ear of guinea pigs. *Eur. Arch. Otorhinolaryngol.*, *266*, 663-667.

Pacenti, M., Dugheri, S., Villanelli, F., Bartolucci, G., Calamai, L., Boccalon, P., Arcangeli, G., Vecchione, F., Alessi, P., Kikic, I. and Cupelli, V. (2008). Determination of organic acids in urine by solid-phase microextraction and gas chromatography-ion trap tandem mass spectrometry previous 'in sample' derivatization with trimethyloxonium tetrafluoroborate. *Biomed. Chromatogr.*, *22*,1155-1163.

Pahl, M.V., Culver, B.D. and Vaziri, N.D. (2005). Boron and the kidney. *J. Ren. Nutr.*, *15*, 362-370.

Pahl, M.V., Culver, B.D., Strong, D.L., Murray, F.J. and Vaziri, N.D. (2001). The effect of pregnancy on renal clearance of boron in humans: a study based on normal dietary intake of boron. *Toxicol. Sci.*, *60*, 252-256.

Patel, A. and Agrawal, S.C. (2002). Sensitivity of ornithophillic fungi to some drugs. *Hindustan Antibiot. Bull.*, *44*, 49-52.

Peters, A.K., Steemans, M., Hansen, E., Mesens, N., Verheyen, G.R. and Vanparys, P. (2008a). Evaluation of the embryotoxic potency of compounds in a newly revised high throughput embryonic stem cell test. *Toxicol. Sci.*, *105*, 342-350.

Peters, A.K., Wouwer, G.V., Weyn, B., Verheyen, G.R., Vanparys, P. and Gompel, J.V. (2008b). Automated analysis of contractility in the embryonic stem cell test, a novel approach to assess embryotoxicity. *Toxicol. In Vitro*, *22*, 1984-1956.

Piemjai, M. and Surakompontorn, J. (2006). Effect of tooth-bleaching on the tensile strength and staining by caries detector solution on bovine enamel and dentin. *Am. J. Dent.*, *19*, 387-392.

Pinto, J.T. and Rivlin, R.S. (1987). Drugs that promote renal excretion of riboflavin. *Drug Nutr. Interact.*, *5*, 143-151.

Pinto, J.T., Huang, Y.P., McConnel, R.J. and Rivlin, R.S. (1978). Increased urinary riboflavin excretion resulting from boric acid ingestion. *J. Lab. Clin. Med.*, *92*, 126-134.

Poller, F., Bauch, T., Sauerwein, W., Bocker, W., Wittig, A. and Streffer, C. (1996). Comet assay study of DNA damage and repair of tumor cells following boron neutron capture irradiation with fast d(14) + Be neutrons. *Int. J. Radiat. Biol.*, *70*, 593-602.

Porter, I.A. and Brodie, J. (1969). Boric acid preservation of urine samples. *Br. Med. J.*, *2*, 353-355.

Prentice, L.H., Tyas, M.J. and Burrow, M.F. (2006). The effects of boric acid and phosphoric acid on the comprehensive strength of glass-ionomer cements. *Dent. Mat.*, *22*, 94-97.

Price, C.J., Strong, P.L., Murray, F.J. and Goldberg, M.M. (1997). Blood boron concentrations in pregnant rats fed boric acid throughout gestation. *Reprod. Toxicol.*, *11*, 833-842.

Price, C.J., Strong, P.L., Murray, F.J. and Goldberg, M.M. (1998). Developmental effects of boric acid in rats related to maternal blood boron concentrations. *Biol. Trace Elem. Res.*, *66*, 359-372.

Price, C.J., Strong, P.L., Marr, M.C., Mayers, C.B. and Murray, F.J. (1996b). Developmental toxicity NOAEL and postnatal recovery in rats fed boric acid during gestation. *Fundam. Appl. Toxicol.*, *32*, 179-193.

Price, C.J., Marr, M.C., Myers, C.B., Seely, J.C., Heindel, J.J. and Schwetz, B.A. (1996a). The developmental toxicity of boric acid in rabbits. *Fundam. Appl. Toxicol.*, *34*, 176-187.

Prutting, S.M. and Cerveny, J.D. (1998). Boric acid vaginal suppositories: a brief review. *Infect. Dis. Obstet. Gynecol.*, *6*, 191-194.

Ray, D., Goswani, R., Dadhwal, V., Goswami, D., Banerjee, U. and Kochupillai, N. (2007a). Prolonged (3-month) mycological cure after boric acid suppositories in diabetic women with vulvovaginal candidiasis. *J. Infect.*, *55*, 374-377.

Ray, D., Goswani, R., Banerjee, U., Dadhwal, V., Goswami, D., Mandal, P., Sreenivas, V. and Kochupillai, N. (2007b). Prevalence of *Candida glabrata* and its response to boric acid vaginal suppositories in comparison with oral fluconazol in patients with diabetes and vulvovaginal candidiasis. *Diabetes Care, 30,* 312-317.

Reddy, K.R. and Kayastha, A.M. (2006). Boric acid and boronic acids inhibition of pigeonpea urease. *J. Enzyme Inhib. Med. Chem.*, *21*, 467-470.

Reilly, W.J. Jr. (2006). Pharmaceutical Necessities. In R. Hendrickson (Ed.), *Remington: The Science and Practice of Pharmacy* (21st edition, pp. 1083-1084, 1089). Philadelphia, PA: Lippincott Williams and Wilkins.

Restuccio, A., Mortensen, M.E. and Kelly, M.T. (1992). Fatal ingestion of boric acid in an adult. *Am. J. Emerg. Med.*, *10*, 545-547.

Richold, M. (1998). Boron exposure from consumer products. *Biol. Trace Elem. Res.*, *66*, 121-129.

Robinson-Fuentes, V.A., Jaime-Sánchez, J.L., García-Aguilar, L., Gómez-Peralta, M., Vázquez-Garcidueñas, M.S. and Vázquez-Marrufo, G. (2008). Determination of alpha- and beta-amanitin in clinical urine samples by Capillary Zone Electrophoresis. *J. Pharm. Biomed. Anal.*, *47*, 913-917.

Robson, H., Craig, D.Q. and Deutsch, D. (2000). An investigation into the release of cefuroxime axetil from taste-masked stearic acid microspheres. II. The effects of buffer composition on drug release. *Int. J. Pharm.*, *195*, 137-145.

Romano, L., Battaglia, F., Masucci, L., Sanguinetti, M., Posteraro, B., Plotti, G., Zanetti, S. and Fadda, G. (2005). In vitro activity of bergamot natural essence and furocoumarin-free and distilled extracts and their associations with boric acid, against clinical yeast isolates. *J. Antimicrob. Chemother.*, *55*, 110-114.

Rowe, R.I., Bouzan, C., Nabili, S. and Eckert, C.D. (1998). The response of trout and zebrafish embryos to low and high boron concentrations is U-shaped. *Biol. Trace Elem. Res.*, *66*, 261-270.

Rust, M.K., Reierson, D.A. and Klotz, J.H. (2004). Delayed toxicity as a critical factor in the efficacy of aqueous baits for controlling Argentine ants (Hymenoptera: Formicidae). *J. Econ. Entomol.*, *97*, 1017-1024.

Salphale, P.S. and Shenoi, S.D. (2003). Contact sensitivity to calcium hypochlorite. *Contact Dermatitis*, *48*, 162.

Saxena, R. and Verma, R.M. (1983). Iodometric microdetermination of boric acid and borax separately or in a mixture. *Talanta*, *30*, 365-367.

Schon, J.S., Jansen, J.A. and Aggerbeck, B. (1984). Human pharmacokinetics and safety of boric acid. *Arch. Toxicol. Suppl.*, *7*, 232-235.

Scorei, R., Ciubar, R., Ciofrangeanu, C.M., Mitran, V., Cimpean, A. and Iordachescu, D. (2008). Comparative effect of boric acid and calcium fructoborate on breast cancer cells. *Biol. Trace Elem. Res.*, *22*, 197-205.

Seigel, E. and Wilson, S. (1986). Boric acid toxicity. *Pediatr. Clin. North Am.*, *33*, 363-367.

Shaposhnikov, V.I. and Zorik, V.V. (2001). Combined treatment of purulent-necrotic lesions of lower extremities in diabetic patients. *Khirurgiia (Mosk.)*, *2*, 46-49.

Sheng, M.H., Taper, L.J., Veit, H., Qian, H., Ritchey, S.J. and Lau, K.H. (2001). Dietary boron supplementation enhanced the action of estrogen, but not that of parathyroid hormone, to improve trabecular bone quality in ovariectomized rats. *Biol. Trace Elem. Res.*, *82*, 109-123.

Singh, S., Sobel, J.D., Bhargava, P., Boikov, D. and Vazquez, J.A. (2002). Vaginitis due to *Candida krusei*: epidemiology, clinical aspects, and therapy. *Clin. Infect. Dis.*, *35*, 1066-1070.

Shinohara, M.S., Peris, A.R., Pimenta, L.A. and Ambrosano, G.M. (2005). Shear bond strength evaluation of composite resin on enamel and dentin after nonvital bleaching. *J. Esthet. Restor. Dent.*, *17*, 22-29.

Shinohara, M.S., Peris, A.R., Rodrigues, J.A., Pimenta, L.A. and Ambrosano, G.M. (2004). The effect of nonvital bleaching on the shear bond strength of composite resin using three adhesive systems. *J. Adhes. Dent.*, *6*, 205-209.

Shomron, N. and Ast, G. (2003). Boric acid reversibly inhibits the second step of pre-mRNA splicing. *FEBS Lett.*, *552*, 219-224.

Sobel, J.D. and Chaim, W. (1997). Treatment of *Torulopsis glabrata* vaginitis: retrospective review of boric acid therapy. *Clin. Infect. Dis.*, *24*, 649-652.

Sobel, J.D., Chaim, W., Nagappan, V. and Leaman, D. (2003). Treatment of vaginitis caused by *Candida glabrata*: use of topical boric acid and flucytosine. *Am. J. Obstet. Gynecol.*, *189*, 1297-1300.

Soine, T.O. and Wilson, C.O. (1967). *Rogers Inorganic Pharmaceutical Chemistry* (8th edition, pp. 125-126). Philadelphia, PA: Lea and Febiger.

Sood, G., Nyirjesy, P., Weitz, M.V. and Chatwani, A. (2000). Terconazole cream for non-*Candida albicans* fungal vaginitis: results of a retrospective analysis. *Infect. Dis. Obstet. Gynecol.*, *8*, 240-243.

Stanley, M.C. and Robinson, W.A. (2007). Relative attractiveness of baits to Paratrechina longicornis (Hymenoptera: Formicidae). *J. Econ. Entomol.*, *100*, 509-516.

Strong, C.A., Koehler, P.G. and Patterson, R.S. (1993). Oral toxicity and repellency of borates to German cockroaches (Dictyoptera: Blattellidae). *J. Econ. Entamol.*, *86*, 1458-1463.

Stuttgen, G., Siebel, T. and Aggerbeck, B. (1982). Absorption of boric acid through skin depending on the type of vehicle. *Arch. Dermatol. Res.*, *272*, 21-29.

Suhara, Y., Takayama, H., Nakane, S., Miyashita, T., Waku, K. and Sugiura, T. (2000). Synthesis and biological activities of 2-arachidonoylglycerol, an endogenous cannabinoid receptor ligand, and its metabolically stable ether-linked analogues. *Chem. Pharm. Bull. (Tokyo)*, *48*, 903-907.

Sweetman, S.C., Ed. (2007). Boric acid. *Martindale: The Complete Drug Reference* (35th edition). London: Pharmaceutical Press, Electronic Version.

Tangermann, R.H., Etzel, R.A., Mortimer, L., Penner, G.D. and Paschal, D.C. (1992). An outbreak of a food related illness resembling boric acid poisoning. *Arch. Environ. Contam. Toxicol.*, *23*, 142-144.

Teixeira, E.C., Hara, A.T., Turssi, C.P. and Serra, M.C. (2003). Effect of non-vital tooth bleaching on microleakage of coronal access restorations. *J. Oral Rehabil.*, *30*, 1123-1127.

Teixeira, E.C., Turssi, C.P., Hara, A.T. and Serra, M.C. (2004). Influence of post-bleaching time intervals on dentin bond strength. *Braz. Oral Res.*, *18*, 75-79.

Temiz, A., Alfredsen, G., Eikenes, M. and Terzier, N. (2008). Decay resistance of wood treated with boric acid and tall oil derivatives. *Bioresour. Technol.*, *99*, 2102-2106.

Teshima, D., Taniyama, T. and Oishi, R. (2001). Usefulness of forced diuresis for acute boric acid poisoning in an adult. *J. Clin. Pharm. Ther.*, *26*, 387-390.

Teshima, D., Morishita, K., Ueda, Y., Futagami, K., Higuchi, S., Komoda, T., Nanishi, F., Taniyama, T., Yoshitake, J. and Aoyema, T. (1992). Clinical management of boric acid ingestion: pharmacokinetic assessment of efficacy of hemodialysis for treatment of acute boric acid poisoning. *J. Pharmacobiodyn.*, *5*, 287-294.

Thierauf, A., Serr, A., Halter, C.C., Al-Ahmad, A., Rana, S. and Weinmann, W. (2008). Influence of preservatives on the stability of ethyl glucuronide and ethyl sulphate in urine. *Forensic. Sci. Int.*, *182*, 41-45.

Timpawat, S., Nipattamanon, C., Kijsamanmith, K. and Messer, H.H. (2005). Effect of bleaching agents on bonding to pulp chamber dentine. *Int. Endod. J.*, *38*, 211-217.

Thongboonkerd, V. and Saetun, P. (2007). Bacterial overgrowth affects urinary proteome analysis: recommendation for centrifugation, temperature, duration, and the use of preservatives during sample collection. *J. Proteome Res.*, *6*, 4173-4181.

Treinen, K.A. and Chapin, R.E. (1991). Development of testicular lesions in F 344 rats after treatment with boric acid. *Toxicol. Appl. Pharmacol.*, *107*, 325-335.

Turkez, H. (2008). Effects of boric acid and borax on titanium dioxide genotoxicity. *J. Appl. Toxicol.*, *28*, 658-664.

Türkez, H., Geyikoğlu, F., Tatar, A., Keleş, S. and Ozkan, A. (2007). Effects of some boron compounds on peripheral human blood. *Z. Naturforsch. C.*, *62*, 889-896.

Turkoglu, S. (2007). Genotoxicity of five food preservatives tested on root tips of Allium cepa L. *Mutat. Res.*, *626*, 4-14.

Tyl, R.W., Chernoff, N. and Rogers, J.M. (2007). Altered axial skeletal development. *Birth Defects Res. B Dev. Reprod. Toxicol.*, *80*, 451-472.

Ulloa-Chacon, P. and Jaramillo, G.I. (2003). Effect of boric acid, tipronil, hydramethylnon, and difluobenzuron baits in colonies of ghost ants (Hymenoptera: Formicidae). *J. Econ. Entomol.*, *96*, 856-862.

United States Pharmacopeia 30 / National Formulary 25 (2007). Rockville, MD: United States Pharmacopeial Convention, Electronic Version.

Usuda, K., Kono, K., Orita, Y., Dote, T., Iguchi, K., Nishuira, H., Tominaga, M., Tagawa, T., Goto, E. and Shirai, Y. (1998). Serum and urinary boron levels in rats after single administration of sodium tetraborate. *Arch. Toxicol.*, *72*, 468-474.

Van Slyke, K.K., Michel, V.P. and Rein, M.F. (1981). Treatment of vulvovaginal candidiasis with boric acid powder. *Am. J. Obstet. Gynecol.*, *141*, 145-148.

VanKessel, K., Assefi, N., Marrazzo, J. and Eckhert, L. (2003). Common complementary and alternative therapies for yeast vaginitis and bacterial vaginosis: a systematic study. *Obstet. Gynecol. Surv.*, *58*, 351-358.

Vaziri, N.D., Oveisi, F., Culver, B.D., Pahl, M.V., Anderson, M.E., Strong, P.L. and Murray, F.J. (2001). The effect of pregnancy on renal clearance of boron in rats given boric acid orally. *Toxicol. Sci.*, *60*, 257-263.

Walmod, P.S., Gravemann, U., Nau, H., Berezin, V. and Bock, E. (2004). Discriminative power of an assay for automated in vitro screening of teratogens. *Toxicol. In Vitro.*, *18*, 511-525.

Wang, C. and Bennett, G.W. (2009). Cost and effectiveness of community-wide integrated pest management for German cockroach, cockroach allergen, and insecticide use reduction in low-income housing. *J. Econ. Entomol.*, *102*, 1614-1623.

Wang, Y., Zhao, Y. and Chen, X. (2008). Experimental study on the estrogen-like effect of boric acid. *Biol. Trace Elem. Res.*, *121*, 160-170.

Werfel, S., Boeck, K., Abeck, D. and Ring, J. (1998). Special characteristics of topical treatment in childhood. *Hautarzt*, *49*, 170-175.

Wery, N., Narotsky, M.G., Pacico, N., Kavlock, R.J., Picard, J.J. and Gofflot, F. (2003). Defects in cervical vertebrae in boric acid-exposed rat embryos are associated with anterior shifts of hox gene expression domains. *Birth Defects Res. A, Clin. Mol. Teratol.*, *67*, 59-67.

Wester, R.C., Hartway, T., Maibacch, H.I., Schell, M.J., Northington, D.J., Culver, B.D. and Strong, P.L. (1998a). In vitro percutaneous absorption of boron as boric acid, borax, and disodium octaborate tetrahydrate in human skin: a summary. *Biol. Trace Elem. Res.*, *66*, 111-120.

Wester, R.C., Hui, X., Maibach, H.I., Bell, K., Schell, M.J., Northington, D.J., Strong, P. and Culver, B.D. (1998b). In vivo percutaneous absorption of boric acid, borax, and disodium octaborate tetrahydrate in humans: a summary. *Biol. Trace Elem. Res.*, *66,* 101-109.

Wester, R.C., Hui, X., Hartway, T., Maibach, H.I., Bell, K., Schell, M.J., Northington, D.J., Strong, E. and Culver, B.D. (1998c). In vitro percutaneous absorption of boric acid, borax, and disodium octaborate tetrahydrate in humans compared to in vitro absorption in human skin from infinite and finite doses. *Toxicol. Sci., 45*, 42-51.

Williams, L.H. and Amburgey, T.L. (1987). Integrated protection against lyctid beetle infestations. IV. Resistance of boron-treated wood (*Virola spp.*) to insect and fungal attack. *Forest. Prod. J., 37*, 10-17.

Wilson, A.D. and Kent, B.E. (1972). A new translucent cement in dentistry. The glass ionomer cement. *Br. Dent. J., 132*, 133-135.

Woods, W.G. (1994). An introduction to boron: history, sources, uses and chemistry. *Environ. Health Perspect., 102*, 5-11.

Xue, R.D. and Barnard, D.R. (2003). Boric acid bait kills adult mosquitoes (Diptera: Culicidae). *J. Econ. Entomol., 96*, 1559-1562.

Xue, R.D., Kline, D.L., Ali, A. and Barnard, D.R. (2006). Applications of boric acid baits to plant foliage for adult mosquito control. *J. Am. Mosq. Control Assoc., 22,* 497-500.

Xue, R.D., Ali, A., Kline, D.L. and Barnard, D.R. (2008). Field evaluation of boric acid- and tipronil-based bait stations against adult mosquitoes. *J. Am. Mosq. Control Assoc., 24*, 415-418.

Yelvigi, M. (2005). Boric acid. In R.C. Rowe, P.J. Sheskey, and S.C. Owen (Eds.), *Pharmaceutical Excipients* (5th edition). London: Pharmaceutical Press, Electronic Version.

Yoshida, M., Watabiki, T. and Ishida, N. (1989). Spectrophotometric determination of boric acid by the curcumin method. *Nihon Hoigaku Zasshi, 43*, 490-496.

Yoshizaki, H., Izumi, Y., Hirayama, C., Fujimoto, A., Kandori, H., Sugitani, T. and Ooshima, Y. (1999). Availability of sperm examination for male reproductive toxicities in rats treated with boric acid. *J. Toxicol. Sci., 24*, 199-208.

Yui, K.C., Rodrigues, J.R., Mancini, M.N., Balducci, I. and Gonçalves, S.E. (2008). Ex vivo evaluation of the effectiveness of bleaching agents on the shade alteration of blood-stained teeth. *Int. Endod. J., 41*, 485-492.

Zhang, Y.C., Perzanowski, M.S. and Chew, C.L. (2005). Sub-lethal exposure of cockroaches to boric acid pesticide contributes to increased Bla g 2 excretion. *Allergy, 60*, 965-968.

Zhao, S., Wang, J., Ye, F. and Liu, Y.M. (2008). Determination of uric acid in human urine and serum by capillary electrophoresis with chemiluminescence detection. *Anal. Biochem., 378*, 127-131.

Zinellu, A., Carru, C., Sotgia, S. and Deiana, L. (2004). Optimization of ascorbic and uric acid separation in human plasma by free zone capillary electrophoresis ultraviolet detection. *Anal. Biochem., 330*, 298-305.

Zurek, L., Gore, J.C., Stringham, S.M., Watson, D.W., Waldvogel, M.G. and Schal, C. (2003). Boric acid dust as a component of an integrated cockroach management program in confined swine production. *J. Econ. Entomol., 96*, 1362-1366.

In: Advances in Medicine and Biology. Volume 20
Editor: Leon V. Berhardt
ISBN 978-1-61209-135-8

Chapter 10

PHAGE DISPLAY TECHNOLOGY: NEW BIOTECHNOLOGICAL APPLICATIONS IN SYNTHETIC BIOLOGY

Santina Carnazza * ***and Salvatore Guglielmino***
Dept. Life Sciences, University of Messina,
sal. Sperone 31, Vill. S. Agata 98166 Messina, Italy

ABSTRACT

Synthetic biology is potentially one of the most powerful emerging technologies today. It is the art of synthesizing and engineering new biological systems that are not generally found in nature, and also of redesigning existing biological molecules, structures and organisms so as to understand their underlying mechanisms. The molecules used in these systems might be naturally occurring or artificially synthesized for a variety of nanotechnological objectives.

Protein evolution in vitro technologies can provide the tools needed to synthetic biology both to develop "nanobiotechnology" in a more systematic manner and to expand the scope of what it might achieve. Proteins designed and synthesized from the scratch – the so-called "never born proteins" (NBPs)– and those endowed with novel functions can be adapted for nano-technological uses.

The phage display technique has been used to produce very large libraries of proteins having no homology with known proteins, and being selected for binding to specific targets. The peptides are expressed in the protein coats of bacteriophage, which provides both a vector for the recognition sequences and a marker that signals binding to the respective target. Specific ligands for virtually any target of interest can be isolated from highly diverse peptide libraries, and many of them selected up till now have shown considerable potential for nanotechnological applications. These involve drug discovery,

* E-mail: santina.carnazza@unime.it

"epitope discovery", design of DNA-binding proteins, source of new materials, antibody phage and recombinant phage probes, next-generation nano-electronics, targeted therapy and phage-display vaccination.

All this can be regarded as a kind of synthetic biology in that it involves the reshaping and redirecting of natural molecular systems, phage, typically using the tools of protein and genetic engineering.

INTRODUCTION

Synthetic biology studies how to build artificial biological systems for engineering applications, using advanced tools of system design, modeling and simulation, as well as the most recent experimental techniques.

There can be various approaches to synthetic biology: engineering of biological systems; redesigning life, by constructing biological systems, aimed to bridge gaps in our current understanding of biology; creating alternative life, by using unnatural molecules in living systems. The focus is often on ways of taking parts of natural biological systems, characterizing and simplifying them, and using them as components of novel, engineered, highly unnatural life forms. The experimental work has a philosophical counterpart, arising in a special way when chemistry, physics and engineering move towards biology.

Biologists are interested in synthetic biology because it provides a complementary perspective from which to consider, analyze, and ultimately understand the living world. Being able to design and build a system is also one very practical measure of understanding. Physicists, chemists and others are interested in synthetic biology as an approach with which to probe the behaviour of molecules and their activity inside living cells. Engineers view biology as a technology; they are interested in synthetic biology because the living world provides an apparently rich yet largely unexplored way for controlling and processing information, materials, and energy.

Protein evolution in vitro technologies, together with protein and genetic engineering, can provide the tools needed for rapid design, fabrication, and testing of systems. Studies of cellular function, discovery of new therapeutic targets, and detailed mechanistic and structural analyses of proteins rely on specific binding reagents. Display techniques are powerful tools to generate, select, and evolve such binding reagents completely in vitro, and they have great potential for biotechnological, medical and proteomic applications.

In particular, phage display technology has become crucial in functional genomics and proteomics. Since it was first reported, it has become an invaluable component of biotechnology. Specific ligands can be isolated from highly diverse peptide libraries against virtually any target of interest, and successfully used in various research fields. Improved library construction approaches —in combination with innovated vector design, display formats and screening methods— have further extended the technology. It seems probable that extremely diverse phage-display libraries contain multiple solutions to most binding problems. Phage display technology aids to explore the links between protein structure and function, and this information will in turn expedite the process of directed molecular evolution; moreover, it extends to the synthesis of artificial proteins with random sequences.

During the past ten years, antibody phage display has evolved into a well-accepted technology, and in a short time has delivered fully human, high-quality antibodies. Phage antibodies have proven their safety and efficacy in clinical trials and are likely to play a great role in the generation of analytical reagents and therapeutic drugs. They offer major advantages in terms of speed and throughput for research and target identification/validation. The greatest challenge for the future will be to translate our ability to create binding sites with tailored size, affinity, valency and sequence, into antibody molecules with improved clinical benefit.

Recombinant phage selected by phage-display may find application also as biosorbent and diagnostic probe in micro- and nano-devices in which antibodies have been used to date. For their properties, phage probes may be exploited for development of bioaffinity sensors, whose essential elements are probes that specifically recognize and selectively bind target structures and, as parts of the analytical platform, generate a measurable signal. For example, they may be used for separation and purification of bacteria prior to their identification with polymerase chain reaction, immunoassays, flow cytometry, or other methods. Otherwise, they can be used to develop rapid real-time diagnostic arrays, by themselves recognizing and binding selectively and specifically the target, with no need of further characterization. The development of highly sensitive and accurate field-usable devices for detection of multiple biological agents could have a number of applications in biomedical field as well as in monitoring of agro-food pathogens and detection of biological warfare agents.

Nanostructured biomaterials represent an ideal system for use in biological detection due to their unique selectivity. Organizing ordered inorganic nanoparticles by using soft materials as templates is essential for constructing electronic devices with new functionalities. The concept of conjugating nanoparticles with biomolecules opens up new possibilities for making functional next-generation electronic devices using biomaterial systems, such as recombinant phage.

Finally, tailored selection processes, in combination with easy-to-engineer phage, open the door wide for novel sophisticated applications of phage display technology in synthetic biology.

SYNTHETIC BIOLOGY AND THE "NEVER BORN PROTEINS"

One of the aims of protein engineering is to design proteins from scratch — for example, new protein-based materials and artificial enzymes. The chemical approach to synthetic biology is concerned with the synthesis of molecular structures and/or multi-molecular organized systems –proteins, nucleic acids, vesicular forms, and other– that do not exist in nature.

Some peptide materials have been successfully designed [McGrath et al. 1992; Urry et al. 1995; Deming 1997], as well as de novo peptides with specified folds [Bryson et al. 1995]. One interesting project belonging to this chemical frontier of synthetic biology concerns the so-called "never born proteins" (NBPs), meaning proteins that have not been produced and/or selected by nature in the course of biological evolution [Luisi 2007]. The proteins existing in nature make only an infinitesimal fraction of the theoretically possible structures, and our life is based on a very limited number of structures. This elicits the question why and how the protein structures existing in our world have been selected out, with the underlying question whether they have something very particular from the structural or thermodynamic point of view that made the selection possible. For example, the few structures selected might be the only ones to be stable (i.e., with the correct folding); or water soluble; or those which have very particular viscosity and/or rheological properties. A second point of view is that "our" proteins have no extraordinary physical properties at all; they have been selected by "chance" among an enormous number of possibilities of quite similar compounds, and it happened that they were capable of fostering cellular life. This last belongs to the so-called "contingency" theory. The NBPs can be produced in laboratory either by chemical synthesis –e.g. fragment condensation of short peptides with selection governed by the contingency of the environmental conditions– or the modern molecular biology techniques –such as the phage display method.

The principle to produce NBPs is simple: if one makes a long string of DNA purely randomly, the probability of hitting an existing sequence in nature is practically zero. If then this DNA is processed by standard recombinant DNA and in vivo expression techniques, a non-existing polypeptide will be obtained, which, when globularly folded, is already a NBP.

In practice, the aim is to produce a very large library of totally random, de novo proteins having no homology with known proteins, and to investigate whether these synthetic biology products are really so different with respect to natural proteins, in terms of stability, solubility, or folding.

Luisi's group [Chiarabelli et al. 2006a, 2006b; De Lucrezia et al. 2006a, 2006b] tackled the question by investigating folding ability of NBPs, considering it a particularly important and stringent criterion, as the prerequisite for the biological activity of proteins determined by their primary structure. The strategy adopted was based on the well-accepted observation that folded proteins are not easily digestible by proteases. It involved the insertion of the tripeptide PRG (proline-arginine-glycine), substrate for the proteolytic enzyme thrombin, in the otherwise totally random protein sequence. In this way, each of the new proteins had the potentiality of being digested by the enzyme, with the expectation, however, that globularly folded NBPs would be protected from digestion. The larger part of the population was rapidly hydrolyzed, but ~ 20% of the NBPs were highly resistant to the action of thrombin, suggesting that folding is indeed a general property, something that arises naturally, even for proteins of medium length. A significant percentage of periodic structure, α-helix in particular, was present, and, furthermore, the globular folding was thermoreversible, indicating to be under thermodynamic control.

It appears possible at this point to state that folding and thermodynamic stability are not properties that are restricted to extant proteins, and that, on the contrary, they appear to be rather common features of randomly created polypeptides. On the basis of this, one is tempted to propose that "our" natural proteins do not belong to a class of polypeptides with privileged physical properties. And, by inference, one could say that this kind of data permit to brake a lance in favour of the scenario of contingency.

Of course, the NBPs may have also bio-technological importance and be very interesting from the structural point of view: they could, for example, display novel catalytic and structural features that have not been observed in natural proteins.

Indeed, a more difficult challenge is achieving novel catalytic function in artificial proteins with an efficacy and specificity similar to that of natural enzymes. Most efforts so far have tended to use the natural combinatorial mechanism of the immune system to develop antibodies with catalytic functions [Tramontano et al. 1986]. Recently, however, advances in computational methods have been exploited to transform a non-catalytic protein receptor into a mimic of a natural enzyme by rationally mutating several residues in the binding site [Dwyer et al. 2004]. Thus, rational protein design does not necessarily have to be conducted wholly de novo: existing protein folds can be used for the "scaffolding", and one can focus simply on retooling the active site. Moreover, proteins can be rationally modified to bind to new, non-natural substrates.

These developments in protein design have been adapted for nanotechnological uses. For example, the versatility of the immune system has been used to generate antibodies that will recognize and bind to fullerenes [Chen et al. 1998], carbon nanotubes [Erlanger et al. 2001], and a variety of crystal surfaces [Izhaky and Addadi 1998]. Synthetic biology, however, could provide the tools and understanding needed to develop "nanobiotechnology" in a more systematic manner, as well as to expand the scope of what it might achieve [Ball 2005].

Synthetic biology includes the broad redefinition and expansion of biotechnology, with the ultimate goals of being able to design and build engineered biological systems that process information, manipulate chemicals, fabricate materials and structures, produce energy, provide food, and maintain and enhance human health and our environment [Chopra and Kamma 2006].

PROTEIN ENGINEERING AND DIRECTED EVOLUTION

Protein engineering is a relatively young discipline, aimed to develop novel useful or valuable proteins with new and uniquely functional attributes [Graff et al. 2004]. Much research is currently taking place into the understanding of the fundamental rules linking a protein's structure to its function, and it involves the application of science, mathematics and economics.

There are two main strategies for protein engineering. The first is known as rational design, in which detailed knowledge of the structure and function of the protein is used to make desired changes. This has the advantage of being generally inexpensive and easy, since site-directed mutagenesis techniques are well-developed. However, there is a major drawback in that detailed structural knowledge of a protein is often unavailable, and even when it is available, it can be extremely difficult to predict the effects of various mutations.

Computational protein design algorithms seek to identify amino acid sequences that have low energies for target structures. While the sequence-conformation space that needs to be searched is large, the most challenging requirement for computational protein design is a fast, yet accurate, energy function that can distinguish optimal sequences from similar suboptimal ones. Using computational methods, a protein with a novel fold has been designed [Yuan et al. 2005], as well as sensors for unnatural molecules [Arnold 1998].

The second strategy is known as directed evolution. This method mimics natural evolution to evolve proteins with desirable properties not found in nature, and generally produces superior results to rational design. Random mutagenesis is applied to a protein, and a selection regime is used to pick out variants that have the desired qualities. Further rounds of mutation and selection are then applied, in order to allow an increase in functional density of the protein of interest, identifying interesting mutants. It is thus possible to use this method to optimize properties that were not selected for in the original organism, including catalytic specificity, thermostability and many others.

A typical directed evolution experiment involves two steps:

1. *Library creation:* The gene encoding the protein of interest is mutated and/or recombined at random to create a large library of gene variants.
2. *Library screening*: The library is screened by the researcher using a high-throughput screen to identify mutants or variants that possess the desired properties. "Winner" mutants identified in this way then have their DNA sequenced to understand what mutations have occurred.

The evolved protein is then characterized using biochemical methods.

The great advantage of directed evolution techniques is that they require no prior structural knowledge of a protein, nor it is necessary to be able to predict what effect a given mutation will have. Indeed, the results of directed evolution experiments are often surprising in that desired changes are often caused by mutations that no one would have expected. The drawback is that they require high-throughput, which is not feasible for all proteins. Large amounts of recombinant DNA must be mutated and the products screened for desired qualities. The sheer number of variants often requires expensive robotic equipment to automate the process. Furthermore, not all desired activities can be easily screened for. New advancements in high-throughput technology will greatly expand the capabilities of directed evolution.

An additional technique known as *DNA shuffling*, or *sexual Polymerase Chain Reaction* (PCR) [Stemmer 1994], mixes and matches pieces of successful variants in order to rapidly propagate beneficial mutations, thus producing better results in a directed evolution experiment. This process mimics recombination that occurs naturally during sexual reproduction and is used to rapidly increase DNA library size.

DNA shuffling is a PCR without synthetic primers. In this process, a family of related genes are first cut with enzymes. The gene fragments then are heated up to separate them into single-stranded templates. Some of these fragments will bind to other fragments that share complementary DNA regions, which in some cases will be from other family members. Regions of DNA that are non-complementary hang over the ends of the templates, and the PCR reaction then treats the complementary regions as primers and builds the new double-helical DNA. But PCR also adds bases to the overhanging piece of the primer, forming a double helix there, too. This ultimately creates a mixed structure or "chimera". In the final step, PCR reassembles these chimeras into full-length, shuffled genes.

Application of these methods to engineer protein cores, active sites and macromolecular interfaces will contribute greatly to our ability to both understand and rationally manipulate the physicochemical properties that drive protein function.

In Vitro and Biological Display Technologies

One of the most powerful strategies to improve the properties of proteins or even create new ones is to imitate the strategy of evolution in the test tube, through an in vitro iteration between diversification and selection, by means of display technologies. The directed evolution of proteins using display methods can be engineered for specific properties and selectivity. A variety of display approaches are employed for the engineering of optimized human antibodies, as well as protein ligands, for such diverse applications as protein arrays, separations, and drug development.

In vitro display technologies, namely ribosome and mRNA display [for a review, Amstutz et al. 2001; Lipovsek and Plückthun 2004], combine two important advantages for identifying and optimizing ligands by evolutionary strategies. First, by obviating the need to transform cells in order to generate and select libraries, they allow much higher library diversity. Second, by including PCR as an integral step in the procedure, they make PCR-based mutagenesis strategies convenient. The resulting iteration between diversification and selection allows true Darwinian protein evolution to occur in vitro. Successful examples of high-affinity, specific target-binding molecules selected by in-vitro display methods include peptides, antibodies, enzymes, and engineered scaffolds, such as fibronectin type III domains [Koide et al. 1998; Xu et al. 2002] and synthetic ankyrins, which can mimic antibody function [Binz et al. 2003, 2004].

Ribosome Display is a technique used to perform in vitro protein evolution to create proteins that can bind to a desired ligand. It was first developed by Mattheakis et al. [1994] for the selection of peptides and further improved for folded proteins [Hanes and Plückthun 1997; He and Taussig 1997, 2007]. A fusion protein is constructed in which the domain of interest is fused to a C-terminal tether, such that this domain can fold while the tether is still in the ribosomal tunnel. This fusion construct lacks a stop codon at the mRNA level, thus preventing release of the mRNA and the polypeptide from the ribosome. The process results in translated proteins that remain associated with ribosome and their mRNA progenitor, which is used, as a non-covalent ternary complex, to bind to an immobilized ligand in a selection step. The mRNA-protein hybrids that bind well are then reverse transcribed to cDNA and their sequence amplified via PCR. The end result is a nucleotide sequence that can be used to create tightly binding proteins.

The complex of mRNA, ribosome, and protein is stabilized with the lowering of temperature and the addition of cations such as Mg^{2+}. During the subsequent panning stages, the complex is introduced to surface-bound ligand in several ways: using an affinity chromatography column with a resin bed containing ligand, a 96-well plate with immobilized surface-bound ligand, or magnetic beads that have been coated with ligand. The complexes that bind well are immobilized. Subsequent elution of the binders, via high salt concentrations, chelating agents, or mobile ligands which complex with the binding motif of the protein, allows dissociation of the mRNA. The mRNA can then be reverse transcribed back into cDNA, and thus, the genetic information of the binding polypeptides is available for analysis, then it can undergo mutagenesis, and iteratively fed into the process with greater selective pressure to isolate even better binders.

By having the protein progenitor attached to the complex, the process of ribosome display skips the microarray/peptide bead/multiple-well sequence separation that is common in assays involving nucleotide hybridization and provides a ready way to amplify the proteins that do bind without decrypting the sequence until necessary. At the same time, this method relies on generating large, concentrated pools of sequence diversity without gaps and keeping these sequences from degrading, hybridizing, and reacting with each other in ways that would create sequence-space gaps.

In addition, as ribosome display is the first method for screening and selection of functional proteins performed completely in vitro, it circumvents many drawbacks of in vivo selection technologies. First, the diversity of the library is not limited by the transformation efficiency of bacterial cells, but only by the number of ribosomes and different mRNA molecules present in the test tube. Second, random mutations can be introduced easily after each selection round, as no library must be transformed after any diversification step. This allows simple directed evolution of binding proteins over several generations.

In ribosome display, the physical link between genotype and phenotype is accomplished by an mRNA–ribosome–protein complex that is used for selection. As this complex is stable for several days under appropriate conditions, several selections can be performed. Ribosome display allows protein evolution through a built-in diversification of the initial library during selection cycles. Thus, the initial library size no longer limits the sequence space sampled.

This technology of directed evolution over many generations is currently being exploited to address fundamental questions of protein structure and stability [Jermutus et al. 2001; Hanes et al. 2000a], catalysis [Amstutz et al. 2002; Cesaro-Tadic et al. 2003], as well as interesting biomedical applications. Recently, the potential of ribosome display for directed molecular evolution was recognised and developed into a rapid and simple affinity selection strategy to obtain scFv fragments of antibodies with affinities in the low picomolar range [Schaffitzel et al. 1999; Hanes et al. 2000b]. The authors selected a range of different scFvs with picomolar affinity from a fully synthetic naïve antibody scFv library using ribosome display. All of the selected antibodies accumulated beneficial mutations throughout the selection cycles. This display method can apply also to other members of the immunoglobulin superfamily; for example single V-domains which have an important application in providing specific targeting to either novel or refractory cancer markers [Irving et al. 2001]. These works demonstrated that ribosome display not only allows the selection of library members but also further evolves them, thereby mimicking the strategy of the immune system.

It was also demonstrated that even those proteins can be selected that cannot be expressed at all in vivo [Schimmele and Plückthun 2005; Schimmele et al. 2005].

Ribosome display systems that are well proven, by the evolution of high affinity antibodies and the optimization of defined protein characteristics, generally use an *Escherichia coli* cell extract for in vitro translation and display of an mRNA library. More recently, a cell-free translation system has been produced by combining recombinant *E. coli* protein factors with purified 70S ribosomes [Villemagne et al. 2006]. Higher cDNA yields are recovered from ribosome display selections when using a reconstituted translation system and the degree of improvement seen is selection specific. These effects are likely to reflect higher mRNA and protein stability and potentially other advantages that may include protein specific improvements in expression. Reconstituted translation systems therefore enable a highly efficient, robust and accessible prokaryotic ribosome display technology.

Competing methods for protein evolution in vitro are mRNA display, yeast display, bacterial display and phage display.

Like other biological display technologies, *mRNA display technology* provides easily accessible coding information for each peptide/protein displayed [Roberts and Szostak 1997; Nemoto et al. 1997]. In mRNA display, mRNA is first translated and then covalently bonded to the nascent polypeptide it encodes, using puromycin as an adaptor molecule. The covalent mRNA–protein adduct is purified from the ribosome and used for selection. Puromycin is an analogue of the 3' end of a tyrosyl-tRNA with a part of its structure mimics a molecule of adenosine, and the other part mimics a molecule of tyrosine. Compared to the cleavable ester bond in a tyrosyl-tRNA, puromycin has a non-hydrolysable amide bond. As a result, puromycin interferes with translation, and causes premature release of translation products.

The protein and the mRNA are thus coupled and are subsequently isolated from the ribosome and purified. In the current protocol, a cDNA strand is then synthesized to form a less sticky RNA–DNA hybrid and these complexes are finally used for selection.

mRNA display technology has many advantages over the other display methods. The biological display libraries (phage, yeast and bacterial) have polypeptides or proteins expressed on the respective microorganism's cell surface, and the accompanying coding information for each polypeptide or protein is retrievable from the microorganism's genome. However, the library size for the in vivo display systems is limited by the transformation efficiency of each organism. For example, the library size for phage and bacterial display is limited to 1-10 × 10^9 different members. The library size for yeast display is even smaller. Moreover, these cell-based display systems only allow the screening and enrichment of peptides/proteins containing natural amino acids. In contrast, mRNA display and ribosome display are totally in vitro selection methods [Roberts 1999]. They allow a library size as large as 10^{14} different members. The large library size increases the probability to select very rare sequences, and also improves the diversity of the selected sequences. In addition, in vitro selection methods remove unwanted selection pressure, such as poor protein expression, and rapid protein degradation, which may reduce the diversity of the selected sequences. Finally, in vitro selection methods allow the application of in vitro mutagenesis and recombination techniques throughout the selection process. Moreover, although both ribosome display and mRNA display are both in vitro selection methods, mRNA display has some advantage over the former. mRNA display utilizes covalent mRNA-polypeptide complexes linked through puromycin; whereas, ribosome display utilizes stalled, noncovalent ribosome-mRNA-polypeptide complexes and selection stringency is limited. This may cause difficulties in reducing background binding during the selection cycle. Also, the polypeptides under selection in a ribosome display system are attached to an enormous rRNA-protein complex, the ribosome itself, and there might be some unpredictable interaction between the selection target and the ribosome, thus leading to a loss of potential binders during the selection cycle. In contrast, the puromycin DNA spacer linker used in mRNA display technology is much smaller comparing to a ribosome, so having less chance to interact with an immobilized selection target. Thus, mRNA display technology is more likely to give less biased results [Gold 2001].

mRNA display has been used to select high affinity reagents from engineered libraries of linear peptides [Barrick and Roberts 2002; Barrick et al. 2001; Wilson et al. 2001; Baggio et al. 2002], constrained peptides [Baggio et al., 2002], single-domain antibody mimics [Xu et al.,2002], variable heavy domains of antibodies and single-chain antibodies [Chen 2003]. In addition, mRNA-display selections from proteomic libraries have identified cellular polypeptides that bind specific signaling proteins [Hammond et al., 2001] and small-molecule drugs, as well as polypeptide substrates of v-abl kinase [Cujec et al., 2002].

In general, in vitro display technologies prove to be valuable tools for many applications other than merely selecting polypeptide binders. They have great potential for directed evolution of protein stability and affinity, the generation of high-quality libraries by in vitro preselection, the selection of enzymatic activities, and the display of cDNA and random-peptide libraries [Amstutz et al. 2001; Lipovsek and Plückthun 2004].

In *Yeast display* (or *yeast surface display*) a protein of interest is displayed as a fusion to the Aga2p protein on the surface of yeast [Boder and Wittrup 1997, 1998]. The Aga2p protein is naturally used by yeast to mediate cell-cell contacts during yeast cell mating. As such, display of a protein via Aga2p projects the protein away from the cell surface, minimizing potential interactions with other molecules on the yeast cell wall. The use of magnetic separation and flow cytometry in conjunction with a yeast display library is a highly effective method to isolate high affinity protein ligands against nearly any receptor through directed evolution.

Advantages of yeast display over other in vitro evolution methods include eukaryotic expression and processing, quality control mechanisms of the eukaryotic secretory pathway, minimal avidity effects, and quantitative library screening through fluorescent-activated cell sorting (FACS) [Feldhaus and Siegel 2004a, 2004b].

Disadvantages include smaller mutagenic library sizes compared to alternative methods and differential glycosylation in yeast compared to mammalian cells. It should be noted that these disadvantages have not limited the success of yeast display for a number of applications, including engineering the highest monovalent ligand-binding affinity reported to date for an engineered protein [Boder et al. 2000].

Similarly, in *Bacterial Display* (or *bacterial surface display*) libraries of polypeptides displayed on the surface of bacteria can be screened using iterative selection procedures (biopanning), flow cytometry or cell sorting techniques [Francisco et al. 1993], thus simplifying the isolation of proteins with high affinity for ligands. Expression of antigens on the surface of non-virulent microorganisms is an attractive approach to the development of high-efficacy recombinant live vaccines [Georgiou et al. 1997]. Finally, cells displaying protein receptors or antibodies are of use for analytical applications and bioseparations.

PHAGE DISPLAY FOR DIRECTED MOLECULAR EVOLUTION

Phage display is a fundamental tool in protein engineering as well as a method for studying protein-protein, protein-peptide, and protein-DNA interactions that utilizes bacteriophage to connect proteins with the genetic information that encodes them [Smith 1985]. This connection between genotype and phenotype enables large libraries of proteins to be screened and amplified in a process of in vitro selection that imitates the strategy of natural evolution in the test tube. Phage display technology involves the expression of random peptides on the surface of a bacteriophage, displayed as a fusion with one of the viral structural protein. The most common bacteriophages used in phage display are M13 and fd filamentous phage [Smith and Petrenko 1997; Kehoe and Kay 2005], though T4 [Ren and Black 1998], T7 [Rosenberg et al. 1996], and λ phage [Santini et al. 1998] have also been used.

Filamentous phages [Marvin 1998] are flexible, thread-like particles approximately 1 μm long and 6 nm in diameter. The bulk of their tubular capsid consists of 2700 identical subunits of the 50-residue major coat protein pVIII arranged in a helical array with a five-fold rotational axis and a coincident two-fold screw axis with a pitch of 3.2 nm. The major coat protein constitutes 87% of total virion mass. Each pVIII subunit is largely αhelical and rod-shaped; about half of its 50 amino acids are exposed to the solvent, the other half being buried in the capsid. At one tip of the particle the capsid is capped with five copies each of minor coat proteins pVII and pIX; five copies each of minor coat proteins pIII and pVI cap the other end. The capsid encloses a single-stranded DNA. Longer or shorter plus strands —including recombinant genomes with foreign DNA inserts— can be accommodated in a capsid whose length matches the length of the enclosed DNA by including proportionally fewer or more pVIII subunits.

In 1985, recombinant DNA techniques were applied to phage to fashion a new type of molecular chimera that underlies today's phage display technology [Smith 1985]. A foreign coding sequence is spliced in-frame into a phage coat protein gene, so that the ''guest'' peptide encoded by that sequence is fused to a coat protein, and thereby displayed on the exposed surface of the virion.

In early examples of M13 filamentous phage display, polypeptides were fused to the amino-terminus of either PIII or PVIII in the viral genome [Scott and Smith 1990; Greenwood et al. 1991]. These systems were severely limited because large polypeptides (>10 residues for PVIII display) compromised coat protein function and so could not be efficiently displayed. The development of phagemid display systems solved this problem because, in such systems, polypeptides were fused to an additional coat protein gene encoded by a phagemid vector [Bass et al. 1990]. Multiple cloning sites are sometimes used to ensure that the fragments are inserted in all three possible frames so that the cDNA fragment is translated in the proper frame. The phagemid is then transformed into *E. coli* bacterial cells such as TG1 or XL1-Blue *E. coli*. The phage particles will not be released from the *E. coli* cells until these are infected with helper phage, which enables packaging of the phagemid DNA in a coat composed mainly of wild-type coat proteins from the helper phage but also containing some fusion coat proteins from the phagemid. In phagemid systems, functional polypeptide display has now been demonstrated with all five M13 coat proteins.

By cloning large numbers of DNA sequences into the phage, display libraries are produced with a repertoire of many billions of unique displayed proteins. A phage display library is, in fact, an ensemble of up to about 10 billion recombinant phage clones, each harboring a different foreign coding sequence, and therefore displaying a different guest peptide on the virion surface.

The foreign coding sequence can derive from a natural source, or it can be deliberately designed and synthesized chemically. For instance, phage libraries displaying billions of random peptides can be readily constructed by splicing degenerate synthetic oligonucleotides, obtained by combinatorial approach into the coat protein gene. Displayed peptides can be linear or disulfide-constrained [McLafferty et al. 1993; Ladner 1995; Saggio and Laufer 1993; Luzzago and Felici 1998], aimed to mimic in minute detail similar natural ligands and epitopes.

Recombinant peptides specifically binding a target of interest can be selected from random peptidic libraries (usually from 8- to 20-mer), by a process of affinity selection known as "biopanning".

By immobilizing the target protein to a solid support of some sort (e.g., on the polystyrene surface of an ELISA well or on a magnetic bead), a phage that displays a protein binding to one of those targets on its surface will be captured on the support and remain there while others are removed by washing. Those that remain —generally a minuscule fraction of the initial phage population— can then be eluted in a solution that loosens target-peptide bonds without destroying phage infectivity, propagated simply by infecting fresh bacterial host cells and so produce a phage mixture that is enriched with relevant binding phage and that can serve as input to another round of affinity selection. Phage eluted in the final step (typically after 2-4 rounds of selection) can be used to infect a suitable bacterial host, from which the phagemids can be collected and the relevant DNA sequenced to identify the interacting proteins or protein fragments.

Recent work published by Chasteen et al. [2006] shows that use of the helper phage can be eliminated by using a novel "bacterial packaging cell line" technology.

In addition, phage selection is not limited to solid support biopanning as described above but has been used also against intact cells for selection of tissue and cell targeting proteins [Samoylova et al. 2003; Spear et al. 2001]. In particular, this technology represents a powerful tool for the selection of peptides binding to specific motifs on whole cells since it is a non-targeted strategy, which also enables the identification of surface structures that may not have been considered as targets or have not yet been identified [Bishop-Hurley et al. 2005].

More recently, in vivo phage display has been used extensively to screen for novel targets of tumor therapy [Schluesener and Xianglin 2004; Li et al. 2006, Lee et al. 2002; Brown C. K. et al. 2000; Zhang et al. 2005].

Phage display is a practical realization of an artificial chemical evolution [Smith and Petrenko 1997]. Using standard recombinant DNA technology, peptides are associated with replicating viral DNAs that include their coding sequences. This confers on them the key properties of evolving organisms: replicability and mutability. Affinity for the target is an artificial analogue to the ''fitness'' that governs an individual's survival in the next generation. Because selection parameters can be designed and controlled, the phage display is an ideal instrument for directed molecular evolution.

The peptide populations so created are managed by simple microbiological methods that are familiar to all molecular biologists, and they are replicable and therefore nearly cost-free after their initial construction or selection. Therefore, phage display has the overwhelming advantage to be cheap and easy.

The proteins displayed range from short amino acid sequences to fragments of proteins [Wang et al. 1995; vanZonneveld et al. 1995], enzymes [Wang et al. 1996], receptors [Gu et al. 1995; Onda et al. 1995; Sche et al. 1999; Fakok et al. 2000], DNA and RNA binding proteins [Wu et al. 1995; Wolfe et al. 1999; Segal et al. 1999; Isalan and Choo et al. 2000; Cheng et al. 1996] and hormones [Cabibbo et al. 1995; Wrighton et al. 1996; Livnah et al. 1996].

Geysen and his colleagues introduced the term "mimotope" to refer to small peptides that specifically bind a receptor's binding site (and in that sense mimic the epitope on the natural ligand) without matching the natural epitope at the amino acid sequence level [Geysen et al. 1986a, 1986b]; the definition includes cases where the natural ligand is non-proteinaceous. Mimotopes are usually of little value in mapping natural epitopes, but may have other important uses, as identifying new receptors and natural ligands for an "orphan receptor" [Houimel et al. 2002; El-Mousawi et al. 2003], peptides that might act as enzyme inhibitors [Hyde-DeRuyscher et al. 2000; Dennis et al. 2000; Maun et al. 2003; Huang et al. 2003; Lunder et al. 2005a, 2005b; Bratkovic et al. 2005; Nakamura et al. 2001] and receptor agonists or antagonists [Skelton et al. 2002; McConnell et al. 1998; Nakamura et al. 2002; Hessling et al. 2003], "epitope discovery" [Wang and Yu 2004; Petit et al. 2003; Leinonen et al. 2002; Coley et al. 2001; Myers et al. 2000; Rowley et al. 2000], design of DNA-binding proteins [Wu et al. 1995; Wolfe et al. 1999; Segal et al. 1999; Isalan and Choo et al. 2000; Cheng et al. 1996], source of new materials [Smith and Petrenko 1997; Souza et al. 2006; Nam et al. 2004]. The proteins so synthesized can indeed be considered as non-extant, and artificial proteins with random sequence can be displayed [Nakashima et al. 2000], which permits the terminology of "never born proteins" (NBPs).

Phage display is an exponentially growing research area, and numerous reviews covering different aspects of it have been published [Felici et al. 1995; Cortese et al. 1995; Smith and Petrenko 1997]. In conclusion, then, it is a useful tool in protein engineering as well as in functional genomics and proteomics, in drug discovery and, we can say, in synthetic biology.

ANTIBODY PHAGE

Over the past decade, phage-displayed antibody fragments have been the subject of intensive research [reviewed in Dall'Acqua and Carter 1998; Griffiths and Duncan 1998]. As a result, antibody phage libraries have become practical tools for drug discovery and several phage-derived antibodies are in advanced clinical trials. Phage display has provided approximately 30% of all human antibodies now in clinical development [Kretzschmar and von Ruden 2002].

Large collection of antibody fragments have been displayed on phage particles, and successfully screened with different antigens [Hust and Dubel 2004]. Since the first demonstration that it was possible to display functional antibody fragments on the surface of filamentous phage [McCafferty et al. 1990; Hoogenboom et al. 1991; Barbas et al. 1991; Breitling et al. 1991; Garrard et al. 1991], the development of this technique has led to the construction of recombinant antibody libraries displayed on the bacteriophage surface. The selection of antibodies by phage display basically relies on several factors: first, the ability to isolate or synthesize antibody gene pools to construct large, highly diverse libraries; second, the possibility to express functional antibody fragments in the periplasmic space of *Escherichia coli* [Better et al. 1988; Skerra and Plückthun 1988]; and third, the efficient coupling of expression and display of the antibody protein with the antibody's genetic information being packaged in the *E. coli* bacteriophage [Smith 1985; McCafferty et al. 1990].

Filamentous bacteriophage such as M13 and its coat protein pIII are most often used for antibody display, although T7 bacteriophage also reportedly allows antibody display [Rosenberg et al. 1996].

Both scFv (single chain Fv fragments, where the heavy and light chain V regions are linked by a linker peptide) [McCafferty et al. 1990] and Fab (Fragment antigen binding dimers) [Hoogenboom et al. 1991; Garrard et al. 1991; Chang et al. 1991] formats have been used successfully in antibody libraries displayed on phage, with the former representing the more popular choice [Carmen and Jermutus 2002]. Large repertoires of heavy and light chain V regions can be obtained through amplification by the polymerase chain reaction from the B cells of an immunized animal (usually extracted from the spleen) [Clackson et al. 1991], or hybridoma cells generated from such an animal [Orlandi et al. 1989; Chiang et al. 1989], or even immunized humans ("immunized libraries") [Persson et al. 1991; Burton et al. 1991; Graus et al. 1997; Cai and Garen 1995]; these repertoires will contain antibodies that are biased towards the immunogen, based on the host's immune response.

Alternative approaches are constituted by the "semi-synthetic libraries", where germ-line heavy and light chain V regions, cloned from human B cells, are assembled and synthetic randomization is used to introduce additional diversity at the CDR3 region to increase the repertoire [Barbas et al. 1992], and the "naïve libraries" heavy and light chain variable regions are amplified from the naïve Ig repertoire of a healthy human donor and randomly combined to produce the phage-displayed library [Carmen and Jermutus 2002; Marks et al. 1991].

An important advance has been the development of high-quality libraries with completely synthetic complementarity-determining regions. Knappik et al. [2000] have constructed a library in which a limited number of human frameworks are used and diversity is introduced by means of synthetic cassettes. Such a system is very amenable to the generation of therapeutic antibodies because preferred frameworks can be used and affinity maturation is aided by the use of defined mutagenic cassettes.

The construction of large, high-quality antibody phage libraries has also been aided by the introduction of improved in vivo recombination systems [Sblattero and Bradbury 2000].

The selection of antibodies from phage libraries consists of two main steps: panning and screening. During panning, library phage preparations are incubated with the antigen of choice, unbound phage are discarded and remaining phage recovered after several washing steps by disrupting the phage–antigen interaction without compromising the phage infectivity (e.g. by applying pH-gradients, competitive elution conditions or proteolytic reactions). Recovered phage subsequently are amplified by infecting *E. coli* and further round(s) of panning are applied, yielding a polyclonal mixture of phage antibodies enriched for antigen-specific binders. The purified antigen can be attached to a solid support (to plastic, by adsorption, or to beads or a column matrix), the phage library run over the support, and antigen-bound phage retrieved after rinsing the support by elution; alternatively, the binding event can take place with antigen in solution, for example, using biotinylated antigen or unlabelled antigen, and the antigen-bound phage might be retrieved by incubation with streptavidin-coated magnetic beads or other ligands that capture the antigen.

The in vitro selection procedure can be performed for function, besides for binding capabilities. Selections can be performed under conditions that mediate the selection of phage antibodies with a particular characteristic, for example, under reducing conditions to retrieve disulfide-free yet stable antibodies [Proba and Wörn 1998], or in the presence of proteases to select for well-folded molecules [Kristensen and Winter 1998]. Antibodies might also be selected with or for a particular functional activity, for example, for receptor cross-linking, signalling, gene transfer or catalysis [Hoogenboom et al. 1998].

The screening process involves subsequently converting the polyclonal mixture obtained by panning into monoclonal antibodies. To this end, *E. coli* cells are infected with the phage pool, plated on selective agar dishes, and single colonies are picked. Thus, highly specific, monoclonal antibody clones are obtained, from which the antibody genes can be readily isolated for further analyses and/or engineering [Rader and Barbas 1997; Griffiths and Duncan 1998; Hoogenboom and Chames 2000; Siegel 2002].

A single phage antibody library can be distributed to thousands of users and serve as the source of cloned antibodies against an unlimited array of antigens. Because selection is based solely on affinity, many toxic and biological threat agents that could not be used to immunize animals without their prior inactivation can nevertheless serve as ''native antigens'' in this artificial immune system. Furthermore, phage display allows selecting of antibodies recognizing unique epitopes on biological agents that may be missed in hybridoma screening [Emanuel et al. 1996]. Another advantage of phage display contrasting it to the hybridoma technique is that the quantity of antigens required for selection of phage antibodies may be surprisingly small [Liu and Marks 2000], and the properties of selected probes can be further improved by affinity maturation and molecular evolution [Chowdhury 2002; Deng et al.1994; Worn and Plückthun 2001]. Thus, for many purposes, this system may well come to replace natural immunity in animals [Liu and Marks 2000]. With phage display, as in the in vivo immune system, antibodies can be affinity-matured in a stepwise fashion, by incorporating mutations and selecting variant cells with decreasing amounts of antigen (reviewed in Hoogenboom 1997]. Various procedures for introducing diversity in the antibody genes have been described, ranging from more-or-less random strategies (e.g. V-gene chain shuffling, error-prone PCR, mutator strains or DNA shuffling), to strategies targeting the CDR regions of the antibody for mutagenesis (e.g. oligonucleotide-directed mutagenesis, PCR). One possible disadvantage of this in vitro process is that the affinity improvement can be accompanied by the appearance of a modified fine-specificity or unwanted cross-reactivity [Ohlin et al. 1996], which the natural immune system might quickly remove. Extensive screening of the in vitro affinity-matured antibodies for changes in specificity is thus required.

There are several examples where phage antibodies selected against various biological threats have been used beneficially in various detection platforms [reviewed in Iqbal et al. 2000] and for therapeutic applications [reviewed in Kretzschmar and von Ruden 2002].

PHAGE AS PROBES IN NANOBIOTECHNOLOGY

Development of systems for monitoring the environment and food for biological threat agents is a challenge because it requires environmentally stable, long lasting, sensitive and specific diagnostic probes capable of tight selective binding of pathogens in unfavorable conditions. To respond to the challenge, large financial investments and extensive collaborative efforts of specialists in different areas of science and technology are required. In the last years, probe technology is being revolutionized by utilizing methods of combinatorial chemistry and directed molecular evolution. In particular, phage display is recently identified [Brigati et al. 2004; Petrenko and Smith 2000; Petrenko and Sorokulova 2004; Petrenko and Vodyanoy 2003] as a new technique for development of diagnostic probes which may meet the strong criteria —fastness, sensitiveness, accuracy, and inexpensiveness— for biological monitoring [Al-Khaldi et al. 2008, 2009].

Recombinant phage probes have been selected from phage display libraries with high specificity and selectivity for a wide range of targets [Kouzmitcheva et al. 2001; Petrenko et al. 1996; Petrenko and Smith 2000; Romanov et al. 2001; Samoylova et al. 2003], including small molecules [Saggio and Laufer 1993], receptors [Balass et al. 1993; Legendre and Fastrez 2002], virus [Gough et al. 1999], bacterial spores [Knurr et al. 2003; Steichen et al. 2003; Turnbough 2003; Brigati et al. 2004], and whole-cell epitopes [Carnazza et al. 2007, 2008; Cwirla et al. 1990; Olsen et al. 2006; Petrenko and Sorokulova 2004; Sorokulova et al. 2005; Stratmann et al. 2002; Yu and Smith 1996].

Affinity-selected landscape phage probes for *Salmonella typhimurium* were demonstrated to possess the specificity, selectivity, and affinity of monoclonal antibodies and used as probes for the detection of *S. typhimurium* [Sorokulova et al. 2005].

Petrenko and Vodyanoy [2003] have discussed the potential of phage-based electronic-based (QCM) biodetectors for threat agents.

Proof-in-concept biosensors were prepared for the rapid detection of *S. typhimurium* in solution, based on affinity-selected filamentous phage prepared as probes physically adsorbed to acoustic wave piezoelectric transducers [Olsen et al. 2006].

At the same time, phage antibody chip strategy proved its efficacy in preliminary diagnostics of cancer or other diseases applications [Hong et al. 2004]; phage arrays were constructed by using five clones, displaying respectively four scFv from mouse and one humanized scFv. The targets were Cy3 fluorescence labeled protein extracts from normal lymphocytes and tumorous HeLa cells. Fluorescence intensity of phage was exploited to indicate overexpression of some proteins in the tumor cell line when compared to normal lymphocytes.

Another similar proof-of-principle experiment for phage antibody chips was also reported from the same research group [Bi et al. 2007]: a protein microarray spotted directly with ninety-six phage-displayed antibody clones, half of them deriving from cell panning with leukocytes from healthy donors, and half from panning with acute myeloid leukemia leukocytes, was created to discriminate between recognition profiles of samples from healthy donors and leukemia patients. The signals of nine of those probes showed significant difference between normal and leukemia samples.

In general, fusion of a peptide to the pIII minor coat protein, located on the tip of the phage capsid, is probably not optimal for obtaining phage probes because this expression format does not take advantage of the avidity effect gained when the binding peptides are displayed multi-valently on the major coat protein pVIII [Liao et al. 2005]. This is because pIII is the minor protein of wild phage and there are only five copies of pIII on the tip of the phage. In the pIII display system, scFv is always expressed mono-valently, and most probably scFv will be either orientated parallel to or in contact with the surface which may restrict the freedom of scFv recognition. On the contrary, the result is surprising in pVIII display system, with a ratio of the positive signal to the negative of 3000:1. This is attributed to the amount and status of pVIII of phage, which forms the tube of phage with about 2700 copies. Using pVIII display system, not only the number of fused scFv is increased, but also the orientation is improved because there is always half of the displayed scFv stretching freely out into solution. Thus, phage antibody chip by pVIII display system seems to be very promising.

The overall data strongly suggest that new generation of selective and inexpensive phage-derived probes will serve as efficient substitutes for antibodies in separation, concentration and detection systems employed for clinical and environmental monitoring, for example by developing rapid diagnostic arrays.

The main advantages of phage probes include: the simplicity of manipulation of the phage libraries, their great variability, high binding affinity, great stability and the possibility to select probes to targets of different nature, also to small molecules or toxic compounds or immunosupressants against which it is difficult to raise natural antibodies.

In fact, while sensitive and selective, antibodies have numerous disadvantages for use as diagnostic biodetectors in biological monitoring, including high cost of production, low availability, great susceptibility to environmental conditions [Shone et al. 1985] and the need for laborious immobilization methods to sensor substrates [Petrenko and Vodyanoy 2003]. The phage probes affinity-selected from random libraries for specific and selective binding to biological targets can be an effective alternative to antibodies [Sorokulova et al. 2005].

They can act as antibody surrogates, possessing distinct advantages including durability, stability, standardization and low-cost production, while achieving equivalent specificity and sensitivity [Petrenko et al. 1996; Petrenko and Vodyanoy 2003]. A selected phage itself can be used as a probe in a detection device, without chemical synthesis of the displayed peptide or fusion to a carrier protein. For example, to be used in an automated fluorescence based sensing assay [Goldman et al. 2000] or FACS [Turnbough 2003], phage, exposing thousands of reactive amino-groups, can be conjugated with fluorescent labels and, in this format, successfully compete with antibody-derived probes.

The use of antibodies as diagnostic probes outside of a laboratory may be problematic because they are often unstable in severe environmental conditions. Environmental monitoring requires stable probes, such as landscape phage, that carry thousands of foreign peptides on their surfaces, are as specific and selective as antibodies, and can operate in non-controlled conditions. Filamentous phages are probably the most stable natural nucleoproteins capable of withstanding high temperatures (up to 80°C), denaturing agents (6–8 mol/L urea), organic solvents [e.g., 50% alcohol or acetonitrile], mild acids (pH 2), and alkaline solutions. The thermostability of a landscape phage probe was recently examined in comparison with a monoclonal antibody specific for the same target [Brigati and Petrenko 2005]. They were both stable for greater than six months at room temperature, but at higher temperatures the antibody degraded more rapidly than the phage probe. Phage retained detectable binding ability for more than 6 weeks at 63°C, and 3 days at 76°C. These results confirm that phage probes are highly thermostable and can function even after exposure to high temperatures during shipping, storage and operation.

Phage-derived probes inherit the extreme robustness of the wild-type phage and, in addition, allow fabrication of bioselective materials by self-assemblage of phages or their composites on metal, mineral, or plastic surfaces [Petrenko and Vodyanoy 2003]. The recombinant phage probes appear to be highly amenable to simple immobilization through physical adsorption directly to the sensor surface, thus providing another engineering advantage while maintaining biological functionality [Carnazza et al. 2007, 2008; Olsen et al. 2006; Sorokulova et al. 2005]. This property allows phage to be used as a recognition element in biosensors, like an inexpensive standard construction material that allows fabrication of bioselective layers by self-assembly of virions or their composites on solid surfaces [Nanduri et al. 2007b].

Protocols for immobilizing bacteriophage particles on solid surfaces have been described since the inception of phage display technology [Smith 1992]. Purified phage particles can be either directly coated to plastic surfaces or anti-phage monoclonal antibodies can be used to tether phage particles onto the surface of multiwell plates directly from crude supernatants of infected bacteria, without any previous purification step [Dente et al. 1994].

With respect to methods relevant for phage probes used in protein chip and biosensor applications, many different technical approaches of immobilization have been exploited, such as physical adsorption, covalent bonding, and molecular recognition. For examples, phage has been immobilized by direct physical adsorption to the gold surface [Nanduri et al. 2007b], by peptide bond between amino residue on phage and carboxyl terminal on surface [Ploug et al 2001], by disulfide bond between one thiol group on phage and another on surface [Dultsev et al. 2001], and by specific recognition between hexahistidine tag on phage and nickel coated surface [Finucane et al. 1999].

In addition, phage probes can functionalize surface with less steric hindrance than antibodies, thus allowing a higher binding avidity for the target per surface unit. In fact, on the same surface unit, a greater number of phage and with a more correct orientation can be patterned in comparison to antibodies.

Therefore, different phage clones could be selected specifically binding to isolated proteins, enzyme or inorganic material, as well as to different microbial species, thus with a single array different targets might be detected at once, by performing in parallel several different assays in real-time, within the same miniaturized substrate. In this way, standardizing data from multiple separate experiments will be unnecessary, and truly meaningful comparisons may be derived. This could ultimately translate to a much lower cost per test. Much of the promise of these microarrays relies in their small dimensions, which reduce sample and reagent requirements and reaction times, while increasing the amount of data available from a single assay.

Furthermore, phage probes may find application as bio-recognition elements of real-time biosensor devices. Recombinant phage selected by phage-display selectively recognize and specifically bind complex target structures, such as bacterial cells, thus they can be used to develop rapid diagnostic arrays. In fact, traditional diagnostic systems usually involve a multi-step detection method with the use of labeled secondary antibody. Phage-displayed detection microsystems could be considered one-step, simultaneously bind and identify the target microorganism, with no need of further characterization steps.

On the other hand, in nanobiotechnology, acquisition of abundant probes and label-free, high sensitive detection now become the important issues. Generally, labeling tends to cause the deactivation of protein, owing to protein complex three-dimensional structure. The combination of phage-displayed probes and the label-free, real-time detection method based on surface plasmon resonance (SPR) technique has been proved to be fit for proteomics research. Lytic phage were used as probes on an SPR platform, SPREETATM (Texas Instrument, US), for detection of *Staphylococcus aureus* [Balasubramanian et al. 2007], and scFv antibody displayed on Lm P4:A8 phage pIII protein was used to detect *Listeria monocytogenes* and its virulence factor ActA [Nanduri et al. 2007a]. More recently, phage display technique and SPR detection method were combined to acquire abundant specific capture molecules and realize a label-free and high-sensitive protein chip [Liu et al. 2008]. A 12-amino acid peptide displayed on phage M13 coat protein pIII was selected as the probe, and it was immobilized on 11-mercapto-undecanoic acid sensor chip to fabricate a reusable phage-displayed protein chip. The interaction between the peptide and the specific ligand protein was detected on the BIAcore3000 (BIAcore AB, Sweden). Experimental results showed that the phage-displayed protein chip can act as a useful tool in proteomics research.

On the whole, this would allow development of analytical methods for detecting and monitoring quantitative changes of agents under any conditions that warrant their recognition, including clinical based diagnostics and biological warfare applications, spanning several potential markets including biomedical and industrial use, monitoring and proteomics research.

VIRUSES AS NEW MATERIALS

Viruses are highly organized supramolecular arrays put together by a combination of non-covalent self-assembly and genetic programming. One of the attractions of viruses as nanostructured materials is that their surface chemistry is highly amenable to fine, site-specific and inheritable modification: the proteins that constitute the coat can be altered by introducing the appropriate sequence into the gene that encodes them.

Viral capsids offer the advantages of being robust and monodisperse, and can exhibit various sizes and shapes. Filamentous bacteriophage [Wilson and Finlay 1998] is a flexible rod, of 6 nm diameter and 800-2000 nm length, depending on the genome length. The capsid is mainly constituted by a protein (pVIII) arranged in a helical array with a 5-fold symmetry axis around a single-stranded DNA molecule. A series of aspartate and glutamate residues ensure a negative potential on the surface, and one tryptophan residue is buried in the hydrophobic region responsible for packing of the capsid.

Landscape phage obtained by phage-display technology might be looked on as a new kind of submicroscopic "fiber" [Smith and Petrenko 1997]. Each phage clone is a type of fiber with unique surface properties. These fibers are not synthesized one by one with some use in mind. Instead, billions of fibers are constructed, propagated all at once in a single vessel and portions of this enormous population are distributed to multiple end-users with many different goals. Each user must devise a method of selecting from this population those fibers that might be suitable for his or her particular applications by affinity selection or whatever other selection principle can be conceived. The phage-display approach provides a physical linkage between the peptide-substrate interaction and the DNA that encodes that interaction.

Localizable or global emergent properties cannot be transferred from the virion surface to another medium; any application that depends on such properties must therefore use phage themselves as the new material.

Given that filamentous phage are resistant to harsh conditions such as high salt concentration, acidic pH, chaotropic agents, and prolonged storage, they are suitable candidate building blocks to meet the challenges of bottom-up nano-fabrication. Moreover, the pIII minor capsid protein of the phage can be easily engineered genetically to display ligand peptides that will bind to and modify the behaviour of target cells in selected tissues. Thus, the tactic of integrating phage display technology with tailored nanoparticle assembly processes offers opportunities for reaching specific nano-engineering and biomedical goals [Giordano et al. 2001; Trepel et al. 2002; Arap et al. 2002; Langer and Tirrell 2004; LaVan et al. 2002].

Recently an approach for fabrication of spontaneous, biologically active molecular networks consisting of phage directly assembled with gold (Au) nanoparticles has been reported [Souza et al. 2006]. In this work, it was shown that such networks are biocompatible and preserve the cell-targeting and internalization attributes mediated by a displayed peptide and that spontaneous organization (without genetic manipulation of the pVIII major capsid protein), and optical properties can be manipulated by changing assembly conditions. By taking advantage of Au optical properties, Au–phage networks were generated that, in addition to targeting cells, could function as signal reporters for fluorescence and dark-field microscopy and near-infrared (NIR) surface-enhanced Raman scattering (SERS) spectroscopy. Notably, this strategy maintains the low-cost, high-yield production of complex polymer units (phage) in host bacteria and bypasses many of the challenges in developing cell-peptide detection tools, such as complex synthesis and coupling chemistry, poor solubility of peptides, the presence of organic solvents, and weak detection signals. These networks can effectively integrate the unique signal reporting properties of Au nanoparticles while preserving the biological properties of phage. Together, the physical and biological features within these targeted networks offer convenient multifunctional integration within a single entity with potential for nanotechnology-based biomedical applications.

Furthermore, the bacteriophage MS2 was used as a potential vector for transporting the anti-tumour drug taxol, by linking it covalently to the acid-labile chemical linker groups attached to the inside of the spherical virion [Kooker et al. 2004].

Of particular value would be methods that could be applied to materials with interesting electronic or optical properties. Organizing ordered inorganic nanoparticles by using biological structures as templates is essential for constructing nano-devices with new functionalities. Nature shows how soluble molecules capable of recognizing and binding to specific materials can be used to shape and control the growth of crystals and other nanostructures. There is no need to rely on the complexity of the immune system in order to conduct combinatorial searches for new peptides of this sort. Although natural evolution has not selected for interactions between biomolecules and such materials, phage-display libraries can be successfully used to identify, develop and amplify binding between organic peptide sequences and inorganic metal and semiconductor substrates.

Peptides with limited selectivity for binding to metal surfaces and metal oxide surfaces have been successfully selected [Brown 1992; Brown 1997]; other researchers have used phage display to select peptides against synthetic polymers such as polystyrene [Adey et al. 1995] and yohimbine-imprinted methacrylate polymer for molecular-imprinted receptors [Berglund et al. 1998].

This approach was then extended and it was shown that combinatorial phage-display libraries can be used to evolve peptides that bind to a range of semiconductor surfaces with high specificity, depending on the crystallographic orientation and composition of the materials used [Whaley et al. 2000].

Phage-display libraries, based on a combinatorial library of random peptides-each containing 12 amino acids-fused to the pIII coat protein of M13, provided 10^9 different peptides that were reacted with crystalline semiconductor structures. Five copies of the pIII coat protein are located on one end of the phage particle, accounting for 10-16 nm of the particle. The experiments utilized different single-crystal semiconductors for a systematic evaluation of the peptide-substrate interactions. In this way, 12-mer peptides could be identified that bind to specific crystal faces of GaAs, as well as to the surfaces of GaN, ZnS, CdS, Fe_3O_4 and $CaCO_3$. Peptide binding selective for the crystal composition (for example, binding to GaAs but not to Si) and crystalline face (for example, binding to GaAs(100), but not to GaAs(111)B) was demonstrated. In addition the preferential attachment of phage to a zinc-blended surface in close proximity to a surface of differing chemical and structural composition was reported, demonstrating the high degree of binding specificity for chemical composition. These recognition peptides might provide selective “glues” for assembling inorganic nanocrystals into complex arrangements, or for attaching them to other biomolecules for labelling or transport.

Subsequently, phage display has been used again in selecting unique peptides against inorganic semiconductor materials [Flynn et al. 2003; Sano and Shiba 2003].

Brown's work on polypeptides that will bind to specific metals [Brown 1997] has been extended by Sarikaya and coworkers [2003] to make so-called GEPIs (genetically engineered polypeptides for inorganics) that bind a host of materials. Again, the peptides are prepared by combinatorial shuffling of sequences, coupled to a phage-display screening process. Some of these GEPIs exhibit the ability to modify crystal growth, for example switching the morphology of gold nanocrystals from cubo-octahedral (the equilibrium form) to flat triangular or pseudo-hexagonal forms [Brown S. et al. 2000].

In addition, M13 bacteriophage surfaces have been engineered with recognition peptides [Whaley et al. 2000] so that they bound ZnS or CdS, acting as templates for the synthesis of polynanocrystalline nanowires [Mao et al. 2003, 2004]. Lee *et al* [Lee et al. 2002] combined recognition peptides with the self-organizing property of rod-shaped M13 bacteriophage to arrange inorganic nanocrystals into an ordered superstructure. The viruses spontaneously packed in concentrated solution into a layered, tilted liquid-crystalline phase. When their coats were tipped with a peptide 9-mer that bound to ZnS, the viruses act as "handles" for arranging ZnS nanocrystals into composite layers with a roughly 700 nm periodicity.

Similar modifications were made to the tubular protein sheath of the tobacco mosaic virus (TMV) so that it can bind metal ions such as cobalt, potentially enabling the virus to template magnetic nanowires and nanotubes [Schlick et al. 2005]. Each TMV tube is 300 nm long, made up of 2100 identical protein subunits. Interestingly, the wedge-shaped proteins were also able to form a variety of other potential template structures, such as shorter tubes and disks, depending on parameters such as pH and ionic strength. Functionalization of these structures with chromophores could provide mimics of the disk-shaped light harvesting complexes of photosynthetic bacteria.

More recently, a new memory effect function was reported in the hybrid system composed of TMV conjugated with platinum nanoparticles (TMV–Pt) [Tseng et al. 2006]. The augmentation of electrical conductivity in this TMV–Pt nanocomposite modifies its properties and makes it a suitable candidate for electronic applications. The function of the TMV is not just to provide a backbone for the organization of discrete nanoparticles (NP). Indeed, the TMV consists of an RNA core with rich aromatic rings, such as guanine, which can behave as charge donors, and of coat proteins, which separate the RNA and Pt NP and act as the energy barrier. These interactions between the RNA and proteins in the TMV with the Pt-NPs are responsible for charge trapping for data storage and tunnelling process in high conductance state, thus creating a conductance switching behaviour and a repeatable memory effect in the TMV–Pt devices.

On the other hand, ring-shaped viruses from genetically modified M13 with two different binding peptides at each end were created [Nam et al. 2004]. When a bifunctional linker molecule binding to the two peptides was added, it secured the flexible rod-shaped viruses into rings about 200 nm in diameter. These were proposed to be used as templates for making nanoscale magnetic rings for magnetic data storage [Ball 2005].

Reviews [Sarikaya et al. 2003; Sarikaya et al. 2004] have highlighted the application of phage display in selecting peptides to functionalize biomaterials such as titanium. More recently, a unique strategy for surface functionalization of an electrically conductive polymer, chlorine-doped polypyrrole (PPyCl), which has been widely researched for various electronic and biomedical applications, has been developed [Sanghvi et al. 2005]. A M13 bacteriophage library was used to screen 10^9 different 12-mer peptide inserts against PPyCl, a binding phage was isolated, and the stability and specificity, strength and mechanism of its binding to PPyCl were assessed. In these studies, phage display was used to select for peptides that specifically bound to an existing biomaterial, PPy, and were subsequently used to modify the surface of PPy. PPy is a conductive synthetic polymer that has numerous applications in fields such as drug delivery [Konturri et al. 1998] and nerve regeneration [Schmidt 1997; Valentini et al. 1992], and has been used in biosensors and coatings for neural probes [Vidal et al. 1999; Cui et al. 2001]. Different dopant ions such as chloride, perchlorate, iodine and poly(styrene sulphonate) can be used during electrochemical synthesis to provide the material with varying properties (for example, conductivity, film thickness and surface topography). PPyCl does not contain a functional group for biomolecule immobilization, making it a suitable model polymer for functionalization using a peptide selected with phage display. Further, the specific peptide selectively binding PPyCl was joined to a cell adhesive sequence and used to promote cell attachment on PPyCl, to serve as a bi-functional linker. The use of the selected peptide for PPyCl by phage display can be extended to encompass a variety of therapies and devices such as PPy-based drug delivery vehicles [Konturri et al. 1998], nerve guidance channel conduits [Schmidt et al. 1997; Valentini et al. 1992], and coatings for neural probes [Cui et al. 2001]. Furthermore, this strategy for surface functionalization can be extended to immobilize a variety of molecules to PPyCl for numerous other applications. In addition, phage display can be applied to other existing polymers (including those that are already approved and/or those polymers that lack functional chemical groups for coupling reactions like PPyCl) to develop bioactive hybrid-materials without altering their bulk properties.

Selection of peptides using phage display thus represents a simple and versatile alternative to methods based on electrostatic and hydrophobic interactions between two moieties to achieve functionalization of surfaces. It is theoretically possible to design bivalent recombinant phage with two-component recognition: such phage have the potential to bind to specific locations on a semiconductor structure by peptides displayed on pIII protein and simultaneously to specific target (molecules or cells to be captured) by peptides displayed on pVIII coat protein.

One of the encouraging messages to emerge from such efforts to use essentially biological structures for nanotechnology is that the potential hurdle of interfacing seems not to be a problem: that is to say, biology is clearly "plastic" enough to accommodate unfamiliar materials from the inorganic world.

Overall, this can be regarded as a kind of synthetic biology in that it involves the reshaping and redirecting of natural molecular systems, typically using the tools of protein and genetic engineering.

Phage Perspectives in Synthetic Biology

In recent years it has been recognized that recombinant bacteriophages have several potential applications in the modern biotechnology industry: they have been proposed, beside for the above described sophisticated design of antibody drugs, detection of pathogenic bacteria and new biomaterials, as alternatives to antibiotics (phage-therapy); as delivery vehicles for protein and DNA vaccines; as gene therapy delivery vehicles; and as tools for screening libraries of proteins, peptides or antibodies. This diversity, with the ease of their manipulation and production, means that they have potential uses in research, therapeutics and manufacturing in both the biotechnology and medical fields [for a review, Clark and March 2006].

Phage-display libraries can be screened in several ways to isolate displayed peptides or proteins with practical applications [Benhar 2001; Willats 2002; Wang and Yu 2004]. For example, it is possible to isolate displayed peptides binding target proteins with affinities similar to those of antibodies, which can then be used as therapeutics that act either as agonists or through the inhibition of receptor–ligand interactions [Ladner et al. 2004].

Furthermore, phage-displayed peptides may be used as signal peptides able to trigger complex cell responses. Overall, phage-display selection of peptides mimicking ligands of cell receptors involved in modulating cell processes such as proliferation, apoptosis and differentiation, opens the door for their potential applications, respectively, in regenerative medicine, anti-tumoral development and stem cell differentiation.

Phages have been used as potential vaccine delivery vehicles in two different ways: by directly vaccinating with phage carrying vaccine antigens on their surface or by using the phage particle to deliver a DNA vaccine expression cassette that has been incorporated into the phage genome [Clark and March 2004a, 2004b]. In phage-display vaccination, phage can be designed to display a specific antigenic peptide or protein on their surface. Alternatively, phage displaying peptide libraries can be screened with a specific antiserum to isolate novel protective antigens or mimotopes – peptides that mimic the secondary structure and antigenic properties of a protective carbohydrate, protein or lipid, despite having a different primary structure [Folgori et al. 1994; Phalipon et al. 1997]. The serum of convalescents can also be used to screen phage-display libraries to identify potential vaccines against a specific disease, without prior knowledge of protective antigens [Meola et al. 1995]. Rather than generating a transcriptional fusion to a coat protein, substances can also be artificially conjugated to the surface of phage after growth, which further increases the range of antigens that can be displayed [Molenaar et al. 2002]. More recently, it has also been shown that unmodified phage can be used to deliver DNA vaccines more efficiently than standard plasmid DNA vaccination [Clark and March 2004a, 2004b; March et al. 2004; Jepson and March 2004]. The vaccine gene, under the control of a eukaryotic expression cassette, is cloned into a standard lambda bacteriophage, and purified whole phage particles are injected into the host. The phage coat protects the DNA from degradation and, because it is a virus-like particle, it should target the vaccine to the antigen presenting cells.

One particularly novel use for phage-displayed peptides is in targeted therapy. One example was in the development of a nasally delivered treatment against cocaine addiction [Dickerson et al. 2005]: whole phage particles delivered nasally can enter the central nervous system where a specific phage-displayed antibody can bind to cocaine molecules and prevent their action on the brain. Theoretically it might also be possible to modify the surface of a bacteriophage by incorporating specific protein sequences to preferentially target the particle to particular cell types, e.g. galactose residues to target galactose-recognizing hepatic receptors in the liver [Molenaar et al. 2002] or peptides isolated by screening phage-display libraries to target dendritic [Curiel et al. 2004] or Langerhans cells [McGuire et al. 2004]. To screen phage for the ability to target specific tissue types, phage-display libraries have been passed through mice several times and at each stage phage were isolated from specific tissues [Rajotte et al. 1998]. A similar in vivo screening strategy was also used to isolate phage displaying peptides that showed increased cytoplasmic uptake into mammalian cells [Ivanenkov and Menon 2000]. Phage-displayed peptides so selected may be used as targeted vehicle for antibiotics or anti-tumorals, or act themselves as targeted anti-bacterials and anti-tumorals. Specific phage-displayed peptides could be used, for example, in anticancer therapy either directly inducing apoptosis processes or targeting anti-tumorals to cancer cells, or also targeting to microorganisms that in turn specifically infect tumoral cells. Recently, in vivo phage display has been used to analyze the structure and molecular diversity of tumor vasculature and to select tumor-specific antigens which have revealed stage- and type-specific markers of tumor blood vessels [Li et al. 2006; Brown C. K. et al. 2000]. Peptides identified by this approach also work as vehicles to transport cargo therapeutic reagents to tumors. These peptides and their corresponding cellular proteins and ligands may provide molecular tools to selectively target the addresses of tumors and their pathological blood vessels and might increase the efficacy of therapy while decreasing side effects.

Finally, phages have also been proposed as potential therapeutic gene delivery vectors [Barry et al. 1996; Dunn 1996]. The phage coat protects the DNA from degradation after injection, and the ability to display foreign molecules on the phage coat also enables targeting of specific cell types, a prerequisite for effective gene therapy. Both artificial covalent conjugation [Larocca et al. 1998] and phage display [Larocca et al. 1999] have been used to display targeting and/or processing molecules on the phage surface. This demonstrates, again, the versatility of phage, showing that tissue targeting can be achieved either by rational design or by the screening of random phage-display libraries.

Conclusion

Synthetic biology could provide the tools and understanding needed both to develop and to expand "nano-biotechnology". Proteins can be rationally modified to bind to new, non-natural substrates, and NBPs can be produced in laboratory by modern molecular biology techniques, being adapted for nano-technological uses. There is no need to rely on the complexity of the immune system in order to conduct combinatorial searches for new proteins of this sort.

Protein engineering and directed evolution in vitro would allow for rapid design, fabrication, and testing of novel and unique systems. Totally in vitro –ribosome and mRNA-display– and biological –yeast, bacterial and phage display– display techniques are powerful tools to generate, select, and evolve such "synthetic" proteins, and they have great potential for biotechnological, medical and proteomic applications.

In particular, phage display technology has become a fundamental tool in functional genomics and proteomics, as well as an invaluable component of biotechnology. It is used for studying protein-protein, protein-peptide, and protein-DNA interactions; moreover, it aids to explore protein structure/function and it extends to the synthesis of artificial proteins with random sequences. During the past ten years, phage display has evolved into a well-accepted technology, and in a short time has delivered both sophisticated, high-quality antibody phage and recombinant phage probes for detection of pathogenic agents.

Phage antibodies are likely to play a great role in the generation of analytical reagents and therapeutic drugs, offering major advantages in terms of speed and throughput for research and target identification/validation.

Recombinant phage selected by phage-display find application also as biosorbents and diagnostic probes in micro- and nano-devices, as effective surrogates of antibodies, used to date. For their properties, phage probes may meet the strong criteria —stability, fastness, sensitiveness, accuracy, and inexpensiveness— for development of bioaffinity sensors for biological monitoring. Highly sensitive and accurate field-usable devices could have a number of applications in biomedical field as well as in environment and food monitoring, and detection of biological warfare agents.

Moreover, important developments have been made in the synthesis of bio-nanostructures with nano-crystals, including protein-shelled viruses modified by metallic or semiconducting nanoparticles. The concept of conjugating nanoparticles with biomolecules opens up new possibilities for making functional electronic devices using biomaterial systems. Phage-display libraries have been successfully used to identify, develop and amplify binding between organic peptide sequences and inorganic metal and semiconductor substrates, in order to provide templates for the synthesis of polynanocrystalline nanowires, nanotubes and nanorings for applications in advanced nanoelectronic devices.

Several other potential applications in the modern biotechnology industry have been recently recognized for recombinant phage –phage-therapy, gene delivery, phage-display vaccination, targeted therapy.

Tailored selection processes, in combination with improved library construction and innovated vector design and display formats, open the door wide for new sophisticated applications of phage display technology in synthetic biology.

REFERENCES

Adey, N. B., Mataragnon, A. H., Rider, J. E., Carter, J. M. and Kay, B. K. (1995). Characterization of phage that bind plastic from phage-displayed random peptide libraries. *Gene, 156,* 27–31.

Al-Khaldi, S. F., Mossoba, M. M., Yakes, B. J., Brown, E., Sharma, D., Carnazza, S. (2008). Recent Advances in microbial discovery using metagenomics, DNA microarray, biosensors, molecular subtyping, and phage recombinant probes. In M. K. Moretti and L. J. Rizzo (Eds.), *Oligonucleotide Array Sequence Analysis* (Chapter 4, pp. 123-147). Nova Science Publishers, Inc.

Al-Khaldi, S. F., Mossoba, M. M., Yakes, B. J., Brown, E., Sharma, D., Carnazza, S. (2009). The biggest winners in DNA and protein sequence analysis: metagenomics, DNA microarray, biosensors, molecular subtyping, and phage recombinant probes. *International Journal of Medical and Biological Frontiers (Nova Science Publishers), 15(3/4),* 4.

Amstutz, P., Pelletier, J. N., Guggisberg, A., Jermutus, L., Cesaro-Tadic, S., Zahnd, C. and Plückthun, A. (2002). In vitro selection for catalytic activity with ribosome display. *J. Am. Chem. Soc., 124,* 9396-403.

Amstutz, P., Forrer, P., Zahnd, C. and Plückthun, A. (2001). In vitro display technologies: novel developments and applications. *Curr. Opin. Biotechnol., 12,* 400–5.

Arap, W., Kolonin, M. G., Trepel, M., Lahdenranta, J., Cardo-Vila, M., Giordano, R. J., Mintz, P. J., Ardelt, P. U., Yao, V. J., Vidal, C. I., et al. (2002). Steps toward mapping the human vasculature by phage display. *Nat. Med., 8,* 121–7.

Arnold, F. H. (1998). Design by directed evolution. *Acc. Chem. Res., 31(3),* 125-31.

Baggio, R., Burgstaller, P., Hale, S. P., Putney, A. R., Lane, M., Lipovsek, D., Wright, M. C., Roberts, R. W., Liu, R., Szostak, J. W. and Wagner, R. W. (2002). Identification of epitope-like consensus motifs using mRNA display. *J. Mol. Recognit., 15,* 126-34.

Balass, M., Heldman, Y., Cabilly, S., Givol, D., Katchalski-Katzir, E. and Fuchs, S. (1993). Identification of a hexapeptide that mimics a conformation-dependent binding site of acetylcholine receptor by use of a phage-epitope library. *Proc. Natl. Acad. Sci. USA, 90,* 10638-42.

Balasubramanian, S., Sorokulova, I. B., Vodyanoy, V. J. and Simonian, A. L. (2007). Lytic phage as a specific and selective probe for detection of *Staphylococcus aureus*. A surface plasmon resonance spectroscopic study. *Biosens. Bioelectr., 22,* 948-55.

Ball, P. (2005). Synthetic biology for nanotechnology. Tutorial. *Nanotechnology, 16,* R1-R8.

Barbas, C. F., Bain, J. D., Hoekstra, D. M. and Lerner, R. A. (1992). Semisynthetic combinatorial libraries: a chemical solution to the diversity problem. *Proc. Natl. Acad. Sci. USA, 89,* 4457–61.

Barbas, C. F., Kang, A. S., Lerner, R. A. and Benkovic, S. J. (1991). Assembly of combinatorial antibody libraries on phage surfaces: the gene III site. *Proc. Natl. Acad. Sci. USA, 88,* 7978-82.

Barrick, J. E. and Roberts, R. W. (2002). Sequence analysis of an artificial family of RNA-binding peptides. *Protein Sci., 11,* 2688-96.

Barrick, J. E., Takahashi, T. T., Balakin, A. and Roberts, R. W. (2001). Selection of RNA-binding peptides using mRNA–peptide fusions. *Methods, 23,* 287-93.

Barry, M. A., Dower, W. J. and Johnston, S. A. (1996). Toward cell-targeting gene therapy vectors: selection of cell-binding peptides from random peptide-presenting phage libraries. *Nat. Med., 2,* 299–305.

Bass, S., Green, R. and Wells, J. A. (1990). Hormone phage: an enrichment method for variant proteins with altered binding properties. *Proteins: Struct. Funct. Genet., 8,* 309-14.

Benhar, I. (2001). Biotechnological applications of phage and cell display. *Biotechnol. Adv., 19,* 1–33.

Berglund, J., Lindbladh, C., Nicholls, I. A. and Mosbach, K. (1998). Selection of phage display combinatorial library peptides with affinity for a yohimbine imprinted methacrylate polymer. *Anal. Commun., 35,* 3–7.

Better, M., Chang, C. P., Robinson, R. R. and Horwitz, A.H. (1988). *Escherichia coli* secretion of an active chimeric antibody fragment. *Science, 240,* 1041-3.

Bi, Q., Cen, X., Wang, W., Zhao, X., Wang, X., Shen, T. and Zhu, S. (2007). A protein microarray prepared with phage-displayed antibody clones. *Biosens. Bioelectron., 22(12),* 3278-82.

Binz, H. K., Amstutz, P., Kohl, A., Stumpp, M. T., Briand, C., Forrer, P., Grütter, M. G. and Plückthun, A. (2004). High-affinity binders selected from designed ankyrin repeat protein libraries. *Nat. Biotechnol., 22,* 575-82.

Binz, H. K., Stumpp, M. T., Forrer, P., Amstutz, P. and Plückthun, A. (2003). Designing repeat proteins: well-expressed, soluble and stable proteins from combinatorial libraries of consensus ankyrin repeat proteins. *J. Mol. Biol., 332,* 489-503.

Bishop-Hurley, S. L., Schmidt, F. J., Erwin, A. L. and Smith, A. L. (2005). Peptides selected for binding to a virulent strain of *Haemophilus influenzae* by phage display are bactericidal. *Antimicrob. Agents Chemother., 49,* 2972-8.

Boder E. T., Midelfort K. S. and Wittrup K. D. (2000). Directed evolution of antibody fragments with monovalent femtomolar antigen-binding affinity. *Proc. Nat. Acad. Sci. USA, 97(20),* 10701-5.

Boder, E. T. and Wittrup, K. D. (1997). Yeast surface display for screening combinatorial polypeptide libraries. *Nat. Biotech., 15,* 553-57.

Boder, E. T. and Wittrup, K. D. (1998). Optimal screening of surface-displayed polypeptide libraries. *Biotechnol. Prog., 14,* 55-62.

Bratkovic, T., Lunder, M., Popovic, T., Kreft, S., Turk, B., Strukelj, B. and Urleb, U. (2005). Affinity selection to papain yields potent peptide inhibitors of cathepsins L, B, H, and K. *Biochem. Biophys. Res. Commun., 332(3),* 897–903.

Breitling, F., Dübel, S., Seehaus, T., Klewinghaus, I. and Little, M. (1991). A surface expression vector for antibody screening. *Gene, 104,* 147-53.

Brigati, J. R. and Petrenko, V. A. (2005). Thermostability of landscape phage probes. *Anal. Bioanal. Chem., 382,* 1346-50.

Brigati, J., Williams, D. D., Sorokulova, I. B., Nanduri, V., Chen, I. H., Turnbough, C. L., Jr. and Petrenko, V. A. (2004). Diagnostic probes for *Bacillus anthracis* spores selected from a landscape phage library. *Clin. Chem., 50,* 1899-906.

Brown, C. K., Modzelewski, R. A., Johnson, C. S. and Wong, M.K. (2000). A novel approach for the identification of unique tumor vasculature bonding peptides using an *E. coli* peptide display library. *Ann. Surg. Oncol. 7(10),* 743-9.

Brown, S. (1992). Engineered iron oxide-adhesion mutants of the *Escherichia coli* phage l receptor. *Proc. Natl Acad. Sci. USA, 89,* 8651-5.

Brown, S. (1997). Metal-recognition by repeating polypeptides. *Nat. Biotechnol., 15,* 269-72.

Brown, S., Sarikaya, M. and Johnson, E. (2000). Genetic analysis of crystal growth. *J. Mol. Biol., 299,* 725-32.

Bryson, J. W., Betz, S. F., Lu, H. S., Suich, D. J., Zhou, H. X., O'Neil, K. T. and DeGrado, W. F. (1995). Protein design: a hierarchic approach. *Science, 270,* 935-41.

Burton, D. R., Barbas, C. F. 3rd, Persson, M. A., Koenig, S., Chanock, R. M. and Lerner, R. A. (1991). A large array of human monoclonal antibodies to type 1 human immunodeficiency virus from combinatorial libraries of asymptomatic individuals. *Proc. Natl. Acad. Sci. USA, 88,* 10134–7.

Cabibbo, A., Sporeno, E., Toniatti, C., Altamura, S., Savino, R., Paonessa, G. and Ciliberto, G. (1995). Monovalent phage display of human interleukin (hIL)-6: Selection of superbinder variants from a complex molecular repertoire in the hIL-6 D-helix. *Gene, 167,* 41-7.

Cai, X. and Garen, A. (1995). Anti-melanoma antibodies from melanoma patients immunised with genetically modified antilogous tumor cells: selection of specific antibodies from single-chain Fv fusion phage libraries. *Proc. Natl. Acad. Sci. USA, 92,* 6537–41.

Carmen, S. and Jermutus, L. (2002). Concepts in antibody phage display. *Brief Funct. Genomic Proteomic, 1(2),* 189-203.

Carnazza, S., Foti, C., Gioffrè, G., Felici, F. and Guglielmino, S. (2008). Specific and selective probes for *Pseudomonas aeruginosa* from phage-displayed random peptide libraries. *Bios. Bioelectron., 23,* 1137-44.

Carnazza, S., Gioffrè, G., Felici, F. and Guglielmino, S. (2007). Recombinant phage probes for *Listeria monocytogenes*. *J. Phys. Condensed Matter, 19,* 395011.

Cesaro-Tadic, S., Lagos, D., Honegger, A., Rickard, J. H., Partridge, L. J., Blackburn, G. M., and Plückthun, A. (2003). Turnover-based in vitro selection and evolution of biocatalysts from a fully synthetic antibody library. *Nat. Biotechnol., 21,* 679-85.

Chang, C. N., Landolfi, N. F. and Queen, C. (1991). Expression of antibody Fab domains on bacteriophage surfaces. *J. Immunology, 147,* 3610–4.

Chasteen, L., Ayriss, J., Pavlik, P. and Bradbury, A. R. (2006). Eliminating helper phage from phage display. *Nucleic Acids Res., 34(21), e145* [Epub 2006 Nov 6].

Chen, B. X., Wilson, S. R., Das, M., Coughlin, D. J. and Erlanger, B. F. (1998). Antigenicity of fullerenes: antibodies specific for fullerenes and their characteristics. *Proc. Natl. Acad. Sci. USA, 95,* 10809-13.

Chen, Y. (2003). *Novel approach to generate human monoclonal antibodies by PROfusion™ Technology*. Cambridge Healthtech Institute 4th Annual Conference on Recombinant Antibodies. Cambridge, MA, USA.

Cheng, X., Kay, B. K. and Juliano, R. L. (1996). Identification of a biologically significant DNA-binding peptide motif by use of a random phage display library. *Gene, 171,* 1–8.

Chiang, Y. L., Sheng-Dong, R., Brow, M. A. and Larrick, J. W. (1989). Direct cDNA cloning of the rearranged immunoglobulin variable region. *Biotechniques, 7,* 360–6.

Chiarabelli, C., Vrijbloed, J. W., Thomas, R. M. and Luisi, P. L. (2006a). Investigation of de novo totally random biosequences. Part I. A general method for in vitro selection of folded domains from a random polypeptide library displayed on phage. *Chem. Biodiv., 3,* 827-39.

Chiarabelli, C., Vrijbloed, J. W., De Lucrezia, D., Thomas, R. M., Stano, P., Polticelli, F., Ottone, T., Papa, E. and Luisi, P. L. (2006b). Investigation of de novo totally random biosequences. Part II. On the folding frequency in a totally random library of de novo proteins obtained by phage display. *Chem. Biodiv., 3,* 840-59.

Chopra, P. and Kamma, A. (2006). Engineering life through Synthetic Biology. *In Silico Biology, 6,* 401-10.

Chowdhury, P. S. (2002). Targeting random mutations to hotspots in antibody variable domains for affinity improvement. In P. M. O'Brien, and R. Aitken (Eds.), *Antibody Phage Display: Methods and Protocols.* (pp. 269–86). Totowa, NJ: Humana Press.

Clackson, T., Hoogenboom, H. R., Griffiths, A. D. and Winter, G. (1991). Making antibody fragments using phage display libraries. *Nature, 352,* 624–8.

Clark, J. R. and March, J. B. (2004a). Bacterial viruses as human vaccines? *Expert Rev. Vaccines, 3,* 463–76.

Clark, J. R. and March, J. B. (2004b). Bacteriophage-mediated nucleic acid immunization. *FEMS Immunol. Med. Microbiol., 40,* 21–6.

Clark, J. R. and March, J. B. (2006). Bacteriophages and biotechnology: vaccines, gene therapy and antibacterials. *Trends Biotech., 24,* 212-8.

Coley, A. M., Campanale, N. V., Casey, J. L., Hodder, A. N., Crewther, P. E., Anders, R. F., Tilley, L. M. and Foley, M. (2001). Rapid and precise epitope mapping of monoclonal antibodies against *Plasmodium falciparum* AMA1 by combined phage display of fragments and random peptides. *Protein Eng., 14,* 691–8.

Cortese, R., Monaci, P., Nicosia, A., Luzzago, A., Felici, F., Galfre, G., Pessi, A., Tramontano, A. and Sollazzo, M. (1995). Identification of biologically active peptides using random libraries displayed on phage. *Curr. Opin. Biotechnol., 6,* 73-80.

Cui, X., Hetke, J. F., Wiler, J. A., Anderson, D. J. and Martin, D. C. (2001). Electrochemical deposition and characterization of conducting polymer polypyrrole/PSS on multichannel neural probes. *Sensors Actuat. A, 93,* 8–18.

Cujec, T. P., Medeiros, P. F., Hammond, P., Rise, C. and Kreider, B. L. (2002). Selection of v-abl tyrosine kinase substrate sequences from randomized peptide and cellular proteomic libraries using mRNA display. *Chem. Biol., 9,* 253-264.

Curiel, T. J., Morris, C., Brumlik, M., Landry, S. J., Finstad, K., Nelson, A., Joshi, V., Hawkins, C., Alarez, X., Lackner, A. and Mohamadzadeh, M. (2004). Peptides identified through phage display direct immunogenic antigen to dendritic cells. *J. Immunol., 172,* 7425–31.

Cwirla, S. E., Peters, E. A., Barrett, R. W. and Dower, W. J. (1990). Peptides on phage: a vast library of peptides for identifying ligands. *Proc. Natl. Acad. Sci. USA, 87,* 6378-82.

Dall'Acqua, W. and Carter, P. (1998). Antibody engineering. *Curr. Opin. Struct. Biol., 8,* 443-50.

De Lucrezia, D., Franchi, M., Chiarabelli, C., Gallori, E. and Luisi, P. L. (2006a). Investigation of de novo totally random biosequences. Part III. RNA foster: a novel assay to investigate RNA folding structural properties. *Chem. Biodiv., 3,* 860-8.

De Lucrezia, D., Franchi, M., Chiarabelli, C., Gallori, E. and Luisi, P. L. (2006b). Investigation of de novo totally random biosequences. Part IV. Folding properties of de novo, totally random RNAs. *Chem. Biodiv., 3,* 869-77.

Deming, T. J. (1997). Polypeptide materials: new synthetic methods and applications. *Adv. Mater., 9,* 299-311.

Deng, S. J., MacKenzie, C. R., Sadowska, J., Michniewicz, J., Young, N. M., Bundle, D. R. and Narang, S. A. (1994). Selection of antibody single-chain variable fragments with improved carbohydrate binding by phage display. *J. Biol. Chem., 269,* 9533–8.

Dennis, M. S., Eigenbrot, C., Skelton, N. J., Ultsch, M. H., Santell, L., Dwyer, M. A., O'Connell, M. P. and Lazarus, R. A. (2000). Peptide exosite inhibitors of factor VIIa as anticoagulants. *Nature, 404,* 465-70.

Dente, L., Cesareni, G., Micheli, G., Felici, F., Folgori, A., Luzzago, A., Monaci, P., Nicosia, A. and Delmastro, P. (1994). Monoclonal antibodies that recognize filamentous phage: tools for phage display technology. *Gene, 148(1),* 7-13.

Dickerson, T. J., Kaufmann, G. F. and Janda, K. D. (2005). Bacteriophage-mediated protein delivery into the central nervous system and its application in immunopharmacotherapy. *Expert Opin. Biol. Ther., 5,* 773–81.

Dultsev, F. N., Speight, R. E., Fiorini, M. T., Blackburn, J. M., Abell, C., Ostanin, V. P. and Klenerman, D. (2001). Direct and quantitative detection of bacteriophage by "hearing" surface detachment using a quartz crystal microbalance. *Anal. Chem., 73(16),* 3935-9.

Dunn, I. S. (1996). Mammalian cell binding and transfection mediated by surface-modified bacteriophage lambda. *Biochimie, 78,* 856–61.

Dwyer, M. A., Looger, L. L. and Hellinga, H. W. (2004). Computational design of a biologically active enzyme. *Science, 304,* 1967-71.

El-Mousawi, M., Tchistiakova, L., Yurchenko, L., Pietrzynski, G., Moreno, M., Stanimirovic, D., Ahmad, D. and Alakhov, V. (2003). A vascular endothelial growth factor high affinity receptor 1-specific peptide with antiangiogenic activity identified using a phage display peptide library. *J. Biol. Chem., 278,* 46681–91.

Emanuel, P., O'Brien, T., Burans, J., DasGupta, B. R., Valdes, J. J. and Eldefrawi, M. (1996). Directing antigen specificity towards botulinum neurotoxin with combinatorial phage display libraries. *J. Immunol. Methods, 193,* 189–97.

Erlanger, B. F., Chen, B.-X., Zhu, M. and Brus, L. (2001). Binding of an anti-fullerene IgG monoclonal antibody to single wall carbon nanotubes. *Nano Lett., 1,* 465-7.

Fakok, V. A., Bratton, D. L., Rose, D. M., Pearson, A., Ezekewitz, R. A. B. and Henson, P. M. (2000). A receptor for phosphatidylserine-specific clearance of apoptotic cells. *Nature, 405,* 85-90.

Feldhaus, M. J. and Siegel, R. W. (2004a). Flow cytometric screening of yeast surface display libraries. *Methods Mol. Biol., 263,* 311-32.

Feldhaus, M. J. and Siegel, R. W. (2004b). Yeast display of antibody fragments: A discovery and characterization platform. *J. Immunol. Methods, 290,* 69-80.

Felici, F., Luzzago, A., Monaci, P., Nicosia, A., Sollazzo, M. and Traboni, C. (1995). Peptide and protein display on the surface of filamentous bacteriophage. In M. R. El-Gewely (Ed.), *Biotechnology Annual Review* (Volume 1, 149-83). Amsterdam, The Netherlands: Elsevier Science B.V.

Finucane, M. D., Tuna, M., Lees, J. H. and Woolfson, D. N. (1999). Core-directed protein design. I. An experimental method for selecting stable proteins from combinatorial libraries. *Biochemistry, 38,* 11604-12.

Flynn, C. E., Mao, C., Hayhurst, A., Williams, J. L., Georgiou, G., Iversona B. and Belcher, A. M. (2003). Synthesis and organization of nanoscale semiconductor materials using evolved peptide specificity and viral capsid assembly. *J. Mater. Chem., 13,* 2414–21.

Folgori, A., Tafi, R., Meola, A., Felici, F., Galfré, G., Cortese, R., Monaci, P. and Nicosia, A. (1994). A general strategy to identify mimotopes of pathological antigens using only random peptide libraries and human sera. *EMBO J., 13,* 2236–43.

Francisco, J. A., Campbell, R., Iverson, B. L. and Georgiou, G. (1993). Production and fluorescence-activated cell sorting of *Escherichia coli* expressing a functional antibody fragment on the external surface. *Proc. Nat. Acad. Sci. USA, 90,* 10444-8.

Garrard, L. J., Yang, M., O'Connell, M. P., Kelley, R. F. and Henner, D. J. (1991). Fab assembly and enrichment in a monovalent phage display system. *Bio/Technology, 9,* 1373-7.

Georgiou, G., Stathopoulos, C., Daugherty, P. S., Nayak, A. R., Iverson, B. L. and Curtis, R. III. (1997). Display of heterologous proteins on the surface of microorganisms: from the screening of combinatorial libraries to live recombinant vaccines. *Nat. Biotech., 15,* 29-34.

Geysen, H. M., Rodda, S. J. and Mason, T. J. (1986a). A priori delineation of a peptide which mimics a discontinuous antigenic determinant. *Mol. Immunol., 23,* 709-15.

Geysen, H. M., Rodda, S. J. and Mason, T. J. (1986b). The delineation of peptides able to mimic assembled epitopes. *Ciba Found. Symp., 119,* 130-49.

Giordano, R. J., Cardo-Vila, M., Lahdenranta, J., Pasqualini, R. and Arap, W. (2001). Biopanning and rapid analysis of selective interactive ligands. *Nat. Med., 7,* 1249–53.

Gold, L. (2001). mRNA display: diversity matters during in vitro selection. *Proc. Natl. Acad. Sci. USA, 98(9),* 4825-6.

Goldman, E. R., Pazirandeh, M. P., Mauro, J. M., King, K. D., Frey, J. C. and Anderson, G. P. (2000). Phage-displayed peptides as biosensor reagents. *J. Mol. Recognit., 13,* 382-7.

Gough, K.C., Cockburn, W. and Whitelam, G.C. (1999). Selection of phage-display peptides that bind to cucumber mosaic virus coat protein. *J. Virol. Methods, 79,* 169–80.

Graff, C. P., Chester, K., Begent, R. and Wittrup, K. D. (2004). Protein Engineering Design and Selection. *Prot. Eng. Des. Sel., 17,* 293-304.

Graus, Y. F., de Baets, M. H., Parren, P. W., Berrih-Aknin, S., Wokke, J., van Breda Vriesman, P. J. and Burton, D. R. (1997). Human anti-nicotinic acetylcholine receptor recombinant Fab fragments isolated from thymus-derived phage display libraries from myasthenia gravis patients reflect predominant specificities in serum and block the action of pathogenic serum antibodies. *J. Immunol., 158,* 1919–29.

Greenwood, J., Willis, A.E. and Perham, R.N. (1991). Multiple display of foreign peptides on a filamentous bacteriophage. *J. Mol. Biol.,220,* 821-7.

Griffiths, A. D. and Duncan, A. R. (1998). Strategies for selection of antibodies by phage display. *Curr. Opin. Biotechnol., 9,* 102-8.

Gu, H. D., Yi, Q. A., Bray, S. T., Riddle, D. S., Shiau, A. K. and Baker, D. (1995). A phage display system for studying the sequence determinants of protein-folding. *Protein Science, 4,* 1108-17.

Hammond, P. W., Alpin, J., Rise, C. E., Wright, M. and Kreider, B. L. (2001). In vitro selection and characterization of Bcl-X(L)-binding proteins from a mix of tissue-specific mRNA display libraries. *J. Biol. Chem., 276,* 20898-906.

Hanes, J. and Plückthun, A. (1997). In vitro selection and evolution of functional proteins using ribosome display. *Proc. Natl. Acad. Sci. USA, 94,* 4937-42.

Hanes, J., Jermutus, L. and Plückthun, A. (2000a). Selecting and evolving functional proteins in vitro by ribosome display. *Methods Enzymol., 328,* 404-30.

Hanes, J., Schaffitzel, C., Knappik, A. and Plückthun, A. (2000b). Picomolar affinity antibodies from a fully synthetic naïve library selected and evolved by ribosome display. *Nat. Biotechnol., 18,* 1287-92.

He, M. and Taussig, M. J. (1997). Antibody-ribosome-mRNA (ARM) complexes as efficient selection particles for in vitro display and evolution of antibody combining sites. *Nucleic Acids Res., 25,* 5132-4.

He, M. and Taussig, M. J. (2007). Eukaryotic Ribosome Display with in situ DNA recovery. *Nat. Methods, 4,* 281-8.

Hessling, J., Lohse, M. J. and Klotz, K. N. (2003). Peptide G protein agonists from a phage display library. *Biochem. Pharmacol., 65,* 961–7.

Hong, L., Liao, W., Zhao, X. S. and Zhu, S. G. (2004). Phage antibody chip for discriminating proteomes from different cells. *Acta Phys.-Chim. Sin., 20,* 1182-5.

Hoogenboom, H. R. (1997). Designing and optimizing library selection strategies for generating high-affinity antibodies. *Trends Biotechnol., 15,* 62–70.

Hoogenboom, H. R. and Chames, P. (2000). Natural and designer binding sites made by phage display technology. *Immunol. Today, 21,* 371-8.

Hoogenboom, H. R., de Bruïne, A. P., Hufton, S. E., Hoet, R. M., Arends, J. W. and Roovers, R. C. (1998). Antibody phage display technology and its applications. *Immunotechnology, 4,* 1–20.

Hoogenboom, H. R., Griffiths, A. D., Johnson, K. S., Chiswell, D. J., Hudson, P. and Winter, G. (1991). Multi-subunit proteins on the surface of filamentous phage: methodologies for displaying antibody (Fab) heavy and light chains. *Nucl. Acids Res., 19,* 4133-7.

Hooker, J. M., Kovacs, E. W. and Francis, M. B. (2004). Interior surface modification of bacteriophage MS2. *J. Am. Chem. Soc., 126,* 3718-9.

Houimel, M., Schneider, P., Terskikh, A., and Mach, J. P. Starovasnik, M. A. and Lowman, H. B. (2002). Stable "zeta" peptides that act as potent antagonists of the high-affinity IgE receptor. *Proc. Natl. Acad. Sci. USA, 99,* 1303–8.

Huang, L., Sexton, D. J., Skogerson, K., Devlin, M., Smith, R., Sanyal, I., Parry, T., Kent, R., Enright, J., Wu, Q. L., Conley, G., DeOliveira, D., Morganelli, L., Ducar, M., Wescott, C. R. and Ladner, R. C. (2003). Novel peptide inhibitors of angiotensin-converting enzyme 2. *J. Biol. Chem., 278,* 15532–40.

Hust, M. and Dubel, S. (2004). Mating antibody phage display with proteomics. *Trends in Biotechnology, 22,* 8-14

Hyde-DeRuyscher, R., Paige, L. A., Christensen, D. J., Hyde-DeRuyscher, N., Lim, A., Fredericks, Z. L., Kranz, J., Gallant, P., Zhang, J., Rocklage, S. M. et al. (2000). Detection of small-molecule enzyme inhibitors with peptides isolated from phage-displayed combinatorial peptide libraries. *Chem Biol, 7,* 17-25.

Iqbal, S. S., Mayo, M. W., Bruno, J. G., Bronk, B. V., Batt, C. A. and Chambers, J. P. (2000). A review of molecular recognition technologies for detection of biological threat agents. *Biosens. Bioelectron., 15,* 549–78.

Irving, R. A., Coia, G., Roberts, A., Nuttall, S. D. and Hudson, P. J. (2001). Ribosome display and affinity maturation: from antibodies to single V-domains and steps towards cancer therapeutics. *J. Immunol. Methods, 248,* 31-45.

Isalan, M. and Choo, Y. (2000). Engineered zinc finger proteins that respond to DNA modification by HaeII and HhaI methyltransferase enzymes. *J. Mol. Biol., 295,* 471-7.

Ivanenkov, V. V. and Menon, A. G. (2000). Peptide-mediated transcytosis of phage display vectors in MDCK cells. *Biochem. Biophys. Res. Commun., 276,* 251–7.

Izhaky, D. and Addadi, L. (1998). Pattern recognition of antibodies for two-dimensional arrays of molecules. *Adv. Mater., 10,* 1009-13.

Jepson, C. D. and March, J. B. (2004). Bacteriophage lambda is a highly stable DNA vaccine delivery vehicle. *Vaccine, 22,* 2413–9.

Jermutus, L., Honegger, A., Schwesinger, F., Hanes, J. and Plückthun, A. (2001). Tailoring in vitro evolution for protein affinity or stability. *Proc. Natl. Acad. Sci. USA, 98,* 75-80.

Kehoe, J. W. and Kay, B. K. (2005). Filamentous phage display in the new millennium. *Chem. Rev., 105(11),* 4056–72.

Knappik, A., Ge, L., Honegger, A., Pack, P., Fischer, M., Wellnhofer, G., Hoess, A., Wolle, J., Plückthun, A. and Virnekas, B. (2000). Fully synthetic human combinatorial antibody libraries (HuCAL) based on modular consensus frameworks and CDRs randomized with trinucleotides. *J. Mol. Biol., 296,* 57-86.

Knurr, J., Benedek, O., Heslop, J., Vinson, R. B., Boydston, J. A., McAndrew, J., Kearney, J. F. and Turnbough, C. L. (2003). Peptide ligands that bind selectively to spores of *Bacillus subtilis* and closely related species. *Appl. Environ. Microbiol., 69,* 6841–7.

Koide, A., Bailey, C. W., Huang, X., Koide, S. (1998). The fibronectin type III domain as a scaffold for novel binding proteins. *J. Mol. Biol., 284,* 1141-51.

Konturri, K., Pentti, P. and Sundholm, G. (1998). Polypyrrole as a model membrane for drug delivery. *J. Electroanal. Chem., 453,* 231–8.

Kouzmitcheva, G. A., Petrenko, V. A. and Smith, G. P. (2001). Identifying diagnostic peptides for lyme disease through epitope discovery. *Clin. Diagn. Lab. Immunol., 8,* 150-60.

Kretzschmar, T. and von Ruden, T. (2002). Antibody discovery: phage display. *Curr. Opin. Biotechnol., 13,* 598-602.

Kristensen, P. and Winter, G. (1998). Proteolytic selection for protein folding using filamentous bacteriophages. *Fold. Des., 3,* 321–8.

Ladner, R. C. (1995). Constrained peptides as binding entities. *Trends Biotechnol., 13,* 426–30.

Ladner, R. C., Sato, A. K., Gorzelany, J. and de Souza, M. (2004). Phage display-derived peptides as therapeutic alternatives to antibodies. *Drug Discov. Today, 9,* 525–9.

Langer, R. and Tirrell, D. A. (2004). Designing materials for biology and medicine. *Nature, 428,* 487–92.

Larocca, D., Kassner, P. D., Witte, A., Ladnera, R. C., Pierce, G. F. and Baird, A. (1999). Gene transfer to mammalian cells using genetically targeted filamentous bacteriophage. *FASEB J., 13,* 727–34.

Larocca, D., Witte, A., Johnson, W., Pierce, G. F. and Baird, A. (1998). Targeting bacteriophage to mammalian cell surface receptors for gene delivery. *Hum. Gene Ther., 9,* 2393–9.

LaVan, D. A., Lynn, D. M. and Langer, R. (2002). Moving smaller in drug discovery and delivery. *Nat. Rev. Drug. Discov., 1,* 77–84.

Lee, L.., Buckley, C., Blades, M. C., Panayi, G., George, A. J. and Pitzalis, C. (2002). Identification of synovium-specific homing peptides by in vivo phage display selection. *Arthritis Rheum., 48(8),* 2109-20.

Lee, S.-W., Mao, C., Flynn, C. E. and Belcher, A. M. (2002). Ordering of quantum dots using genetically engineered viruses. *Science, 296,* 892-5.

Legendre, D. and Fastrez, J. (2002). Construction and exploitation in model experiments of functional selection of a landscape library expressed from a phagemid. *Gene, 290,* 203–15.

Leinonen, J., Wu, P. and Stenman, U. H. (2002). Epitope mapping of antibodies against prostate-specific antigen with use of peptide libraries. *Clin. Chem., 48,* 2208–16.

Li, X. B., Schluesener, H. J. and Xu, S. Q. (2006). Molecular addresses of tumors: selection by in vivo phage display. *Arch. Immunol. Ther. Exp. (Warsz), 54(3),* 177-81.

Li, X. B., Schluesener, H. J. and Xu, S. Q. (2006). Molecular addresses of tumors: selection by in vivo phage display. *Arch. Immunol. Ther. Exp. (Warsz), 54(3),* 177-81.

Liao, W., Hong, L., Wei, F., Zhu, S. G. and Zhao, X. S. (2005). Improving phage antibody chip by pVIII display system. *Acta Physico-Chimica Sinica, 21,* 508-11.

Lipovsek, D. and Plückthun, A. (2004). In-vitro protein evolution by ribosome display and mRNA display. *J. Imm. Methods, 290,* 51-67.

Liu, B. and Marks, J. D. (2000). Applying phage antibodies to proteomics: selecting single chain Fv antibodies to antigens blotted on nitrocellulose. *Anal. Biochem., 286,* 119–28.

Liu, F., Luo, Z., Ding, X., Zhu, S. and Yu, X. (2008). Phage-displayed protein chip based on SPR sensing. *Sens. Actuators B: Chem., doi:10.1016/j.snb.2008.11.031.*

Livnah, O., Stura, E. A., Johnson, D. L., Middleton, S. A., Mulcahy, L. S., Wrighton, N. C., Dower, W. J., Jolliffe, L. K. and Wilson, I. A. (1996). Functional mimicry of a protein hormone by a peptide agonist: the EPO receptor complex at 2.8 Å. *Science, 273,* 464–71.

Luisi, P. L. (2007). Chemical aspects of synthetic biology. *Chemistry and Biodiversity, 4,* 603-21.

Lunder, M., Bratkovic, T., Doljak, B., Kreft, S., Urleb, U., Strukelj, B. and Plazar, N. (2005a). Comparison of bacterial and phage display peptide libraries in search of target-binding motif. *Appl. Biochem. Biotechnol., 127(2),* 125–31.

Lunder, M., Bratkovic, T., Kreft, S. and Strukelj, B. (2005b). Peptide inhibitor of pancreatic lipase selected by phage display using different elution strategies. *J. Lipid Res., 46(7),* 1512–6.

Luzzago, A. and Felici, F. (1998). Construction of disulfide-constrained random peptide libraries displayed on phage coat protein VIII. *Methods Mol. Biol., 87,* 155-64.

Mao, C., Flynn, C. E., Hayhurst, A., Sweeney, R., Qi, J., Georgiou, G., Iverson, B. and Belcher, A. M. (2003). Viral assembly of oriented quantum dot nanowires. *Proc. Natl Acad. Sci. USA, 100,* 6946-51.

Mao, C., Solis, D. J., Reiss, B. D., Kottmann, S. T., Sweeney, R. Y., Hayhurst, A., Georgiou, G., Iverson, B., Belcher, A. M. (2004). Virus-Based toolkit for the directed synthesis of magnetic and semiconducting nanowires. *Science, 303,* 213-7.

March, J. B., Clark, J. R. and Jepson, C. D. (2004). Genetic immunization against hepatitis B using whole bacteriophage lambda particles. *Vaccine, 22,* 1666–71.

Marks, J. D., Hoogenboom, H. R., Bonnert, T. P., McCafferty, J., Griffiths, A. D. and Winter, G. (1991). By-passing immunization: human antibodies from V-gene libraries displayed on phage. *J. Mol. Biol., 222,* 581–97.

Marvin, D.A. (1998). Filamentous phage structure, infection and assembly. *Curr. Opin. Struct. Biol., 8,* 150-8.

Mattheakis, L. C., Bhatt, R. R. and Dower, W.J. (1994). An in vitro polysome display system for identifying ligands from very large peptide libraries. *Proc. Natl. Acad. Sci. USA, 91,* 9022-6.

Maun, H. R., Eigenbrot, C. and Lazarus, R. A. (2003). Engineering exosite peptides for complete inhibition of factor VIIa using a protease switch with substrate phage. *J. Biol. Chem., 278,* 21823–30.

McCafferty, J., Griffiths, A. D., Winter, G. and Chiswell, D. J. (1990). Phage antibodies: filamentous phage displaying antibody variable domains. *Nature, 348,* 552-4.

McConnell, S. J., Dinh, T., Le, M. H., Brown, S. J., Becherer, K., Blumeyer, K., Kautzer, C., Axelrod, F. and Spinella, D. G. (1998). Isolation of erythropoietin receptor agonist peptides using evolved phage libraries. *Biol. Chem., 379,* 1279–86.

McGrath, K. P., Fournier, M. J., Mason, T. L. and Tirrell, D. A. (1992). Genetically directed synthesis of new polymeric materials. Expression of artificial genes encoding proteins with repeating -(AlaGly)3ProGluGly- elements. *J. Am. Chem. Soc., 114,* 727-33.

McGuire, M. J., Sykes, K. F., Samli, K. N., Timares, L., Barry, M. A., Stemke-Hale, K., Tagliaferri, F., Logan, M., Jansa, K., Takashima, A., Brown, K. C. and Johnston, S. A. (2004). A library-selected, Langerhans cell targeting peptide enhances an immune response. *DNA Cell Biol., 23,* 742–52.

McLafferty, M. A., Kent, R. B., Ladner, R. C., and Markland, W. (1993). M13 bacteriophage displaying disulfide-constrained microproteins. *Gene, 128,* 29–36.

Meola, A., Delmastro, P., Monaci, P., Luzzago, A., Nicosia, A., Felici, F., Cortese, R. and Galfre, G. (1995). Derivation of vaccines from mimotopes. Immunologic properties of human hepatitis B virus surface antigen mimotopes displayed on filamentous phage. *J. Immunol., 154,* 3162–72.

Molenaar, T. J., Michon, I., de Haas, S. A. M., van Berkel, T. J. C., Kuiper, J. and Biessen, E. A. L. (2002). Uptake and processing of modified bacteriophage M13 in mice: implications for phage display. *Virology, 293,* 182–91.

Myers, M. A., Davies, J. M., Tong, J. C., Whisstock, J., Scealy, M., Mackay, I. R. and Rowley, M. J. (2000). Conformational epitopes on the diabetes autoantigen GAD65 identified by peptide phage display and molecular modelling. *J. Immunol., 165,* 3830–8.

Nakamura, G. R., Reynolds, M. E., Chen, Y. M., Starovasnik, M. A. and Lowman, H. B. (2002). Stable "zeta" peptides that act as potent antagonists of the high-affinity IgE receptor. *Proc. Natl. Acad. Sci. USA, 99,* 1303–8.

Nakamura, G. R., Starovasnik, M. A., Reynolds, M. E. and Lowman, H. B. (2001). A novel family of hairpin peptides that inhibit IgE activity by binding to the high-affinity IgE receptor. *Biochemistry, 40,* 9828–35.

Nakashima, T., Ishiguro, N., Yamaguchi, M., Yamauchi, A., Shima, Y., Nozaki, C., Urabe, I. and Yomo, T. (2000). Construction and characterization of phage libraries displaying artificial proteins with random sequences. *J. Biosci. Bioeng., 90,* 253–9.

Nam, K. T., Peelle, B. R., Lee, S..-W and Belcher, A. M. (2004). Genetically driven assembly of nanorings based on the M13 virus. *Nano Lett., 4,* 23-7.

Nanduri, V., Bhunia, A. K., Tu, S.-I., Paoli, G. C. and Brewster, J. D. (2007a). SPR biosensor for the detection of *L. monocytogenes* using phage-displayed antibody. *Biosens. Bioelectr., 23,* 248-52.

Nanduri, V., Sorokulova, I. B., Samoylov, A. M., Simonian, A. L., Petrenko, V. A. and Vodyanoy, V. (2007b). Phage as a molecular recognition element in biosensors immobilized by physical adsorption. *Biosens. Bioelectron., 22,* 986-92.

Nemoto, N., Miyamoto-Sato, E., Husimi, Y. and Yanagawa, H. (1997). In vitro virus: bonding of mRNA bearing puromycin at the 3′-terminal end to the C-terminal end of its encoded protein on the ribosome in vitro. *FEBS Lett., 414,* 405-8.

Ohlin, M., Owman, H., Mach, M. and Borrebaeck, C. A. (1996). Light chain shuffling of a high affinity antibody results in a drift in epitope recognition. *Mol. Immunol., 33,* 47–56.

Olsen, E. V., Sorokulova, I. B., Petrenko, V. A., Chen, I. H., Barbaree, J. M. and Vodyanoy, V. J. (2006). Affinity-selected filamentous bacteriophage as a probe for acoustic wave biodetectors of *Salmonella typhimurium*. *Biosens. Bioelectron., 21,* 1434-42.

Onda, T., LaFace, D., Baier, G., Brunner, T., Honma, N., Mikayama, T., Altman, A. and Green, D. R. (1995). A phage display system for detection of T cell receptor-antigen interactions. *Mol. Immunol., 32,* 1387-97.

Orlandi, R., Güssow, D. H., Jones, P. T. and Winter, G. (1989). Cloning immunoglobulin variable domains for expression by the polymerase chain reaction. *Proc. Natl. Acad. Sci. USA, 86,* 3833–7.

Persson, M.A., Caothien, R. H. and Burton, D. R. (1991). Generation of diverse high-affinity human monoclonal antibodies by repertoire cloning. *Proc. Natl. Acad. Sci. USA, 88,* 2432–6.

Petit, M. A., Jolivet-Reynaud, C., Peronnet, E., Michal, Y. and Trepo, C. (2003). Mapping of a conformational epitope shared between E1 and E2 on the serum-derived human hepatitis C virus envelope. *J. Biol. Chem., 278,* 44385–92.

Petrenko, V. A. and Smith, G. P. (2000). Phages from landscape libraries as substitute antibodies. *Protein. Eng., 13,* 589-92.

Petrenko, V. A. and Sorokulova, I. B. (2004). Detection of biological threats. A challenge for directed molecular evolution. *J. Microbiol. Methods, 58,* 147-68.

Petrenko, V. A. and Vodyanoy, V. J. (2003). Phage display for detection of biological threat agents. *J. Microbiol. Methods, 53,* 253-62.

Petrenko, V. A., Smith, G. P., Gong, X. and Quinn, T. (1996). A library of organic landscapes on filamentous phage. *Protein. Eng., 9,* 797-801.

Phalipon, A., Folgori, A., Arondel, J., Sgaramella, G., Fortugno, P., Cortese, R., Sansonetti, P. J. and Felici, F. (1997). Induction of anti-carbohydrate antibodies by phage library-selected peptide mimics. *Eur. J. Immunol., 27,* 2620–5.

Ploug, M., Østergaard, S., Gårdsvoll, H., Kovalski, K., Holst-Hansen, C., Holm, A., Ossowski, L. and Danø, K. (2001). Peptide-derived antagonists of the urokinase receptor. Affinity maturation by combinatorial chemistry, identification of functional epitopes, and inhibitory effect on cancer cell intravasation. *Biochemistry, 40(40),* 12157-68.

Proba, K., Wörn, A., Honegger, A. and Plückthun, A. (1998). Antibody scFv fragments without disulfide bonds made by molecular evolution. *J. Mol. Biol., 275,* 245–53.

Rader, C. and Barbas, C. F. (1997). Phage display of combinatorial antibody libraries. *Curr. Opin. Biotechnol., 8,* 503-8.

Rajotte, D., Arap, W., Hagedorn, M., Koivunen, E., Pasqualini, R. and Ruoslahti, E. (1998). Molecular heterogeneity of the vascular endothelium revealed by in vivo phage display. *J. Clin. Invest., 102,* 430–7.

Ren, Z. and Black, L. W. (1998). Phage T4 SOC and HOC display of biologically active, full-length proteins on the viral capsid. *Gene, 215,* 439-44.

Roberts, R.W. (1999). Totally in vitro protein selection using mRNA-protein fusions and ribosome display. *Curr. Opin. Chem. Biol., 3(3),* 268-73.

Roberts, R. W. and Szostak, J. W. (1997). RNA-peptide fusions for the in vitro selection of peptides and proteins. *Proc. Natl. Acad. Sci. USA, 94(23),* 12297-302.

Romanov, V. I., Durand, D. B. and Petrenko, V. A. (2001). Phage display selection of peptides that affect prostate carcinoma cells attachment and invasion. *Prostate, 47,* 239-51.

Rosenberg, A., Griffin, K., Studier, F. W., McCormick, M., Berg, J., Novy, R. and Mierendorf, R. (1996). T7Select® phage display system: a powerful new protein display system based on bacteriophage T7. *Innovations (Newsletter of Novagen, Inc.), 6,* 1-6.

Rowley, M. J., Scealy, M., Whisstock, J. C., Jois, J. A., Wijeyewickrema, L. C. and Mackay, I. R. (2000). Prediction of the immunodominant epitope of the pyruvate dehydrogenase complex E2 in primary biliary cirrhosis using phage display. *J. Immunol., 164,* 3413–9.

Saggio, I. and Laufer, R. (1993). Biotin binders selected from a random peptide library expressed on phage. *Biochem. J., 293(3),* 613-6.

Samoylova, T. I., Petrenko, V. A., Morrison, N. E., Globa, L. P., Baker, H. J. and Cox, N. R. (2003). Phage probes for malignant glial cells. *Mol. Cancer Ther., 2,* 1129-37.

Sanghvi, A. B., Miller, K. P.-H., Belcher, A. M. and Schmidt, C. E. (2005). Biomaterials functionalization using a novel peptide that selectively binds to a conducting polymer. *Nature Materials, 4,* 496-502.

Sano, K. and Shiba, K. (2003). A hexapeptide motif that electrostatically binds to the surface of titanium. *J. Am. Chem. Soc., 125,* 14234–5.

Santini, C., Brennan, D., Mennuni, C., Hoess, R. H., Nicosia, A., Cortese, R. and Luzzago, A. (1998). Efficient display of an HCV cDNA expression library as C-terminal fusion to the capsid protein D of bacteriophage lambda. *J. Mol. Biol., 282,* 125-35.

Sarikaya, M., Tamerler, C., Jen, A. K.-Y., Schulten, K. and Baneyx, F. (2003). Molecular biomimetics: nanotechnology through biology. *Nat. Mater., 2,* 577-85.

Sarikaya, M., Tamerler, C., Schwartz, D. T. and Baneyx, F. (2004). Materials assembly and formation using engineered polypeptides. *Annu. Rev. Mater. Res., 34,* 373–408.

Sblattero, D. and Bradbury, A. (2000). Exploiting recombination in single bacteria to make large phage antibody libraries. *Nat. Biotechnol., 18,* 75-80.

Schaffitzel, C., Hanes, J., Jermutus, L. and Plückthun, A. (1999). Ribosome display: an in vitro method for selection and evolution of antibodies from libraries. *J. Immunol. Methods, 231,* 119-35.

Sche, P. P., McKenzie, K. M., White, J. D. and Austin, D. J. (1999). Display cloning: functional identification of natural receptors using cDNA-phage display. *Chem. Biol., 6,* 707-16.

Schimmele, B., and Plückthun, A. (2005). Identification of a functional epitope of the Nogo receptor by a combinatorial approach using ribosome display. *J. Mol. Biol., 352,* 229-41.

Schimmele, B., Gräfe N., and Plückthun, A. (2005). Ribosome display of mammalian receptor domains. *Protein Eng. Des. Sel., 18,* 285-94.

Schlick, T. L., Ding, Z., Kovacs, E. W. and Francis, M. B. (2005). Dual-surface modification of the tobacco mosaic virus. *J. Am. Chem. Soc., 127,* 3718–23.

Schluesener, H. J. and Xianglin, T. (2004). Selection of recombinant phages binding to pathological endothelial and tumor cells of rat glioblastoma by in vivo display. *J. Neurol. Sci., 224(1-2),* 77-82.

Schmidt, C. E., Shastri, V. R., Vacanti, J. P. and Langer, R. (1997). Stimulation of neurite outgrowth using an electrically conducting polymer. *Proc. Natl Acad. Sci. USA, 94,* 8948–53.

Scott, J. K. and Smith, G. P. (1990). Searching for peptide ligands with an epitope library. *Science, 249,* 386-90.

Segal, D. J., Dreier, B., Beerli, R. R. and Barbas, C. F. III (1999). Toward controlling gene expression at will: selection and design of zinc finger domains recognizing each of the 5'-GNN-3' DNA target sequences. *Proc. Natl. Acad. USA, 96,* 2758-63.

Shone, C., Wilton-Smith, P., Appleton, N., Hambleton, P., Modi, N., Gatley, S. and Melling, J. (1985). Monoclonal antibody-based immunoassay for type A *Clostridium botulinum* toxin is comparable to the mouse bioassay. *Appl. Environ. Microbiol., 50,* 63-7.

Siegel, D. L. (2002). Recombinant monoclonal antibody technology. *Transfus. Clin. Biol., 9,* 15-22.

Skelton, N. J., Russell, S., de Sauvage, F. and Cochran, A. G. (2002). Amino acid determinants of beta-hairpin conformation in erythropoeitin receptor agonist peptides derived from a phage display library. *J. Mol. Biol., 316,* 1111–25.

Skerra, A. and Plückthun, A. (1988). Assembly of a functional immunoglobulin Fv fragment in *Escherichia coli*. *Science, 240,* 1038-41.

Smith, G. P. (1985). Filamentous fusion phage: novel expression vectors that display cloned antigens on the virion surface. *Science, 228(4705)*, 1315-7.

Smith, G. P. (1992). Cloning in fUSE vectors. Available from: Prof. G. P. Smith, Division of Biological Sciences, University of Missouri, URL: http://www.biosci.missouri.edu/smithgp/PhageDisplayWebsite/PhageDisplayWebsiteIndex.html.

Smith, G. P. and Petrenko, V. A. (1997). Phage Display. *Chem. Rev., 97,* 391-410.

Sorokulova, I. B., Olsen, E. V., Chen, I. H., Fiebor, B., Barbaree, J. M., Vodyanoy, V. J., Chin, B. A. and Petrenko, V. A. (2005). Landscape phage probes for *Salmonella typhimurium*. *J. Microbiol. Methods, 63,* 55-72.

Souza, G. R., Christianson, D. R., Staquicini, F. I., Ozawa, M. G., Snyder, E. Y., Sidman, R. L., Miller, J. H., Arap, W. and Pasqualini, R. (2006). Networks of gold nanoparticles and bacteriophage as biological sensors and cell-targeting agents. *PNAS, 103,* 1215-20.

Spear, M. A., Breakefield, X. O., Beltzer, J., Schuback, D., Weissleder, R., Pardo, F. S., and Ladner, R. (2001). Isolation, characterization, and recovery of small peptide phage display epitopes selected against viable malignant glioma cells. *Cancer Gene Ther., 8,* 506–11.

Steichen, C., Chen, P., Kearney, J. F. and Turnbough, C. L. Jr. (2003). Identification of the immunodominant protein and other proteins of the *Bacillus anthracis* exosporium. *J. Bacteriol., 185,* 1903–10.

Stemmer, W. P. (1994). Rapid evolution of a protein in vitro by DNA shuffling. *Nature, 370(6488),* 389-91.

Stratmann, J., Strommenger, B., Stevenson, K. and Gerlach, G. F. (2002). Development of a peptide-mediated capture PCR for detection of *Mycobacterium avium* subsp. *paratuberculosis* in milk. *J. Clin. Microbiol., 40,* 4244-50.

Tramontano, A., Janda, K. D. and Lerne, R. A. (1986). Catalytic antibodies. *Science, 234,* 1566-70.

Trepel, M., Arap, W. and Pasqualini, R. (2002). In vivo phage display and vascular heterogeneity: implications for targeted medicine. *Curr. Opin. Chem. Biol., 6,* 399–404.

Tseng, R. J., Tsai, C., Ma, L., Ouyang, J., Ozkan, C. S. and Yang, Y. (2006). Digital memory device based on tobacco mosaic virus conjugated with nanoparticles. *Nature Nanotechnology, 1,* 72-7.

Turnbough, C. L. Jr. (2003). Discovery of phage display peptide ligands for species-specific detection of *Bacillus* spores. *J. Microbiol. Methods, 53,* 263-71.

Urry, D. W., McPherson, D. T., Xu, J., Gowda, D. C. and Parker, T. M. (1995). In C. Gebelein, and C. E. Carraher (Eds.), *Industrial Biotechnological Polymers* (pp. 259-281). Lancaster, PA: Technomic.

Valentini, R. F., Vargo, T. G., Gardella, J. A. Jr. and Aebischer, P. (1992). Electrically conductive polymeric substrates enhance nerve fibre outgrowth in vitro. *Biomater., 13,* 183–90.

vanZonneveld, A. J., vandenBerg, B. M. M., vanMeijer, M., and Pannekoek, H. (1995). Identification of functional interaction sites on proteins using bacteriophage-displayed random epitope libraries. *Gene, 167,* 49-52.

Vidal, J. C., Garcia, E. and Castillo, J. R. (1999). In situ preparation of a cholesterol biosensor: entrapment of cholesterol oxidase in an overoxidized polypyrrole film electrodeposited in a flow system: Determination of total cholesterol in serum. *Anal. Chim. Acta., 385,* 213–22.

Villemagne, D., Jackson, R. and Douthwaite, J. A. (2006). Highly efficient ribosome display selection by use of purified components for in vitro translation. *J. Immunol. Methods, 313,* 140-8.

Wang, C. I., Yang, Q. and Craik, C. S. (1996). Phage display of proteases and macromolecular inhibitors. *Combinatorial Chemistry, 267,* 52-68.

Wang, L. F. and Yu, M. (2004). Epitope identification and discovery using phage display libraries: applications in vaccine development and diagnostics. *Curr. Drug Targets, 5,* 1–15.

Wang, L. F., Duplessis, D. H., White, J. R., Hyatt, A.D. and Eaton, B. T. (1995). Use of a gene-targeted phage display random epitope library to map an antigenic determinant on the Bluetongue Virus outer capsid protein Vp5. *J. Immunol. Methods, 178,* 1-12.

Weaver-Feldhaus, J. M., Lou, J., Coleman, J. R., Siegel, R. W., Marks, J. D. and Feldhaus, M.J. (2004). Yeast mating for combinatorial Fab library generation and surface display. *FEBS Lett., 564(1-2),* 24-34.

Whaley, S. R., English, D. S., Hu, E. L., Barbara, P. F. and Belcher, A. M. (2000). Selection of peptides with semiconductor binding specificity for directed nano-crystal assembly. *Nature, 405,* 665-8.

Willats, W. G. (2002). Phage display: practicalities and prospects. *Plant Mol. Biol., 50,* 837–54.

Wilson, D. R. and Finlay, B. B. (1998). Phage display: application, innovations, and issues in phage and host biology. *Can. J. Microbiol., 44,* 313–29.

Wilson, D. S., Keefe, A. D. and Szostak, J. W. (2001). The use of mRNA display to select high-affinity protein-binding peptides. *Proc. Natl. Acad. Sci. USA, 98,* 3750-5.

Wolfe, S. A., Greisman, H. A., Ramm, E. I. and Pabo, C.O. (1999). Analysis of zinc fingers optimized via phage display: evaluating the utility of a recognition code. *J. Mol. Biol., 285,* 1917-34.

Worn, A. and Plückthun, A. (2001). Stability engineering of antibody single-chain Fv fragments. *J. Mol. Biol., 305,* 989–1010.

Wrighton, N. C., Farrell, F. X., Chang, R., Kashyap, A. K., Barbone, F. P., Mulcahy, L. S., Johnson, D. L., Barrett, R. W., Jolliffe, L. K. and Dower, W. J. (1996). Small peptides as potent mimetics of the protein hormone erythropoietin. *Science, 273,* 458–64.

Wu, H., Yang, W. P., and Barbas, C. F. (1995). Building zinc fingers by selection - toward a therapeutic application. *Proceedings of the National Academy of Sciences of the United States of America, 92,* 344-8.

Xu, L., Aha, P., Gu, K., Kuimelis, R. G., Kurz, M., Lam, T., Lim, A. C., Liu, H., Lohse, P. A., Sun, L., Weng, S., Wagner, R. W. and Lipovsek, D. (2002). Directed evolution of high-affinity antibody mimics using mRNA display. *Chem. Biol., 9,* 933-42.

Yu, J. and Smith, G. P. (1996). Affinity maturation of phage-displayed peptide ligands. *Methods Enzymol., 267,* 3-27.

Yuan, L., Kurek, I., English, J. and Keenan, R. (2005). Laboratory-directed protein evolution. *Microbiol. Mol. Biol. Rev., 69(3),* 373-92.

Zhang, L., Hoffman, J. A. and Ruoslahti, E. (2005). Molecular profiling of heart endothelial cells. *Circulation, 112(11),* 1601-11.

In: Advances in Medicine and Biology. Volume 20
Editor: Leon V. Berhardt
ISBN 978-1-61209-135-8

Chapter 11

ANALYTICAL METHODS FOR THE QUANTITATIVE DETERMINATION OF OXYTOCIN

Faith A. Chaibva and Roderick B. Walker*

ABSTRACT

Oxytocin is a clinically important nonapeptide that is used for the induction and/or augmentation of labor and is normally administered as a slow intravenous infusion diluted with normal saline or Ringer's lactate solution. Oxytocin is also indicated for use in the prevention and treatment of post partum hemorrhage and may be administered via either the intramuscular or intravenous routes in order to increase uterine tone and/or reduce bleeding. The analysis of oxytocin in different media has evolved over the past 30 years with the result that more sophisticated, selective and sensitive techniques are used for the determination of the compound. A variety of techniques have been applied to the determination of oxytocin in different matrices ranging from simple paper chromatography to hyphenated liquid chromatographic such as liquid chromatography coupled with mass-spectrometry. Additionally enzyme linked immuno-sorbent assays (ELISA) and radio immuno-assays (RIA) are used for the determination of low concentrations of oxytocin in biological matrices. This manuscript provides a systematic survey of the analytical methods that have been reported for isolation and quantitation of oxytocin in different matrices.

INTRODUCTION

Oxytocin (Figure 1) is a nonapeptide that is synthesized in the cell bodies of the paraventricular and supraoptic nuclei of the hypothalamus [1-3].

* E-mail: R.B.Walker@ru.ac.za

Cys – Tyr – Ile – Gln – Asn – Cys – Pro – Leu – Gly – NH_2

Figure 1. Amino acid sequence of oxytocin.

The amino acid moieties in the endogenous peptide exist in the L-configuration and a disulphide bridge links the two (2) cysteine residues resulting in a six (6)-membered ring linked to a tripeptide residue that is amidated at the carboxy terminal [4]. The chemical structure of OT is depicted in Figure 2 [4].

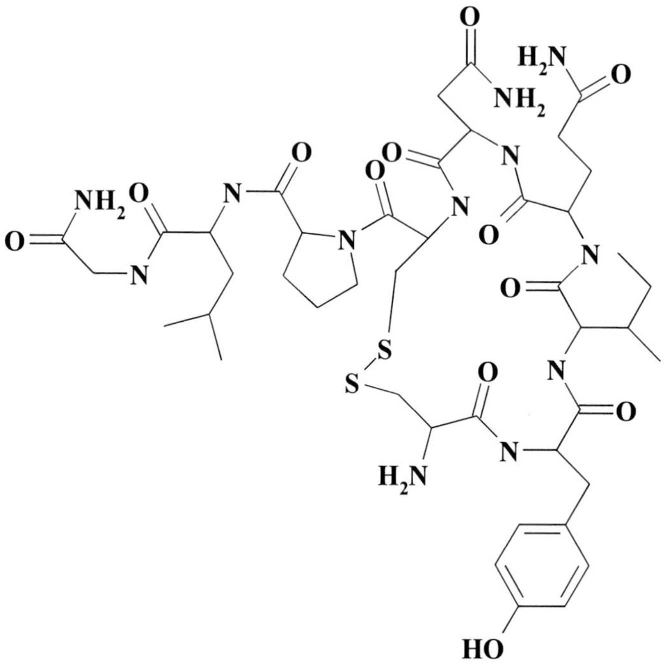

Figure 2. The chemical structure of OT ($C_{43}H_{66}N_{12}O_{12}S_2$) MW = 1007.23 [4].

Following synthesis in the hypothalamus, oxytocin is attached to the carrier protein, neurohypophysin and is subsequently transported via axonal processes to the posterior lobe of the pituitary gland, where it is stored until required for use [1, 2].

Oxytocin activity in mammals include both uterogenic and galactogenic effects that result in uterine contractions and milk letdown respectively [4-7]. The extensive expression of oxytocin receptors in organs other than the uterus and mammary glands suggests that oxytocin has diverse pharmacological effects. Indeed it has been shown that oxytocin plays a role in the stimulation of prostaglandin production in the endometrium, T-cell function, bone and muscle formation, secretion of prolactin, luteolysis among many other diverse biological functions [8, 9].

Analytical methods for the determination of oxytocin need to be sensitive and precise for several reasons. Specifically, commercially available pharmaceutical preparations of oxytocin contain 5 and 10 IU/ml [10], which with reference to the 4th International Standard [11] is equivalent to only 8.4 and 16.8 μg/ml, respectively. In addition oxytocin solutions are often diluted with large volumes of intravenous fluids such as normal saline, Ringer's lactate and Ringer's lactate/dextrose solutions resulting in the delivery of low doses of the drug over extended periods of time. [7]. Furthermore, oxytocin is rapidly cleared from the systemic circulation via liver metabolism and kidney excretion [12] and therefore plasma levels of circulating oxytocin are likely to be low, necessitating the use of highly sensitive analytical techniques for its determination in biological matrices. Sensitivity may be achieved by use of selective and efficient extraction procedures to isolate oxytocin from different matrices or by use of analytical methods with highly sensitive detection systems or that are selective for oxytocin.

Several methods reporting the analysis of oxytocin from different matrices have been published and this review provides an overview of these methods and their diverse application. A summary of the different analytical methods that have been reported and an overview of the subsequent progress made in the determination and quantitation of oxytocin is included.

BIOASSAY

The use of biological assay methods are the earliest techniques reported for the determination of oxytocic activity [13-15]. These techniques involved the measurement of the depression of blood pressure in a chicken or the assessment of uterine contractions in comparison to those produced from a known standard [4, 16]. As recently as the early 1990's, official compendia such as the 1990 United States Pharmacopeia [17] and the 1993 British Pharmacopoeia [18] recommended the use of such assays to quantitate the activity of oxytocin.

Bioassays however have a number of limitations including a lack of precision and selectivity and hence they have limited applicability for use as quality control procedures and their use for the determination of low concentrations of analyte is problematic [19]. Furthermore, the high cost of analysis, total analysis time, limited sensitivity and poor reproducibility of bioassays further restrict their use [20]. In addition bioassay methods cannot provide adequate information about the presence or absence and/or levels of related substances or degradation products in a sample [20] further limiting their possible use for routine analysis and therefore other techniques are required for the quantitation of oxytocin.

PARTITION AND THIN LAYER CHROMATOGRAPHY

The use of classic partition chromatography with a Sephadex® G-25 separation system and a mobile phase consisting of butanol, acetic acid, pyridine and water was successfully used to separate oxytocin from its diastereoisomers [21].

Thin-layer chromatography (TLC) has been successfully used for the purposes of identifying oxytocin during synthetic procedures [22, 23] and TLC was found to be adequate for the identification of oxytocin during solid phase and solution based synthetic procedures [24-26]. In general the retention factor (R_f) of a known standard in a particular mobile phase system was used to compare TLC plates developed following synthesis of oxytocin. The mobile phase compositions used included the use of butanol-acetic acid-water (4:1:1 (v/v/v)) [22, 23] or methanol-chloroform-acetic acid-water (38:62:2:2(v/v/v/v)) with a Pauly and/or chlorine-o-toluidine color reagent for oxytocin identification [27].

The separation of oxytocin from other nonapeptides, including [8-lysine]-vasopressin, [8-ornithine]-vasopressin, [2-phenylalanine,8-lysine]-vasopressin and [des-1-amino]-oxytocin was achieved using commercially available silica gel plates with a mobile phase consisting of methanol-chloroform-acetic acid-water (30/70/1/6 (v/v/v/v)) [28]. TLC was also used for the separation of oxytocin and related nonapeptides with a mobile phase comprised of chloroform-isopropanol-water (2:8:1 (v/v/v)) or acetone-ethyl acetate-methanol-water (3:2:1:1 (v/v/v/v)) and included a derivatization reaction with fluorescamine for the identification process [29].

An advantage of TLC when used for the purposes of identification of oxytocin during synthetic procedures is that the technique permits a rapid qualitative assessment of the compounds and analytes of interest [30]. However the technique is not appropriate for the accurate quantitation of the exact amount of oxytocin present in reaction mixtures [30]. Therefore more accurate and selective methods of analysis using techniques such as high performance liquid chromatography (HPLC) for the determination of oxytocin in different media are required.

Electrophoretic Separations

Electrophoretic methods of separation have also been applied to the identification of oxytocin during synthesis [23, 27]. Cellulose-coated plates in an electrolyte solution of pyridine-acetic acid-water (1:10:90) at pH 4.6 with a gradient of approximately 23 V/cm for 45 – 60 minutes was found to be suitable for the identification of oxytocin produced during such procedures [23]. In addition a thin-layer electrophoretic method developed using an electrolyte solution comprised of 0.1 N pyridine-acetic acid buffer at pH 5.6 and an applied voltage of 400 V over a 2 hour period was also found to be suitable for the identification of oxytocin during synthetic procedures [27].

Sutcliffe and Corran [31] compared the selectivity of capillary zone electrophoresis (CZE), micellar capillary zone electrophoresis (MCZE) and HPLC and established that successful separations were possible using MCZE and HPLC whereas complete separation was not achieved using CZE. However MCZE did not offer an advantage over HPLC with respect to speed of analysis and therefore the method was not recommended as an alternative to HPLC for the analysis and identification of oxytocin.

Reversed Phase High Performance Liquid Chromatography

The advent of HPLC and the subsequent development of speciality chemically modified stationary phases have facilitated the use of this technique for the routine analysis of peptides during synthesis and for the purposes of quality control. Furthermore the selectivity of HPLC makes it possible to acquire accurate quantitative data for peptides of interest [30]. It is clear that HPLC is considered the method of choice for the analysis of oxytocin as the majority of published and compendial methods for the determination of oxytocin us the technique due to its relative simplicity, wide applicability and sensitivity. Several studies [32-37] have shown that it is possible to establish a good quality correlation between results generated using HPLC analysis with those obtained using bioassays in which the contraction of the rat uterus or drop in blood pressure of chickens were monitored. Furthermore the standard error of estimates for HPLC methods (< 1%) is considerably better than those observed when using bioassay techniques (approximately 7%) [35]. A summary of the HPLC methods that have been reported for the determination of oxytocin in different matrices is shown in Table 1. One of the earliest applications of HPLC to monitor the synthesis of oxytocin was achieved using a LiChrosorb® RP-8 5 µm, 3.2 mm i.d. x 150 mm stationary phase and a mobile phase comprised of sodium phosphate buffer and acetonitrile [30]. This method was selective since it was possible to separate oxytocin from its immediate precursor, i.e. a reduced nonapeptideamide which contains reduced cysteine groups as compared to oxytocin which contains oxidized residues that form a disulphide bridge [30]. Furthermore the method was found to be suitable for the quantitative determination of oxytocin and its synthetic intermediates with a limit of detection in the nanogram range.

Reversed-phase HPLC methods that have been used for the separation of oxytocin and its diastereoisomers have selected various stationary phases including µBondapak® C_{18} columns with mobile phase compositions of 10% v/v tetrahydrofuran in ammonium acetate buffer or 18% v/v acetonitrile in ammonium acetate buffer [21]. The separation of oxytocin from 2-D-tyrosineoxytocin, 4-D-glutaminyloxytocin and 2-D-tyrosine-4-D-glutaminyloxytocin has also been achieved using reversed-phase HPLC in which both gradient and isocratic separations were successful [33]. By use of a gradient system the resolution of oxytocin from its diastereoisomers was better than that achieved using an isocratic separation method. Isocratic separation was achieved using a LiChrosorb® RP 8 (5 or 10 µm) column with a mobile phase consisting of 12% v/v acetonitrile in a phosphate buffer of pH 3 or 7 whereas the gradient separation was achieved using an octadecylsilanized silica gel (5 µm) stationary phase with phosphate buffer (pH 2.3) initially and a buffer-acetonitrile (1:1) gradient to separate oxytocin and similar nonapeptides over the analysis time [33]. However, the isocratic method was found to be suitable for monitoring the quality and stability of synthetic oxytocin [33, 34].

The determination of oxytocin and other nonapeptides in liquid dosage forms such as intravenous solutions and concentrated oxytocin products has also been achieved using reversed-phase HPLC [38]. Binary mixtures of 20% v/v acetonitrile in phosphate buffer were used as the mobile phase for the quantitation of oxytocin in the aforementioned products. The limit of detection for oxytocin was reported to be 0.88 and 1.17 ng/µl at wavelengths of 210 and 215 nm, respectively. The methods were found to be reproducibile for the analysis of oxytocin in ampoules and concentrates with relative standard deviation values of between 1 and 1.5%, respectively [38].

The separation of oxytocin from other peptide hormones using gradient elution HPLC with a Hypersil® ODS (5 mm x 100 mm) stationary phase and 0.1 M phosphate buffer at pH 2.1 and acetonitrile as the primary and secondary solvents, respectively, has been reported [43]. The method applied a gradient of 0 – 60% v/v acetonitrile over a 50 minute period with oxytocin eluting after approximately 19.5 minutes. The eluant was monitored by UV detection at a wavelength of 225 nm [43].

Table 1. HPLC methods for the analysis of oxytocin

Column	Mobile phase	Elution	Flow rate (ml/min.)	Detection	Retention time(min.)	Ref
L1 packing, 5 µm, 4.6 mm x 120 mm	A: 0.1 M monobasic sodium phosphate buffer B: 50% acetonitrile in water Linear gradient from 70% mobile phase A to 50% mobile phase B over 20 minutes	Gradient	1.5	UV 220 nm	~ 10	[5]
ODS, 5 µm, 4.6 mm x 125 mm	A: 15.6 g/l sodium dihydrogen phosphate B: 50% acetonitrile in water Linear gradient from 70% mobile phase A to 40% over 30 minutes, and then switch to 70% over 0.1 minute then 70% mobile phase A for 15 minutes	Gradient	1.0	UV 220 nm	~7.5	[6]
LiChrosorb® RP-8 (E. Merck), 5 µm, 3.2 mm i.d. x 150 mm	18% acetonitrile in 0.15 M phosphate buffer, pH = 3.0	Isocratic	0.7	UV 215 nm	~ 7.5	[30]
Merck® RP 8, 10 µm, 25 cm x 3 mm i.d.	20% v/v acetonitrile in phosphate buffer, pH = 7	Isocratic	3.0	UV 215 nm	~ 2	[16]
Nucleosil® C_8 5 µm, 15 cm x 0.4 mm i.d.	20% v/v acetonitrile in phosphate buffer, pH = 7	Isocratic	2.0	UV 210 nm	~ 5	[16]
Spherisorb® S5 ODS, 5 µm, 7.5 cm x 3 mm i.d.	17.5% v/v acetonitrile in borate buffer, pH = 10	Isocratic	1.0	UV 220 nm	~ 2.5	[16]
Nucleosil® C_{18}, 10 µm, 25 cm x 4 mm i.d.	20% v/v acetonitrile in 267 mM phosphate buffer, pH = 7	Isocratic	4.0	UV 210 nm	~ 4	[38]
Nucleosil® C_8, 5 µm, 15 cm x 4 mm i.d.	20% v/v acetonitrile in 0.67 M phosphate buffer, pH = 7	Isocratic	1.8	UV 220 nm	~ 7	[38]
Merck® RP 8, 10 µm, 25 cm x 3 mm i.d.	20% v/v acetonitrile in 0.67 M phosphate buffer, pH = 7	Isocratic	3.0	UV 215 nm	~ 2.5	[38]
RP-C18, 12.5 cm x 4.6 mm i.d. e.g. Shandon Hypersil®	A: 50% acetonitrile B: 0.1 M sodium dihydrogephosphate Gradient: 30% A to 60% A in 30 minutes	Gradient	1.0	UV 220 nm	~ 9	[35]
LiChrosopher® 60 RP-select, 5 µm	18% v/v acetonitrile in phosphate buffer pH = 2.1	Isocratic	1.0	UV 220 nm	10	[32]
Alltech Hypersil® ODS, 5 µm, 120 X 4.6 mm Beckman Ultrasphere® ODS, 5 µm, 150 X 4.6 mm	A: 100mM sodium phosphate monobasic with pH varied from pH = 3.1 to pH = 4.5 B: 50% v/v acetonitrile	Gradient	1.5	UV 220 nm	10	[39]
C_{18} micro bond-a-clone 10 µm column (Phenomenex®)	20% acetonitrile in a 0.1 M potassium dihydrogen phosphate buffer at pH 7.0.	Isocratic	1.6 – 1.8	UV 210 nm	7.4	[40]
Phenomenex® Hypersil C_{18} 5 µm, 4.6 mm x 150 mm	20% acetonitrile in 0.08 M phosphate buffer, pH = 5.0	Isocratic	1.5	UV 220 nm	5.0	[41]
Agilent Zorbax® SB-C18, 5 µm, 2.1 mm x 150 mm	50% v/v acetonitrile in 0.05% v/v formic acid	Isocratic	0.25	Mass spectrometry	4.51	[42]

Several other authors have recommended the use of HPLC for the determination of oxytocin in pharmaceutical preparations [32, 36, 39, 41, 44]. Dudkiewicz et al., [32] used a 5 µm LiChrosopher® 60 RP-select stationary phase with a mobile phase of 18% v/v acetonitrile in phosphate buffer at pH = 2.1 at a flow rate of 1.0 ml/minute and a detection wavelength 220 nm to determine the concentration of oxytocin in pharmaceutical dosage forms. A similar method reported by Ohta et al., [36] made use of a 5 µm, 4.6 mm i.d. x 250 mm Zorbax® TMS column maintained at 40 °C and a mobile phase comprised of 18% v/v acetonitrile with a 50 mM phosphate buffer at pH 5.0 and a flow rate of 1.0 ml/minute for the determination of oxytocin in pharmaceutical preparations.

The determination of oxytocin in parenteral formulations was achieved using a validated stability indicating assay that used reversed-phase gradient chromatography [39]. The stationary phases used in these studies were 5 µm, 120mm X 4.6 mm i.d. Alltech Hypersil® ODS or 150 mm X 4.6 mm i.d. Beckman Ultrasphere® ODS with mobile phase A consisting of 100 mM monobasic sodium phosphate of varying pH between 3.1 and 4.5 and mobile phase B consisting of 50% v/v acetonitrile in water. The flow rate was set at 1.5 ml/minute and detection was achieved at 220 nm. The resultant retention time for oxytocin was approximately 10.2 minutes and chlorobutanol, the preservative used in the formulations eluted at approximately 21.1 minutes. Stress studies were conducted by exposing Oxytocin Injection USP and synthetic oxytocin containing 10 units/ml to thermal, acidic, basic, oxidative and fluorescent light conditions. In all cases oxytocin was well resolved from any degradation products and the percent degradation was calculated from the peak area response of samples relative to a calibration curve [39].

An analytical method for the assay a combination formulation of oxytocin and ergometrine has been reported and a Nucleosil® C_{18} column was used as the stationary phase [44]. The mobile phase used to achieve the separation was comprised of 35% v/v acetonitrile in a buffer containing 0.05% sodium tetradecyl sulfate and 0.83 mM phosphoric acid buffer adjusted to pH 5 with triethylamine. The flow rate for the analysis was 2.5 ml/minute and the analytes of interest were detected using UV detection with a retention time for oxytocin of approximately 8 minutes.

A simple isocratic stability indicating HPLC method for the determination of oxytocin in ampoules was developed by Chaibva and Walker [41]. Stress studies conducted during validation revealed that the method was stability indicating as oxytocin was found to be well separated from any degradation products. Separation was achieved on a 5 µm, 4.6 mm i.d. X 150 mm Phenomenex® C_{18} Hypersil®, stationary phase using a mobile phase consisting of 20% v/v acetonitrile in an 80 mM phosphate buffer at a pH of 5. The limits of quantitation and detection were reported to be 0.4 IU/ml and 0.1 IU/ml respectively with a maximum of 4.84% RSD indicating that the method was precise.

The stability of oxytocin in fluids commonly used for intravenous adminsitration was evaluated using a C_{18} bond-a-clone 10 stationary phase with ethylphychoxybenzoate as an internal standard [40]. Two vials of oxytocin (10 IU/ml) were injected into 1000 ml of an intravenous solution after which samples were periodically removed from the solutions for analysis. Samples were concentrated prior to analysis that was conducted using a mobile phase of 20% v/v acetonitrile in a 10 mM potassium dihydrogen phosphate buffer at pH 7.0 at a flow rate of 1.6 to 1.8 ml/minute with UV detection at 210 nm. The retention time of oxytocin under these conditions was reported to be approximately 7.4 minutes [40]. Solid phase extraction has also been used for concentrating oxytocin from dilute Ringer's Lactate solutions [45]. A Supelco® C_8 (5 μm, 150 mm x 4.6 mm) reversed-phase column attached to a pellicular guard column with a mobile phase consisting of 20% v/v acetonitrile in a 50 mM potassium dihydrogen orthophosphate buffer at pH 7 at a flow rate of 1.25 ml/minute with detection at 220 nm was used for sample analysis. Under these conditions oxytocin and internal standard were eluted at approximately 8.5 minutes and 17 minutes respectively. The lower limit of quantitation was found to be 0.0075 IU/ml.

Bridges et al, [46] reported the use of a simple isocratic HPLC technique for the determination of neurohypophyseal hormones including oxytocin with UV detection. The method was reported to be highly sensitive and levels of oxytocin as low as 200 fmol were analyzed.

UV Detection

The ultraviolet (UV) spectrum of oxytocin reveals that oxytocin has a λ_{max} absorbance at approximately 275 nm with an additional peak at approximately 280 nm with an additional region of increased absorbance occurring at between 200 – 240 nm [4]. Therefore it is not surprising that the majority of analytical methods summarized in Table 1 reveal the use of UV as a method of detection at a wavelength of approximately 220nm. The use of UV detection for the analysis of oxytocin may is a relatively simple procedure without the need for sample derivatization. However, the major limitation of UV detection is its relatively low sensitivity when applied to samples in which particularly low concentrations of oxytocin or other synthetic by-products or degradation products of the molecule are present.

Fluorescence Methods with Derivatization

The use of post column derivatization methods prior to detection to improve the sensitivity of the analytical methods of oxytocin has been reported [20, 47-49]. It has been shown [48] that spectrofluorometric methods of detection are more sensitive when compared to the use of UV detection at 210 nm with an improvement in sensitivity of between 2- and 5-fold for optimized reaction conditions [47].

The application of post column derivatization with Fluram® has been reported for the HPLC analysis of injection solutions of oxytocin [47, 48]. The method was found to be reproducible with a retention time of less than 10 minutes for oxytocin. The high reproducibility and sensitivity of the reported method permits the accurate characterization and quantitation of oxytocin and related substances in pharmaceutical formulations and is also applicable for the determination and quantitation of by-products that contain free amino functional groups at low concentrations [47].

Derivatization with fluorescamine followed by HPLC has also been applied to the determination of oxytocin in large volume intravenous fluids [20]. Samples were concentrated using an on-line trap C_{18} pre-column concentrator/guard column and switching valve. Separation was achieved using a 5 μm, 4.6 mm i.d. x 125 mm Whatman® C_{18} column and a mobile phase consisting of 21% v/v acetonitrile in 0.1% phosphoric acid. Detection was performed with excitation at 250 nm and measured through a 418 nm cut off filter [20]. The method was used for determining concentrations as low as 40 parts per billion and was validated for selectivity, reproducibility, accuracy and precision.

Fluorometric detection following derivatization was one of the earliest methods applied to the determination of oxytocin (and vasopressin) in biological tissues [50]. Gruber et al., [50] described the extraction of oxytocin from rat pituitary glands and subsequent derivatization with fluorescamine. The peptide reacts with fluorescamine through the free amino functional group to produce a fluorophor. Separation in this case was achieved using a Partisil® ODS reversed-phase bonded column which had been equilibrated with a mobile phase containing 15% v/v acetone, 0.03% ammonium formate and 0.01% thiodiglycol. A 55 minute linear gradient from 15 to 50% v/v acetone with both solutions containing ammonium formate and thiodiglycol at the aforementioned concentration and run at a flow rate of 0.25 ml/minute was used to elute the fluorophors. The retention time for oxytocin under these conditions was approximately 30 minutes and the method was applied to the assay of peptides in tissue samples. The limits of quantitation and detection were in the picomole range with a concentration of 15 pmol of the oxytocin derivative giving a peak with a signal to noise ratio of 15:1.

PHOTODIODE ARRAY DETECTION

Rao et al., [51] used a simple isocratic HPLC method with photodiode-array detection (PDA) for the simultaneous detection of oxytocin, lysine vasopressin and arginine vasopressin. Sample concentration was performed using a solid phase extraction technique and analysis was achieved using HPLC following after reconstitution with the mobile phase. A Dynamax® 3009-A C_8 column was used with a mobile phase consisting of 20% v/v acetonitrile with 0.1% trichloroacetic acid, 50 mM heptanesulfonic acid and 30 mM triethylamine in water at pH 2.5. The retention time of oxytocin under these conditions was reported to be 4.6 minutes [51].

COULOMETRIC DETERMINATION

Although UV detection has been the primary method of choice for the analysis of oxytcon in all sample matrices the use of coulometric detection for the determination of oxytocin in biological samples has also been considered. Samples were prepared using solid phase extraction with an antibody immuno-affinity purification used to extract the oxytocin prior to analysis using HPLC coupled with coulometric detection [52]. The use of dual-electrode coulometric detection permits the oxidation of electro-active amino acids (such as tyrosine and tryptophan) by use of an upstream electrode thereby enhancing the detection capability of the downstream electrode. Consequently the sensitivity of the method is enhanced (up to 5 fold) permitting the analysis of extremely low concentrations of analytes in biological matrices [52]. This method was found to be highly sensitive with the lower limit of detection reported to be 40 pg/ml The retention time for this method was found to be 9.72 minutes.

LC MS Methods

The use of mass spectrometry has recently been reported for the detection of very low concentrations of oxytocin [42]. The use of liquid chromatography with mass spectrometry provides additional selectivity and sensitivity thereby which eliminating the need for time consuming sample preparation methods that are usually required for concentrating dilute samples. The separation and characterization of oxytocin (and other peptides) using reversed-phase liquid chromatography – mass spectrometry has been performed successfully, indicating the potential usefulness of this technique for the detection of oxytocin [53] Furthermore, the degradation of oxytocin (and other peptides) has also been studied [54] using mass spectrometry and shows the application of this method to monitor oxytocin levels in pharmaceutical dosage forms. The fragmentation patterns of oxytocin were used to identify oxytocin and degradation products [42]. Karbiwnyk et al., [42] used LC-MS to determine the concentration of oxytocin in dilute intravenous solutions. An LC-MS ion trap instrument with an electrospray ionization interface in a positive ion mode was used for the analysis. The isocratic method used an Agilent Zorbax® SB C_{18}, 5μm, 150 mm x 2.1 mm i.d., stationary phase with a mobile phase of 50% acetonitrile (v/v) and water containing 0.05% formic acid at a flow rate of 0.25 ml/minute. Under these conditions the limits of quantitation and detection were 7 and 2ng/ml, respectively.

Cation Exchange Chromatography

Radhakrishnan et al., [49] used cation-exchange chromatography with a Partisil® SCX cation exchange resin, volatile pyridine acetate buffers and an automated fluorescamine column monitoring system to separate oxytocin from related peptides. Separation was achieved using a 50 minute gradient with a mobile phase was varying from 5 x 10^{-3}M pyridine at pH 3.0 to 5 x 10^{-2} M pyridine at pH 4.0 and from which oxytocin was eluted between 40 and 50 minutes.

Ion exchange chromatography has also been reported to be applicable for the analysis of the degradation products from enzymatic degradation of oxytocin [55]. Degradation products from enzymatic degradation were synthesized and were separated on a Partisil® SCX, 10 μm (4.6 mm i.d. x 250 mm) columns. Typical degradation products that were isolated included the peptides Asn and Gln, dipeptides such as Gln-Asn, Leu-Gly, aminated tripeptides including Pro-Leu-Gly-NH2, and other peptides. The mobile phase consisted of 10% v/v methanol in 0.02 M aqueous potassium dihydrogen phosphate of pH 5 and the flow rate was set at 0.5 ml/minute. Samples were monitored with UV detection at 209 nm and the retention time of oxytocin under these conditions was approximately 18 minutes.

RADIO-IMMUNOASSAY

It has been reported that a coupling of ion-pair HPLC and post column detection by RIA may be used for the determination of nonapeptides, including oxytocin in the pineal and pituitary glands [56].

The majority of methods used for the determination of oxytocin levels in pharmacological applications have used radio-imunno assay (RIA) techniques [57-62]. The major advantage of RIA is that very low levels of analyte can be determined and the limits of quantitation and detection are in the sub picogram range [60].

ENZYME-LINKED IMMUNO-SORBENT ASSAY

A simple Enzyme-Linked ImmunoSorbent Assay (ELISA) method for the determination of oxytocin levels and changes of oxytocin levels during parturition in monkeys was described by Kawasaki et al., [63]. The method was considered to be effective since it had a short run time, permitting multiple analyses to be carried out expediently. Furthermore, the method was sensitive and it was possible to determine concentrations of oxytocin as low as 8 pg/ml in serum.

CONCLUSION

Oxytocin is an important neurohypophyseal hormone with a diverse pharmacological profile and important clinical function in preventing and controlling post partum hemorrhage. The analysis of oxytocin has been performed using a diverse range of methods including bioassay methods that measure the drop in the blood pressure of chickens or the effect of oxytocin on uterine contractions. However with advances in analytical technology methods for the assay of oxytocin have become complex and range from classic partition chromatography to non-derivatized HPLC with ultraviolet detection and HPLC with derivatization and fluorescence detection. The need for sensitive methods that are applicable to the detection of very low levels of analyte in dosage forms, solutions for use and biological matrices has resulted in the use of alternate detection methods for analysis. Spectrofluometric and coulometric detectors have been successfully applied to the determination of oxytocin as has the use of HPLC with mass spectrometry to facilitate the detection of extremely low concentrations of oxytocin in samples of interest.

REFERENCES

[1] Pituitary hormones and their hypothalamic releasing factors. In: Hardman JG, Limbird LE, Goodman Gilman A, eds. *Goodman and Gilman's the Pharmacological Basis of Therapeutics*. New York: McGraw Hill Medical Publishing Division, 2001; 1541-62.

[2] Hruby VJ, Chow MS, Smith DD. Conformational and structural considerations in oxytocin-receptor binding in biological activity. *Annual Review of Pharmacology and Toxicology* 1990; 30: 501-34.

[3] Sokol HW. Evidence for oxytocin synthesis after electrolytic destruction of the paravenricular nucleus in rats with hereditary hypothalamic diabetes insipidus. *Neuroendocrinology* 1970; 6: 90-7.

[4] Nachtmann F, Krummen K, Maxl F, Riemer E. Oxytocin. *Analytical Profiles of Drug Substances* 1981; 10: 563-600.

[5] United States Pharmacopeia/National Formulary. 31, 2897-2898. 2008. Rockville, MD, United Stated Pharmacopeial Conventionn Inc. Twinbrook Parkway.

[6] British Pharmacopoeia. Volume 1, 1635-1636. 2009. London, The Stationary Office.

[7] Oxytocin. Sweetman, S. Electronic Version. 2009. London, Pharmaceutical Press. Martindale: The Complete Drug Reference. 1-27-2009.

[8] Gimpl G, Fahrenholz F. The oxytocin receptor system: Structure, function, and regulations. *Physiological Reviews* 2001; 81: 629-83.

[9] Zingg HH, Laporte SA. The oxytocin receptor. *Trends in Endocrinology and Metabolism* 2003; 14: 222-7.

[10] Syntocinon. *Monthly Index of Medical Specialities (MIMS)* 2008; 48: 239.

[11] WHO International Biological Reference Preparations. http://www.who.int/biologicals/reference_preparations/catalogue_no/en/index.html . 2009. Access date: 13 May 2009.

[12] Fjellestad-Paulsen A, Lundin S. Metabolism of vasopressin, oxytocin and their analogues [Mpa1, -Arg8]-vasopressin (dDAVP) and [Mpa1, -Tyr(Et)2, Thr4, Orn8]-oxytocin (antocin) in human kidney and liver homogenates. *Regulatory Peptides* 1996; 67: 27-32.

[13] Deptula S. Evaluation of oxytocin determination methods. *Acta Poloniae Pharmaceutica (Abstract on Science Direct®)* 1966; 23: 51-4.

[14] Houvenaghel A. Biological determination of plasma oxytocin. *Memoires de l'Academie Royale de Medecine de Belgique (Abstract on Science Direct®)* 1979; 7: 78.

[15] Van Dongen CG, Hays RL. A sensitive in vitro assay for oxytocin. *Endocrinology* 1966; 78: 1-5.

[16] Krummen K, Frei RW. The separation of nonapeptides by reversed-phase high-performance liquid chromatography. *Journal of Chromatography A* 1977; 132: 27-36.

[17] United States Pharmacopeia. 22, 1005-1006. 1990. Twinbrook Parkway, Rockville MD, United States Pharmacopeial Convention, Inc.

[18] British Pharmacopoeia. Volume 1, 475-478. 1993. London, The Pharmaceutical Press.

[19] Andre M. Effects of mobile phase and stationary phase on the quantitative determination of oxytocin. *Journal of Chromatography A* 1986; 351: 341-5.

[20] Brown DS, Jenke DR. Determination of trace levels of oxytocin in pharmaceutical solutions by high-performance liquid chromatography. *Journal of Chromatography A* 1987; 410: 157-68.

[21] Larsen B, Fox BL, Burke MF, Hruby VJ. The separation of peptide hormone diasteroisomers by reverse phase high pressure liquid chromatography. Factors affecting separation of oxytocin and its diastereoisomers - structural implications. *International Journal of Peptide and Protein Research* 1979; 13: 12-21.

[22] Hase S, Walter R. Symmetrical disulphide bonds as sulphur-protecting groups and their cleavage by dithiothreitol. Synthesis of oxytocin with high biological activity. *International Journal of Peptide and Protein Research* 1973; 5: 283-8.

[23] Mühlemann M, Titov MI, Schwyzer R, Rudinger J. The use of intermediates with preformed disulphide bridge for the synthesis of oxytocin and deamino-oxytocin. *Helvetica Chimica Alta* 1972; 55: 2854-60.

[24] Khan SA, Sivanandaiah KM. Solid-phase synthesis of oxytocin, desaminooxytocin and 4-Thr-oxytocin using active esters in the presence of 1-hydroxybenzotriazole. *International Journal of Peptide and Protein Research* 1978; 12: 164-9.

[25] Live DH, Agosta WC, Cowburn D. A rapid, efficient synthesis of oxytocin and [8-arginine]-vasopressin. Comparison of benzyl, p-methoxybenzyl, and p-methylbenzyl as protecting groups for cysteine. *Journal of Organic Chemistry* 1977; 42: 3556-61.

[26] Spatola AF, Cornelius DA, Hruby VJ, Blomquist AT. Synthesis of oxytocin and related diastereomers deuterated in the half-cystine positions. Comparison of solid phase and solution methods. *Journal of Organic Chemistry* 1974; 39: 2207-12.

[27] Flouret G, Terada S, Kato T, Gualtieri R, Lipkowski A. Synthesis of oxytocin using iodine for oxidative cyclization and silica gel adsorption chromatography for purification. *International Journal of Peptide and Protein Research* 1979; 13: 137-41.

[28] SANDOZ AG, Basel Switzerland. Quality control. 1981. In, Nachtmann F, Krummen K, Maxl F, Riemer E. Oxytocin. *Analytical Profiles of Drug Substances* 1981; 10: 563-600.

[29] Nakamura H, Pisano JJ. Derivatization of compounds at the origin of thin-layer plates with fluorescamine. *Journal of Chromatography* 1976; 121: 33-40.

[30] Nachtmann F. High-performance liquid chromatography of intermediates in the oxytocin synthesis. *Journal of Chromatography A* 1979; 176: 391-7.

[31] Sutcliffe N, Corran PH. Comparison of selectivities of reversed-phase high-performance liquid chromatography, capillary zone electrophoresis and micellar capillary electrophoresis in the separation of neurohypophyseal peptides and analogues. *Journal of Chromatography A* 1993; 636: 95-103.

[32] Dudkiewicz W, Snycerski A, Tautt J. HPLC method for the determination of oxytocin in pharmaceutical dosage form and comparison with biological method. *Acta Poloniae Pharmaceutica - Drug Research* 2000; 57: 403-6.

[33] Krummen K, Maxl F, Nachtmann F. The use of HPLC in the quality control of oxytocin. *Pharmaceutical Technology International* 1979; 2: 37-43.

[34] Krummen K, Maxl F, Nachtmann F. The use of high-performance liquid chromatography (HPLC) in the quality control of oxytocin. *Pharmaceutical Technology* 1979; 3: 77-83.

[35] Maxl F, Siehr W. The use of high-performance liquid chromatography in the quality control of oxytocin, vasopressin and synthetic analogues. *Journal of Pharmaceutical and Biomedical Analysis* 1989; 7: 211-6.

[36] Ohta M, Fukuda H, Kimura T, Tanaka A. Quantitative analysis of oxytocin in pharmaceutical preparations by high-performance liquid chromatography. *Journal of Chromatography A* 1987; 402: 392-5.

[37] Pask-Hughes RA, Hartley RE, Gaines Das RE. A comparison of high performance liquid chromatographic assays with the current pharmacopoeial assays for the combined

formulation of ergometrine and oxytocin. *Journal of Biological Standardization* 1983; 11: 13-7.

[38] Krummen K, Frei RW. Quantitative analysis of nonapeptides in pharmaceutical dosage forms by high-performance liquid chromatography. *Journal of Chromatography A* 1977; 132: 429-36.

[39] Wang G, Miller RB, Melendez L, Jacobus R. A stability-indicating HPLC method for the determination of oxytocin acetate in oxytocin injection, USP, synthetic. *Journal of Liquid Chromatography and Related Technologies* 1997; 20: 567-81.

[40] Gard JW, Alexander JM, Bawdon RE, Albrecht JT. Oxytocin preparation stability in several common obstetric intravenous solutions. *American Journal of Obstetrics and Gynecology* 2002; 186: 496-8.

[41] Chaibva FA, Walker RB. Development and validation of a stability-indicating analytical method for the quantitation of oxytocin in pharmaceutical dosage forms. *Journal of Pharmaceutical and Biomedical Analysis* 2007; 43: 179-85.

[42] Karbiwnyk CM, Faul KC, Turnipseed SB, Andersen WC, Miller KE. Determination of oxytocin in a dilute IV solution by LC-MS. *Journal of Pharmaceutical and Biomedical Analysis* 2008; 48: 672-7.

[43] O'Hare MJ, Nice EC. Hydrophobic high-performance liquid chromatography of hormonal polypeptides and proteins on alkylsilane bonded silica. *Journal of Chromatography* 1979; 171: 209-26.

[44] Pask-Hughes RA, Corran PH, Calam DH. Assay of the combined formulation of ergometrine and oxytocin by high-performance liquid chromatography. *Journal of Chromatography A* 1981; 214: 307-15.

[45] Kumar V, Madabushi R, Derendorf H *et al.* Development and validation of an HPLC method for oxytocin in Ringer's Lactate and its application in stability analysis. *Journal of Liquid Chromatography and Related Technologies* 2006; 29: 2353-65.

[46] Bridges TE, Marino V. A rapid and sensitive method for the identification and quantitation of sub-picomolar amounts of neurohypophysial peptides. *Life Sciences* 1987; 41: 2815-22.

[47] Frei RW, Michel L, Santi W. Post-column fluorescence derivatization of peptides : Problems and potential in high-performance liquid chromatography. *Journal of Chromatography* 1976; 126: 665-77.

[48] Frei RW, Michel L, Santi W. New aspects of post-column derivatization in high-performance liquid chromatography. *Journal of Chromatography* 1977; 142: 261-70.

[49] Radhakrishnan AN, Stein S, Licht A, Gruber KA, Udenfriend S. High-efficiency cation-exchange chromatography of polypeptides and polyamines in the nanomale range. *Journal of Chromatography A* 1977; 132: 552-5.

[50] Gruber KA, Stein S, Brink L, Radhakrishnan A, Udenfriend S. Flourometric assay of vasopressin and oxytocin: A general approach to the assay of peptides in tissues. *Proceedings of the National Academy of Sciences of the United States of America* 1976; 73: 1314-8.

[51] Rao PS, Weinstein GS, Wilson DW, Rujikarn N, Tyras DH. Isocratic high-performance liquid chromatography--photodiode-array detection method for determination of lysine- and arginine-vasopressins and oxytocin in biological samples. *Journal of Chromatography A* 1991; 536: 137-42.

[52] Kukucka MA, Misra HP. Determination of oxytocin in biological samples by isocratic high-performance liquid chromatography with coulometric detection using C18 solid-phase extraction and polyclonal antibody-based immunoaffinity column purification. *Journal of Chromatography B: Biomedical Sciences and Applications* 1994; 653: 139-45.

[53] Toll Hr, Oberacher H, Swart R, Huber CG. Separation, detection, and identification of peptides by ion-pair reversed-phase high-performance liquid chromatography-electrospray ionization mass spectrometry at high and low pH. *Journal of Chromatography A* 2005; 1079: 274-86.

[54] Huck CW, Pezzei V, Schmitz T, Bonn GK, Bernkop-Schnürch A. Oral peptide delivery: Are there remarkable effects on drugs through sulfhydryl conjugation? *Journal of Drug Targeting* 2006; 14: 117-25.

[55] Heping W, Pacáková, Stulík K, Barth T. Ion-exchange high-performance liquid chromatographic analysis of the products of the enzymatic degradation of oxytocin. *Journal of Chromatography A* 1990; 519: 244-9.

[56] Fisher LA, Fernstrom JD. Measurement of nonapeptides in pineal and pituitary using reversed-phase, ion-pair liquid chromatography with post-column detection by radioimmunoassay. *Life Sciences* 1981; 28: 1471-81.

[57] Janaky T, Szabo P, Kele Z *et al.* Identification of oxytocin and vasopressin from neurohypophyseal cell culture. *Rapid Communications in Mass Spectrometry* 1999; 12: 1765-8.

[58] Mori M, Vigh S, Miyata A, Yoshihara T, Oka S, Arimura A. Oxytocin is the major prolactin releasing factor in the posterior pituitary. *Endocrinology.* 1990; 126: 1009-13.

[59] Rosenblum LA, Smith ELP, Altemus M *et al.* Differing concentrations of corticotropin-releasing factor and oxytocin in the cerebrospinal fluid of bonnet and pigtail macaques. *Psychoneuroendocrinology* 2002; 27: 651-60.

[60] Schams D. Oxytocin determination by radioimmunoassay. III. Improvement to subpicogram sensitivity and application to blood levels in cyclic cattle. *Acta Endocrinology(Copenh).* 1983; 103: 180-3.

[61] Vecsernyes M, Torok A, Jojart I, Laczi F, Penke B, Julesz J. Specific radioimmunoassay of oxytocin in rat plasma. *Endocrinology Regulation.* 1994; 28: 145-50.

[62] Bosch OJ, Kromer SA, Brunton PJ, Neumann ID. Release of oxytocin in the hypothalamic paraventricular nucleus, but not central amygdala or lateral septum in lactating residents and virgin intruders during maternal defence. *Neuroscience* 2004; 124: 439-48.

[63] Kawasaki K, Mitsui Y, Ono T *et al.* Simple method for assaying serum oxytocin and changes in serum oxytocin level during parturition in cynomolgus monkeys. *Experimental Animals* 2002; 51: 181-5.

In: Advances in Medicine and Biology. Volume 20
Editor: Leon V. Berhardt

ISBN 978-1-61209-135-8

Chapter 12

PLANT CELL WALL FUNCTIONAL GENOMICS: NOVELTIES FROM PROTEOMICS

Rafael Pont-Lezica[a], Zoran Minic[b], David Roujol[a], Hélène San Clemente[a] and Elisabeth Jamet*[a]

[a]Surfaces Cellulaires et Signalisation chez les Végétaux, UMR 5546 CNRS - UPS - Université de Toulouse, Pôle de Biotechnologie Végétale, 24 chemin de Borde-Rouge, BP 42617 Auzeville, 31326 Castanet-Tolosan, France

[b]Department of Chemistry, University of Saskatchewan, 110 Science Place, Saskatoon, SK S7N 5C9, Canada

ABSTRACT

Proteomics has become an important contributor to the knowledge of plant cell wall structure and function by allowing the identification of proteins present in cell walls. This chapter will give an overview on recent development in the cell wall proteomic field. Results from proteomics show some discrepancies when compared to results from transcriptomics obtained on the same organ. It suggests that post-transcriptional regulatory steps involve an important proportion of genes encoding cell wall proteins (CWPs). Proteomics thus complements transcriptomic. The cell wall proteome of *Arabidopsis thaliana* is the most completely described at the moment with about one third of expected CWPs identified. CWPs were grouped in functional classes according to the presence of predicted functional domains to allow a better understanding of main functions in cell walls. The second best-described cell wall proteome is that of *Oryza sativa.* Same functional classes were found, with different compositions reflecting the differences in polysaccharide structure between dicot and monocot cell walls. All these proteomic data were collected in a new publicly accessible database called *WallProtDB*

* E-mail: jamet@scsv.ups-tlse.fr

(http://www.polebio.scsv.ups-tlse.fr/WallProtDB/). In conclusion, some perspectives in plant cell wall proteomics are discussed.

INTRODUCTION

Several years after the launching of systematic programs of genome sequencing, the challenge of gene function discovery is still enormous, especially in the case of genes encoding cell wall proteins (CWPs). To date, only about 10% of them have a characterized function. Plant cell walls are mainly composed of networks of polysaccharides which represent up to 95% of cell wall mass. After completion of growth, secondary walls reinforce primary walls around cells. Models of primary cell wall structure describe the arrangement of their components into two structurally independent but interacting networks, embedded in a pectin matrix [9, 16]. Cellulose microfibrils and hemicelluloses constitute the first network; the second one is formed by structural proteins among which extensins [38]. CWPs only represent 5 to 10% of the cell wall mass [10], but they are playing many roles especially in polysaccharide network remodelling during plant development and in response to environmental stresses [25].

Many experimental approaches were developed to understand cell wall structure and function. Transcriptomics has greatly contributed to the understanding of gene regulation and to the identification of candidate genes for biogenesis of cell walls, especially secondary walls [19, 39, 55, 73]. Biochemistry of cell wall polysaccharides and proteins has been particularly studied, allowing a good knowledge of the cell wall structure [25]. Proteomics has recently been a newcomer in the field with the first significant results obtained since 2002. Since then, a large number of studies were performed, especially on *Arabidopsis thaliana*, the dicot model plant which genome was the first to be completely sequenced [1]. They led to the identification of about 500 CWPs, representing one third of the expected CWPs [31]. Moreover, improvement of proteomic tools allowed comparative and quantitative studies between different physiological stages or in response to stresses [13, 68, 75]. Bioinformatics strongly helped in prediction of gene structure and function and allowed the building of sophisticated databases collecting all this information (http://www.ncbi.nlm.nih.gov/; http://www.arabidopsis.org/). In this chapter, three questions will be addressed: (i) What is the interest of proteomics as compared to transcriptomics? (ii) What is the current picture of *A. thaliana* cell wall proteome? (iii) What is expected from the knowledge of a monocot cell wall proteome?

IS PROTEOMICS REDUNDANT WITH REGARD TO TRANSCRIPTOMICS?

Transcript profiling is one of the most widespread methods to identify genes involved in a developmental process or in response to environmental changes. Although it is widely accepted that the rate of synthesis and degradation determines transcript abundance, the majority of studies only measure steady state transcript levels, largely due to the technical simplicity with which they are determined [48]. It is generally assumed that a high level of transcripts at a particular physiological stage means that the genes play an important role in the process studied, allowing their selection for in-depth studies. However, gene expression can be regulated at different levels: transcription, post-transcription, translation, post-translation, biological activity of the encoded protein, and its degradation.

The proteome is the full complement of proteins expressed by a genome at a specific time [70]. The development of proteomics during the last 20 years allowed the large scale identification of the working end of the cell: the protein machinery [52]. However, proteomics encounters great difficulties:

(i) All proteins cannot be extracted by a simple method because they can have different properties like pI ranging from 2 to 12, and hydrophobicity/hydrophilicity.
(ii) The dynamic range of proteins in the cell may be orders of magnitude different since one protein can be expressed as 10.000 copies and another as 10 copies only. Thus, since there is no polymerase chain reaction (PCR) equivalent for replicating proteins, for the vast majority of proteomic analyses, it is only the most abundant 10% to 20% of proteins which are monitored.
(iii)The proteome of each living cell is dynamic, altering in response to the individual cell metabolic state and perception of intracellular and extracellular signal molecules.
(iv)Many proteins undergo post-translational modifications (PTMs) which can interfere with their separation prior to mass spectrometry (MS) analysis, and/or their identification through standard MS protocols.

In this part, we will present results of transcriptomic and proteomic studies performed on the same *A. thaliana* organs to show that both approaches are not redundant, but rather complementary. Mature stems and dark-grown hypocotyls were studied [29, 33, 45, 46]. In the transcriptomic studies, around 3.000 genes coding for proteins predicted to be targeted to the secretory pathway (SPGs, for secretory pathway genes) were selected. Fifty-eight percent of them showed detectable level of transcripts, but only genes with high and moderate level of transcripts (*i.e.* 4- to 64-fold the background level) were compared to the proteomic data. The rationale for this choice was that given the restrictions of proteomics, only the more abundant proteins were identified. If the idea that highly expressed genes produce the most abundant proteins is correct, the results should fit. One hundred ninety-three and 433 SPGs were selected on the basis on their transcript levels in mature stems and etiolated hypocotyls respectively. Among those genes, only 23 (11.9%) and 48 (11%) were also identified in the respective proteomic studies [29, 45]. This is illustrated in Figure 1A in the case of etiolated hypocotyls. This means that many CWPs escape proteomic analyses, but these results are in the low end of the normal results for proteomics [50]. In addition to the difficulties encountered by proteomics mentioned above, two features are specific to cell wall proteomics: (i) extraction of CWPs from the polysaccharide matrix of the wall can be difficult because they can be insolubilized within the matrix by several types of linkages [6, 62]; (ii) heavily glycosylated proteins are not easily identified. The latter case is illustrated in Table 1 with the arabinogalactan protein (AGP) gene family. These proteins are hydroxyproline-rich glycoproteins (HRGPs) which undergo many PTMs: most Pro residues are hydroxylated and *O*-glycosylated [61], thus preventing their identification by classical methods. Their identification requires a specific deglycosylation procedure using hydrogen fluoride (HF) [64,

74]. Indeed, no AGP has been identified although sixteen AGP genes show significant levels of transcripts in *A. thaliana* etiolated hypocotyls. However, there are some other cases where none of the proposed explanation is valid. As illustrated in Table 1, this is the case of the xyloglucan endotransglucosylase/hydrolase (XTH) gene family (http://labs.plantbio.cornell.edu/xth/) . Five XTHs were identified in the cell wall proteomic study performed on *A. thaliana* etiolated hypocotyls. There is one case where the protein is detected when the level of transcripts of the gene is below background (AtXTH33), and twelve cases where transcript levels are above background with no protein identified. One has to assume that post-transcriptional events contribute to the regulation of such genes.

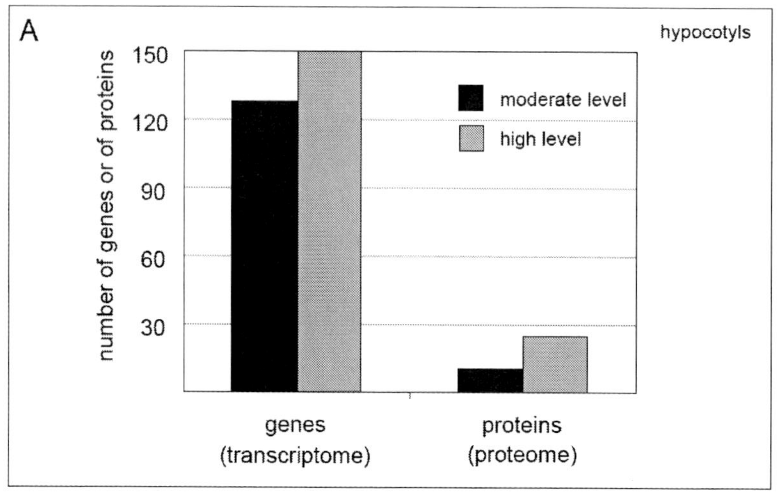

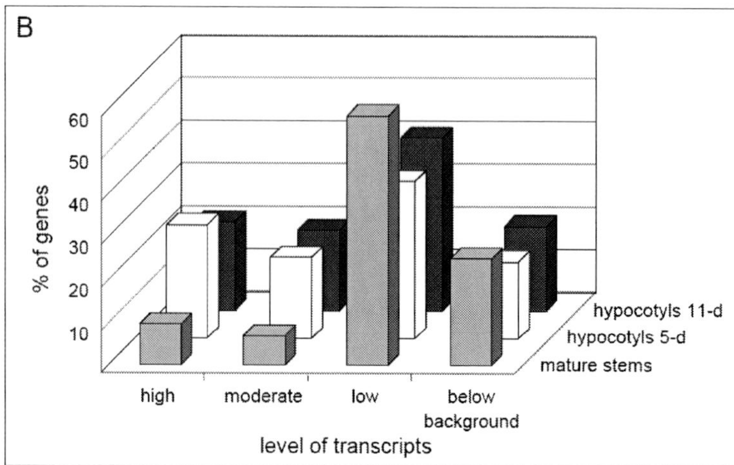

Figure 1. Transcriptomics *vs* proteomics. A. Distribution of SPGs having moderate or high levels of transcripts in *A. thaliana* etiolated hypocotyls: comparison between results of transcriptomics and proteomics [29, 33]. B. Levels of transcripts of *A. thaliana* genes encoding CWPs identified by proteomics in stems [45] and in 5- (5-d) and 11-day-old (11-d) etiolated hypocotyls [29]. Percentages of genes falling in the three following categories are represented: high: corresponds to log_2 values of the mean signal intensity higher than 10; moderate: values between 9 and 10; low: values between background and 9; values under the background level [33, 46].

Table 1. Comparison of transcriptomic and proteomic data obtained with etiolated hypocotyls of *A. thaliana* for two gene families: XTHs and AGPs. Transcriptomic data were obtained on CATMA microarrays with 5- and 11-day-old etiolated hypocotyls [33]. They are expressed as log_2 of mean signal intensity. The background of the experiment was estimated to 6.83. Cell wall proteomic data are from [29]: + means that the protein has been identified; - means that the protein has not been found. Transcriptomic results which are consistent between transcriptomic and proteomic studies are in bold characters

XTH family	transcriptomics		proteomics		AGP family	transcriptomics	
	5 days	11 days	5 days	11 days		5 days	11 days
AtXTH4	13.27	12.62	+	+	AtAGP9	12.43	12.55
AtXTH31	10.10	8.76	+	-	AtAGP15	12.04	12.70
AtXTH5	9.02	8.63	+	-	AtAGP31	11.98	11.89
AtXTH32	8.82	7.91	-	+	AtAGP12	11.93	12.97
AtXTH33	6.47	6.47	+	+	AtAGP4	11.76	11.86
AtXTH15	13.61	12.31	-	-	AtAGP22	9.98	10.94
AtXTH19	12.37	12.15	-	-	AtAGP1	9.62	9.86
AtXTH30	11.24	11.17	-	-	AtAGP26	9.07	9.62
AtXTH8	10.48	10.34	-	-	AtAGP25	8.68	8.80
AtXTH27	10.42	10.34	-	-	AtAGP18	8.64	8.41
AtXTH7	9.89	8.65	-	-	AtAGP10	8.18	8.88
AtXTH24	9.88	10.14	-	-	AtAGP19	7.94	7.93
AtXTH28	9.72	9.73	-	-	AtAGP5	7.34	7.49
AtXTH16	8.17	7.89	-	-	AtAGP30	6.95	6.57
AtXTH20	7.93	8.12	-	-	AtAGP17	6.94	6.84
AtXTH14	7.47	7.43	-	-	AtAGP41	6.84	6.93
AtXTH10	7.07	7.08	-	-			

Conversely, we looked at the level of transcripts of genes encoding CWPs identified through proteomics in etiolated hypocotyls (137 proteins) [29] and mature stems (87 proteins) [45]. The level of transcripts of some of the genes were not found in the microarray experiments (about 20%) since some have no gene specific tags (GSTs) or were eliminated because of poor signals of hybridization to the RNA probe. The results in Figure 1B are normalized plotting the percentage of genes for each level of transcripts. The big surprise was that most of the CWPs identified by proteomics originate from genes which level of transcripts was low (between 37 and 58%) or below the background (between 18 and 25%). Considering the limitations of proteomics, the identified proteins were expected to be the products of most abundant transcripts, but it was not the case. It suggests that transcripts could have short half-lives and/or that CWPs could have a low turnover. Post-transcriptional regulation seems to be important for more than 56% of CWPs in mature stems and etiolated hypocotyls. Only 16 to 44 % of the identified CWPs are the product of genes with moderate or high expression level, suggesting that transcription is the main regulatory step for that group of proteins.

These results are in agreement with previous ones obtained in yeast [28, 40], *A. thaliana* [34], and *Brassica napus* [75], showing that the quantification of transcripts does not always reflect the actual level of protein. We should consider that most of the genes encoding CWPs have a post-transcriptional regulation. This means that transcriptomics and proteomics are complementary, and not redundant [41].

A Summary of Present Proteomic Results Obtained on Arabidopsis Thaliana

Nearly 500 *A. thaliana* CWPs were identified in seventeen studies on different organs using various strategies. A new database (*WallProtDB*) collecting all this information was set up (http://www.polebio.scsv.ups-tlse.fr/WallProtDB/). *WallProtDB* allows searching for CWPs identified in the seventeen cell wall proteomes already published, and looking for the presence of proteins or protein families of interest in several proteomes. An exportation format is offered to download results of queries in the Microsoft Office Excel format (http://www.microsoft.com/france/office/2007/programs/excel/overview.mspx). Each proteomic study is described through a simplified flowchart showing its different steps from plant material to protein identification. For each proteomic study, the used strategy is highlighted in color.

As illustrated in Figure 2, two types of methods can be used to prepare a CWP fraction. Non-destructive methods leave the cells alive and allow elution of CWPs from cell walls using different buffered solutions (*e.g.* [4, 5]).

Destructive methods start with tissue grinding thus mixing CWPs and intracellular proteins (*e.g.* [2, 14, 21]). The CWP fraction needs to be fractionated to allow identification of proteins by mass spectrometry (MS). Proteins can be directly submitted to enzymatic digestion with appropriate proteases such as trypsin or to chemical treatment to get peptides of appropriate mass (usually between 750 and 4000 Da). Alternatively, proteins are separated prior to cleavage into peptides.

Cell wall proteomic analysis of *Arabidopsis thaliana* 11-day-old etiolated hypocotyls

A new picture of cell wall protein dynamics in elongating cells of *Arabidopsis thaliana*: confirmed actors and newcomers.
M Irshad, H Canut, G Borderies, R Pont-Lezica and E Jamet
BMC Plant Biology, 8: 94, 2008.

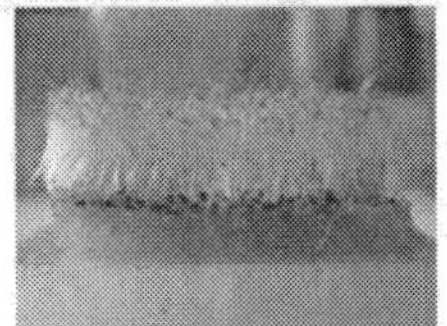

11-day-old *A. thaliana* etiolated seedlings grown *in vitro*

Strategy for the preparation of proteins

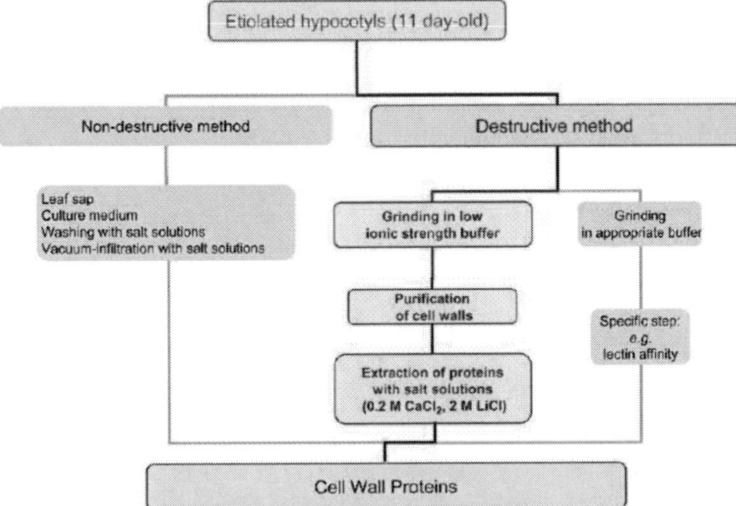

Strategy for the separation and the identification of proteins

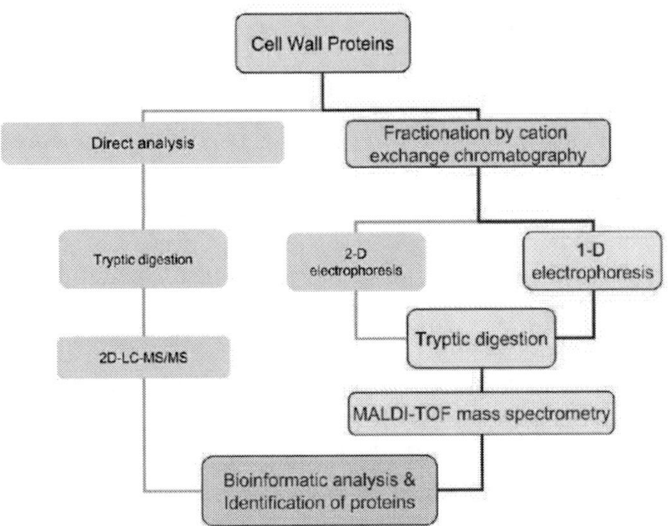

Figure 2. A strategy for cell wall proteomics as illustrated in *WallProtDB* (http://www.polebio.scsv.ups-tlse.fr/WallProtDB/index.php). For each experimental work, a description of the procedure is provided as a simplified flowchart showing its main steps for the preparation of the CWP sample, the separation of CWPs, and their identification using MS. All strategies were put on the same scheme that is customized for each experiment. The example shown is taken from [21], with the used strategy highlighted in color.

Since most CWPs are basic glycoproteins poorly resolved by bi-dimensional electrophoresis (2D-E), the most efficient ways to separate them are mono dimensional electrophoresis (1D-E) (*e.g.* [5]) or cationic exchange chromatography followed by 1D-E of protein fractions eluted with a salt gradient [29]. Identification of proteins can be done either by peptide sequencing by liquid-chromatography (LC) coupled to MS (LC-MS/MS) or by peptide mass mapping using matrix-assisted laser desorption/ionization-time of flight (MALDI-TOF) MS. Bioinformatic analysis of results is required both for protein identification, prediction of sub-cellular localization, and presence of functional domains [57]. To facilitate the interpretation of data, a database called *ProtAnnDB* dedicated to structural and functional annotation of *A. thaliana* proteins was recently built up [57]. *ProtAnnDB* also provides a link to the NCBI reference protein sequences (RefSeq) containing curated sequences (http://www.ncbi.nlm.nih.gov/RefSeq/) [54]. Proteins present in *WallProtDB* were linked to their annotation in *ProtAnnDB*.

The *Arabidopsis* CWPs were classified in nine groups on the basis on their predicted functional domains or known function [31]. The main class is represented by proteins acting on cell wall carbohydrates (26.1%) such as expansins (http://www.bio.psu.edu/expansins/), and glycoside hydrolases (GHs) among which xyloglycan endotransglucosylases/hydrolases (XTHs) [56], polygalacturonases (http://cellwall.genomics.purdue.edu/families/4-3-3.html), and β-1,4-glucanases (http://cellwall.genomics.purdue.edu/families/4-3-2-1.html). These proteins are assumed to modify polysaccharide networks *in muro* during growth and development or in response to environmental constraints. As the second most important group of CWPs (13.9%), oxido-reductases comprise peroxidases [51], multicopper oxidases [30], blue copper binding proteins [49] and berberine bridge oxido-reductases. Proteases (http://merops.sanger.ac.uk/) represent 11.6% of the CWPs. They can play roles in protein degradation, protein maturation and peptide signaling [69].

CWPs having interacting domains with polysaccharides and/or with proteins (11.1%) comprise proteins homologous to lectins, proteins with leucine-rich repeats (LRRs) domains, and enzyme inhibitors [17, 35, 67].

Proteins involved in signaling (6.7%) are arabinogalactan proteins (AGPs) [61] and receptor protein kinases which have transmembrane domains and intracellular kinase domains [63]. Considering the high number of CWPs (5.9%) related to lipid metabolism identified in cell wall proteomic studies, this group of proteins was recently introduced in our classification. They are thought to be involved in cuticle synthesis [42, 53]. Proteins of yet unknown function represent 12.2% of CWPs, and can have conserved structural domains or domains common to many proteins called domains of unknown function (DUFs). Such proteins or protein families are also called uncharacterized conserved proteins (UCPs) or uncharacterized protein families (UPFs).

Finally, miscellaneous proteins are diverse proteins which cannot be assigned to any of the previous group of CWPs. In particular, they comprise pathogenesis-related proteins (PR-proteins) [18], purple acid phosphatases [37], proteins homologous to strictosidine synthases [66], and proteins homologous to phosphate induced proteins like EXORDIUM [59]. Although we have now an overview of CWPs present in plant cell walls, the exact function of most of these proteins *in muro* is still unknown.

WHY INVESTIGATING ADDITIONAL CELL WALL PROTEOMES?

As discussed above, plant cell wall proteomic studies have encountered several limitations. The choice of a strategy for a cell wall proteomic study leads to the selection of sub-proteomes. Elution of proteins around living cells limits the use of drastic treatments to extract CWPs that would be more tightly interacting with cell walls [4]. Conversely, purification of cell walls prior to extraction of CWPs leads to the loss of proteins weakly interacting with cell wall polysaccharides [21]. In *A. thaliana*, two thirds of the predicted CWPs remain to be identified. Additional physiological stages can be studied, as well as interactions with environmental factors. Proteomic strategies can also be customized for specific protein families, e.g. [60]. Cell walls from monocots have only recently started to be analyzed with first papers published in 2008 on *Oryza sativa* and the identification of about 300 CWPs [12, 15, 36]. Among monocots, *O. sativa* is the model plant for cereals, whereas *Brachypodium distachyon* will probably become the model plant for herbaceous grasses [26]. *WallProtDB* was extended to *O. sativa*, thus collecting nearly 800 CWPs from twenty studies. Due to differences in cell wall composition and structure, there is interest in characterizing monocot cell wall proteomes. Differences between available *A. thaliana* and *O. sativa* cell wall proteomes will be discussed mainly for proteins acting on cell wall polysaccharides.

In a recently published review [47], the families of GHs involved in the modification and /or degradation of cell wall polysaccharides of *A. thaliana* and *O. sativa* were selected on the basis of their classification using the CAZY (http://www.cazy.org/) and TAIR (http://www.arabidopsis.org/) databases, predicted cellular localization and experimentally-determined substrate specificities in various plants and microorganisms. A total of 200 genes were selected in the *A. thaliana* genome. They belong to thirteen different families (GH 1, 3, 5, 9, 10, 16, 17, 27, 28, 31, 35, 43, and 51). Similarly, 174 genes were selected in the *O. sativa* genome. Putative substrates *in muro* for these enzymes are listed in Table 2. The comparison between the *A. thaliana* and *O. sativa* GH genes shows differences in the number of genes in some GH families. For example, the number of GH28 genes differs significantly between *A. thaliana* and *O. sativa*, particularly for the genes encoding polygalacturonases [44, 72]. This is in agreement with the different amounts of pectins present in the cell walls of these two species. The cell walls of commelinoid monocots, such as *O. sativa*, have a low content in pectins in contrast to those of *A. thaliana*, a dicot, which are rich in this constituent. In addition, dicots contain higher levels of xyloglucans than monocots [20, 24]. The total number of genes encoding GH16 and GH31 which are involved in the hydrolysis of xyloglucans is slightly higher in *A. thaliana* than in *O. sativa*. In contrast, the *O. sativa* genome contains more genes in the GH3 and GH17 families. These GH families are known to be involved in hydrolysis of (1,3)(1,4)-β-D-glucans [23] which are, in addition to glucuronoarabinoxylans, major polysaccharides in monocot cell walls.

Comparison of the cell wall proteomic analyses of *O. sativa* [12, 15, 36] and *A. thaliana* [31] leads to the conclusion that proteins acting on polysaccharides constitute the major functional class in both plants. Inside this class, GHs represent the largest number of proteins. Figure 3 presents a comparison between the number of GH families identified by cell wall proteomics in *A. thaliana* and *O. sativa* from leaves, cell suspension culture and culture medium of cell suspension cultures. Although the plant materials and the strategies used for proteomic analyses in *A. thaliana* and *O. sativa* were not wholly the same, these analyses give useful information on the relative abundance of GH families in the cell walls of both plants. For example, two GH5 enzymes were identified in the cell wall of rice. They might be involved in modification of mixed glycan polymers, only found in monocot cell walls [44]. In contrast, a higher number of GH16 enzymes, which are XTHs involved in the modification of the structure of xyloglucans [56], were identified in cell walls of *A. thaliana*. Similarly, a higher number of GH28 and GH35 enzymes were found in cell walls of *A. thaliana*. As mentioned above, such enzymes are assumed to be involved in modification of pectins. In addition, a higher number of carbohydrate esterases (CE8 and CE13) were identified in cell walls of *A. thaliana* cell suspension cultures and leaves respectively. These families comprise respectively pectin methylesterases and pectin acylesterases which modify pectins [43, 71].

Table 2. CAZy GH gene families and their putative substrates in plant cell walls. GH families are named according to the CAZy nomenclature (http://www.cazy.org/) [8]. Putative substrates *in muro* are given according to [47]

GH family	Putative substrates *in muro*
GH1	glucan/cellulose xyloglucan
GH3	xylan arabinan arabinoxylan
GH5	glucan/cellulose mannan
GH9	glucan/cellulose
GH10	xylan
GH13	1,4-α-glucosidic linkages
GH16	xyloglucan
GH17	1,3-β-glucan (1,3)(1,4)-β-glucan
GH18	GlcNAc linkages
GH19	GlcNAc linkages
GH27	galactomannan
GH28	homogalacturonan
GH31	xyloglucan
GH35	galactan glycoproteins (AGPs)
GH43	xylan
GH51	arabinoxylan xylan arabinan
GH79	glycoproteins (AGPs)

Most of the identified *O. sativa* GH families are also found in *A. thaliana*. However, GH13 enzymes, which are predicted to be α-amylases, were only found in *O. sativa* cell walls. α-amylases are endo-amylolytic enzymes which hydrolyze the 1,4-α-glucosidic linkages of starch. They are found in most storage tissues during periods of starch mobilization [3, 22]. Eight of them were predicted to be extracellular proteins (http://psort.ims.u-tokyo.ac.jp/form.html). Subcellular localization of α-amylases in *O. sativa* cell suspension cultures revealed that these enzymes are localized in cell walls as well as in starch granules within amyloplasts [11].

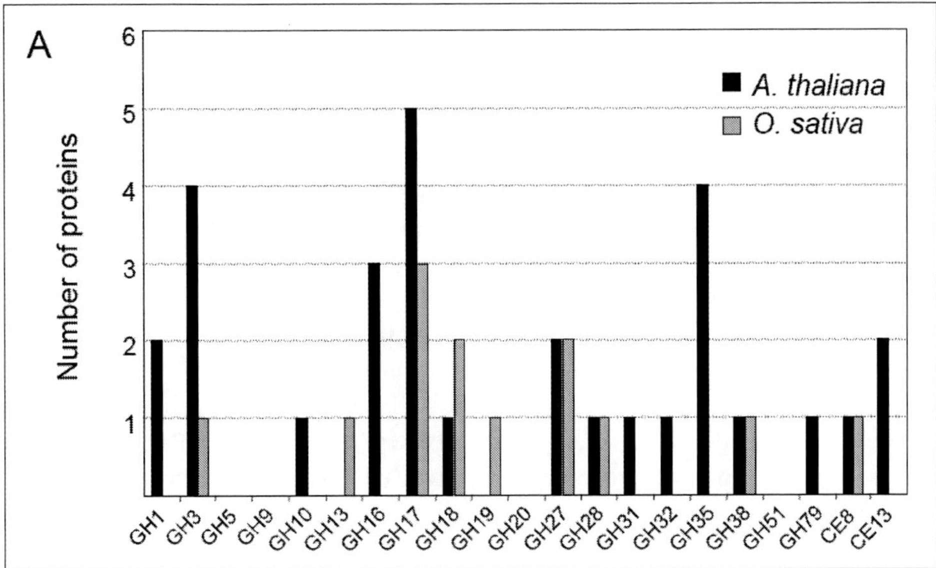

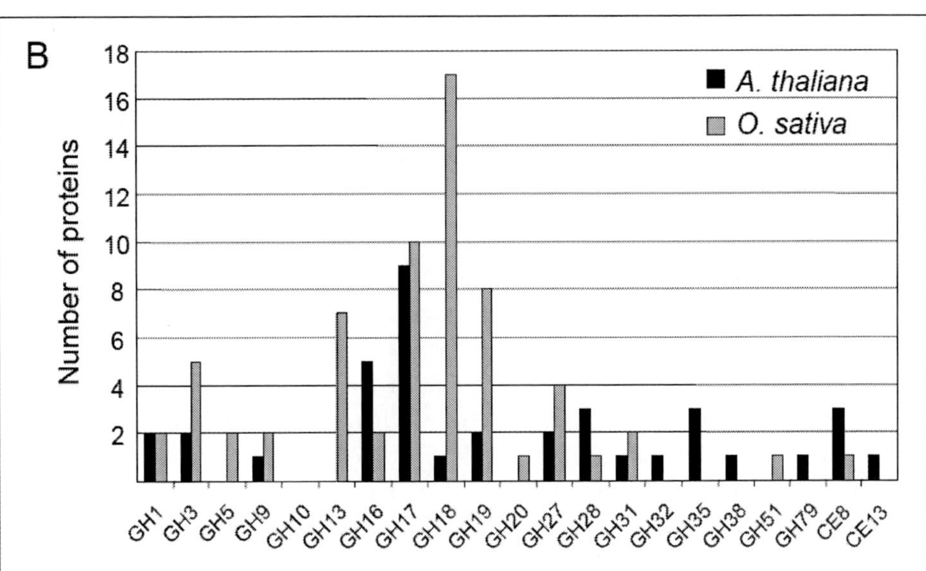

Figure 3. Comparison of *A. thaliana* and *O. sativa* GHs identified by proteomics. GH families were identified by cell wall proteomics either in leaves (A) or in cell suspension cultures and culture medium of cell suspension cultures (B). GHs families are numbered according to the CAZy database (http://www.cazy.org/). The list of proteins was obtained from *WallProtDB* (http://www.polebio.scsv.ups-tlse.fr/WallProtDB/).

The dual localization of α-amylases in rice is in agreement with the mobilization of starch in the endosperm by secreted amylases, and the mobilization of chloroplastic starch in leaves during the dark periods by chloroplastic α-amylases [65].

Important difference in the number of identified enzymes between *A. thaliana* and *O. sativa* were found for the GH18 and GH19 families. The number of proteins identified in *O. sativa* is much higher than in *A. thaliana*, especially in cell suspension cultures which are stressed cells. GH18 and GH19 are predicted to be chitinase-like enzymes. Chitinases catalyze the hydrolysis of N-acetyl-β-D-glucosaminide 1,4-linkages in chitins and chitodextrins which are not found in plants [58]. Thus, they are assumed to be involved in the protection of plants against pathogens [7, 58]. Hence, chitinases can inhibit fungal growth and kill fungi presumably by degradation of their cell walls made of chitin. Phylogenetic analysis of identified chitinases by proteomic analyses reveals that many of them are predicted to belong to class III, according to their primary structures [36]. Although precise functions of class III chitinases cannot be predicted, it is known that GH18 enzymes of class III also have lysozyme activity. These lysozymes/chitinases show higher activity on bacterial cell walls peptidoglycan (murein) [7].

On the other hand, some class III chitinases, of the TAXI-type (*Triticum aestivum* xylanase inhibitor), were found to be inhibitors of fungal and bacterial xylanases (GH11) in cereals [27]. However, chitinases were shown to be induced in response to abiotic stress and during development. It was suggested that class III chitinases could act on GlcNAc-containing glycolipids or glycoproteins, and be involved in signal transduction [58].

In conclusion, comparative genomic and proteomic analyses between the *A. thaliana* dicot plant, and the *O. sativa* monocot plant, show some similarity in number of cell wall enzymes involved in cell wall polysaccharide modifications, but also some differences with regard to the presence/absence of some GH families. These features could be related with variations in their cell wall compositions, stages of development, and environmental factors.

CONCLUSION

Plant cell wall proteomics has brought a new vision of CWPs and cell wall functions. New questions now need to be addressed. A comprehensive understanding of gene regulation requires all steps from gene transcription to protein degradation to be taken into account. Indeed, comparisons of transcriptomics and proteomics data show that the amount of an mRNA is not always strictly correlated with that of the translated protein. Major biological roles for proteolytic activities were only recently demonstrated in maturation of enzymes or production of extracellular peptide signals [69]. But there is still no information about CWP turnover which might be of critical importance in the regulation of extracellular functions. We are still far from the exhaustive description of cell wall proteomes. Additional information on CWPs could be obtained using specific methods for extracting CWPs strongly bound to cell wall components. Efforts should be made to extract CWPs that are physically-linked to cell wall components, such as polysaccharides or lignins [32]. On the other hand, the cell wall proteomes of monocots are only begun to be described. Many new CWPs having predicted enzymatic activities toward cell wall polysaccharides specific to monocots should be found [44]. A complete description of CWP biological functions will require complementary approaches including genetics, biochemistry, and the study of patterns of gene expression. Special attention should be drawn on the numerous CWPs (12.2%) with yet unknown predictable function. Unexpected cell wall functions during plant development and response to environmental factors will certainly arise from such studies.

ACKNOWLEDGMENTS

This work was supported by the Université Paul Sabatier in Toulouse and Centre National de la Recherche Scientifique (France). The authors wish to thank the *Plate-forme de Bioinformatique GenoToul Midi-Pyrénées* in Toulouse (France) for providing calculation facilities for the *O. sativa* sequences (http://www.bioinfo.genotoul.fr).

REFERENCES

[1] Arabidopsis Genome Initiative. (2000). Analysis of the genome sequence of the flowering plant *Arabidopsis thaliana*. Nature, 408, 796-815.

[2] Bayer, E. M., Bottrill, A. R., Walshaw, J., Vigouroux, M., Naldrett, M. J., Thomas, C. L. and Maule, A. J. (2006). *Arabidopsis* cell wall proteome defined using multidimensional protein identification technology. Proteomics, 6, 301-11.

[3] Beck, E. and Ziegler, P. (1989). Biosynthesis and degradation of starch in higher plants. *Annu. Rev. Plant Physiol Plant Mol. Biol.*, 40, 95-117.

[4] Borderies, G., Jamet, E., Lafitte, C., Rossignol, M., Jauneau, A., Boudart, G., Monsarrat, B., Esquerré-Tugayé, M. T., Boudet, A. and Pont-Lezica, R. (2003). Proteomics of loosely bound cell wall proteins of *Arabidopsis thaliana* cell suspension cultures: A critical analysis. *Electrophoresis,* 24, 3421-3432.

[5] Boudart, G., Jamet, E., Rossignol, M., Lafitte, C., Borderies, G., Jauneau, A., Esquerré-Tugayé, M.-T. and Pont-Lezica, R. (2005). Cell wall proteins in apoplastic fluids of *Arabidopsis thaliana* rosettes: Identification by mass spectrometry and bioinformatics. *Proteomics*, 5, 212-221.

[6] Brady, J. D., Sadler, I. H. and Fry, S. C. (1996). Di-isodityrosine, a novel tetrametric derivative of tyrosine in plant cell wall proteins: a new potential cross-link. *Biochem. J.,* 315, 323-7.

[7] Brunner, F., Stintzi, A., Fritig, B. and Legrand, M. (1998). Substrate specificities of tobacco chitinases. *Plant J.,* 14, 225-234.

[8] Cantarel, B., Coutinho, P., Rancurel, C., Bernard, T., Lombard, V. and Henrissat, B. (2009). The Carbohydrate-Active EnZymes database (CAZy): an expert resource for Glycogenomics. *Nucleic Acids Res.*, 37, D233-D238.

[9] Carpita, N. C. and Gibeaut, D. M. (1993). Structural models of primary cell walls in flowering plants: consistency of molecular structure with the physical properties of the walls during growth. *Plant J.*, 3, 1-30.

[10] Cassab, G. I. and Varner, J. E. (1988). Cell wall proteins. *Annu. Rev. Plant Physiol.* Plant *Mol. Biol.*, 39, 321-353.

[11] Chen, M., Huang, L., Li, H., Chen, Y. and Yu, S. (2004). Signal peptide-dependent targeting of a rice alpha-amylase and cargo proteins to plastids and extracellular compartments of plant cells. *Plant Physiol.*, 135, 1367-1377.

[12] Chen, X., Kim, S., Cho, W., Rim, Y., Kim, S., Kim, S., Kang, K., Park, Z. and Kim, J. (2008). Proteomics of weakly bound cell wall proteins in rice calli. *J. Plant Physiol.*, doi:10.1016/j.jplph.2008.09.010.

[13] Cheng, F.-Y., Blackburn, K., Lin, Y.-M., Goshe, M. and Wiliamson, J. (2009). Absolute protein quantification by LC/MS for global analysis of salicylic acid-induced plant protein secretion responses. *J. Proteome Res.*, 8, 82-93.

[14] Chivasa, S., Ndimba, B. K., Simon, W. J., Robertson, D., Yu, X.-L., Knox, J. P., Bolwell, P. and Slabas, A. R. (2002). Proteomic analysis of the *Arabidopsis thaliana* cell wall. *Electrophoresis*, 23, 1754-1765.

[15] Cho, W., Chen, X., Chu, H., Rim, Y., Kim, S., Kim, S., Kim, S.-W., Park, Z.-Y. and Kim, J.-Y. (2009). The proteomic analysis of the secretome of rice calli. *Physiol. Plant*, 135, 331-341.

[16] Cosgrove, D. J. (2005). Growth of the plant cell wall. *Nat. Rev. Mol. Cell Biol.*, 6, 850-61.

[17] Di Matteo, A., Federici, L., Mattei, B., Salvi, G., Johnson, K., Savino, C., De Lorenzo, G., Tsernoglou, D. and Cervone, F. (2003). The crystal structure of polygalacturonase-inhibiting protein (PGIP), a leucine-rich repeat protein involved in plant defense. *Proc. Natl. Acad. Sci. USA*, 100, 10124-10128.

[18] Edreva, A. (2005). Pathogenesis-related proteins: research progress in the last 15 years. *Gen. Appl. Plant Phys.*, 31, 105-124.

[19] Ehlting, J., Mattheus, N., Aeschliman, D., Hamberger, B., Cullis, I., Zhuang, J., Kaneda, M., Mansfield, S., Samuels, L., Ritland, K. et al. (2005). Global transcript profiling of primary stems from *Arabidopsis thaliana* identifies candidate genes for missing links in lignin biosynthesis and transcriptional regulators of fiber differentiation. *Plant J.*, 42, 618-640.

[20] Farrokhi, N., Burton, R., Brownfield, L., Hrmova, M., Wilson, S., Bacic, A. and Fincher, G. (2006). Plant cell wall biosynthesis: genetic, biochemical and functional genomics approaches to the identification of key genes. *Plant Biotechnol. J.*, 4, 145-167.

[21] Feiz, L., Irshad, M., Pont-Lezica, R. F., Canut, H. and Jamet, E. (2006). Evaluation of cell wall preparations for proteomics: a new procedure for purifying cell walls from *Arabidopsis* hypocotyls. *Plant Methods*, 2, 10.

[22] Fincher, G. (1989). Molecular and cellular biology associated with endosperm mobilization in germinating cereal grains. *Annu. Rev. Plant Physiol. Plant Mol. Biol.*, 40, 305-346.

[23] Fincher, G. (2009). Exploring the evolution of (1,3;1,4)-beta-D-glucans in plant cell walls: comparative genomics can help! *Curr. Opin. Plant Biol.*, 12, 140-147.

[24] Fincher, G. (2009). Revolutionary times in our understanding of cell wall biosynthesis and remodeling in the grasses. *Plant Physiol*, 149, 27-37.

[25] Fry, S. C. (2004). Primary cell wall metabolism: tracking the careers of wall polymers in living plant cells. *New Phytol.*, 161, 641-675.

[26] Garvin, D., Gu, Y.-Q., Hasterok, R., Hazen, S., Jenkins, G., Mockler, T., Mur, L. and Vogel, J. (2008). Development of genetic and genomic resources for *Brachypodium distachyon*, a new model for grass crop research. *Crop. Sci.*, 48, S69-S84.

[27] Gebruers, K., Brijs, K., Courtin, C., Fierens, K., Goesaert, H., Rabijns, A., Raedschelders, G., Robben, J., Sansen, S., Sørensen, J. et al. (2004). Properties of TAXI-type endoxylanase inhibitors. *Biochim. Biophys. Acta*, 1696, 213-221.

[28] Gygi, S., Rochon, Y., Franza, B. and Aebersold, R. (1999). Correlation between protein and mRNA abundance in yeast. *Mol. Cell Biol.*, 19, 1720-1730.

[29] Irshad, M., Canut, H., Borderies, G., Pont-Lezica, R. and Jamet, E. (2008). A new picture of cell wall protein dynamics in elongating cells of *Arabidopsis*: confirmed actors and newcomers. *BMC Plant. Biol.*, 8:94.

[30] Jacobs, J. and Roe, J. L. (2005). *SKS6*, a multicopper oxidase-like gene, participates in cotyledon vascular patterning during *Arabidopsis thaliana* development. *Planta,* 222, 652-666.

[31] Jamet, E., Albenne, C., Boudart, G., Irshad, M., Canut, H. and Pont-Lezica, R. (2008). Recent advances in plant cell wall proteomics. *Proteomics*, 8, 893-908.

[32] Jamet, E., Canut, H., Boudart, G. and Pont-Lezica, R. F. (2006). Cell wall proteins: a new insight through proteomics. *Trends in Plant Sci.,* 11, 33-39.

[33] Jamet, E., Roujol, D., San Clemente, H., Irshad, M., Soubigou-Taconnat, L., Renou, J.-P. and Pont-Lezica, R. (2009). Cell wall biogenesis of *Arabidopsis thaliana* elongating cells: transcriptomics complements proteomics. (submitted).

[34] Jones, A. M., Thomas, V., Truman, B., Lilley, K., Mansfield, J. and Grant, M. (2004). Specific changes in the *Arabidopsis* proteome in response to bacterial challenge: differentiating basal and R-gene mediated resistance. *Phytochemistry,* 65, 1805-16.

[35] Juge, N. (2006). Plant protein inhibitors of cell wall degrading enzymes. *Trends Plant Sci.*, 11, 359-367.

[36] Jung, Y..-H., Jeong, S.-H., Kim, S., Singh, R., Lee, J.-E., Cho, Y.-S., Agrawal, G., Rakwal, R. and Jwa, N.-S. (2008). Systematic secretome analyses of rice leaf and seed callus suspension-cultured cells: Workflow development and establishment of high-density two-dimensional gel reference maps. *J. Proteome Res.*, 7, 5187-5210.

[37] Kaida, R., Hayashi, T. and Kaneko, T. (2008). Purple acid phosphatase in the walls of tobacco cells. *Phytochemistry*, 69, 2546-2551.

[38] Kieliszewski, M. J. and Lamport, D. T. (1994). Extensin: repetitive motifs, functional sites, post-translational codes, and phylogeny. *Plant J.*, 5, 157-72.

[39] Ko, J., Han, K., Park, S. and Yang, J. (2004). Plant body weight-induced secondary growth in Arabidopsis and its transcription phenotype revealed by whole-transcriptome profiling. *Plant Physiol.*, 135, 1069-1083.

[40] Kolkman, A., Daran-Lapujade, P., Fullaondo, A., Olsthoorn, M., Pronk, J., Slijper, M. and Heck, A. (2006). Proteome analysis of yeast response to various nutrient limitations. *Mol. Syst. Biol.*, 2, 2006.0026.

[41] Kolkman, A., Daran-Lapujade, P., Fullaondo, A., Olsthoorn, M. M., Pronk, J. T., Slijper, M. and Heck, A. J. (2006). Proteome analysis of yeast response to various nutrient limitations. *Mol. Syst. Biol.,* 2, 2006.0026.

[42] Kurdyukov, S., Faust, A., Nawrath, C., Bär, S., Voisin, D., Efremova, N., Franke, R., Schreiber, L., Saedler, H., Métraux, J. et al. (2006). The epidermis-specific extracellular

BODYGUARD controls cuticle development and morphogenesis in Arabidopsis. *Plant Cell*, 18, 321-339.

[43] Micheli, F. (2001). Pectin methylesterases: cell wall enzymes with important roles in plant physiology. *Trends Plant Sci.*, 6, 414-419.

[44] Minic, Z. (2008). Physiological roles of plant glycoside hydrolases. *Planta,* 227, 723-740.

[45] Minic, Z., Jamet, E., Negroni, L., der Garabedian, P. A., Zivy, M. and Jouanin, L. (2007). A sub-proteome of *Arabidopsis thaliana* trapped on Concanavalin A is enriched in cell wall glycoside hydrolases. *J. Exp. Bot.*, in press.

[46] Minic, Z., Jamet, E., San Clemente, H., Pelletier, S., Renou, J.-P., Rihouey, C., Okinyo, D., Proux, C., Lerouge, P. and Jouanin, L. (2009). Transcriptomic analysis of Arabidopsis developing stems: a close-up on cell wall genes. *BMC Plant Biol.*, 9, 6.

[47] Minic, Z. and Jouanin, L. (2006). Plant glycoside hydrolases involved in cell wall polysaccharide degradation. *Plant Physiol. Biochem.*, 44, 435-449.

[48] Narsai, R., Howell, K., Millar, A., O'Toole, N., Small, I. and Whealan, J. (2007). Genome-wide analysis of mRNA decay rates and their determinants in *Arabidopsis thaliana. Plant Cell,* 19, 3418-3436.

[49] Nersissian, A. M. and Shipp, E. L. (2002). Blue copper-binding domains. *Adv. Protein Chem.*, 60, 271-340.

[50] Newton, R., Brenton, A., Smith, C. and Dudley, E. (2004). Plant proteome analysis by mass spectrometry: principles, problems, pitfalls and recent developments. *Phytochemistry,* 65, 1449-1485.

[51] Passardi, F., Penel, C. and Dunand, C. (2004). Performing the paradoxical: how plant peroxidases modify the cell wall. *Trends Plant Sci*, 9, 534-540.

[52] Peck, S. (2005). Update on proteomics in Arabidopsis. Where do we go from here? *Plant Physiol.*, 138, 591-599.

[53] Pollard, M., Beisson, F., Li, Y. and Ohlrogge, J. (2008). Building lipid barriers: biosynthesis of cutin and suberin. *Trends Plant Sci*, 13, 236-246.

[54] Pruitt, K., Tatusova, T. and Maglott, D. (2007). NCBI reference sequences (RefSeq): a curated non-redundant sequence database of genomes, transcripts and proteins. *Nucleic Acids Res.*, 34.

[55] Ranik, M., Creux, N. and Myburg, A. (2006). Within-tree transcriptome profiling in wood-forming tissues of a fast-growing Eucalyptus tree. *Tree physiol*, 26, 365-375.

[56] Rose, J., Braam, J., Fry, S. and Nishitani, K. (2002). The XTH family of enzymes involved in xyloglucan endotransglucosylation and endohydrolysis: current perspectives and a new unifying nomenclature. *Plant Cell Physiol.*, 43, 1421-1435.

[57] San Clemente, H., Pont-Lezica, R. and Jamet, E. (2009). Bioinformatics as a tool for assessing the quality of sub-cellular proteomic strategies and inferring functions of proteins: plant cell wall proteomics as a test case. *Bioinform Biol. Insights*, 3, 15-28.

[58] Sasaki, C., Vårum, K., Itoh, Y., Tamoi, M. and Fukamizo, T. (2006). Rice chitinases: sugar recognition specificities of the individual subsites. *Glycobiology*, 16, 1242-1250.

[59] Schröder, F., Lisso, J., Lange, P. and Müssig, C. (2009). The extracellular EXO protein mediates cell expansion in Arabidopsis leaves. *BMC Plant Biol*, 13, 9-20.

[60] Schultz, C. J., Ferguson, K. L., Lahnstein, J. and Bacic, A. (2004). Post-translational modifications of arabinogalactan-peptides of *Arabidopsis thaliana. J. Biol. Chem.*, 279, 455103-45511.

[61] Seifert, G. and Roberts, K. (2007). The biology of arabinogalactan proteins. *Annu. Rev. Plant Biol.*, 58, 137-161.

[62] Shah, K., Penel, C., Gagnon, J. and Dunand, C. (2004). Purification and identification of a Ca^{2+}-pectate binding peroxidase from *Arabidopsis* leaves. *Phytochemistry*, 65, 307-312.

[63] Shiu, S. H. and Bleecker, A. B. (2001). Receptor-like kinases from *Arabidopsis* form a monophyletic gene family related to animal receptor kinases. *Proc. Natl. Acad. Sci. USA,* 98, 10763-8.

[64] Shpak, E., Leykam, J. and Kieliszewski, M. (1999). Synthetic genes for glycoprotein design and the elucidation of hydroxyproline-O-glycosylation codes. *Proc. Natl. Acad. Sci. USA*, 21, 14736-14741.

[65] Smith, A., Zeeman, S. and Smith, S. (2005). Starch degradation. *Annu. Rev. Plant Biol.*, 56, 73-98.

[66] Sohani, M., Schenk, P., Schultz, C. and Schmidt, O. (2009). Purple acid phosphatase in the walls of tobacco cells. *Plant Biol.*, 11, 105-117.

[67] Spadoni, S., Zabotina, O., Di Matteo, A., Mikkelsen, J., Cervone, F., De Lorenzo, G., Mattei, B. and Bellincampi, D. (2006). Polygalacturonase-inhibiting protein interacts with pectin through a binding site formed by four clustered residues of arginine and lysine. *Plant Physiol.*, 141, 557-564.

[68] Thelen, J. and Peck, S. (2007). Quantitative proteomics in plants: choices in abundance. *Plant Cell*, 19, 3339-3346.

[69] van der Hoorn, R. (2008). Plant proteases: from phenotypes to molecular mechanisms. *Annu. Rev. Plant Biol.*, 59, 191-223.

[70] Wasinger, V., Cordwell, S., Cerpa-Poljak, A., Yan, J., Gooley, A., Wilkins, M., Duncan, M., HArris, R., Williams, K. and Humphery-Smith, I. (1995). Progress with gene-product mapping of the molliculites: *Mycoplasma genitalium*. *Electrophoresis*, 16, 1090-1094.

[71] Willats, W., McCartney, L., Mackie, W. and Knox, J. (2001). Pectin: cell biology and prospects for functional analysis. *Plant Mol. Biol.*, 47, 9-27.

[72] Yokoyama, R. and Nishitani, K. (2004). Genomic basis for cell-wall diversity in plants. A comparative approach to gene families in rice and Arabidopsis. *Plant Cell Physiol.*, 45, 1111-1121.

[73] Yokoyama, R. and Nishitani, K. (2006). Identification and characterization of *Arabidopsis thaliana* genes involved in xylem secondary cell walls. *J. Plant Res.*, 119, 189-194.

[74] Zhao, Z., Tan, L., Showalter, A., Lamport, D. and Kieliszewski, M. (2002). Tomato LeAGP-1 arabinogalactan-protein purified from transgenic tobacco corroborates the Hyp contiguity hypothesis. *Plant J.*, 31, 431-444.

[75] Zhu, M., Dai, S., McClung, S., Yan, X. and Chen, S. (2009). Functional differentiation of *Brassica napus* guard cells and mesophyll cells revealed by comparative proteomics. *Mol. Cell Proteomics*, 8, 752-766.

In: Advances in Medicine and Biology. Volume 20
Editor: Leon V. Berhardt
ISBN 978-1-61209-135-8

Chapter 13

ISOLATED RAT HEARTS PRESERVED FOR 24-120 HOURS IN THE PRESENCE OF CARBON MONOXIDE AND CARBON DIOXIDE, RESUSCITATION AND HETEROTOPIC TRANSPLANTATION

Yu Yoshida*[*], *Naoyuki Hatayama and Kunihiro Seki
Kanagawa University, Faculty of Science, 2946, Tsuchiya, Hiratsukashi, Japan

ABSTRACT

Currently, the primary technique for preserving isolated human organs for transplant is maintaining them at low temperature, but the time limit of this method is from 4 hr to 24 hr depending on the type of organs. If organs could be preserved long-term like blood cells or microorganisms, then the problem of organ shortage could be considerably alleviated. New techniques for the long-term preservation of organs still have to be developed and they remain eagerly awaited [4].

Seki *et al.* (1998) focused on cryptobiosis which enables living things to adapt themselves to extreme environment such as severe desiccation or very low temperature by decreasing the water content of the organism, which leads to a slow down of the metabolism. In this context, Seki *et al.* succeeded in resuscitating tardigrades placed under extremely high pressure (600 MPa) [22]. Since 1998, using liquid perfluorocarbon (PFC) for organ preservation, Seki *et al.* decreased the water contained in isolated rat hearts and preserved them for resuscitation. This occasionally led to good results, but the reproducibility was poor. Taking advantage of the anesthetic and metabolic inhibitory actions of carbon dioxide (CO_2) gas on organisms, Seki *et al.* (1998) repeatedly

[*] E mail: yoshiy412@hotmail.co.jp Tel: 0463-59-4111 Post 2513 Fax: 0463-58-9684

conducted experiments on rat hearts whose water content was considerably decreased and were exposed to high partial pressure of PCO_2=200 hPa during preservation and good results were obtained with reproducibility [21]. After having thus established organ preservation method by means of desiccation with CO_2, Yoshida *et al.* successfully extended the preservation time of isolated rat hearts to 72 hr by supplying CO_2 (PCO_2=200 hPa) to liquid PFC in the form of bubbles [23].

Various combinations of gas mixtures were tried. For example, in place of the usual CO_2, CO which is in reversible relation with oxygen (O_2) was introduced. Isolated rat hearts were placed in a hyperbaric environment of 2 ATA where high partial pressure between 200 hPa and 1000 hPa was chosen for the CO gas and the remaining pressure was complemented by the partial pressure of O_2. The rat hearts preserved 24 hr in this hyperbaric environment were heterotopically transplanted and were resuscitated. This successfully demonstrated the usefulness of CO for the preservation of organs and in search of longer preservation period, the partial pressure of CO was varied. For example, when the predetermined partial pressure of CO was between 1000 hPa and 5000 hPa, O_2 gas was supplied until the total pressure reached 7 ATA. After 48 hr of heart preservation by desiccation, heterotopic transplantation of the rat heart was performed and it was resuscitated. The reproducibility of this series of experiments was good.

In addition, CO_2 was introduced into the hyperbaric environment of 7 ATA. At the same time, to prevent CO_2 from causing decompression sickness, helium (He), an inert gas, was used to create a hyperbaric environment of 7 ATA. Consequently in the hyperbaric environment of PCO=400 hPa + PCO_2=100 hPa + PO_2=900 hPa + PHe=5600 hPa, isolated rat hearts were desiccated and preserved for 72 hr and were heterotopically transplanted and resuscitated successfully. Significant reproducibility was obtained. By using a mixture of CO_2, CO, O_2 and He, reproducible techniques have been developed allowing the desiccation and preservation of isolated rat hearts for a maximum of 120 hr, followed by their heterotopical transplantation and resuscitation. Therefore, the new scientific field of semibiology had been created.

Keywords: CO, CO_2, Isolated rat heart, perfluorocarbon (PFC), desiccation, preservation, resuscitation, heterotrophic transplantation

INTRODUCTION

Transplantation of human lungs, hearts, livers and kidneys is routinely practiced as clinical therapy [1]. However, the shortage of organs is getting serious, as the number of patients waiting for transplantation is increasing every year. Even in a country as advanced as the USA, over 70% of the patients waiting for the transplantation have to die without undergoing the operation. The supply of organs for transplantation is deficient because organs cannot be preserved for a long time. If organs could be preserved long-term like blood cells or microorganisms, the problem of organ shortage could be considerably alleviated. New techniques for the long-term preservation of organs have been eagerly anticipated [4].

At present, the primary technique of preserving isolated human organs for transplant is maintaining them at low temperature, but the time limit of this method is from 4 hr to 24 hr depending on the type of organs. The organs cannot be preserved longer, because the cell walls in the organs thus preserved are damaged by the low temperature of 4°C and ischemia [9] [16] [17].

The following methods yielded positive results. The simple immersion of isolated organs in the University of Wisconsin solution (UWS) is the widely used clinical method of organ preservation. The isolated hearts of rats, rabbits, baboons and humans can be kept at low temperature for resuscitation, but the time limit for revival ranges from 4 hr to 18 hr [12] [28]. Even at low temperature, metabolism of tissues continues at a reduced rate producing waste. To further extend the duration of preservation time, some have supplied O_2 to either the preservation solution or a perfusate and subsequently longer preservation time was reported [26]. Kuroda *et al.* filled rat hearts with UWS and immersed them in liquid PFC which is an inert liquid capable of absorbing a great amount of O_2. To this liquid, mixed gas containing PCO_2=50 hPa+PO_2=950 hPa was constantly supplied. The rat hearts thus immersed were successfully preserved from 24 hr (100%) to 48 hr (four of five rats) [11].

In this context, Seki *et al.* (1998), focused on cryptobiosis which enables living things to adapt themselves to an extreme environment such as severe desiccation or very low temperature by decreasing water contained in their tissues, which leads to a slow-down of their metabolism. Seki showed that it was possible to resuscitate tardigrades which were desiccated via immersion in liquid PFC and were placed in an extremely high hydrostatic environment of 600 MPa [22]. The success of this procedure stimulated Seki *et al.* to embark on a series of experiments on organ preservation. If desiccation of living things and their exposure to high pressure allows them resuscitate later, this phenomenon may be applied for long-term organ preservation. If long-term preservation of organs becomes a reality, like the storage of blood, the shortage of organs for transplantation may be resolved, so in 1998 Seki *et al.* started experiments on organ preservation.

Isolated rat hearts have been investigated regarding their preservation of from 24 hr to 72 hr, their heterotopic transplantation and their resuscitation. These experiments are divided into 3 groups: the first group (Section 1) includes those experiments using CO_2 for desiccation and preservation; the second group (Section 2) used CO for desiccation and preservation; the third group (Section 3) using a mixture of CO_2, CO, O_2 and He for the same purpose.

SECTION 1: DESICCATION AND PRESERVATION WITH CARBON DIOXIDE

Based on cryptobiosis, Seki *et al.* (1999) desiccated isolated rat hearts with silica gel to preserve them for 10-26 days and resuscitated them by means of Langendorff method [17]. He also undertook experiments with liquid PFC to decrease the water content of isolated rat hearts and then resuscitated them using Langendorff method. These experiments yielded occasional satisfactory results, but the reproducibility was poor. Taking advantage of the anesthetic and metabolic inhibitory effects which are characteristics of CO gas [4] [13] [19], Seki *et al.* (2005) placed isolated rat hearts in the dry environment containing PCO_2=200 hPa prior to transplantation and resuscitation. This series of experiments produced reproducible results [21].

The rats used in this experiment as donors and recipients were both inbred strains (LEW/SsN Slc; male, 6 weeks old) developed by Japan SLC, Inc. especially for organ transplantation. Donor rats were prepared as follows; Each rat was administered an anesthetic of diethyl ether and 1.0 ml of physiological saline containing heparin sodium (100 U) was injected intravenously. After the chest was shaved and disinfected with 75% alcohol, a midline longitudinal incision was made. Both the right and left ribs were cut toward the back and the chest walls were rolled back toward the head to expose the heart. After the heart was removed and was placed in a vessel filled with heparin-containing physiological saline (room temperature), the ascending aorta and pulmonary arteries were both transected as far as possible from their respective origins. Heparin- containing physiological saline was infused into the heart via the transected root of the ascending aorta to remove blood by means of perfusion. After perfusion, both the inferior and superior vena cava were litigated and transected. The pulmonary veins and bronchial tubes were litigated together in one lump and lungs were removed. After the weight of the harvested heart was measured, it was filled with Krebs-Henseleit (KH) solution and then was placed in a hyperbaric chamber (2 ATA) where it was exposed to the pressurized mixture of PO_2 =1800 hPa+PCO_2=200 hPa. It was later moved to a refrigerator set at 4°C. Antibiotics and warfarin were added to the KH solution, both of which were dissolved beforehand. A beaker filled with distilled water was placed inside the hyperbaric chamber, to keep the humidity inside the hyperbaric chamber over 90% (Figure 1).

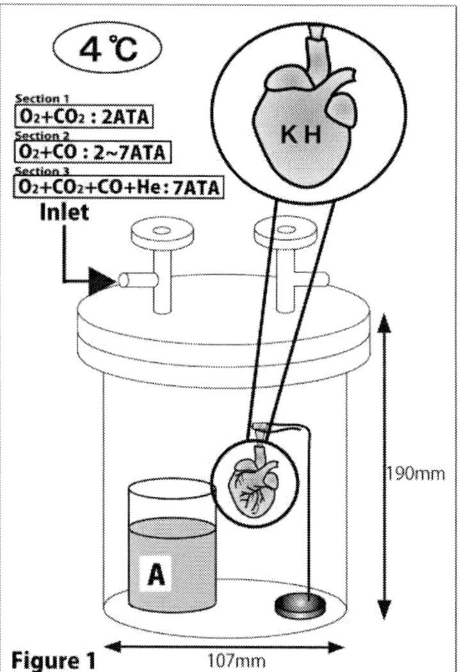

Figure 1. A hyperbaric chamber was used for the preservation of isolated heart. A beaker filled with distilled water was placed inside the chamber to maintain the humidity at over 90%. The heart was filled with KH solution into which antibiotics and warfarin sodium were dissolved and it was hung as shown in the illustration.
[Section 1] The mixture of O_2 and CO_2 was pressurized up to 2 ATA and the chamber was put in a refrigerator and stored at 4°C and for 24 hr.
[Section 2] The mixture of CO and O_2 was pressurized from 2 ATA to 7 ATA and the chamber was put in a refrigerator and stored at 4°C for 48 hr.
[Section 3] The mixture of O_2, CO_2, CO and He was pressurized up to 7 ATA and the chamber was put in a refrigerator and stored at 4°C for 72 hr.

When the isolated heart had been preserved for 24-72 hr, the recipient rat was put under anesthesia with the inhalation of diethyl ether and the right neck of the rat was shaved widely and disinfected with 75% alcohol. Next, a longitudinal 3 cm-long incision was made and the connective tissue under the skin was cauterized and eliminated with electrocautery to expose the external jugular vein and common carotid artery. After the proximal side of the external jugular vein was blocked with a micro-clip, the distal side of the external jugular vein was tied and cut off. Thereafter, the same procedures were repeated with the proximal side of the common carotid artery. After blocking its proximal side with a micro-clip, cuffs fitting the external jugular vein and the common carotid artery were put on respectively and heparin-containing physiological saline was flushed through the blood vessels. As the recipient rat was thus prepared, the preserved heart of the donor rat was taken out of the refrigerator and was immersed temporarily in physiological saline.

After a while, the donor rat heart was taken out and its aorta and recipient rat's common carotid artery as well as the donor rat heart's external jugular vein and the recipient rat's pulmonary artery underwent end-to-end anastomosis. To allow for the passage of blood, micro-clips clamped to the external jugular vein and common carotid artery were removed and the blood again flowed through the transplanted heart. When its pulsation was confirmed, the incision of the neck was sutured. For the next 10 weeks, the recipient rat was given drinking water into which antibiotics were dissolved beforehand and post-operative observation was conducted in the animal room. Ten weeks later, the pulsation of the transplanted heart was recorded with ECG. With regard to the isolated rat hearts preserved 24 hr with the mixture of PO_2=1800 hPa+PCO_2=200 hPa, of 5 hearts heterotopically transplanted, heartbeats were confirmed in all 5 immediately after the operation and all 5 transplanted hearts kept on beating for more than 10 weeks. Therefore, this experiment turned out to be 100% reproducible.

During the 24 hr preservation, a reduction of water contained in the heart and inhibitory effects of CO_2 on metabolism were both observed. However, to ameliorate the O_2 supply to tissues, Yoshida *et al.* adopted the method of Kuroda *et al.* but by changing the CO_2 concentration, the mixture described above yielded better results. Kuroda chose the partial pressure of 50 hPa for the CO_2 bubbled into liquid PFC, but PCO_2=100 hPa was selected in the current study and it successfully increased the preservation time up to 72 hr ($P<0.01$) [23]. After the 72 hr preservation, the donor rat heart was harvested following the same procedures used for the 24 hr preservation. The harvested heart was exposed to the mixture of PCO_2=100 hPa+PO_2=900 hPa and then was immersed in the liquid PFC at 4°C. During the preservation, the isolated heart was constantly exposed to bubbling CO_2 at the rate of 35 ml/min (Figure 2). After the preservation for 72 hr, the isolated rat heart was transplanted heterotopically to the right neck of the recipient rat and was resuscitated. Twelve weeks later the pulsation of the donor rat heart transplanted to the recipient rat was recorded with ECG (Figure 3).

After being preserved in liquid PFC from 48 hr to 120 hr, an isolated heart of a donor rat underwent end-to-end anastomosis with the heart of the recipient rat, the latter's common carotid artery was attached to the former's aorta and the latter's external jugular vein with the former's pulmonary artery and the isolated rat heart was resuscitated. Of 20 heterotopically transplanted hearts, pulsation of 11 transplanted hearts could be recorded with ECG immediately after the operation. In the case of 72 hr preservation, at 10 weeks after the transplantation, the pulsation of 4 of 5 of the transplanted hearts was confirmed with ECG (Figure 4).

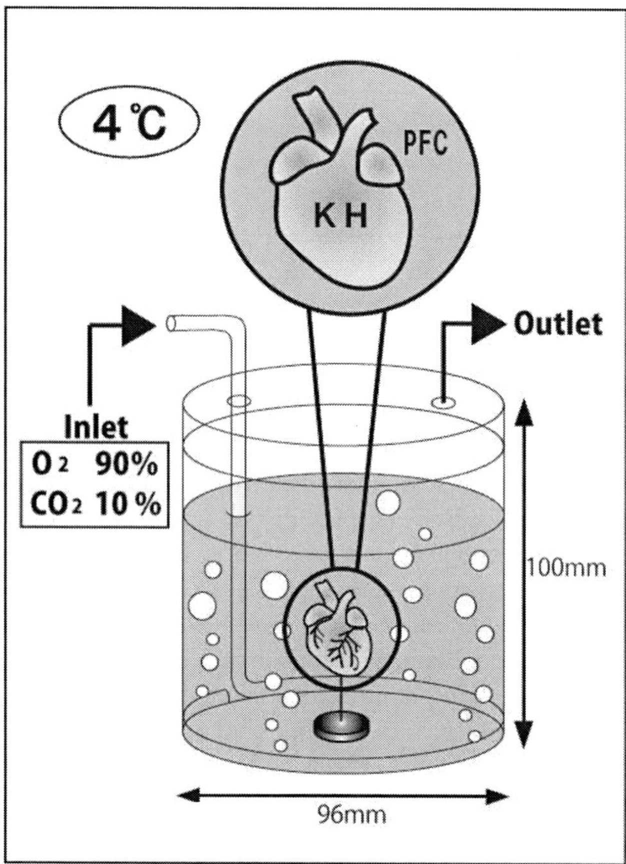

Figure 2. An isolated rat heart was placed in a container and it was kept in a refrigerator at 4°C. The isolated rat heart was preserved in inert liquid PFC. During the preservation, the mixture of O_2 and CO_2 was constantly supplied at the rate of 35ml/min. (A: glass bottle)

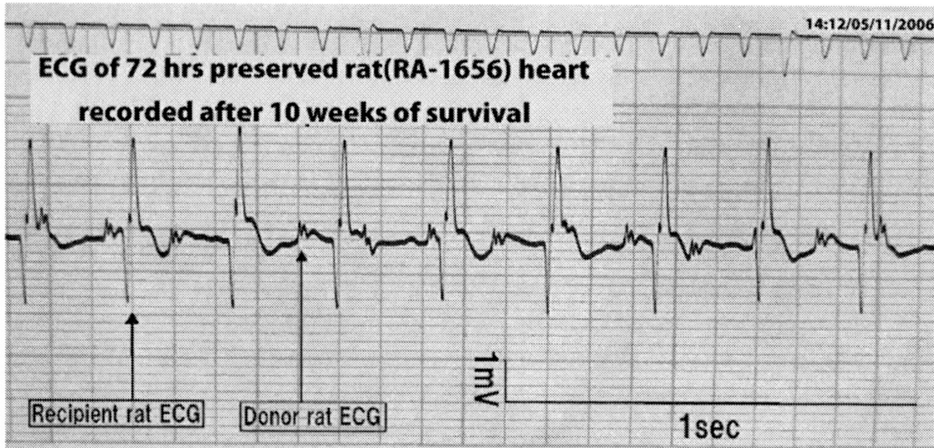

Figure 3. An isolated rat heart was preserved for 72 hr at PCO_2=100 hPa. After the preservation, a heterotopic transplantation was performed at the neck of a recipient rat and the transplanted heart was resuscitated. This is the ECG of the donor rat heart and recipient rat heart recorded on November 5, 2006, at 12 weeks after the heterotopic transplantation was performed.

Group	Preservation method	Preservation time(hr)	The number of hearts	Resuscitation rate(%)	Ten weeks survival rate(%)
1A	KH/PFC+CO2(100hPa)	48	5	5/5 (100)	5/5 (100)
1B	KH/PFC+CO2(100hPa)	72	5	5/5 (100)	5/5 (80)
1C	KH/PFC+CO2(100hPa)	96	5	1/5 (20)	0/5 (0)
1D	KH/PFC+CO2(100hPa)	120	5	0/5 (0)	0/5 (0)
2	control	0	10	10/10 (100)	10/10 (100)

Figure 4. The isolated rat hearts were preserved in liquid PFC for 48-120 hr which was exposed constantly to the mixture of PCO_2=100 hPa+PO_2=900 hPa. The resuscitation rate immediately after the heterotopic transplantation as well as the survival rate 10 weeks after the operation is shown.

Using liquid PFC, UW solution and O_2, Kuroda *et al.* reported that they preserved isolated rat hearts for a maximum of 48 hr after which they were heterotopically transplanted. Of 5 rats thus transplanted 4 were resuscitated and survived 6 weeks (4/5=80%) [11]. Increasing the CO_2 partial pressure allowed the preservation of the isolated rat hearts for 72 hr. In addition, a perfusion method was adopted to allow for the metabolic activity of the tissues at low temperature and to eliminate the waste produced by the tissues. The adoption of perfusion method [17] resulted in better preservation time in comparison to the simple immersion method and demonstrated that after 72 hr preservation and heart transplantation, heartbeats could be recorded for more than 10 weeks. However, with preservation for 96-120 hr, though the hearts pulsated temporarily after the operation, they didn't survive and within 24 hr after transplantation, they all ceased to pulsate.

Carbon dioxide gas was used for the long-term preservation of isolated rat hearts because it is known that depolarizing CO_2 gas or any inert gas dissolved in water causes hydrophobic hydration to the surrounding water thus suppressing the thermal activity of water molecules. When depolarizing CO_2 is dissolved into free water, it increases the formation of structured water and turns it into hydrophobic material [24]. This suggests that when CO_2 is dissolved in water, the formation of structured water contained in cells is enhanced and biological activity is inhibited. Since depolarizing CO_2 gas diminishes the metabolic activity of each cell in the organs, it is possible for cells to stay in a state of dormancy so that when the resuscitation is performed, the tissue resumes its activities. All these observations seem to suggest the efficacy of CO_2 for the preservation of organs.

Before these experiments in tissue preservation by means of CO_2, Ohta *et al.* used CO_2-added Modified Euro-Collins solution and obtained good results for the preservation of dogs' veins [14] [15]. However, although the same experiment was attempted with rat hearts, no resuscitation could be obtained after preservation for less than 48 hr. This may be because myocardial muscles require a greater O_2 supply and when the O_2 supply was ameliorated with the use of PFC, it was possible to extend the preservation time of rat hearts. Though it is difficult to measure the concentration of free water contained in tissues, it is conceivable that the immersion of rat hearts in liquid PFC, aerated continually with O_2 and CO_2, tends to diminish the amount of water in the tissues. A possible connection between a decrease in free water and an extension of preservation time is suggested.

Another method of preservation for organisms involves eliminating as much intracellular water as possible. Many bacteria enter into a dormant state when the relative humidity falls below 60%. This phenomenon of the loss of intracellular water is put into practical use as a preservation method by dehydration. However, the 72 hr preservation of isolated mammalian organs followed by their resuscitation was never tried till the current experiments. The results of the experiment undertaken by Yoshida *et al.* (2008) suggested the future application of this method for the preservation and resuscitation of human and other mammalian organs. Besides changing the CO_2 partial pressure, other modifications brought about to the method of Kuroda *et al.* was the use of KH solution in place of UWS. The 72-hr-long preservation was repeated using UWS, but no heartbeat was observed following resuscitation. On the other hand, better results were obtained using KH solution s. However, there is no reason to consider KH solution as the best preservation solution and we are still in search of the most appropriate solution for preservation. The high viscosity of UWS seems to make it unsuitable for the preservation by perfusion.

In the desiccation and preservation with CO_2 followed by transplantation and resuscitation, the isolated heart continued to beat for more than 12 weeks after the transplantation. This demonstrated that in spite of the temporary arrest of biological activity for 72 hr on the part of isolated hearts, they could resuscitate after transplantation and their biological activity could be resumed. These experiments also suggest that further extension of preservation time could be obtained by choosing right level of CO_2 partial pressure which favors the dormant state of myocardial muscles and by performing delicate adjustments and maintenance of the intracellular and extracellular structured water.

SECTION 2: DESICCATION AND PRESERVATION WITH CARBON MONOXIDE

After establishing the method of preservation by desiccation, the composition of mixed gas was changed. Instead of CO_2, CO which has reversible relationship with O_2 was introduced. Carbon monoxide is extremely toxic. For humans, inhalation of air containing 0.01% of CO (100 ppm) can cause toxic symptoms. When exposed to the air containing 0.08-0.12% of CO, humans may fall into coma, suffer from respiratory failure and heart failure. When the CO concentration is 0.08%, 50% of body's hemoglobin can be converted into carboxyhemoglobin. When the CO concentration reaches 0.15%, it may be lethal and the concentration of 0.19% (1900 ppm) leads to prompt death [25].

In recent studies, isolated rat hearts were exposed to CO gas with a partial pressure of 200-1000 hPa. The former is equal to 200000 ppm in atmospheric pressure and is 100 times over the level lethal to humans. To the CO gas with predetermined partial pressure of 200-1000 hPa, compressed O_2 gas was added until the total pressure reached 2 ATA.

Group	Preservation method	Preservation time(hr)	The number of hearts	Resuscitation rate(%)	Ten weeks survival rate(%)
1A	O2(1800hPa)+CO(200hPa)	24	5	3/5 (60)	3/5 (60)
1B	O2(1600hPa)+CO(400hPa)	24	5	5/5 (100)	5/5 (100)
1C	O2(1400hPa)+CO(600hPa)	24	5	5/5 (100)	3/5 (60)
1D	O2(1200hPa)+CO(800hPa)	24	5	1/5 (20)	1/5 (20)
1E	O2(1000hPa)+CO(1000hPa)	24	5	1/5 (20)	0/5 (0)
2	control	0	10	10/10 (100)	10/10 (100)

Figure 5. An isolated rat heart was preserved in a hyperbaric chamber of 2 ATA of a mixture of CO and O_2, with predetermined level of CO partial pressure. After 24 hr preservation, it was heterotopically transplanted. This table indicates the rate of resuscitation immediately after the transplantation and the rate of survival 10 weeks later.

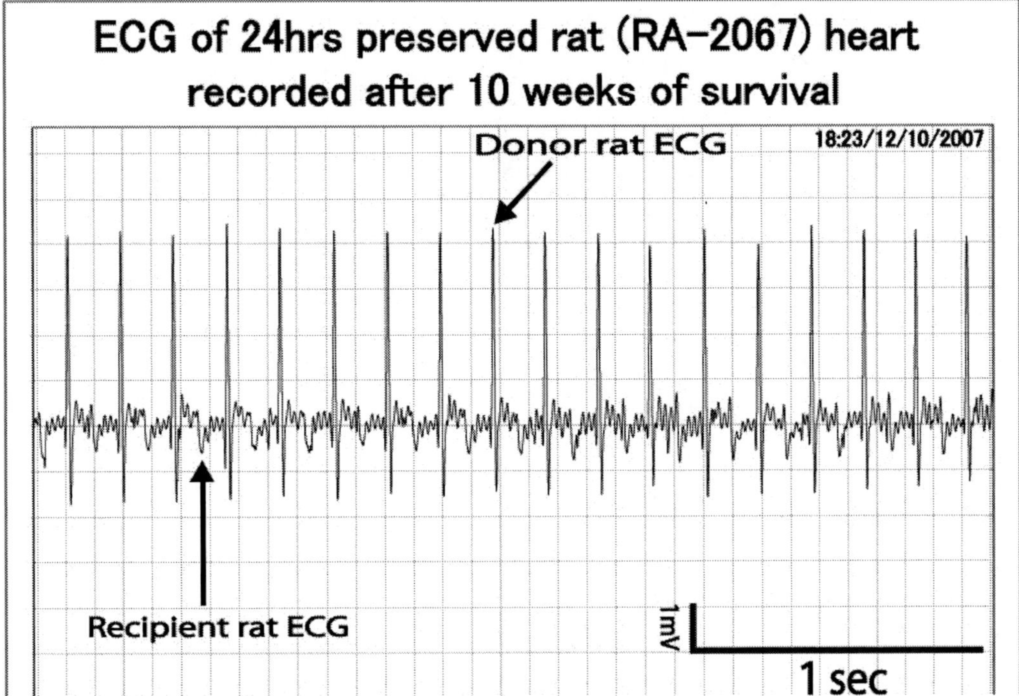

Figure 6. An isolated rat heart was put into a hyperbaric chamber which was pressurized to 2 ATA with PO_2=1600 hPa+PCO=400 hPa. After 24 hr of preservation, the preserved heart was heterotopically transplanted to the neck of a recipient rat and was resuscitated. Ten weeks later on October 12, 2007 this ECG was recorded for the donor rat heart and the recipient rat heart.

After the preservation for 24 hr at PCO=200-1000 hPa, isolated rat hearts were transplanted heterotopically and were resuscitated. Carbon monoxide turned out to be useful for the preservation of organs and various levels of CO's partial pressure were assessed. For example, isolated rat hearts were desiccated and preserved at 7 ATA with the CO partial pressure of 1000-5000 hPa, the rest being provided by O_2. After the preservation for 48 hr in such a hyperbaric environment, isolated rat hearts were heterotopically transplanted and resuscitated with significant reproducibility.

Group	Preservation method	Preservation time(hr)	The number of hearts	Resuscitation rate(%)	Ten weeks survival rate(%)
1A	O2(3000hPa)+CO(4000hPa)	48	5	5/5 (100)	4/5 (80)
1B	O2(2500hPa)+CO(4500hPa)	48	4	4/4 (100)	3/4 (75)
2	control	0	10	10/10 (100)	10/10 (100)

Figure 7. Isolated rat heart was preserved 48 hr in a hyperbaric chamber of 7 ATA pressurized by CO and O_2 with their respective level of partial pressure. The resuscitation rate immediately after transplantation and the survival rate 10 weeks after the operation are indicated here.

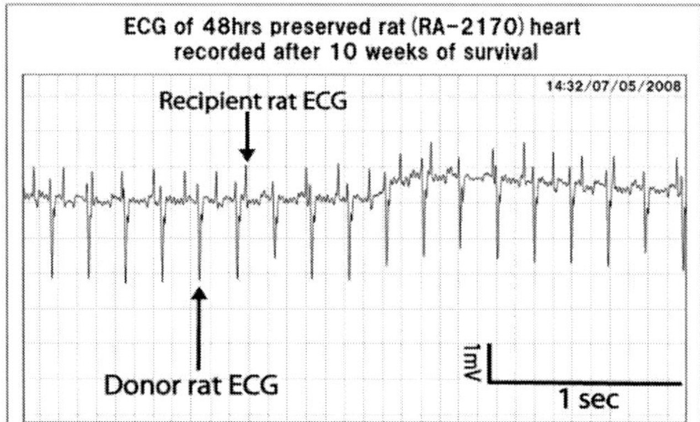

Figure 8. Isolated rat heart was put in a hyperbaric chamber and kept there for 48 hr at 7 ATA (PO_2=3000 hPa + PCO=4000 hPa) and then was transplanted heterotopically to the neck of a recipient rat and was resuscitated. This ECG was recorded on May 7, 2008 from the donor heart and recipient heart.

With this series of experiments, the heart harvested from a donor rat was hung in a hyperbaric chamber into which mixed gas of pressurized O_2 and CO was supplied until the total pressure reached 2-7 ATA. The chamber was stored at 4°C. Antibiotics and warfarin potassium were added to the KH solution, both of which were dissolved beforehand. A beaker filled with distilled water was placed inside the hyperbaric chamber to keep the humidity there over 90% (Figure 1). After the preservation for 48 hr, the isolated rat heart was heterotopically transplanted to a recipient rat and blood flow to the preserved heart was allowed to resume. When the resumption of cardiac activity of the donor heart was confirmed (Figure 5, 7) the incision at the neck of the recipient rat was sutured. The recipient rat was given drinking water containing antibiotics and post-operative observation was conducted in the animal room. Ten weeks later, the pulsation of the recipient rat heart and of the donor rat heart were both recorded by ECG (Figures 6, 8).

Heart pulsation was confirmed immediately after transplantation in all 5 of the hearts preserved 24 hr with mixed gas pressurized to PO_2=1600 hPa+PCO=400 hPa. At 10 weeks after transplantation, all 5 hearts continued to pulsate. These findings are therefore considered to be of good results. Heart pulsation was confirmed in all 5 hearts after 48 hr preservation with the gas mixture of PO_2=3000 hPa+PCO=4000 hPa, and transplantation. Ten weeks later the pulsation of all 5 hearts was verified, and this was considered to be of good result.

Section 3: Desiccation and Preservation with Carbon Monoxide, Carbon Dioxide, Helium and Oxygen

The good results obtained with 48 hr preservation with CO, indicated that a hyperbaric environment of 7 ATA was effective for the long-term cardiac preservation. Subsequently the CO_2 described in Section 1 and the hyperbaric environment of 7 ATA as described in Section 2 were combined in the hope of achieving longer preservation time. However, the experiment led to the formation of bubbles in the preserved heart. In other words, decompression sickness developed and heart preservation was badly affected. Carbon dioxide is extremely soluble in water. With water as cold as 4°C, CO_2 is 45 times more soluble than CO. It can be presumed that CO_2 being dissolved in water and forming minute bubbles damaged the nervous system, blocked fine capillaries and caused necrosis of the tissues. To take advantage of effectiveness of CO_2 for preservation while maintaining the environment of 7 ATA, He, an inert gas, was added to the gas mixture of O_2 and CO_2. Like CO_2, inert gas has anesthetic [6] and metabolic inhibitory effects [3] and is used for the storage of plants [29] and is known to structure water molecules in living organism [24]. The solubility of He is about one fourth that of CO so that it does not easily cause depression sickness. Isolated rat heart was preserved for 72 hr with the mixture of 3 different gasses of PCO_2=100 hPa+PO_2=900 hPa+PHe=6000 hPa, in a hyperbaric chamber. After the lapse of 72 hr, cardiac pulsation of transplanted heart was confirmed for the first time, though it ceased to pulsate at the end of 1 hr.

Adding CO to this gas mixture of CO_2, O_2 and He, organ preservation was thought to yield synergistic effects resulting from the combination of CO and CO_2 both of which contribute respectively to organ preservation. Therefore, the next series of experiments employed a gas mixture of 4 different gasses. The heart was taken out of a donor rat as described above and was filled with KH solution to be hung in a hyperbaric chamber. Each gas was pressurized to the respectively predetermined partial pressure level and the total pressure in the chamber was 7 ATA. The chamber was kept in a refrigerator of 4°C (Figure 1). Seventy-two hr later, at the end of heterotopic transplantation, blood started to flow again through the preserved heart and the resumption of cardiac activity was checked. After the heart started to pump, the incision at the neck was sutured. The recipient rat was given drinking water containing antibiotics and post-operative observation was conducted in the animal room. Ten weeks later, the pulsation of the recipient rat's heart and of the transplanted heart were both recorded by ECG (Figure 9).

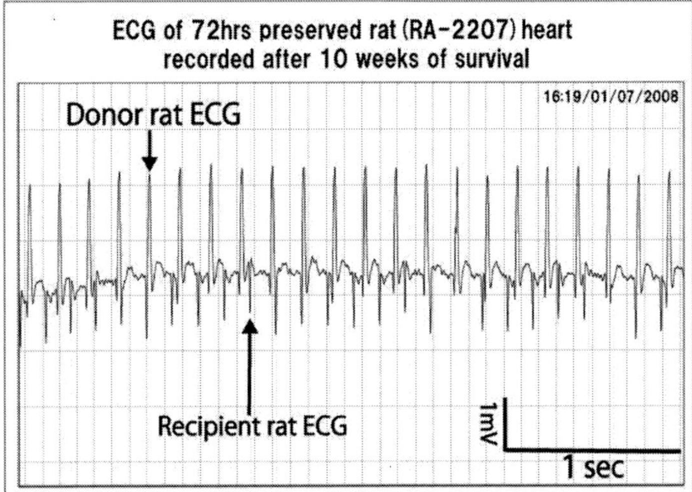

Figure 9. Isolated rat heart was placed in a hyperbaric chamber and was compressed with PCO=400 hPa+PCO_2=100 hPa+PO_2=900 hPa+PHe=5600 hPa. After 72 hr of preservation, the heart was transplanted heterotopically to the neck of the recipient rat and was resuscitated. Ten weeks after resuscitation, on July 1, 2008, this ECG was recorded from the donor rat heart and the recipient rat heart.

Various levels of partial pressure for each gas were tested and the best result was obtained with PCO=400 hPa+PCO_2=100 hPa+PO_2=900 hPa+PHe=5600 hPa. Of 10 preserved and transplanted hearts, both the ventricle and atrium of 8 hearts pulsated and 3 hearts successfully survived. In order to identify the best preservation method via desiccation, experimentation continues while changing the partial pressure of each component gas.

Discussion

Seki *et al.* (1998) successfully resuscitated tardigrades that had been dehydrated and exposed to the high hydrostatic pressure of 600 MPa. This success suggested that the same physiological mechanism of preservation and resuscitation could be applied to the preservation of mammalian organs. A tardigrade is a multicellular organism made up of about 40,000 cells including nerve cells. Among the cells within mammals, only cells as blood, sperm and ovum can be preserved long-term and resuscitated. As for tissues which are aggregates of cells and organs which are composed of a number of tissues, their transplantation must be performed within 24 hr at the maximum after their removal [10].

With regard to the preservation of mammalian organs, after preservation, the survival of tissues within the preserved organs must be verified. For the verification, several methods exist such as the histological anatomy method, transplantation method and electrophysiology method. Whichever method is chosen, however, the survival of the cells within each tissue must be verified. In the experiments described in this review, heterotopic transplantation of isolated rat heart to the neck of a recipient rat was chosen as a means to verify cardiac resuscitation. Though the heart thus transplanted does not contribute to the outflow of blood from the recipient's heart, it is easy to see if the transplanted heart is still alive. The merit of this verification is it can be done objectively by means of visual inspection or ECG.

Preservation by desiccation has been put into practical use for the storage of unicellular organisms and tissues of plants and animals. However, for the preservation and resuscitation of mammalian organs, no other research groups have attempted the technique of preservation by desiccation. When isolated rat hearts are exposed to the hyperbaric environment, they gradually lose water and dry up. It turned out that the resuscitation rate of hearts depends upon the kind of gas used for desiccation. Taking advantage of CO_2's anesthetic and metabolic inhibitory actions on organisms, cardiac dehydration was combined with hyperbaric environment of PCO_2=200 hPa and reproducible results were obtained [21][23].

This organ preservation and resuscitation technique has given birth to a new academic field of semibiology wherein vital functions can be stopped and resumed at will. These experiments were designed to restrain metabolic activity of tissues by means of decreasing in their water content and by the use of CO_2. One of the objectives of these experiments was place tissues in a state of artificial hibernation. Using hypothermic exposure to mixed gas and inactivation of the enzymes of the electron transport system, biological activity itself can be temporarily suspended to be resuscitated later. This phenomenon is called "suspended animation" and some researchers consider it applicable to organ preservation. The aim of the efforts is to allow passage between the world of life activities and the world of non-life activities.

In order to prolong the preservation time of isolated rat hearts, $CO_2$was replaced with CO. In addition, they were exposed for 24 hr to the pressurized mixed gas with a partial pressure of CO of 400 hPa. The isolated rat heart was heterotopically transplanted to the right neck of recipient rat. After resuscitation, the transplanted heart survived more than 3 months. Carbon monoxide gas is known to bind with Fe^{2+} in cytochrome oxidase which is an enzyme indispensable to living organisms for converting glucose into energy. Combined with this enzyme, CO gas suppresses the enzyme's activity [27]. According to the law of increasing entropy, decaying components decompose much earlier than they should be so as to compose again before clutter accumulates and all the molecules are replaced rapidly in dynamic equilibrium. This is known as the theory of dynamic equilibrium. The same mechanism seems to be working in establishing dynamic equilibrium between $O_2$and CO within the cells and tissues of isolated hearts preserved in the hyperbaric environment. It can be presumed that CO gas prevents necrosis by binding with Fe^{2+} in cytochrome oxidase and that this binding inhibits the activity of the enzyme, thus restraining intracellular metabolism and leading to the prevention of necrosis [7] [8] [18].

To further prolong preservation time, a variety of partial pressures were tried. For 48- hr-long preservation, the best resuscitation rate could be obtained in the hyperbaric environment of 7 ATA with PCO=4000 hPa. This result was obtained because of exposure to the mixed gas in the hyperbaric environment of 7 ATA, dynamic equilibrium seems to have been established between $O_2$and CO within their cells and tissues, which led to the increase in the ratio of cells bound with Fe^{2+} of cytochrome oxidase and at the same time led to the increase in the number of cells saved from necrosis as a result of the restrained intracellular metabolic activity. This experiment seems to suggest that hyperbaric environment of 7 ATA itself is highly effective for the long-term cardiac preservation. Consequently while maintaining a hyperbaric environment of 7 ATA, $CO_2$was incorporated. At the same occasion, to prevent decompression sickness to be caused by CO_2, the inert gas He was also introduced, thereby contributing to the creation of a hyperbaric environment. Like CO_2, inert gas has an anesthetic effect [6] and inhibitory effect on metabolism [3] and is used for the storage of plants [29]. Inert gas is known to somehow form a structured aggregate of water molecules contained in living organisms [24].

To the gas mixture of CO, $O_2$and He, $CO_2$was added to provide its preservative properties. Diverse levels of partial pressure for each constituent gas were tested and a good result with 72 hr preservation and resuscitation was obtained with a mixture of PCO=400 hPa+PCO_2=100 hPa+PO_2=900 hPa+PHe=5600 hPa. Ongoing experiments are presently investigating a diversity of partial pressures for each component gas. In the scientific world of physics and chemistry, CO is used daily, but in the field of biology CO is known as poisonous and has not been used. The present experiments proved that CO could be used profitably for the preservation and resuscitation of organs.

When automobiles break down, they can be repaired and driven again, because there are blueprints and specifications, spare parts and repair techniques. In the case of humans, like automobiles there are blueprints (anatomical charts), repair techniques (surgical operations), but spare parts (organs) are not easily available. They are provided by brain dead donors, but they can be preserved for only 24 hr at the maximum. The technique of organ preservation and resuscitation with the use of CO_2 gas extended this preservation time from 24 hr to 72 hr. With the development of the semibiology techniques, it may some day be possible to preserve human organs for more than 1 year, and therefore human life could in the future become semi-permanent in a manner similar to that of automobiles.

In the natural world, every year plants and animals go into a desiccation-induced dormant state and wake up the following spring. The mechanism of this natural phenomenon has been applied to the cells and tissues of plants and animals for practical use. With these experiments, it may be possible to demonstrate that the mechanism of this natural phenomenon can be applied to the preservation and resuscitation of mammalian organs. In order to apply this technique of organ preservation and resuscitation to human organs, organs of large-sized mammals such as pigs and monkeys are being tested. Thanks to the techniques of organ preservation developed here, in the near future a storehouse of organs may thus be achieved, which will supply organs anytime, anywhere and to anyone.

Conclusion

Using various combinations of gas mixtures, significantly reproducible techniques of organ preservation were established which allowed the preservation of isolated rat hearts for up to a maximum of 120 hr. Therefore, a new scientific field of semibiology has thus been established.

References

[1] Behringer W, Safar P, Wu X, Kentner R, Radovsky A, Kochanek PM, Dixon CE, Tisherman SA. Survival without brain damage after clinical death of 60-120 min in dogs using suspended animation by profound hypothermia. *Crit Care Med* 2003;31:1523-1531.

[2] Blackstone E, Morrison M, Roth M.B. H_2S induces a suspended animation-like state in mice, *Science* 2005;308:518.

[3] Bruemmer JH, Brunetti BB, Schreiner HR. Effects of helium group gases and nitrous oxide on Hela cells. *J Cell Physiol.* 1967 Jun;69(3):385-92.

[4] Cooper JD, Patterson GA, Trulock EP, et al. Results of single and bilateral lung transplantation in 131 consencutive recipients. *J Thorac Cardiovasc Surg* 1994;107: 460-471.

[5] Dean JB, Mulkey DK, Garcia AJ, Putnam RW, Henderson RA. Neuronal sensitivity to hyperoxia, hypercapnia, and inert gases at hyperbaric pressures. *J Appl Physiol.* 2003 Sep;95(3):883-909.

[6] Goto T, Nakata Y, Morita S. How does xenon perspective from electrophysiological studies. *International Anesthesiology Clinics*. 2001 Spring;39(2):85-94.

[7] Gorman D, Drewry A, Huang YL, Sames C. The clinical toxicology of carbon monoxide. *Toxicology* 2003 May 1;187(1):25-38.

[8] Guggenheim KY. Rudolf Schoenheimer and the concept of the dynamic state of body constituents. *J Nutr.* 1991 Nov;121(11):1701-1704.

[9] Heffner JE, Pepine JE. Pulmonary strategies of antioxidant defence. *Am. Rev. Pespir.* 1989 Dis; 140:531-554.

[10] Kalayoglu M, Sollinger HW, Stratta RJ, et al. Extended preservation of the liver for clinical transplantation. *Lancet* 1988;2:617-619.

[11] Kuroda Y, Kawamura T, Tanioka T, et al. Heart preservation using a cavitary two-layer (University of Winsconsin Solution/Perfluorochemical) cold storage method. *Transplantation* 1995;59:699-701.

[12] Makowka I, Zerbe TR, Champman F, et al. Prolonged rat cardiac preservation with UW lactobionate solution. *Transplant Proc.* 1989;21:1350-1352.

[13] Nunn JF. The effects of changes in the carbon dioxide tension. *In Nunn's Applied respiratory physiology*. 3rd ed. London: Butterworth-Heinemann Ltd. 1987; 460-470.

[14] Ohta T, Yasuda A, Mitsuda H. Effect of carbon dioxide on conservation of physiological activities of animal tissue, *Proc, Japan Acad* 1987; 63B:340-343.

[15] Ohta T, Yasuda A, Mitsuda H: Effect of carbon dioxide on conservation of vasoactivity of canine mesenteric arteries,*Transplantation* 1989 Apr;47(4):740-742.

[16] Oz MC, Pinsky DJ, Koga S, et al. Novel preservation solution permits 24-hours presevation in rat and baboon cardiac transplant model. *Circulation* 1993;88:291-297.

[17] Pegg DE. Organ presevation. *Surg. Clin. North Am,* 1986;66:617-632.

[18] Prockop LD, Chichkova RI. Carbon monoxide intoxication, an updated review. *J Neurol Sci.* 2007 Nov 15; 262(1-2): 122-130.

[19] Schaefer KE, Messier AA, Morgan C, Baker GT. Effect of chronic hypercapnia on body temperature regulation. *J Appl Physiol* 1975;38:900-906.

[20] 20 . Seki K. Preservation and resuscitation of rat isolated heart for 10-26 days in perfluorocarbon and silicagel.*36th annual meeting of the society for Cryobiology* 1999 July; No.101.

[21] Seki K, Yoshida Y Mizushima Y et al. Preservation of rat heart under hypercapnic and hyperbaric condition (in Japanese). *The japaneses journal of hyperbaric medicine* 2005;Vol.40 No.3,26(P183).

[22] Seki K, Toyoshima M. Preserving tardigarades under pressure. *Nature* 1998 29 October;Vol.395:6705:853-854.

[23] Yoshida Y, Hatayama N, Sekino H, Seki K. Heterotopic transplant of an isolated rat heart preserved for 72 h in perfluorocarbon with CO_2. *Cell Transplant.* 2008;17(1-2):83-89.

[24] Tanaka H, Nakanishi K. Hydrophobic hydration of inert gases. Thermodynamic properties, inherent structures, and normal-mode analysis. *J. Chem Phys* 1991;95(5): 3717-3727.

[25] Thomas M. D. Investigating Carbon Monoxide Poisonings. *Carbon Monoxide Poisoning* CRC Press, Taylor & Francis Group, NY 2007; 287-303.

[26] Tsutsumi H, Oshima K, Mohara J, et al. Successful orthotopic cardiac transplantation following 24-hr preservation using a hypothermic perfusion apparatus in canine hearts. *J Heart Lung Transplant* 2000;19:60-61.

[27] White SR. Treatment of carbon monoxide poisoning. *Carbon Monoxide Poisoning* CRC Press, Taylor & Francis Group, NY 2007;341-374.

[28] Yeh T, Hanan SA, Johson DE, et al. Superior myocardial preservation with modified UW solution after prolonged ischemia in the rat heart. *Ann. Thorac. Surg.* 1990;49:932-939.

[29] Yoshino T, Sotome I, Ohtani T, Isobe S, Oshita S, Maekawa T. Observations of xenon gas-treated barley cells in solution by atomic force microscopy. *Journal of Electron Microscopy* 2000;(49)3:483-486.

In: Advances in Medicine and Biology. Volume 20
Editor: Leon V. Berhardt

ISBN 978-1-61209-135-8

Chapter 14

SYSTEM SPECIFIC CHAPERONES FOR MEMBRANE REDOX ENZYME MATURATION IN BACTERIA

Raymond J. Turner*, Tara M. L. Winstone, Vy A. Tran and Catherine S. Chan

Department of Biological Sciences, University of Calgary, Calgary, Alberta T2N 1N4, Canada

ABSTRACT

A group of bacterial system specific chaperones are involved with the maturation pathway of redox enzymes that utilize the twin-arginine protein translocation (Tat) system. These chaperones are referred collectively as REMPs (*R*edox *E*nzyme *M*aturation *P*rotein). They are proteins involved in the assembly of a complex redox enzyme which itself does not constitute part of the final holoenzyme. These proteins have been implicated in coordinating the folding, cofactor insertion, subunit assembly, protease protection and targeting of these complex enzymes to their sites of physiological function. The substrates of REMPs include respiratory enzymes such as N- and S-oxide oxidoreductases, nitrate reductases, and formate dehydrogenases, which contain at least one of a range of redox-active cofactors including molybdopterin (MoPt), iron sulfur [Fe-S] clusters, and b- and c-type haems. REMPS from *Escherichia coli* include TorD, DmsD, NarJ/W, NapD, FdhD/E, HyaE, HybE and the homologue YcdY. The biochemical, structural and functional information on these REMPs are reviewed in detail here.

* Corresponding author: turnerr@ucalgary.ca

INTRODUCTION

Coordinating the folding, assembly, and targeting of complex enzymes to their sites of physiological function is an important feature of all biological systems. In bacteria, the generation of energy by respiratory electron transport chains involves the cytoplasmic membrane. The redox-active protein components of electron transport chains are often embedded in this membrane to interact with the quinone/menaquinone pool and significant numbers are located on the extra-cytoplasmic side. Extra-cytoplasmic respiratory enzymes such as N- and S-oxide oxidoreductases, periplasmic nitrate reductases, and formate dehydrogenases contain at least one of a range of redox-active cofactors including molybdopterin, [Fe-S] clusters, and b- and c-type haems. The assembly of such proteins is recognized as one of the most complex processes in the field of bioinorganic chemistry. With the notable exception of the c-type cytochromes, such cofactors are assembled into the protein in the cytoplasm prior to enzyme export from the cell.

The acquirement of molybdenum and the biosynthetic pathway for the molybdopterin cofactor (MoPt) or the more complex bis-(molybdopterin guanosine dinucleotide) (MGD) cofactor is a complex multi-step process requiring numerous enzymes (Schwarz, 2005). In bacteria, high-affinity molybdate transporters acquire the molybdate anion form of molybdenum. Molybdate is catalytically inactive within the cell until complexed with a tricyclic pterin cofactor (sometimes referred to as a molybdopterin, or molybdenum cofactor or Moco). Molybdopterin biosynthesis involves numerous enzymatic steps that can be abbreviated into 3 general steps: (1) Synthesis of precursor from guanosine-5'-triphosphate; (2) Incorporation of 2 sulfur atoms to form the tricyclic pyranopterin; and (3) Insertion of the molybdenum atom (each Mo atom can have one or two pyranopterin molecules associated) (Schwarz 2005). Many bacterial molybdoenzymes utilize a fourth step in which a guanosine dinucleotide is added to the molybdenum-complexed molybdopterin (Schwarz, 2005). As the MoPt and MGD are very labile they must be protected in the protein at all times during the biosynthesis. This raises the question of their insertion into their final host proteins polypeptide chain, with consideration that this step occurs concurrent with protein folding during the conversion from apo to cofactor holo form.

THE TWIN ARGININE TRANSLOCASE (TAT) SYSTEM

Studies of the model prokaryote *E. coli* have established that many of these cofactor-containing exported proteins are synthesized with amino-terminal leader signal peptides containing the distinctive S*RR*xFLK "twin-arginine" amino acid sequence motif (Berks, 1996). The leader contains 3 regions similar to Sec dependent leaders; the n, h, and c region. The twin-arginine motif is typically found between the n and h regions (Berks, 1996) (Figure 1). The "n-region" is a small charged region of the signal peptide that precedes the twin-arginine motif and is unique to each individual protein. The h-region is a hydrophobic stretch of 7-30 amino acids. The c-region contains a cleavage recognition site which is thought to be processed by leader peptidase I. Preproteins bearing these twin-arginine-leaders are secreted post-translationally and most interestingly post-folding, across the cytoplasmic membrane by a unique protein translocase initially termed the membrane targeting and translocation (Mtt) system (Weiner et al., 1998) or the ΔpH translocase (Settles et al., 1997). The system is now most frequently referred to as the twin-arginine translocase (Tat) (Sargent et al., 1998; 2002; 2007a; Berks et al., 2000a; 2000b; 2003; Lee et al., 2006).

A most remarkable feature of the Tat system is its substrates from both bacteria and plant chloroplasts are required to be fully folded before successful translocation can occur (e.g. DeLisa et al., 2003). Additionally, the system appears to translocate protein complexes where only one protein in the complex may contain a twin-arginine-leader (see reviews: Berks et al., 2003; Sargent 2007a; b). Biochemical studies have shown that the integral membrane proteins TatA, TatB, and TatC form the core components of the *E. coli* Tat system (reviewed by Berks et al., 2003). The TatBC unit is believed to form the signal recognition module, while TatA forms a very large oligomeric ring-structure presumed to be the protein-conducting channel itself (Berks et al., 2003).

	N- n-region SRRxFLK h-region c-region	
DmsA	--MKTKIPDAVLAAEVSRRGLVKT----TAIG-G-LAMASSALTLPFSRI----------	42
FdnG	-----------MD--VSRRQFFK-----ICAG-G--MAGTTVAALGFAPKQ---------	30
FdoG	-----------MQ--VSRRQFFK-----ICAG-G--MAGTTAAALGFAPSV---------	30
HyaA	MNNEETFYQAMRRQGVTRRSFLK-----YCSL----AATSLGLGAGMAPKI---------	42
HybO	MTGDNTL---IHSHGINRRDFMK-----LCAA--------LAATMGLSSK----------	34
NapA	-----------MK--LSRRSFMK------ANA---VAAAAAAAGLSVPGV----------	28
TorA	MNNND-----LFQ--ASRRRFLA------QLG-GLTVAGMLGPSLLTPRR---------	36
NarG	-----------MSKFLDRFRYFKQKGETFADGHGQLL-NTNRDWEDGYRQRWQHDKIVRSTH	50
NarZ	-----------MSKLLDRFRYFKQKGETFADGHGQVM-HSNRDWEDSYRQRWQFDKIVRSTH	50

Figure 1. Twin-arginine leader sequences of Tat system substrates. Typical structure of leader peptides consists of a basic n-region (red), followed by the twin-arginine motif (purple), a variable length hydrophobic h-region (grey), and a short polar c-region (blue). The leader peptidase I cleavage site follows the c-region. The peptides of NarG and NarZ do not contain these structures as their proteins are not translocated and are attached from the cytoplasmic side of the membrane. Sequence alignment of all the peptides demonstrate the architecture of these structures and highlight the vestige motifs of NarG and NarZ.

The majority of work to date on the Tat system has been in the model bacterial system *Escherichia coli, which* has ~27 Tat targeted proteins (Tullman-Erick et al., 2007). The majority are cofactor-containing enzymes, most of which acquire their prosthetic groups in the cytoplasm prior to export. Additionally, assembly with signal-less partner subunits must occur prior to translocation and these partners must also have any prosthetic groups preloaded prior to translocation. Once the discovery of this Tat pathway was reported, the system became a focus of many excellent research groups (B. Berks, T. Palmer, C. Robinson, J. Weiner, L.-F. Wu, K. Cline, M. Müller and many newly inspired groups). Their focus has been on proteins that use the translocase, and the translocase itself, with some interest in comparing the twin – arginine leader to that of Sec-dependent leaders.

DISCOVERY OF REMPs

A key question was asked early on for this system; is there a twin-arginine leader binding protein for the Tat system? This was asked as the general secretory system utilizes at least two chaperones such as SecB and the signal recognition particle (SRP). The question was answered with the discovery of DmsD (Oresnik et al., 2001). This experiment utilized the twin-arginine signal peptide from the Dimethyl sulfoxide reductase catalytic subunit (DmsA) fused to the N-terminus of glutathione-S-transferase and pre-bound to a Glutathione-sepharose resin over which anaerobic *E. coli* lysate was poured. Two proteins were found to bind to this peptide, the chaperone DnaK and a product from an uncharacterized reading frame YnfI, now labled as DmsD. Subsequent studies described here identified other system specific chaperones. As the result of a bioinformatics study of the similar oxidoreductase systems, the collective term REMP (*R*edox *E*nzyme *M*aturation Protein) was proposed to refer to any protein involved in the assembly of a complex redox enzyme which itself does not constitute part of the final holoenzyme (Turner et al., 2004). Different REMPs in *E. coli* are listed in Table 1 but are also found in other organisms where similar redox enzymes are found. The need for these system specific chaperones in microbial physiology has become increasingly accepted in the past few years. Many potential roles for these systems specific REMPs have been suggested and include:

Foldase chaperone – promoting assembly of the enzyme.
Unfoldases – correcting mistakes in folding.
Sec-avoidance chaperone – preventing incorrect targeting during assembly.
Cofactor chaperone – maintaining the apoenzyme in a competent state for prosthetic group loading.
Cofactor binding protein – binds the cofactor to transfer to the apoenzyme.
Cofactor insertase – directly involved in prosthetic group insertion.
Targeting protein – involved in specific targeting of their substrate to locations in the cell (e.g. prosthetic group biosythesis, membrane, Tat translocase). This has also been referred to as an 'escort protein'.
Tat system Proof reading – suppressing transport until the assembly process is complete.
Control "policing" protein – verification of 'hitchhiker' proteins that are associated prior to translocation. Hitchhiker proteins are co-translocated with twin-arginine-leader containing proteins, yet do not have such a signal sequence themselves. May or may not be a step in proof reading.
Protection from proteases – preventing degradation during assembly.

TMAO Reductase (TorD)

In *E. coli*, trimethylamine N-oxide (TMAO) reduction is important during anaerobic respiration. TMAO reduction is dependent on the gene products of the *torCAD* operon (Mejean *et al.*, 1994). The first gene product, TorC is a membrane anchored pentahemic c-type cytochrome whose proposed role is to feed electrons to the catalytic subunit of the reductase, TorA (Mejean *et al.*, 1994, Gon *et al.*, 2001). TorA, encoded by the second gene of the operon, is a periplasmic molybdoenzyme which bares the characteristic twin-arginine motif and is thus a substrate of the Tat system (Voorduow, 2000). Characteristic to other Tat substrates in *E. coli*, TorA obtains its molybodpterin cofactor from the cytoplasm prior to transport (Voorduow, 2000). The third gene of the *tor* operon encodes for a 22kDa protein,

TorD, the now established REMP for TorA. Homologues to the *tor* operon have also been identified in *Shewanella massilia* (*torECAD*) as well as *Rhodobacter capsulatus* (*dorCDA*) and *Rhodobacter sphaeroides* (*dmsCBA*). DorD and DmsB are the TorD homologues in the *Rhodobacter* species, and displays 26-27% sequence identity with *E. coli* TorD while *S. massilia* TorD shares 34% identity to *E. coli* TorD (Pommier *et al.*, 1998). Although TMAO reduction by TorC and TorA is well studied, the specific function of TorD has not been completely elucidated. However, of the chaperones involved in redox enzyme maturation, the most information is available on TorD. Multiple roles for TorD have been proposed, all of which are centered on its association with TorA to aid in the successful maturation of a fully functional protein.

Initially, it was thought that TorD was a membrane-associated protein, likely acting as another player in the electron transport pathway for TMAO reduction. Recognizing two hydrophobic sections in the amino and carboxy ends of the protein, it was proposed that these regions could anchor the protein to the membrane (Mejean *et al.*, 1994). However it was later shown that TorD is found primarily in the cytoplasm and interacts with an unfolded form of TorA reductase (Pommier *et al.*, 1998). TorD was then coined as a specific chaperone protein of TorA. There is some evidence that the fraction associated with the membrane may be the result of interactions with the Tat translocase itself (Papish et al., 2003).

Table 1. Tat-dependent redox enzyme systems in *E. coli*

Enzyme	Twin arginine leadercontaining subunit	Predicted REMP	Catalyticcofactor
Biotin sulfoxide reductase 1	BisC	YcdY[a]	MoPt
Biotin/TMAO reductase TMAO reductase	BisZ/TorZ TorA	YcdY/TorD[a] TorD	MoPt MoPt
DMSO reductase	DmsA	DmsD	MoPt
Putative DMSO reductase	YnfE	DmsD[a]	MoPt[c]
Putative DMSO reductase	YnfF	DmsD[a]	MoPt[c]
DMSO/TMAO reductase Formate dehydrogenase Formate dehydrogenase Hydrogenase 1 Hydrogenase 2 Nitrate reductase (periplasmic) Nitrate reductase (cytoplasmic) Nitrate reductase (cytoplasmic) Hypothetical protein Hypothetical protein Hypothetical Peroxidase (periplasmic)	YedY FdnG FdoG HyaA HybO NapA NarG NarZ YfhG YcdO YcdB	DmsD/TorD FdhD/FdhE FdhD/FdhE HyaE HybE NapD NarJ NarW[a] ? b ? b ? b	MoPt[c] MoPt MoPt Ni Ni MoPt MoPt MoPt None predicted Heme ? Heme cofactor

[a] Predicted REMP for enzyme in question based on predicted function/role of the enzyme and its relatedness to other enzymes with an appropriate REMP. However, no experimental evidence to support this.

[b] No predicted REMP for enzyme in question due to limited knowledge of its function.

[c] Predicted catalytic cofactor of molybdopterin (MoPt) based on the presence of a signature domain in its sequence.

ENZYME SYSTEMS AND THEIR COGNATE REMP CHAPERONES

Below each enzyme system that requires a REMP system specific chaperone for maturation will be briefly introduced followed by a discussion of their system specific chaperone.

Proposed TorD functions

Over the past decade, many studies have suggested that TorD is involved in the maturation of TorA. This has been attributed to its protective role and its proposed ability to induce apoTorA into a cofactor competent state in a variety of conditions (Pommier *et al.*, 1998; Genest *et al.*, 2005; Genest *et al.*, 2006a; 2006b; Ilbert *et al.*, 2003; Shaw *et al.*, 1999). TorD was first implicated in a protective role for TorA in an *E. coli* strain deficient in molybdenum cofactor synthesis (Pommier *et al.*, 1998). In the absence of TorD, apoTorA levels were significantly lower than that of wild type levels, yet when TorD was overexpressed, apoTorA levels were restored. This was also observed during anaerobic growth in the presence of nitrate and TMAO creating a condition where the cofactor was limiting (Genest *et al.*, 2006). In these conditions, a strain devoid of TorD showed minimal levels of apoTorA. These data suggest that in the presence of TorD, apoTorA is either protected or stabilized against degradation. And, upon the availability of the MoPt cofactor, apoTorA protected by TorD could undergo maturation, suggesting that TorD may also be able to induce apoTorA into a cofactor accepting conformation.

TorD also appears to be capable of protecting apoTorA from temperature induced misfolding or degradation (Genest *et al.*, 2005). At high temperatures, the absence of TorD was accompanied by markedly decreased levels of TorA and a substantially decreased TMAO reductase activity. However, in the presence of TorD, the maturation of TorA was unabated.

Similarly, in *R. capsulatus*, the absence of DorD (TorD homologue) caused a complete loss of dimethylsulfoxide (DMSO) activity, likely due to the degradation of DorA in the absence of its chaperone (Shaw *et al.*, 1999). The group also found data implicating DorD involvement in the insertion of the MoPt cofactor into apoDorA. This is complemented by data suggesting that TorD could modify the conformation of apoTorA in order to aid in cofactor insertion (Ilbert *et al.*, 2003). When TorD was preincubated with apoTorA, an increase in activation of apoTorA upon the availability of the MoPt cofactor is observed, suggesting that TorD acts on the apoprotein prior to cofactor insertion in order to induce apoTorA into a cofactor receptive state.

In a 2006 investigation (Genest *et al.*, 2006), TorD was shown to specifically protect the signal peptide from degradation. TorD co-purified with the apoTorA form that contained a preserved signal peptide. On the other hand, no co-purification was observed with a truncated form of apoTorA, missing regions of the signal peptide suggesting that TorD has a high affinity to the signal peptide. In the absence of TorD, only a truncated form of TorA was seen corresponding to a loss of a 36 residue N-terminal region of the apoprotein. Even a cofactor loaded TorA had a truncated signal peptide in the absence of TorD. It is thus possible that the binding of TorD to the signal peptide protects a certain region that is a target for proteolysis.

The degradation of TorA could likely begins at the N-terminal signal peptide and the data to date suggest that a role of TorD may be to bind and protect the twin-arginine-peptide of apoTorA. However, it is not clear if this is the primary function or simply a consequence of its primary role in maintaining the nascent polypeptide chain of TorD in a Moco competent state. It cannot be ruled out that other general chaperones are involved in TorA maturation, but it is convincing that TorD plays a more prominent role than any other protein alone.

TorD Interactions with TorA

TorD is implicated in signal peptide binding (Genest *et al.*, 2006) and it is likely that TorD also binds the core region of apoTorA in order to aid in co-factor insertion. It is also proposed that the binding of a REMP to its substrate is a "proofreading event" to prevent the premature export of the protein before cofactor loading and/or dimerization in cases of multi-subunit enzyme complexes with only one binding partner containing the signal peptide. Here, the specifics of these binding events will be explored.

In a 2004 investigation (Jack *et al.*, 2004) it was demonstrated that REMPs recognize the twin-arginine signal peptide of their specific molybdoenzyme substrate. Using bacterial two-hybrid screening, TorD was shown to specifically bind the cleavable signal peptide baring the twin arginine motif. Interactions were only found when the entire signal peptide was used. Furthermore, there was only a 1:1 ratio of TorD to the signal peptide of TorA, showing no evidence of TorD binding the signal peptide as a dimer or trimer. Despite the lack of interaction seen via bacterial two hybrid with a truncated form of apoTorA, a fusion protein experiment demonstrated that TorD may also bind regions in the core of TorA (Jack *et al.*, 2004) in order to produce a functional mature protein. When the signal peptide of TorA was replaced with that of another Tat substrate, TorD was still required for TorA maturation, although it could not bind the foreign signal peptide.

Specific regions where TorD binds the ssTorA was elucidated using isothermal calorimetry and synthetic regions of the TorA signal peptide (Hatzixanthis *et al.*, 2005). The hydrophobic h-region as well as a positively charged n- and c-region is needed for binding. Interestingly, a mutation of the RR residues to KK in the n-region did not change the binding affinity of TorD to the peptide, thus suggesting that the twin-arginine motif is not the primary source for peptide recognition. Thus, the n-region containing the twin-arginine motif is not sufficient for binding but is necessarily required for efficient binding in combination with the core region of the peptide (Jack *et al.*, 2004; Hatzixanthis *et al.*, 2005). It is possible that TorD recognizes other residues in the n-region in order to distinguish from other Tat leader peptides.

Individual residues within a leucine rich portion of TorA signal peptide h-region was identified to be involved in TorD binding both *in vivo* and *in vitro* (Buchanan *et al.*, 2008). A TorD variant was also identified that displayed increased affinity for both wild type and variant TorA signal peptides. Residues within the leucine rich region were systematically mutated to glutamine residues, and subjected to bacterial two-hybrid screening to evaluate binding *in vivo*. Binding of TorD to L31Q mutant was undetectable, along with other leucine and glycine residues towards the c-terminus of the h-region. However, the L31Q mutation did not appear to have a physiological effect on TMAO reductase activity or periplasmic targeting of TorA, suggesting that binding to the core of apoTorA was uninhibited. Through the two-hybrid screen of a library of random mutants of TorD, the Q7L TorD mutant was identified with an increased affinity for wild type and the L31Q variant of the signal peptide. This mutant maintained its proofreading activity and could also complement a strain devoid of its endogenous *torD* allowing for the restoration of TMAO reductase activity. However, despite its increased binding, it was not co-transported by the Tat system.

Through random mutagenesis, two mutants with single amino acid mutations were identified in the fifth helix of TorD that were important for TorA maturation (Genest *et al.*, 2008). The residues identified were in less conserved regions of the REMP family, yet the protein still maintained the α-helical structure. These mutants also maintained their ability to bind the signal peptide with dissociation constants that were comparable to wild type TorD. On the other hand, when the interaction of the mutant TorD with apoTorA devoid of its signal sequence was probed via bacterial two-hybrid, no interaction between the mutant TorD proteins was seen with apoTorA. Additionally, TMAO reductase activity was significantly reduced in strains expressing the mutant TorD forms, suggesting that despite the ability to bind the signal peptide of TorA, cofactor insertion and thus TorA maturation was hindered. This supports the idea that twin-arginine-leader interactions with TorD are not sufficient for TorA maturation and that binding of TorD to the core of TorA is required and it is this role that takes precedence.

Other TorD substrates and binding partners

The binding event of TorD to the TorA signal peptide appears to be a primary event, which occurs with a KD in the micromolar range. However, the events that trigger signal peptide release is not understood. It can be speculated that, like other general chaperones such as DnaK/DnaJ, GroEL/GroES as well as the Sec transport system, the binding and subsequent hydrolysis of nucleotides provides the triggering for the binding and release of protein substrates. Thus, the nucleotide binding ability of TorD was investigated (Hatzixanthis *et al.*, 2005). Weak binding was seen with GTP using a tryptophan fluorescence based assay. Minimal changes to the dissociation constant were observed with GDP, GMP and cGMP, which suggest that the guanosine group is what governs the binding. When TorA signal peptide was bound to TorD, the binding to GTP was stronger. Also, the Q7 equivalent in *S. typhimurium* DmsD was predicted to be involved in GTP binding which was shown to be associated with signal peptide binding as discussed above. With this data in mind, the binding of signal peptide could cause a conformational change in TorD so to induce and enhance the binding of GTP. Although weak binding was observed, hydrolysis of GTP was not detected with the assays used. This does not rule out the possibility that hydrolysis could occur, but other binding partners may be necessary if TorD lacks an intrinsic GTPase.

Since it has been shown that TorD interactions with the core of TorA is important for TorA maturation, it was important to investigate if TorD had the capabilities to bind components of the MoPt cofactor. Most recently, experimental data showed a 1:1 ratio for TorD binding with the cofactor intermediate Mo-molybdopterin (MoCo) as well as the final cofactor form, molybdopterin-guanine dinucleotide (MGD) (Genest *et al.*, 2008). These results coincide with the binding studies done with GTP (Hatzixanthis *et al.*, 2005) as it was proposed that the guanosine moiety was directly involved in binding. It is also likely that

TorD could bind both GTP and MoCo to help facilitate the synthesis of MGD. Finally it was found that MobA, the enzyme responsible for synthesizing MGD from MoCo and GTP, associates with TorD. This interaction was shown by cross-linking as well as surface plasmon resonance (Genest *et al.*, 2008).

TorD structure

S. massilia TorD was seen to self associate into various oligomeric forms that retained the ability to bind TorA (Tranier *et al.*, 2002). The purified monomeric and dimeric species were stable and did not interconvert between the two forms unless they were exposed to pH 3.0 for a period of time. Binding capabilities of the monomeric and dimeric forms of TorD to TorA was explored via surface plasmon resonance. Both forms showed binding, however, the dimeric form bound TorA more efficiently than the monomeric form suggesting a physiological relevance for the dimeric form of this REMP. Stable trimer forms of TorD were also thought to occur. Furthermore, circular dichroism revealed that the secondary structure of the monomeric or dimeric species were very similar.

The solved crystal structure of *S. massilia* TorD (pdb:1n1c) has an N-terminal domain and a C-terminal domain that is attached via a hinge region, and when interacting as a dimer, domain swapping occurs such that the N domain from one monomer interacts with the C-domain of the other (Tranier *et al.*, 2003) (Figure 2A). A possibility exists for each domain to serve different roles when binding to TorA (Jack *et al.*, 2004). When the N and C domains were expressed separately in a *torD* mutant strain, the phenotype could not be rescued. However, when both domains were co-expressed, TMAO reductase activity was restored to near wild type levels. Thus, each individual domain is functional and allows for proper assembly and maturation of TorA when co-expressed, despite not being covalently linked. Furthermore, two conserved residues were identified, one each in the C and N domain that are important for MoCo insertion into TorA. On the other hand, the separated TorD domains were incapable of Tat proofreading providing evidence that the hinge region is important for this potential function. Indeed, two residues identified in the hinge region were identified to be important for signal peptide recognition, while the two N and C domain residues were not required. It is then possible that the monomeric and dimeric forms are providing different roles when bound to TorA.

Concluding Remarks for TorD

Taken altogether, TorD binds apoTorA in two regions, possibly in a cooperative manner to protect the molydoenzyme from degradation as well as to help in MoCo insertion and the successful maturation of TorA. Cooperativity may be why binding is not observed when the twin-arginine-leader signal peptide is lacking or mutated as this first binding event provides the catalyst for binding to the core of TorA. This is also consistent with the prediction that TorD first binds the signal peptide with high affinity to prevent the initiation of proteolysis of apo-TorA (Genest *et al.*, 2006). With the most current information on other binding partners of TorD, it is possible that TorD acts as a platform to connect all the necessary components for the final step of MGD synthesis in order to produce a fully functional TorA that is ready for transport by the Tat system.

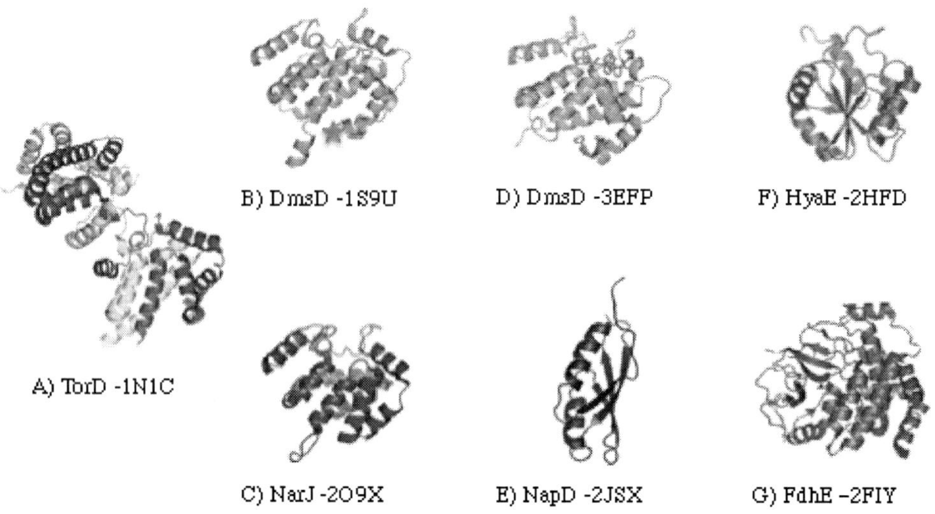

Figure 2. Structures of REMPs. A) *Shewanella massilia* X-ray structure (2.40Å). B) *Salmonella typhimurium* LT2 X-ray structure (1.38Å). A modeled site of GTP binding site is indicated as the star. C) *Archaeoglobus fulgidus* NarJ (3.40Å). D) E. *coli* DmsD (2.00Å). RR-leader binding site residues highlighted in red. E) *E. coli* NapD NMR structure. F) *E. coli* HyaE NMR structure. G) *Pseudomonas aeruginosa* FdhE X ray structure (2.10Å).

DMSO Reductase (DmsD)

E. coli is able to grow anaerobically on dimethyl sulfoxide (DMSO) due to the presence of the membrane-localized DMSO reductase enzyme (Bilous and Weiner 1985) encoded for by the *dmsABC* operon (Bilous et al. 1988). DMSO reductase (DmsABC) is a membrane-associated [Fe-S]-molybdoenzyme. During anaerobic respiration, electron flow proceeds from the electron donor menaquinol pool in the membrane bilayer through to DMSO, the terminal electron acceptor. The integral membrane protein, DmsC, is believed to be composed of 8 transmembrane helices and contains a high-affinity binding site for menaquinol (Geijer and Weiner 2004; Zhao and Weiner 1998). DmsC serves to anchor the soluble DmsAB complex (Weiner et al. 1993). The DmsB subunit contains 4 [4Fe-4S] clusters; one of which interacts closely with the menaquinol oxidation site of DmsC (Rothery and Weiner 1996; Zhao and Weiner 1998). Electrons are passed to another [4Fe-4S] center in DmsB before moving to the molybdenum cofactor of the DmsA catalytic subunit (Rothery et al. 1999; Rothery and Weiner 1996). DMSO is reduced to DMS (dimethyl sulfide) at the active site of DmsA which contains a bis-(pyranopterin guanosine dinucleotide) molybdenum cofactor (MGD). A predominantly negatively charged funnel-like active site entrance lined with aromatic residues may give DMSO reductase its more broad substrate specificity, accepting both S- and N-oxides (Kisker et al. 1997; McAlpine et al., 1998; Schindelin et al., 1996; Schneider et al. 1996), whereas the TorA active site is more specific for the N-oxide TMAO.

The active site of the DmsA enzyme contains a bis-(molybdopterin guanosine dinucleotide) cofactor (Schwarz 2005). The depth to which the MGD cofactor is buried within the DmsA protein, would suggest MGD biosynthesis and insertion into DmsA involves a co-folding mechanism (Schindelin et al. 1996; Schneider et al. 1996). In addition to requiring the MGD association and complete folding prior to translocation by the Tat system, DmsA must also associate with folded DmsB, which must acquire four [4Fe-4S] centres. This hetero-dimeric complex is targeted to the Tat translocase via the twin-arginineleader on DmsA. The DmsA twin-arginine-leader is cleaved and DmsAB associates with the integral membrane protein DmsC. The chaperone DmsD plays a critical role in the maturation of the DMSO reductase enzyme.

DmsD Interactions with DmsA

DmsD was originally identified by its ability to interact with the DmsA twin-arginine-signal peptide (Oresnik et al. 2001). Prior to this study the DmsD protein was uncharacterized in *E. coli* and known as YnfI, part of the *ynfEFGHI* operon. DmsD was shown to interact with the premature forms of both DmsA and TorA in an affinity chromatography technique in which a hexahistidine tagged DmsD was coupled to a Ni-NTA resin and exposed to anaerobically grown *E. coli* lysate (Oresnik et al. 2001). DmsD was shown to be required for biogenesis of DMSO reductase, likely mediated via the interaction with the DmsA twin - arginine leader (Oresnik et al. 2001; Ray et al. 2003). Further proof of the interaction between DmsD and the DmsA twin-arginine leader sequence (DmsAL) has been shown through an *in vitro* far-Western assay (Sarfo et al. 2004; Winstone et al. 2006) in which the DmsAL was fused at the N-terminus of GST. The dissociation constant (K_d) of the DmsD interaction with the DmsAL was determined to be 0.2 μM using isothermal calorimetry (ITC) (Winstone et al. 2006). These ITC experiments used the same hexahistidine tagged DmsD and DmsAL:GST fusions mentioned above in which the first 43 (of a total 45) residues of the DmsAL were fused to the N-terminus of GST. Interestingly, a GST:DmsAL fusion protein (in which the DmsA signal peptide was fused to the C-terminus of GST) was unable to interact with DmsD *in vitro* suggesting that N-terminal display may be important for interaction (Winstone et al. 2006). DmsD binding to DmsAL:GST was also characterized with SEC-FPLC in which monomeric DmsD bound DmsAL to form a 1:1 complex (Winstone et al. 2006).

Amino acid residues conserved in DmsD homologous proteins were subjected to site-directed mutagenesis, the variant proteins were expressed, purified and assayed for binding the DmsAL:GST fusion protein. More than 20 single amino acids in DmsD were substituted and 9 variant proteins exhibited reduced binding relative to the wild type protein (Chan et al. 2008). A DmsD structural model was made and the residues most affected (W72, L75, F76, P86, W87, P124, D126, H127) were found to cluster together on the surface of the DmsD model as a "hot pocket" (Chan et al. 2008) (Figure 3).

DmsD is capable of binding four different twin-arginine-leaders from different enzymes in *E. coli* (DmsA, TorA, YnfE and YnfF) (Chan et al. 2009). The specificity may originate from a four residue hydrophobic motif (-LAMA-) present downstream of the twin-arginine motif, that may be important for interactions with DmsD.

DmsD Interactions with the Tat System and other Proteins

DmsD localization studies were performed in a variety of wild type and Tat system deletion *E. coli* strains. DmsD was shown to localize to the membrane in the presence of the Tat components (Papish et al., 2003). The membrane localization of DmsD in anaerobically grown cells was shown to be dependant on the presence of the TatB and TatC subunits, suggesting that DmsD was capable of handing off the DmsA substrate to the translocase via an interaction with this TatBC receptor complex (Papish et al., 2003). Other studies investigated the localization of green fluorescent protein fused to DmsA and TorA leader peptides in *dmsD* deletion *E. coli* cells and showed that localization was similar to that of wild-type cells (Ray et al., 2003). The anaerobic growth of Δ*dmsD*, Δ*tatABCDE* and wild-type *E. coli* supplemented with DMSO, showed that DmsD is essential for the biogenesis of the DMSO reductase enzyme (Ray et al., 2003). However, there remained a basal level of activity in Δ*dmsD*. Together these observations suggest that the REMP is not absolutiely required but is a critical helperin the biogenesis.

A variety of *in vitro* and *in vivo* protein-protein interaction techniques have been used to probe the DmsD interactome in a variety of cellular/genetic backgrounds. DmsD was shown to interact with a variety of other proteins in *E. coli* including the general chaperones DnaK and GroEL as well as molybdopterin biosynthetic enzymes MoeA, MoeB and MobB (Li, submitted work and unpublished data). Additionally, preliminary data suggest that DmsD also binds NTP. These data suggest a parallel activity role to that of TorD.

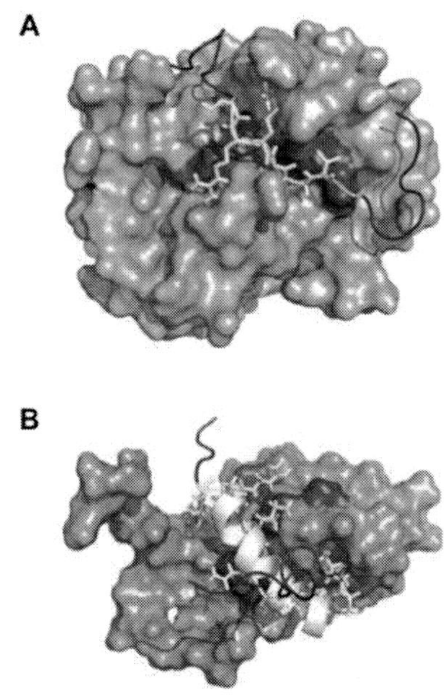

Figure 3. Twin-arginine signal peptide binding sites. A. *E. coli* DmsD and DmsA signal peptide residues 1-29 (3EFP.pdb; Stevens et al. 2009). B. EcNapD and NapA signal peptide residues 1-35 (2PQ4.pdb). DmsD and NapD surface representations are shown with residues implicated in binding colored red. DmsA and NapA peptides are colored dark grey or white (residues aligned with binding sites). The N-terminus of each peptide begins at the top and follows down to the C-terminus. The twin-arginine leader motif SRRxFLK is colored white in both structural representations.

DmsD Structure

DmsD has been shown to exist in multiple oligomeric and folded forms (Sarfo et al., 2004). Purification of *E. coli* hexahistidine tagged DmsD protein with an imidazole gradient has made it possible to separate monomers from dimers. Exposure to low pH induces a conformational change that is visualized as a 'ladder' on a native polyacrylamide gel suggesting that the protein may be aggregating or forming multimers (Sarfo et al., 2004). However, when the low-pH induced form of DmsD was characterized with SEC-FPLC the protein eluted as a single peak beyond the void volume of the column (Sarfo et al., 2004). This elution behavior suggests that the protein is present in solution as a single 'form' that was interacting with the column matrix. The secondary structure (alpha helical content) of the DmsD low pH form was almost identical to that of the native monomer and dimer when characterized by circular dichroism spectroscopy (Sarfo et al., 2004). The low pH induced form of the homologous protein *Shewanella masillia* TorD bound significant amounts of the ANS (1-anilino naphtalene-8-sulfonic acid) fluorophore while the native monomers, dimers and even guanidium hydrochloride (GdnHCl) denatured protein did not (Tranier et al., 2002). ANS binding is indicative of the presence of exposed hydrophobic surfaces (Semisotnov et al. 1991). Subsequently, Sarfo et al., (2004) found that if the low-pH induced form of *E. coli* DmsD was treated with 8 M urea (followed by its removal) the protein exhibited a similar native PAGE profile to that of the monomeric protein, suggesting re-folding to the native folded form. The significance of this low-pH folded form of DmsD is unclear; however, it should be noted that it is capable of binding the DmsA twin arginine leader peptide (Sarfo et al. 2004).

Intrinsic fluorescence spectroscopy has been used to characterize pH and denaturation profiles of *E. coli* DmsD. The protein has multiple pH dependant conformations (Sarfo et al. 2004). The nature of the significance and relevance of the different folding forms at different pHs has not been further explored by researchers in the field; however, as the folding transition occurs at pHs in the physiologically relevant range, such observations must be considered. Very little conformational change occurs when *E. coli* DmsD is exposed to urea concentrations at or below 5 M (Winstone et al., 2006). For significant conformational changes to occur greater than 7 M urea was necessary. GdnHCl denaturation of DmsD appeared to have 2 phases. These observations imply that this protein may have different folding forms.

The protein structure of the DmsD homologue from *Salmonella typhimurium* homologue was solved (pdp;1S9U) (Qiu et al., 2008). The protein showed an all alpha-helical structure (12 helices) and similar organization to a monomeric unit of the previously solved *S. masillia* TorD homologue (pdb;1N1C) (Tranier et al., 2003). With the *E. coli* DmsD structure solved to 2.0 Å resolution (pdb;3EFP) it was possible to perform docking and molecular dynamics simulations with a peptide composed of the 29 N-terminal residues of the DmsA twin-arginine leader peptide (Stevens et al. 2009) which showed support for the previously determined binding site (Chan et al., 2008) (Figure 3). Overall, the DmsD structures are very similar to each other.

NITRATE REDUCTASES (NARJ, NARW, NAPD)

Respiration using nitrate is performed by a set of redox enzymes in bacteria that catalyze the reduction of nitrate (NO^3) to nitrite (NO^2). Other enzymes also carry out further denitrification until NO^2 is converted to NO, N^2O, and finally N2 (summarized in Gonzalez et al., 2006). Three membrane-bound nitrate reductases have been identified in bacteria in addition to the soluble assimilatory nitrate reductases located in the cytoplasm, all of which coordinate a molybdenum-*bis*-molybdopterin guanine dinucleotide (MGD) cofactor in the catalytic site (Potter et al. 2001). *E. coli* contains two cytoplasmically-anchored enzymes commonly referred to as nitrate reductase A (NRA) and nitrate reductase Z (NRZ), or collectively as Nar. The last one is periplasmically localized and is commonly termed the periplasmic nitrate reductase, nitrate reductase P, or Nap. Several reviews provide extensive information into the structure, function, and mechanism of all three (Potter et al., 2001; González et al., 2006; Martinez-Espinosa et al., 2007).

Cytoplasmic Nitrate Reductases (NarJ and NarW)

All bacterial cytoplasmic nitrate reductases isolated to date are heterotrimeric enzymes anchored to the membrane through an integral membrane subunit (González et al. 2006; Martinez-Espinosa et al. 2007). Nitrate reductase A and nitrate reductase Z are no exception, consisting of NarGHI and NarZYV subunits, respectively. NarG is the catalytic subunit with MGD and 1 [4Fe-4S] cluster. NarH connects NarG to the membrane anchor NarI and contains 1 [3Fe-4S] and 3 [4Fe-4S] clusters. Finally, NarI is the membrane anchor subunit consisting of five transmembrane helices coordinating two *b*-type haems and connects to the menaquinone pool in the membrane. A structural overview of the NarGHI complex from *E. coli* is provided in Bertero et al., (2003).

Nitrate reductase Z was shown to exhibit similar physical and chemical characteristics to nitrate reductase A (Iobbi et al., 1987). Its gene products have high homology with the nitrate reductase A components (Blasco et al., 1990), where sequence comparisons identified conserved regions in NarZ similar to known molybdoproteins, NarY contained four cysteine clusters in the same order as in NarH, and NarV has five putative transmembrane segments and haem motifs as organized in NarI. These suggest that the three subunits have highly similar biochemical roles and functional organization as NarGHI. Unfortunately, few further studies have been performed on NarZYV.

Early studies demonstrated the absolute requirement of the accessory proteins NarJ and NarW for biogenesis of nitrate reductases A and Z (Sodergren et al., 1988; Blasco et al., 1992; Dubourdieu and DeMoss 1992). NarJ was not found to be associated with the final holoenzyme of NarGHI, thus little understanding of its role in enzyme maturation was understood at the time other than it was required for functional biogenesis of nitrate reductase A (Sodergren et al., 1988). Further studies demonstrated that NarJ is required for MGD cofactor assembly into NarG, a step that must occur prior to membrane attachment. This led Blasco and colleagues to propose that NarJ is a specific chaperone that binds to NarG to keep it in a cofactor-accepting conformation (Palmer et al., 1996; Blasco et al., 1998a). Additional experiments have also shown the requirement of NarJ for insertion of the [4Fe-4S] cluster into NarG and indirectly affecting haem insertion of NarI by preventing premature attachment of NarGH to NarI (Lanciano et al., 2007).

Functional characterization of NarJ reveals that it accumulates at lower stoichiometric levels than the other three subunits and co-purifies with NarGH or NarG by itself (Liu and DeMoss 1997). Studies provide contradicting evidence on the cellular localization of NarJ (exclusive cytoplasmic versus exclusive membrane localization) that may be attributed to subtle experimental differences. However, it was evident that the exclusive membrane localization was dependent on the presence of the Tat complex (Chan et al., 2006). NarG from membrane fractions was isolated and identified using purified NarJ as bait. These observations implicate the previously accepted role that NarJ is a chaperone that only functions for nitrate reductase A biogenesis in the cytoplasm and suggests that nitrate reductase A targeting is dependent on the Tat system. Evidence further supporting this is indicated in recent studies demonstrating interactions of NarJ and NarW with the vestige Tat motif-containing N-terminal portion of NarG and NarZ and that the interaction with NarG *in vivo* has Tat-dependence (Chan et al. 2009; Li and Turner 2009).

The homologous NarJ from *Thermus thermophilus* was found to be required for membrane attachment to the NarCI complex along with its requirement for the maturation of NarG (Zafra et al., 2005). In *E. coli*, NarJ did not appear to have an effect on membrane attachment of NarGH (Blasco et al. 1992; Vergnes et al., 2006). Its role appeared to be more important for maintaining the apoenzyme in a soluble and competent cofactor-accepting conformation while preventing premature membrane-anchoring to NarI (Vergnes et al., 2006). Recent studies have demonstrated that NarW cross-interacts with the NarG N-terminus that is important for membrane-attachment (Chan et al., 2009), an observation that was attributed to a potential rescue mechanism. Thus NarJ may still be required for NarGH membrane attachment in *E. coli,* but its role may have been substituted by NarW. The two share 56% sequence identity and 71% sequence similarity, which was the highest amongst the ten *E. coli* REMP chaperones (Chan et al., 2009). Both NarJ and NarW have been identified as part of the DmsD family of chaperones including TorD and YcdY as members (Turner et al., 2004).

Periplasmic Nitrate Reductase (NapD)

The periplasmic nitrate reductase Nap was demonstrated to support anaerobic growth in a strain lacking *narG* and *narZ*, providing evidence that a third nitrate reductase exists in *E. coli* (Potter et al., 2000; Stewart et al., 2002). Nap appeared to be selectively preferred under low nitrate conditions (less than 0.1 mM in the medium), suggesting that it provides a selective advantage under limited nitrate conditions (Potter et al., 1999; Stewart et al., 2002). Nap consists of the NapABC subunits where NapA is the catalytic subunit with a twin-arginine motif and contains a MGD cofactor and [4Fe-4S] cluster (Thomas et al., 1999; Brondijk et al., 2004). NapB contains two *c*-type haems and a conventional Sec-targeting signal (Thomas et al., 1999). NapC contains four *c*-type haems but unlike NarI or NarV which are multi-transmembrane integral membrane proteins, it contains a single transmembrane helix that inserts into the membrane from the periplasmic side (Roldan et al. 1998), hence the earlier description that Nap is membrane attached and not anchored.

Early studies established an essential role of NapD for nitrate reductase activity (Potter and Cole 1999). NapD has been shown in two independent studies to interact with the twin-arginine motif-containing N-terminus of NapA. This interaction is very tight with a dissociation constant determined to be 7 nM (0.007 μM) using the technique of ITC (Maillard et al. 2007; Chan et al. 2009). The tight association can even suppress transport by Tat, which may be a control mechanism for proper folding of NapA (Maillard et al., 2007). The structure of NapD was determined by NMR spectroscopy to have four β-strands and two α-helices, which are unique, compared to the other twin-arginine leader binding proteins with structural information currently available; DmsD and TorD exhibit all-alpha folds (Tranier et al., 2003; Maillard et al., 2007; Qiu et al., 2008) (Figure 2). A recent study also demonstrated a highly selective *in vivo* interaction between NapD and the signal peptide of HybO of the hydrogenase enzyme, which also bears a twin-arginine motif (Chan et al., 2009).

Formate Dehydrogenase (FdhD, FdhE)

E. coli has three formate dehydrogenase (FDH) membrane-bound isoenzymes that catalyze oxidation of formate (HCOO) to carbon dioxide (CO_2). Two FDH's are anchored to the cytoplasmic membrane from the periplasm while the other from the cytoplasm. The two periplasmic FDH's are transported across the cytoplasmic membrane via the Tat system (Stanley et al. 2002). The first periplasmic isoenzyme is commonly referred to FDH-N where the N-denotation stemmed from observations that its expression was induced by nitrate anaerobically. The isoenzyme, FDH-O or FDHZ, is also induced by nitrate in the media but also in the presence of oxygen (Sawers et al., 1991; Abaibou et al., 1995).

FDH-N consists of the heterotrimeric FdnGHI subunits. FdnG is the MGD and [4Fe-4S] containing subunit that catalyzes the oxidation of formate. It contains a single selenocysteine residue (Berg et al. 1991). It is also the twin-argnine motif-containing subunit. FdnH is the electron conduit subunit that accepts two electrons from FdnG and shuttles them through its four [4Fe-4S] clusters to FdnI, which is the membrane anchor subunit consisting of four transmembrane helices and contains two *b*-type haems. FDH-N forms a complete redox loop with nitrate reductase A (NarGHI) that contributes to the proton gradient across the membrane. The crystal structure of FdnGHI demonstrates where the MGD cofactor and Fe-S clusters are coordinated and provide a molecular view of how the redox loop is completed during electron shuttle (Jormakka et al., 2002).

FDH-O/Z consists of the FdoGHI subunits that are highly similar to FdnGHI in terms of size, composition, and structural fold (Abaibou et al., 1995). It likely forms a redox loop with nitrate reductase Z in the same manner that FDH-N does with nitrate reductase A (reviewed in Sawers, 1994). FdoG contains the twin-arginine motif, and requires molybdenum and selenocysteine incorporation, and is considered the catalytic subunit (Sawers et al., 1991; Pommier et al., 1992; Abaibou et al., 1995). FdoH and FdoI likely perform similar functions as FdnH and FdnI as they were found to be of similar size and have similar structural folds based on recognition by FdnHI-specific antibodies (Abaibou et al., 1995).

Early studies on *E. coli* mutants deficient in formate dehydrogenase activity isolated two classes of mutants, one involving *fdhD* and the other *fdhE*, coding for two accessory proteins of the FDH's (Mandrand-Berthelot et al. 1988). The two proteins were required for FdnGHI activity but not cytochrome incorporation, indicating a role in the cytoplasmic biogenesis of the FdnGH subunits. FdhD and FdhE do not appear to control transcription of *fdnGHI* (Stewart et al. 1991). FdhD/E does not have a direct role in incorporation of the selenocysteinyl residue as incorporation occurs during mRNA transcription (Schlindwein et al. 1990). FdhE was recently demonstrated to interact with the catalytic subunits FdnG and FdoG (Lüke et al., 2008; Chan et al., 2009). An interesting observation in the study by Lüke et al., (2008) showed that the interaction of FdhE with FdnG excluded the interaction between FdnG and FdnH, suggesting that the interacting sites may overlap or induce a conformation in FdnG to occlude its binding site to FdnH. From this observation, the authors propose that the interaction between FdhE and FdnG prevents premature association of FdnGH prior to complete folding of FdnG.

Biochemical studies show cytoplasmic localization of the FdhD and FdhE and they are the largest of the known REMP chaperones (Schlindwein et al., 1990), and share 11% and 18% sequence identity and similarity, respectively (Chan et al., 2009). Less information is available about FdhD, but a recent study showed that FdhE adopts monomeric and homodimeric forms *in vitro* while dimerization *in vivo* appeared to be stabilized under anaerobic conditions (Lüke et al., 2008). FdhE was also shown to bind a ferric iron through conserved cysteine residues, but the function of the ligated iron is still not understood (Lüke et al. 2008).

Hydrogenase (HyaE HybE)

Three hydrogenases have been identified in *E. coli* and are denoted hydrogenase-1, -2, and -3. Hydrogenases-1 and -2 catalyze the oxidation of hydrogen (H2) to protons (H^+) whereas hydrogenase-3 catalyzes the opposite reduction reaction. Hydrogenases-1 and -2 are membrane-anchored heterodimers that are transported by the Tat pathway to the periplasm (reviewed in Wu et al., 2000). Both enzymes consist of a large and small subunit that contain nickel and iron-sulfur cofactors as with other dimeric hydrogenases in other bacteria (Przybyla et al., 1992). Hydrogenase-1 consists of HyaAB whereas hydrogenase-2 consists of HybOC. The small subunits HyaA and HybO contain the consensus twin-arginine motif at their N-terminus and a C-terminal membrane anchor helix (Dubini et al., 2002). Studies have demonstrated co-targeting and co-translocation of the motif-lacking Hyb as a 'hitchhiker' with HybO, suggesting that this mechanism applies to both hydrogenases (Rodrigue et al., 1999). The large subunits HyaB and HybC are the catalytic centers for hydrogen oxidation. Crystal structure of hydrogenase from *Desulfovibrio gigas* shows nickel coordination by the large subunit and one [3Fe-4S] and two [4Fe-4S] centers in the small subunit (Volbeda et al., 1995).

The solution structure of the *E. coli* accessory protein HyaE was determined by NMR to have a thioredoxin fold, but lacked the canonical C-X-X-C motif typically found in the active site of thioredoxin proteins (Parish et al. 2008) (Figure 2F). HyaE was shown to interact with the twin-arginine leader of HyaA and potentially HybO (Dubini and Sargent 2003). Conserved Glu and Asp residues on the surface of the recently determined HyaE structure were postulated for binding to the positively charged arginines in the leaders (Parish et al. 2008). There is no structure currently available for HybE but it was demonstrated to interact with the leader of HybO and HybC (Dubini and Sargent 2003; Chan et al. 2009). In the study by Dubinin and Sargent, HyaE and HybE also appeared to interact with themselves as well as each other, which have been suggested that they have cooperative functions to work together for maturation of the hydrogenases in a large complex. Despite increasing evidence to suggest that HyaE and HybE participate through 'cross-talk' for maturation of all three hydrogenases, they have low sequence homology at 12% and 19% identity and similarity, respectively (Chan et al. 2009). They also do not share any similar motifs, which raises the question of how specificity of cross-recognition towards the three hydrogenases is conferred (Turner et al. 2004).

REMP Classification and Structural Analysis

Bioinformatics

In 2004, two studies examined the amino acid sequences of accessory proteins that bind to twin-arginine signal peptides (Ilbert et al., 2004; Turner et al., 2004). At the time these proteins were loosely referred to as the TorD family. The Turner et al study proposed that any protein involved in the maturation process of a complex redox enzyme but not part of the holo-enzyme itself be called a Redox Enzyme Maturation Protein or REMP, and classified a large number of proteins as such. This study took the known *E. coli* twin-arginine peptide binding protein sequences, which were "Blasted" (BLASTp) against the databases and the sequence similarities were analyzed. Based on the cofactor contained within the REMP substrate or functionality of the REMP substrate, the REMPs were classified into either a hydrogenase or molybdoprotein super-family. REMPs found to bind molybdoprotein substrates were classified into three families (based on the *E. coli* protein included): DmsD, NapD and FdhE. The NapD and FdhE families were not sub-classified further; however, the DmsD family was further classified into 3 clades (based on the *E. coli* protein included): DmsD, TorD and NarJ. The protein sequences within each clade were aligned and analyzed for defined signature sequence motifs and the *E. coli* protein sequence motifs were highlighted. *E. coli* DmsD family members were predicted to be mostly alpha-helical by secondary structure prediction algorithms and included DmsD, YcdY, TorD, NarJ and NarW (Turner et al. 2004). When looking at the *E. coli* REMP phylogenetic tree and considering the classifications above; the molybdoenzyme REMPs NapD and FdhE are located on the same side of the phylogenetic tree as the hydrogenase REMPs HyaE and HybE, opposite all *E. coli* DmsD family members. Interestingly, these same 4 proteins (NapD, FdhE, HyaE and HybE) were predicted to contain beta structure in addition to alpha helix, contrasting the all alpha-helical nature of the proteins classified to DmsD family members (Turner et al., 2004).

In the study by Ilbert et al. (2004) only protein sequences with pre-determined biochemical information and those proteins found to bind to molybdoenzyme twin-arginine leader peptides were analyzed, thus a much smaller number of proteins were classified as members of the TorD family.

These proteins [sequences] were then classified based on the characteristics (and classification) of the molybdoenzyme that the protein interacted with, and represented an extension of their group's previous definition of the TorD Family (Tranier et al., 2003). This group divided the TorD family into 4 clades, according to the molybdoenzyme classification types (I, II or III) of the DMSO reductase family, which is based on cofactor composition, and organization of the protein domain around the MGD cofactor (McEwan et al., 2002). The newly defined fourth clade (IV) contained TorD homologues that could not be related to any specific molybdoenzyme family member and were exemplified by the *E. coli* protein YcdY and shown to have the closest relationship to Type II TorD homologues (ie. DmsD). Types I and III TorD family members contain the molybdoenzyme itself and the TorD family member (TorD homologue) are located within the same operon (ie. *torCAD*, type III), while Types II and IV have separate operons for the molybdoenzyme and the gene of the TorD homologue.

The accessory proteins found to bind the twin-arginine leader peptides of molybdoenzymes can be correlated between the two papers. The TorD clade (Turner et al. 2004) can be correlated to Group III proteins of the DMSO reductase family (Ilbert et al. 2004) while types II and IV (Ilbert et al., 2004) would both be included within the DmsD clade (Turner et al., 2004). The NarJ clade, within the DmsD family, identified by Turner et al., (2004) is not mentioned by Ilbert et al., (2004). Both papers defined sequence motifs, slightly different amongst each clade/type. Overall, the Turner paper includes a much more diverse set of sequences than in the Ilbert paper focusing only on the TorD Family.

It is worth noting that there are REMP substrate molybdoenzymes of each clade/type that are not translocated (Sargent 2007). The enzyme substrates are assembled first, then targeted and translocated later. Therefore, at this time it is not always possible to extrapolate the type of molybdoenzyme with its localization within the cell, and likely not with the corresponding function of each REMP.

Structures

There are currently 7 REMP protein structures available (Figure 2 and 3). Almost all the different REMP protein families are represented. Structures from X-ray diffraction of crystals have been solved representing 3 different clades of the DmsD family – TorD, DmsD and NarJ. *Shewanella massilia* TorD was the first REMP structure published. It was found to contain a unique fold that is primarily helical, with two domains that undergo domain swapping (Tranier et al. 2003). *Archaeoglobus fulgidus* NarJ (Kirillova et al. 2007) and *E. coli* DmsD (Stevens et al., 2009) crystallized as dimers with no domain swapping and *S. typhimurium* DmsD (Qiu et al. 2008) was monomeric. Each monomeric domain has a very similar architecture and is displayed in figure 2.

The *E. coli* NapD solution structure was determined using nuclear magnetic resonance and is significantly different from the DmsD/TorD family members (Maillard et al. 2007). NapD is a significantly smaller protein and belongs to a distinct protein family and adopts a mixed alpha-beta structure (Figure 2E). The NapD protein interacted specifically with a region of the NapA signal peptide (Maillard et al. 2007). The NapD residues most affected by NapA leader-peptide binding were also examined and included S9, V11, E33, A35, S37, Q43, E49, L74 and Y76 (Maillard et al. 2007). These residues are highlighted in Figure 3 and show the twin-arginine leader forming a helix lying along a beta-sheet surface. For comparison the modeled twin-arginine leader peptide bound to DmsD is also shown in Figure 3. The peptide in this case is in an extended conformation lying across a groove in the protein.

The REMPs HyaE (pdb:2HFD) from *E. coli* and *Pseudomonas aeroginosa* FdhE (pdb:2FIY) display very different structures. Despite their assumed similar function, these proteins adopt independent folds where only DmsD, NarJ and TorD have similar architecture, which was suggested by the bioinformatic analysis establishing the family (Turner et al. 2004). Regardless of whether the twin-arginine-leaders from each substrate enzyme have a final teleological purpose to target to the Tat system or to facilitate cofactor loading, the structure of the twin-arginine signal peptides must be fine tuned to fit only one type of structure and REMP.

Concluding Comments

Of the REMPs DmsD and TorD are the most thoroughly studied. DmsD research has focused on its ability to bind the DmsA twin-arginine leader peptide (Chan et al., 2008; Oresnik et al., 2001; Sarfo et al., 2004; Winstone et al., 2006). Characterization of DmsD has shown it to exist in multiple folded forms when aerobically expressed and purified (Sarfo et al. 2004). Both *E. coli* and *S. massilia* TorD were also shown to exist in multiple folded conformations (Tranier et al. 2003; Tranier et al. 2002). TorD and DmsD have both been shown to bind to a molybdopterin biosynthetic enzyme which provides the hypothesis that the REMPs may form a scaffold for folding and cofactor loading of the reductase; ie. involved in co-binding of the twin-arginine-leader substrate upon translation in concert with interacting with Moco biosynthetic machinery.

Early studies suggested that the interaction between TorD and TorA, unlike DmsD, was not mediated by the twin-arginine signal peptide but with the mature protein sequence (Pommier et al. 1998). TorD has been reported to interact specifically with the unfolded form of TorA within the cytoplasm and assist with the insertion of the molybdopterin cofactor (Ilbert et al. 2003; Pommier et al., 1998). Similarly, NarJ was shown to recognize the nascent unfolded topology of its molybdoenzyme NarG (Blasco et al., 1998). Additionally, NarJ was required for the insertion of the molydopterin cofactor into NarG (Blasco et al., 1998). NarJ was recently shown to interact with a NarG N-terminal leader peptide *in vitro* with both an *in vitro* far-western analysis and ITC (Chan et al. ,2006). NarG and NarZ also have a molybdopterin cofactor but the N-terminal region (vestige signal sequence) is not cleaved (Blasco et al., 1990). The structure of the nitrate reductase complex (NarGHI) was solved (Bertero et al., 2003) and confirmed that the N-terminal region (a vestige twin-arginine-leader) was not cleaved. These results suggest that the REMP has multiple interactions with their substrates and with other enzymatic events leading to their substrates maturation.

A model of the pathway of maturation of the example *E. coli* redox enzyme DMSO reductase is cartooned in figure 4. It has been hypothesized that the REMP may interact with the twin-arginine-leader immediately after it appears from the ribosome tunnel, remaining bound throughout the folding and cofactor loading of its cognate substrate and continue to be associated until immediately prior to translocation. This interaction is likely important for multiple functions within the cell providing protease protection of the twin-arginine-leader, maintinaing the substrate in a cofactor competent state, insertion of cofactor, proofreading of final folding and stability to the substrate – hitchhiker couple, providing targeting to the membrane via TatBC recognition complex and finally release to allow translocation to occur. The specificity and strength of this interaction would likely need to vary during the protein folding cycle of the substrate (and DmsB association) to allow these multiple functions to take place and provide for various interaction partners to aid in this maturation pathway.

With the concept of REMPs now accepted in the field and with several groups actively researching these proteins it is worth reviewing the type of questions being perused about their function and their relationship to the Tat system. Such questions include: What is the physiology of 'REMP' mutants? Is there a level of gene regulation and if so how does it occur? How common are REMPs within an organism? How common are REMPs in other bacterial species? What are the key residues in the REMPs that are involved in leader peptide binding? How specific is the interaction between the REMPs and their target proteins? When do REMPs bind the leader? What are the residues in the twin-arginine-leader that play a role? How do REMPs distinguish between twin-arginine-leaders and Sec-dependent leaders or does it? What is the structure of the twin-arginine-leader bound to the REMP? At what stage in the process does the REMP release the substrate peptide?

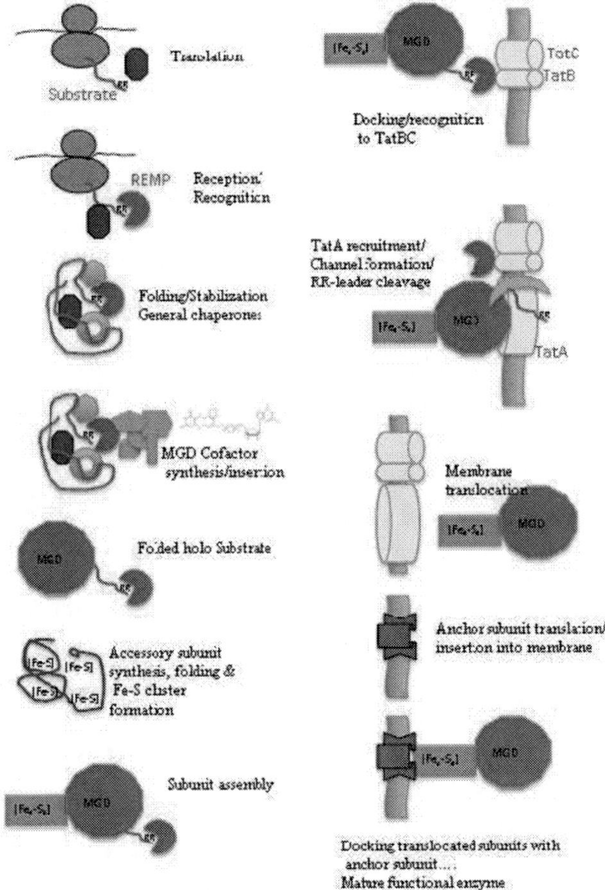

Figure 4. Model for an example maturation pathway (DMSO Reductase (DmsABC)). At the point of translation of the *dmsA* transcript, the system specific chaperone DmsD interacts to mediate the maturation pathway to target to general chaperones and molybdopterin cofactor biogenesis. Subsequent steps require the biogenesis of DmsB with iron-sulfur cluster formation and its interaction with the folded form of DmsA. The DmsAB complex must be targeted to the membrane and to the Tat system for twin arginine leader processing and translocation. The translocated DmsAB in the periplasm must then locate and dock to the membrane anchor subunit DmsC to interface with the quinone pool.

What is the energetics of peptide binding and release? Do REMPs interact with the Tat translocase and if so what is the nature of this interaction? What other proteins are involved in the functionality pathway? This list demonstrates how little we know about these chaperones, yet their answers will ultimately lead to a more complete understanding of the physiology and biochemistry of organisms that respire with alternative oxido reductase enzymes.

Acknowledgments

Research on REMPs is supported by a Grant from the Canadian Institutes of Health Research to RJ Turner. TML Winstone and CS Chan were supported by Graduate student scholarships from the Natural Sciences and Engineering Research Council of Canada.

References

Abaibou, H., Pommier, J., Benoit, S., Giordano, G., and Mandrand-Berthelot, M.A. 1995. Expression and characterization of the *Escherichia coli fdo* locus and a possible physiological role for aerobic formate dehydrogenase. *J. Bacteriol.* 177: 7141-49.

Alami, M., Luke, I., Deitermann, S., Eisner, G., Koch, H.G., Brunner, J. and Muller, M. 2003. Differential interactions between a twin arginine signal peptide and its translocase in *Escherichia coli*. *Mol. Cell* 12: 937-46.

Berg, B.L., Li, J., Heider, J., and Stewart, V. 1991. Nitrate-inducible formate dehydrogenase in *Escherichia coli* K-12. I. Nucleotide sequence of the *fdnGHI* operon and evidence that opal (UGA) encodes selenocysteine. *J. Biol. Chem.* 266: 22380-85.

Berks, B.C., Palmer, T., and Sargent, F. 2003. The Tat protein translocation pathway and its role in microbial physiology. *Adv. Microbiol. Physiol.* 47: 187-254.

Berks, B.C., Sargent, F., De Leeduw, E., Hinsely, A.P., Stanley, N.R., Jack, R.L., Buchanan, G., and Palmer, T. 2000a. A novel protein transport system involved in the biogenesis of bacterial electron transfer chains. *Biochim. Biophys. Acta* 1459: 325-30.

Berks, B.C., Sargent, F., and Palmer, T. 2000b. The Tat protein export pathway. *Mol. Microbiol.* 35: 260-74.

Berks, B.C. 1996. A common export pathway for proteins binding complex redox cofactors. *Mol. Microbiol* 22: 393-404.

Bertero, M.G., Rothery, R.A., Palak, M., Hou, C., Lim, D., Blasco, F., Weiner, J.H., and Strynadka, N.C. 2003. Insights into the respiratory electron transfer pathway from the structure of nitrate reductase. *A. Nat Struct Biol.* 10: 681-7.

Bilous, P.T., Cole, S.T., Anderson, W.F., and Weiner, J.H. 1988. Nucleotide sequence of the dmsABC operon encoding the anaerobic dimethylsulphoxide reductase of *Escherichia coli*. *Mol. Microbiol* 2: 785-95.

Bilous, P.T., and Weiner, J.H. 1985. Dimethyl sulfoxide reductase activity by anaerobically grown *Escherichia coli* HB101. *J Bacteriol* 162: 1151-5.

Blasco, F., Dos Santos, J.P., Magalon, A., Frixon, C., Guigliarelli, B., Santini, C.L., and Giordano, G. 1998. NarJ is a specific chaperone required for molybdenum cofactor assembly in nitrate reductase A of *Escherichia coli*. *Mol. Microbiol.* 28: 435-47.

Blasco, F., Pommier, J., Augier, V., Chippaux, M., and Giordano, G. 1992. Involvement of the *narJ* or *narW* gene product in the formation of active nitrate reductase in *Escherichia coli*. *Mol. Microbiol.* 6: 221-30.

Blasco, F., Iobbi, C., Ratouchniak, J., Bonnefoy, V., and Chippaux, M. 1990. Nitrate reductases of *Escherichia coli*: sequence of the second nitrate reductase and comparison with that encoded by the *narGHJI* operon. *Mol Gen Genet* 222: 104-11.

Brondijk, T.H.C., Nilavongse, A., Filenko, N., Richardson, D.J., and Cole, J.A. 2004. NapGH components of the periplasmic nitrate reductase of *Escherichia coli* K-12: location, topology and physiological roles in quinol oxidation and redox balancing. *Biochem. J.* 379: 47-55.

Buchanan, G., Maillard, J., Nabuurs, S. B., Richardson, D. J., Palmer, T., and Sargent, F. 2008. Features of a twin-arginine signal peptide required for recognition by a Tat proofreading chaperone. *Fed Eur Biochemical Soc Lett* 582: 3979-84.

Chan, C.S., Chang, L., Rommens, K.L., and Turner, R.J. 2009. Differential interactions between Tat-specific redox enzyme peptides and their chaperones. *J Bacteriol.*: 2091-101.

Chan, C.S., Winstone, T.M.L., Chang, L., Stevens, C.M., Workentine, M.L., Li, H., Wei, Y., Ondrechen, M.J., Paetzel, M., and Turner, R.J. 2008. Identification of residues in DmsD for twin-arginine leader peptide binding, defined through random and bioinformatics-directed mutagenesis. *Biochemistry* 47: 2749-59.

Chan, C.S., Howell, J.M., Workentine, M.L., and Turner, R.J. 2006. Twin-arginine translocase may have a role in the chaperone function of NarJ from *Escherichia coli*. *Biochem Biophys Res Commun* 343: 244-51.

DeLisa, M.P., Tullman, D., and Georgiou, G. 2003. Folding quality control in the export of proteins by the bacterial twin-arginine translocation pathway. *PNAS.* 100: 6115-6120.

Dubini, A., Pye, R.L., Jack, R.L., Palmer, T., and Sargent, F. 2002. How bacteria get energy from hydrogen: a genetic analysis of periplasmic hydrogen oxidation in *Escherichia coli*. *Int. J. Hydrogen Energ.* 27: 1413-20.

Dubini, A., and Sargent, F. 2003. Assembly of Tat-dependent [NiFe] hydrogenases: identification of precursor-binding accessory proteins. *FEBS Lett.* 549: 141-6.

Dubourdieu, M., and DeMoss, J.A. 1992. The *narJ* gene product is required for biogenesis of respiratory nitrate reductase in *Escherichia coli*. *J. Bacteriol.* 174: 867-72.

Geijer, P., and Weiner, J.H. 2004. Glutamate 87 is important for menaquinol binding in DmsC of the DMSO reductase (DmsABC) from *Escherichia coli*. *Biochim Biophys Acta* 1660: 66-74.

Genest, O., Neumann, M., Seduk, F., Stocklein, W., Mejean, V., Leimkuhler, S., and lobbi-Nivol, C. 2008. Dedicated metallochaperone connects apoenzyme and molybdenum cofactor biosynthesis components. *J. Biol. Chem.* 283: 21433-40.

Genest, O., Seduk, F., Ilbert, M., Mejean, V., and Iobbi-Nivol, C. 2006a. Signal peptide protection by specific chaperone. *Biochem Biophys Res Commun* 339: 991-5.

Genest, O., Seduk, F., Theraulaz, L., Mejean, V., and Iobbi-Nivol, C. 2006b. Chaperone protection of immature molybdoenzyme during molybdenum cofactor limitation. *FEMS Microbiol Lett* 265: 51-5.

Genest, O., Ilbert, M., Mejean, V., and Ilbert, M. 2005. TorD, an essential chaperone for TorA molybdoenzyme maturation at high temperature. *J. Biol. Chem.* 280: 15644-48.

Gon, S., Giudici-Orticonis, M. T., Mejean, V., and Iobbi-Nivol, C. 2001. Electron Transfer and Binding of the c-type cytochrome TorC to the trimethylamine N-oxide reductase in *Escherichia coli*. *J. Biol. Chem.* 276: 11545-51.

González, P.J., Correia, C., Moura, I., Brondino, C.D., and Moura, J.J.G. 2006. Bacterial nitrate reductases: Molecular and biological aspects of nitrate reduction. *J. Inorg. Biochem.* 100: 1015-23.

Hatzixanthis, K., Clarke, T. A., Oubrie, A., Richardson, D. J., Turner, R. J., and Sargent, F. 2005. Signal peptide-chaperone interactions on the twin-arginine protein transport pathway. *PNAS* 102: 8460-5.

Ilbert, M., Mejean, V., and Iobbi-Nivol, C. 2004. Functional and structural analysis of members of the TorD family, a large chaperone family dedicated to molybdoproteins. *Microbiology* 150: 935-43.

Ilbert, M., Mejean, V., Giudici-Orticoni, M.T., Samama, J.P., and Iobbi-Nivol, C. 2003. Involvement of a mate chaperone (TorD) in the maturation pathway of molybdoenzyme TorA. *J Biological Chemistry*278: 28787-92.

Iobbi, C., Santini, C.-L., Bonnefoy, V., and Giordano, G. 1987. Biochemical and immunological evidence for a second nitrate reductase in *Escherichia coli* K12. *Eur. J. Biochem.* 168: 451-9.

Jack, R. L., Buchanan, G., Dubini, A., Hatzixanthis, K., Palmer, T., and Sargent, F. 2004. Coordinating assembly and export of complex bacterial proteins. *EMBO Journal* 23: 3962-72.

Jormakka, M., Tornroth, S., Byrne, B., and Iwata, S. 2002. Molecular basis of proton motive force generation: structure of formate dehydrogenase-N. *Science* 295: 1863-68.

Kirillova, O., Chruszcz, M., Shumilin, I.A., Skarina, T., Gorodichtchenskaia, E., Cymborowski, M., Savchenko, A., Edwards, A., and Minor, W. 2007. An extremely SAD case: structure of a putative redox-enzyme maturation protein from Archaeoglobus fulgidus at 3.4 A resolution. *Acta Crystallogr D Biol Crystallogr* 63: 348-54.

Kisker, C., Schindelin, H., and Rees, D.C. 1997. Molybdenum-cofactor-containing enzymes: structure and mechanism. *Annu Rev Biochem* 66: 233-67.

Lanciano, P., Vergnes, A., Grimaldi, S., Guigliarelli, B., and Magalon, A. 2007. Biogenesis of a respiratory complex is orchestrated by a single accessory protein. *J. Biol. Chem.*282: 17468-74.

Lee, P.A., Tullman-Ercek, D., and Georgiou, G. 2006 The bacterial twin-arginine translocation pathway. *Annu. Rev. Microbiol.* 60: 373-95.

Li, H., and Turner, R.J. 2009. *In vivo* associations of *Escherichia coli* NarJ with a peptide of the first 50 residues of nitrate reductase catalytic subunit NarG. *Can. J. Microbiol.* 55: 179-88.

Liu, X., and DeMoss, J.A. 1997. Characterization of NarJ, a system-specific chaperone required for nitrate reductase biogenesis in *Escherichia coli*. *J. Biol. Chem.* 272: 24266-71.

Lüke, I., Butland, G., Moore, K., Buchanan, G., Lyall, V., Fairhurst, S., Greenblatt, J., Emili, A., Palmer, T., and Sargent, F. 2008. Biosynthesis of the respiratory formate dehydrogenases from *Escherichia coli*: characterization of the FdhE protein. *Arch Microbiology* 190: 685-96.

Maillard, J., Spronk, C., Buchanan, G., Lyall, V., Richardson, D.J., Palmer, T., Vuister, G.W., and Sargent, F. 2007. Structural diversity in twin-arginine signal peptide-binding proteins. *PNAS* 104: 15641-46.

Mandrand-Berthelot, M.A., Couchoux-Luthaud, G., Santini, C.L., and Giordano, G. 1988. Mutants of *Escherichia coli* specifically deficient in respiratory formate dehydrogenase activity. *J. Gen. Microbiol.* 134: 3129-39.

Martinez-Espinosa, R.M., Dridge, E.J., Bonete, M.J., Butt, J.N., Butler, C.S., Sargent, F., and Richardson, D.J. 2007. Look on the positive side! The orientation, identification and bioenergetics of 'Archaeal' membrane-bound nitrate reductases. *FEMS Microbiol. Lett.* 276: 129-39.

McAlpine, A.S., McEwan, A.G., and Bailey, S. 1998. The high resolution crystal structure of DMSO reductase in complex with DMSO. *J Mol Biol* 275: 613-23.

McEwan, A.G., Ridge, J.P., McDevitt, C.A., and Hugenholtz, P. 2002. The DMSO Reductase Family of Microbial Molybdenum Enzymes; Molecular Properties and Role in the Dissimilatory Reduction of Toxic Elements. *Geomicrobiology Journal* 19: 3-21

Mejean, V., Iobbi-Nivol, C., Lepelletier, M., Giordano, G., Chippaux, M., and Pascal, M. (1994). TMAO anaerobic respiration in *Escherichia coli* involvement of the *tor* operon. *Mol. Microbiol.* 11: 1169-79.

Oresnik, I.J., Ladner, C.L., and Turner, R.J. 2001. Identification of a twin-arginine leader-binding protein. *Mol. Microbiol.* 40: 323-31.

Palmer, T., Santini, C.-L., Iobbi-Nivol, C., Eaves, D.J., Boxer, D.H., and Giordano, G. 1996. Involvement of the *narJ* and *mob* gene products in distinct steps in the biosynthesis of the molybdoenzyme nitrate reductase in *Escherichia coli. Mol. Microbiol.* 20: 875-84.

Papish, A.L., Ladner, C.L., and Turner, R.J. 2003. The twin-arginine leader-binding protein, DmsD, interacts with the TatB and TatC subunits of the *Escherichia coli* twin-arginine translocase. *J. Biol. Chem.* 278: 32501-6.

Parish, D., Benach, J., Liu, G., Singarapu, K., Xiao, R., Acton, T., Su, M., Bansal, S., Prestegard, J., Hunt, J., Montelione, G.T. and Szyperski, T. 2008. Protein chaperones Q8ZP25_SALTY from *Salmonella typhimurium* and HYAE_ECOLI from *Escherichia coli* exhibit thioredoxin-like structures despite lack of canonical thioredoxin active site sequence motif. *J. Struct. Funct. Genomics* 9: 41-9.

Pommier, J., Mejean, V., Giordano, G., and Iobbi-Nivol, C. 1998. TorD, a cytoplasmic chaperone that interacts with the unfolded trimethylamine N-oxide reductase enzyme (TorA) in *Escherichia coli. J. Biol. Chem.* 273: 16615-20.

Pommier, J., Mandrand, M.A., Holt, S.E., Boxer, D.H., and Giordano, G. 1992. A second phenazine methosulphate-linked formate dehydrogenase isoenzyme in *Escherichia coli. Biochem. Biophys. Acta* 1107: 305-13.

Potter, L., Angove, H., Richardson, D., and Cole, J. 2001. Nitrate reduction in the periplasm of gram-negative bacteria. *Adv. Microb. Physiol.* 45: 51-112.

Potter, L.C., Millington, P.D., Thomas, G.H., Rothery, R.A., Giordano, G., and Cole, J.A. 2000. Novel growth characteristics and high rates of nitrate reduction of an *Escherichia coli strain*, LCB2048, that expresses only a periplasmic nitrate reductase. *FEMS Microbiol. Lett.* 185: 51-7.

Potter, L.C., and Cole, J.A. 1999. Essential roles for the products of the *napABCD* genes, but not *napFGH*, in periplasmic nitrate reduction by *Escherichia coli* K-12. *Biochem. J.* 344: 69-76.

Potter, L.C., Millington, P., Griffiths, L., Thomas, G.H., and Cole, J.A. 1999. Competition between *Escherichia coli* strains expressing either a periplasmic or a membrane-bound nitrate reductase: does Nap confer a selective advantage during nitrate-limited growth? *Biochem. J.* 344: 77-84.

Przybyla, A.E., Robbins, J., Menon, N., and Peck, H.D.J. 1992. Structure-function relationships among the nickel-containing hydrogenases. *FEMS Microbiol. Lett.* 88: 109-36.

Qiu, Y., Zhang, R., Binkowski, T.A., Tereshko, V., Joachimiak, A., and Kossiakoff, A. 2008. The 1.38 A crystal structure of DmsD protein from Salmonella typhimurium, a proofreading chaperone on the Tat pathway. *Proteins* 71: 525-33.

Ray, N., Oates, J., Turner, R.J., and Robinson, C. 2003. DmsD is required for the biogenesis of DMSO reductase in Escherichia coli but not for the interaction of the DmsA signal peptide with the Tat apparatus. *FEBS Lett* 534: 156-60.

Rodrigue, A., Chanal, A., Beck, K., Muller, M., and Wu, L.F. 1999. Co-translocation of a periplasmic enzyme complex by a hitchhiker mechanism through the bacterial Tat pathway. *J. Biol. Chem.* 274: 13223-28.

Roldan, M.D., Sears, H.J., Cheesman, M.R., Ferguson, S.J., Thomson, A.J., Berks, B.C., and Richardson, D.J. 1998. Spectroscopic characterization of a novel multiheme *c*-type cytochrome widely implicated in bacterial eectron transport. *J. Biol. Chem.* 273: 28785-90.

Rothery, R.A., Trieber, C.A., and Weiner, J.H. 1999. Interactions between the molybdenum cofactor and iron-sulfur clusters of Escherichia coli dimethylsulfoxide reductase. *J. Biol. Chem.* 274: 13002-9.

Rothery, R.A., and Weiner, J.H. 1996. Interaction of an engineered [3Fe-4S] cluster with a menaquinol binding site of *Escherichia coli* DMSO reductase. *Biochemistry* 35: 3247-57.

Sarfo, K.J., Winstone, T.L., Papish, A.L., Howell, J.M., Kadir, H., Vogel, H.J., and Turner, R.J. 2004. Folding forms of Escherichia coli DmsD, a twin-arginine leader binding protein. *Biochem Biophys Res Commun* 315: 397-403.

Sargent, F. 2007a. The twin-arginine transport system: moving folded proteins across membranes. *Biochem Soc Trans.* 35: 835-47.

Sargent, F. 2007b. Constructing the wonders of the bacterial world: biosynthesis of complex enzymes. *Microbiology* 153: 633-51.

Sargent, F., Berks, B.C., and Palmer, T. 2002. Assembly of membrane-bound respiratory complexes by the Tat protein-transport system. *Arch Microbiology* 178: 77-84.

Sargent, F., Bogsch, E.G., Stanley, N.R., Wexler, M., Robinson, C., Berks, B.C., and Palmer, T. 1998. Overlapping functions of components of a bacterial Sec-independent protein export pathway. *EMBO J.* 17: 3640-50.

Sawers, G. (1994) The hydrogenases and formate dehydrogenases of *Escherichia coli. Antonie Van Leeuwenhoek* 66; 57-88

Sawers, G., Heider, J., Zehelein, E., and Bock, A. 1991. Expression and operon structure of the *sel* genes of *Escherichia coli* and identification of a third selenium-containing formate dehydrogenase isoenzyme. *J. Bacteriol.* 173: 4983-93.

Schindelin, H., Kisker, C., Hilton, J., Rajagopalan, K.V., and Rees, D.C. 1996. Crystal structure of DMSO reductase: redox-linked changes in molybdopterin coordination. *Science* 272: 1615-21.

Schlindwein, C., Giordano, G., Santini, C.L., and Mandrand, M.A. 1990. Identification and expression of the *Escherichia coli fdhD* and *fdhE* genes, which are involved in the formation of respiratory formate dehydrogenase. *J. Bacteriol.* 172: 6112-21.

Schneider, F., Lowe, J., Huber, R., Schindelin, H., Kisker, C., and Knablein, J. 1996. Crystal structure of dimethyl sulfoxide reductase from Rhodobacter capsulatus at 1.88 A resolution. *J Mol Biol* 263: 53-69.

Schwarz, G. 2005. Molybdenum cofactor biosynthesis and deficiency. *Cellular and Molecular Life Sciences* 62: 2792-810.

Semisotnov, G.V., Rodionova, N.A., Razgulyaev, O.I., Uversky, V.N., Gripas, A.F., and Gilmanshin, R.I. 1991. Study of the "molten globule" intermediate state in protein folding by a hydrophobic fluorescent probe. *Biopolymers* 31: 119-28.

Settles, A.M., Yonetani, A., Baron, A., Bush, D.R., Cline, K., and Martienssen, R. 1997. Sec-independent protein translocation by the Maize Hcf106 protein. *Science* 278: 1467-70.

Shaw, A. L., Leimkuhler, S., Klipp, W., Hanson, G. R., and McEwan, A. G. 1999. Mutational analysis of the dimethylsulfoxide respiratory (*dor*) operon of *Rhodobacter capsulatus. Microbiology* 145: 1409-20.

Sodergren, E.J., Hsu, P.Y., and DeMoss, J.A. 1988. Roles of the *narJ* and *narI* gene products in the expression of nitrate reductase in *Escherichia coli. J. Biol. Chem.* 263: 16156-62.

Stanley, N.R., Sargent, F., Buchanan, G., Shi, J., Stewart, V., Palmer, T., and Berks, B.C. 2002. Behaviour of topological marker proteins targeted to the Tat protein transport pathway. *Mol. Microbiol* 43: 1005-21.

Stevens, C.M., Winstone, T.L. Turner, R.J., and Paetzel, M. (2009) Structural analysis of a monomeric form of the twin-arginine leader peptide binding chaperone *Escherichia coli* DmsD *J. Molecular Biology*. 389;124-133.

Stewart, V., Lu, Y., and Darwin, A.J. 2002. Periplasmic Nitrate Reductase (NapABC Enzyme) Supports Anaerobic Respiration by *Escherichia coli* K-12. *J. Bacteriol.* 184: 1314-23.

Stewart, V., Lin, J.T., and Berg, B.L. 1991. Genetic evidence that genes *fdhD* and *fdhE* do not control synthesis of formate dehydrogenase-N in *Escherichia coli* K-12. *J. Bacteriol.* 173: 4417-23.

Thomas, G., Potter, L., and Cole, J.A. 1999. The periplasmic nitrate reductase from *Escherichia coli*: a heterodimeric molybdoprotein with a double-arginine signal sequence and an unusual leader peptide cleavage site. *FEMS Microbiol. Lett.* 174: 167-71.

Tranier, S., Iobbi-Nivol, C., Birck, C., Ilbert, M., Mortier-Barriere, I., Mejean, V., and Samama, J.P. 2003. A novel protein fold and extreme domain swapping in the dimeric TorD chaperone from Shewanella massilia. *Structure* 11: 165-74.

Tranier, S., Mortier-Barriere, I., Ilbert, M., Birck, C., Iobbi-Nivol, C., Mejean, V., and Samama. J.P. 2002. Characterization and multiple molecular forms of TorD from Shewanella massilia, the putative chaperone of the molybdoenzyme TorA. *Protein Sci* 11: 2148-57.

Tullman-Ereck, D., Delisa, M.P., Kawarasaki, Y., Iranour, P., Ribnicky, B., Palmer, T., and Georgiou, G. 2007. Export pathway selectivity of *Escherichia coli* twin-arginine translocation signal peptides. *J. Biol. Chem.* 282: 8309-16.

Turner, R.J., Papish, A.L., and Sargent, F. 2004. Sequence analysis of bacterial redox enzyme maturation proteins (REMPs). *Can. J. Microbiol.* 50: 225-38.

Vergnes, A., Pommier, J., Toci, R., Blasco, F., Giordano, G., and Magalon, A. 2006. NarJ chaperone binds on two distinct sites of the aponitrate reductase of *Escherichia coli* to coordinate molybdenum cofactor insertion and assembly. *J. Biol. Chem.* 281: 2170-76.

Volbeda, A., Charon, M.-H., Piras, C., Hatchikian, E.C., Frey, M., and Fontecilla-Camps, J.C. 1995. Crystal structure of the nickel-iron hydrogenase from *Desulfovibrio gigas*. *Nature* 373: 580-7.

Voorduow, G. 2000. A universal system for the transport of redox proteins: early roots and latest developments. *Biophysical Chemistry* 86: 131-40.

Weiner, J.H., Bilous, P.T., Shaw, G.M., Lubitz, S.P., Frost, L., Thomas, G.H., Cole, J.A., and Turner, R.J. 1998. A novel and ubiquitous system for membrane targeting and secretion of proteins in the folded state. *Cell* 93: 93-101.

Weiner, J.H., Shaw, G., Turner, R.J., and Trieber, C.A. 1993. The topology of the anchor subunit of dimethyl sulfoxide reductase of *Escherichia coli*. *J Biol Chem* 268: 3238-44.

Winstone, T.L., Workentine, M.L., Sarfo, K.J., Binding, A.J., Haslam, B.D., and Turner, R.J. 2006. Physical nature of signal peptide binding to DmsD. *Arch Biochem Biophys* 455: 89-97.

Wu, L.-F., Chanal, A., and Rodrigue, A. 2000. Membrane targeting and translocation of bacterial hydrogenases. *Arch. Microbiology* 173: 319-24.

Zafra, O., Cava, F., Blasco, F., Magalon, A., and Berenguer, J. 2005. Membrane-associated maturation of the heterotetrameric nitrate reductase of *Thermus thermophilus*. *J. Bacteriol.* 187: 3990-96.

Zhao, Z., and Weiner, J.H. 1998. Interaction of 2-n-heptyl-4-hydroxyquinoline-N-oxide with dimethyl sulfoxide reductase of *Escherichia coli*. *J Biol Chem* 273: 20758-63.

In: Advances in Medicine and Biology. Volume 20
Editor: Leon V. Berhardt

ISBN 978-1-61209-135-8

Chapter 15

PHENOTYPIC MODIFICATION OF STAPHYLOCOCCAL CELL WALL CHARACTERISTICS MEDIATED BY DHEA

Balbina J. Plotkin[1,*] and Monika I. Konaklieva[2,†]

Department of Microbiology and Immunology[1] Midwestern University, Downers Grove, IL 60515; Department of Chemistry, American University, Washington, D.C. 20016, USA

ABSTRACT

Vancomycin treatment failures clinically account for 23-40% of patients with *S. aureus* infections; this occurs in the absence of laboratory-documented vancomycin resistance. Resistance of methicillin-resistant *S. aureus* (MRSA) clinical isolates to vancomycin can be phenotypically increased by exposure to dehydroepiandosterone (DHEA), an androgen with a chemical structure analogous to that of cholesterol. This study describes a phenotypic increase in resistance to the host cationic defense peptide, β-1 defensin, as well as vancomycin and other antibiotics that have a positive charge, in response to DHEA. The DHEA-mediated alteration in cell surface architecture appears to correlate with increased resistance to vancomycin. DHEA-mediated cell surface changes include alterations in: cell surface charge, surface hydrophobicity, capsule production, and carotenoid production. In addition, exposure to DHEA results in decreased resistance

1 * Corresponding Author: Balbina J. Plotkin, Ph. D., Department of Microbiology and Immunology, Midwestern University, Downers Grove, IL 60515, (630)515-6163, (630)515-7245 (FAX), bplotk@midwestern.edu

2 † Co-Author: Monika Konaklieva, Ph.D. Department of Chemistry, American University, 4400 Massachusetts Ave, NW Washington DC, 20016-8014 (202)885-1777, FAX: (202) 885-1752 e-mail: mkonak@american.edu

to lysis by Triton X-100 and lysozyme, indicating activation of murien hydrolase acivity. We propose that DHEA is an interspecies quorum-like signal that triggers innate phenotypic host survival strategies in *S. aureus* including but not limited to increased carotenoid production. A side effect of this phenotypic change is increased vancomycin resistance. Furthermore this DHEA-mediated survival system may share the cholesterol-squalene pathway shown to be statin sensitive.

Keywords: testosterone, dehydrotestosterone, DHT, dehydroepiandosterone, DHEA, *Staphylococcus aureus,* antimicrobial susceptibility, vancomycin resistance

INTRODUCTION

A pathogen's survival in the host is dependent on its capability to differentially express virulence factors and survive host defenses. This fundamental ability to survive in multiple host habitats can be triggered by recognition of host substances which function as interkingdom communication signals [1-3]. Niche localization, be it mucosal surface, tissue, or circulation could be informed by a microbes ability to recognize host signals e.g., insulin, and sterols [4-11]. Receipt of these chemical signals would warn the organism of potential exposure to host immune defenses including reactive oxygen intermediates; this could subsequently trigger expression of phenotypes more adapted for survival in that hostile environment. In addition, to host hormones acting as signaling compounds they could also serve as substrates for production of essential survival factors, e.g. carotenoids. An alteration in *S. aureus* that promotes its survival is its ability to alter its surface architecture and relative electronegativity [4-5, 11-14]. We previously reported that exposure of clinical isolates of *S. aureus* from multiple body sites to DHEA results in an increased MIC for vancomycin [17]. This change was DHEA concentration specific and associated with a change in colony morphology. The presence of DHEA, a C3 hydroxylated steroid *in vivo* may contribute to treatment failure. An intrinsic alteration in the physiochemical nature of *S. aureus* cell surfaces also appears to play a role in resistance to both drugs and host defense molecules.

The negative surface charge on staphylococci is determined by anionic phospholipids and by the teichoic acids and lipoteichoic acids, which are highly negatively-charged polymers of the cell wall and membrane complex of staphylococci [5, 13]. *S. aureus* increases its surface positive charge by incorporating D-alanine into teichoic and lipoteichoic acids and introducing L-lysine-modified phosphatidylglycerol, the only phospholipid with a positive net charge, into the cell membrane [5, 11, 14, 16]. Recently, two surface charge mutants of *S. aureus*, *dltABCD* and *mprF*, have been described that are highly susceptible to human defensins, as well as to other highly cationic peptides. Interestingly, *mprF*$^-$ mutants also exhibit an increased susceptibility to vancomycin [14, 16]. Taken together, these findings suggest that changes in surface charge directly affect sensitivity to vancomycin. It is not known what environmental condition regulates expression of this surface charge phenotype. What can be postulated, is that the signal that triggers this phenotype must communicate to the organism that exposure to host immune factors, e.g., highly cationic defense peptides, is imminent. An essential characteristic of such a signal would be its consistent presence at sites where exposure to harmful substances, e.g. defensins, is inevitable. From an evolutionary perspective it is also essential that these signals be present irrespective of age, circadian cycle or gender. DHEA, an androgen, fits this criterion [26-28].

DHEA is a naturally occurring steroid hormone in humans, whose levels decline with aging [18, 20, 23-24]. This decline appears to parallel immunosenescence [24-27]. Experimental evidence supports a protective role for DHEA(S) in reversing immunosensecence, reducing heart disease, controlling obesity, insulin resistance, memory loss, and vascular narrowing [27-30]. Both middle-aged adults and seniors take DHEA, which they obtain over the counter, to supplement normal endogenous androgen production in the hopes of improving their general health and resistance to disease [27, 29, 31]. Thus, *S. aureus* is exposed to various levels of DHEA, throughout the life of the host, which fluctuate depending on host age and consumption of nutritional supplements. The focus of this study was to determine whether DHEA affected *S. aureus* cell surface architecture and how the changes may correlate to antibiotic susceptibility.

Materials and Methods

Bacterial Stains and Growth Conditions

S. aureus ATCC 25923, a highly stable quality control strain, was grown in Mueller-Hinton medium with and without physiologic levels of steroid as previously described, unless otherwise noted [10, 17, 27].

Chemicals

Steroids used were testosterone, estrone, estradiol, pregnenolone, cortisol dehydrotestosterone, hydrocortisone, dehydroepiandrosterone (DHEA), cholesterol, and progesterone (Sigma-Aldrich Chemical Co.). All steroids except cholesterol were prepared in 95% (v/v) ethanol then diluted at least 100 fold into the manufacturer supplied diluent (Pluronic®) or Mueller-Hinton (MH) broth (0.05% v/v highest final ethanol concentration) as described [10]. Cholesterol was initially dissolved in chloroform which was diluted at least 100 fold in disulfamethoxazole (DMSO) to a final maximum DMSO concentration of 0.5% v/v. All steroids were tested at 0.1, 0.5,1.0 and 5.0 μM. Controls consisted of parallel assays done in the presence of the highest concentration of steroid diluent (0.5% ethanol or DMSO-cholesterol only). Lysostaphin and lysozyme were obtained from Sigma-Aldrich. β-1 defensin was obtained from Peptides International.

Antibiotic Sensitivity

Susceptibility in the presence of physiologically relevant concentrations of DHEA (0.1, 0.5, 1.0, 5.0 μM) was screened for using MicroScan® Gram positive Breakpoint Combo panels (Dade Behring Inc, West Sacramento, CA). The kits were used according to manufacturer's instructions [17] with results confirmed by microdilution methodology per

NCCLS procedures [32-33]. All hormones (Aldrich Chemical Co.) were prepared in 95% ethanol then diluted at least 100 fold in broth. Bacteria were suspended in the manufacturer's diluent with or without DHEA (0.1, 0.5, 1.0, 5.0 μM) prior to inoculation of the microtiter plates. Breakpoints were read by visual examination (24 and 48 hr). Samples (0.1 ml) from wells exhibiting no growth were inoculated onto Muller-Hinton agar for cidal concentration determinations. Negative controls included the highest concentration of hormone diluent, 0.05% ethanol and manufacturer's diluent alone. To confirm that the quality control strain (ATCC 25923) had not acquired a cryptic penicillinase that could be activated by DHEA, the presence of β-lactamase was tested for in all β-lactam antibiotic containing wells and growth control wells using nitrocefin disks (Cefinase; BBL Microbiology Systems) [34-40]. The positive control was a known β-lactamase producing *S. aureus* strain. No β-lactamase was detected in ATCC 25923. The specificity of DHEA-induced alteration of antibiotic sensitivity was determined by also testing testosterone (differs from DHEA by a -OH group), and dehydroxy-testosterone (DHT), at physiologically relevant concentrations (0.1, 0.5, 1.0, 5.0 μM) [10, 36, 37]. In addition, the steroids hydrocortisone, dexamethasone, and cortisone were also tested at physiologically relevant concentrations (1, 20, 50 μM) [10]. Antibiotic testing was done as described above for DHEA.

Binding of Radiolabeled [^{3}H]DHEA

Direct evidence that DHEA interacts specificity with *S. aureus* was obtained from radiometric binding assays, done as previously described [38]. Mid-logarithmic growth phase bacteria (10^8CFU/ml) were incubated (37^0 C, 30 min; 5^0 C, 18 hr) with a constant concentration of [^{3}H]DHEA (2.94 Ci/mmol; 20 nM) and increasing levels of cold DHEA (0.2 to 100μM). After incubation the cells were harvested by centrifugation, washed four times and cell-associated radioactivity determined by scintillation count. [^{3}H]DHEA associated with heat-killed cells was subtracted from that measured for live cell levels under homologous incubation conditions. All experiments were done in triplicate and repeated once. Data was analyzed and plotted using Prism/EnZ Plot (GraphPad).

QSARIS

Due to the fact that DHEA affects bacterial response to molecules from different antimicrobial classes regardless of their mechanism of drug action, QSARIS™ a program for quantitative structure-activity or structure-property modeling, was used to determine what, if any, similarities exist between the antimicrobials whose MICs were or weren't affected by the presence of DHEA. The similarity search algorithm assigns each molecule a correspondent vector in the multidimensional space formed by molecular descriptors. Similarity of molecules is expressed in terms of either the distance between these vectors in the multidimensional space, or the similarity coefficient based on the same distances. The similarity search was performed with DHEA as the pharmacophore for the 32 antimicrobials tested in the antibiotic sensitivity assays using QSARIS program Version 1.2, copyright 2001, by SciVision (www.scivision.com). The molecules similarity was based on their 3D structures using the cosine coefficient where x and y are the compared vectors, their components being $\{xi\}$ and $\{yi\}$; n is a dimension in space (the total number of descriptors). In this method, one starts from a molecule, such as a receptor ligand or an enzyme substrate, whose pharmacophoric groups are known for a particular target. Then a search of a database of compounds is performed to determine which ones have a similar 3D structure as pharmacophore. The top virtual hits are inspected and tested.

$$S(x,y) = \frac{\sum_{i=1}^{n} x_i y_i}{\left[\sum_{i=1}^{n} (x_{i})^2 \sum_{i=1}^{n} (y_{i})^2 \right]^{1/2}}$$

Carotenoid Production

Carotenoids were extracted from cells grown in MH broth with the steroids (0.1, 0.5, 1.0, and 5.0 μM) listed above, or in medium alone essentially as described by Morikawa, et al. [39]. After growth in media with and without steroid for 24 hr (shaking, 37 ^{0}C) the cells were harvested from 850μl of culture, washed with distilled deionized water and the pellet resuspended in 200 μl of methanol which was heated to 55^{0}C for 3 min. After the supernatant was removed, the cell pellet was re-extracted with methanol (200 μl). The extracted cell supernantant was placed in a quartz microcuvette with absorbance read at 465nm.

Surface Hydrophobicity/Capsule Production

Plastic adherence was used to measure surface hydrophobicity, as previously described [40]. Bacteria were grown (overnight, Mueller Hinton broth) in flat bottom uncoated microtiter plates in the presence and absence of DHEA. After extensive washing, the plates were stained with crystal violet (50% v/v). Stain was dissolved in 300μl of 95% ethanol and absorbance read in an EIA microtiter plate reader (Dynatech Laboratories, Inc., Chantilly, Virginia) at 590 nm. Duplicate plates stained with Alcian blue (1% PBS, 30 min, room temperature) were used for capsule production determinations. Stain was dissolved in 300μl pyridine (Sigma). Optical density was read in an EIA reader at 450 nm.

Surface Electronegativity

Adherence to glass was used to measure surface charge (glass = negative surface charge) [41]. Bacteria were grown (overnight) in Muller-Hinton broth with and without DHEA (flat bottom, 24 well plates containing glass coverslips). After extensive washing the coverslips were stained with crystal violet. The stain was washed off the coverslips with 300ml of 95% ethanol and absorbance (510nm) read in a Beckman Spectrophotometer.

Autolytic Assay

Triton X-100- Autolysis induced by Triton X-100 was measured as previously described with some modifications [42]. Mid-logarithmic cells grown in Muller-Hinton broth, with and without DHEA, were washed (4X; cold distilled water) and resuspended in buffer (0.05M Tris-HCL 7.2 pH, 0.05% Triton X- 100). Cell lysis was measured as a decrease in O.D (600nm) over time. Each data point represents the mean of six data points +/- S.D.

β-1 Defensin- Resistance to human β-1 defensin (Peptide International) was measured in LB broth without salts essentially as described by Peschel, et al. [18] with the exception of the addition of DHEA (0.5μM).

Coagulase Production

Coagulase production was performed according to manufacturers' (DIFCO) directions. *S. aureus* grown in the presence and absence of DHEA was pelleted and resuspended to a starting concentration of 10^9 CFU/ml. The suspension was serial diluted then tested for coagulase production; 1:100 dilution at 4 hr, 37^0C was the lowest bacterial concentration to result in coagulation

Statistical Analysis

Unless otherwise indicated experiments were performed in triplicate and repeated at least twice. All samples were coded, and performed in a blinded fashion. A one-way analysis of variance was used to determine differences between growth conditions. If statistical significance was found, a Tukey-Kramer post-hoc analysis was applied (Instat, GraphPad Software, San Diego, CA).

Results and Discussion

1. DHEA's Effect on Antibiotic Activity

Co-incubation of *S. aureus* with DHEA adversely impacts *S. aureus* sensitivity to vancomycin [17]. DHEA (0.1μM) exposure resulted in a 32-fold increase in vancomycin

MIC (0.25 μg/ml, unexposed to 8μg/ml, exposed) which now classifies the organism as a vancomycin-intermediate resistant *S. aureus* or VISA. In addition, DHEA (0.5μM) increased vancomycin's MIC from 0.25μg/ml vancomycin to 4μg/ml; this represents a MIC that is 1μg/ml below the recommended pharmacotherapeutic trough level in patients [43-44].

DHEA's effect on *S. aureus'* response to antimicrobials was not limited to vancomycin (Table 1). Furthermore, as with vancomycin, increased MICs were only measured in the presence of DHEA despite repeated serial transfers through medium with and without DHEA and antibiotic [17]. This indicates that DHEA's effect on antimicrobial resistance regardless of drug class is phenotypic and not the result of mutant selection.

Table 1. Effect of DHEA on the antibiotic resistance of *Staphylococcus aureus* ATCC 25923 (MSSA)

Antibiotic[b]	Control (medium alone)	DHEA			
		0.1 μM	0.5 μM	1.0 μM	5.0 μM
Cell wall synthesis inhibitors					
Penicillin	≤ 0.03 (S)[a]	< 0.03 (S)	8 **(R)**	≤ 0.03 (S)	≤ 0.03 (S)
Oxacillin	0.5 (S)	< 0.5 (S)	>2 **(R)**	≤0.5 (S)	≤0.5 (S)
Ampicillin	≤ 0.25 (S)	< 0.25 (S)	>8 **(R)**	≤ 0.25 (S)	≤ 0.25 (S)
Ampicillin/sulbactam	≤8/4 (S)	16/8 **(I)**	16/8 **(I)**	≤8/4 (S)	≤8/4 (S)
Piperacilllin/tazobactam	≤1 (S)	< 1 (S)	>8 **(R)**	≤1 (S)	≤1 (S)
Cefepime	≤2 (S)	>16 **(R)**	< 2 (S)	≤2 (S)	≤2 (S)
Imipenem	≤1 (S)	< 1 (S)	>8 **(R)**	≤1 (S)	≤1 (S)
Meropenem	≤4 (S)	>8 **(R)**	>8 **(R)**	≤4 (S)	≤4 (S)
Amoxicillin/K Clavulanate	≤4/2 (S)	>4/2 **(R)**	< 4/2 (S)	≤ 4/2 (S)	≤ 4/2 (S)
Vancomycin	0.25 (S)	8 **(I)**	4 (S)	0.25 (S)	0.25 (S)
Protein synthesis inhibitors					
Erythromycin	0.5 (S)	1 **(I)**	0.5 (S)	0.5 (S)	0.5 (S)
Clindamycin	≤0.25 (S)	< 0.25 (S)	0.5 (S)	≤ 0.25 (S)	≤ 0.25 (S)
Nitrofurantoin	≤ 32 (S)	>64 **(R)**	>64 **(R)**	≤ 32 (S)	≤ 32 (S)

[a] MIC in μg/ml (S= sensitive, I = intermediate, R = resistant - Interpretive Breakpoints as indicated in NCCLS document M100-S7).

[b] Antibiotics not affected by DHEA: chloramphenicol, cefazolin, cefuroxime, ceftriaxone, cephalothin, amoxicillin/K clavulanate, levofloxacin, gentamicin, tetracycline, ciprofloxacin, amikacin, cefotaxime, rifampin, norfloxacin, trimethoprim/sulfamethoxazole.

DHEA's effect on antibiotic resistance patterns was drug and hormone concentration specific with antimicrobials from multiple classes affected. The cross-class induction of multiple-drug resistance affected MIC values for members of the penicillin, cephalosporin, macrolide, aminoglycoside and nitrofurantoin classes. The cell wall active agents affected by DHEA were either zwitterionic (e.g., cefepine), or reported to have lower concentrations of bound cations than others of the same class [45-47]. For the drugs that affect protein synthesis, the ability of DHEA to alter drug activity also appeared to be related to the amount of drug-associated cations, which are required for the drug's transport across cytoplasmic membranes [48-49]. Additional structure-function analysis of the drugs with QSARIS modeling indicated that drugs affected by DHEA contained functionality's which have 60% or greater similarity with respect to overall charge (amphiphilic to positive) and hydrophobicity (logP value). Those antibiotics affected, within specific classes, exhibited the most positive charge. For example, resistance to cefipine, the most positively charged of the cephalosporins, was increased (Table 1). Additional QSARIS analysis of the antibiotics, using cosine coefficient as one of the similarity coefficients, showed that a compound having less than 60% similarity with the structure of the DHEA was not affected by DHEA. This QSARIS similarity search which was performed independently (blinded) correlated with the experimental data. For example, cefepime (cos. coeff 0.80) was affected by the DHEA while the rest of the cephalosporines tested were not (0.18-0.36). This DHEA-mediated alteration of response to antibiotics appeared to be specific for DHEA, and not the result of nonspecific perturbation of cell wall architecture, since a chemically similar steroid, testosterone, as well as other steroids (dehydrotestosterone, dexamethasone, hydrocortisone, cortisone) (Figure 1) had no affect on either *S. aureus* antibiotic susceptibility, or growth [10, 17]. Thus, antibiotic structure plays a role in DHEA's effect on drug response. However, as previously reported, ESI-MS studies show that the effect of DHEA is not the result of steroid-drug interaction [50].

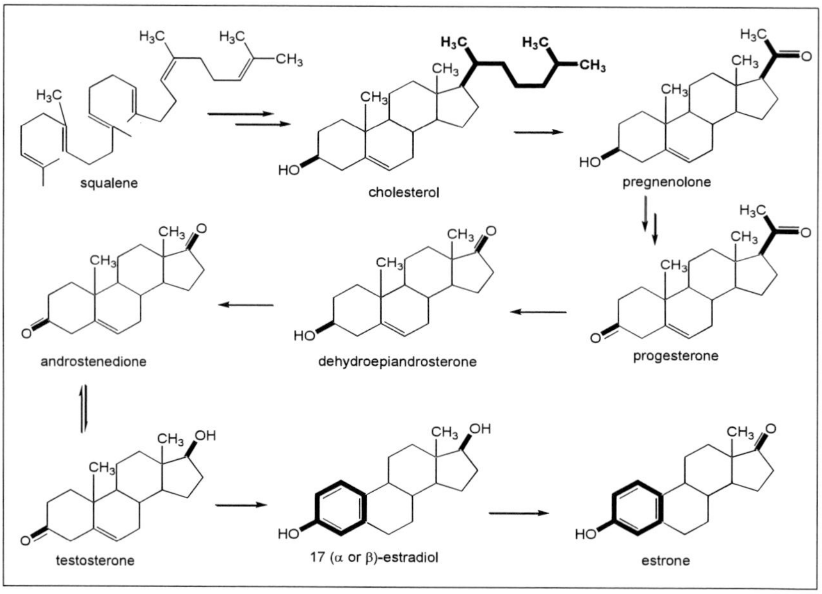

Figure 1. Biosynthetic relations of some steroids. Many pathways are abbreviated and intermediates not shown. Cholesterol, pregnenolone, and DHEA have a hydroxyl group at the third position in their cyclohexane ring.

2. DHEA Interaction with *S. aureus*

Direct evidence that DHEA interacts specifically with *S. aureus* was obtained from radiometric binding assays. $[^3H]$DHEA interaction with *S. aureus* at 5^0C, a temperature permissive for binding but inhibitory for uptake, indicates that $[^3H]$DHEA binds in a saturable manner (Figure 2) with an estimated 783 $\pm$ 30 binding sites per bacterial cell. At 37^{0}C, a permissive temperature for uptake, the data indicate that DHEA can also be taken up by *S. aureus*. Procaryotes normally contain no sterols, instead they contain polyterpenes [6, 51]. An important functional structure, regardless of the phylogenetic ranking of organisms, common to all membrane sterols or sterol-like compounds is the presence of a single hydroxyl group at the C-3 position. This is a characteristic of the structure of DHEA that may relate to its ability to interact with *S. aureus.* However, the identity of a specific receptor remains unclear. BLAST amino acid sequence analysis and genomic sequence analysis appear to rule out the presence in *S. aureus* of sigma 1, the putative cell surface human receptor for DHEA, as well as the mammalian intracellular steroid nuclear receptor superfamily [52]. This does leave open the possibility that DHEA does not bind to a specific receptor but intercalates into the membrane as reported for some mammalian cells [53].

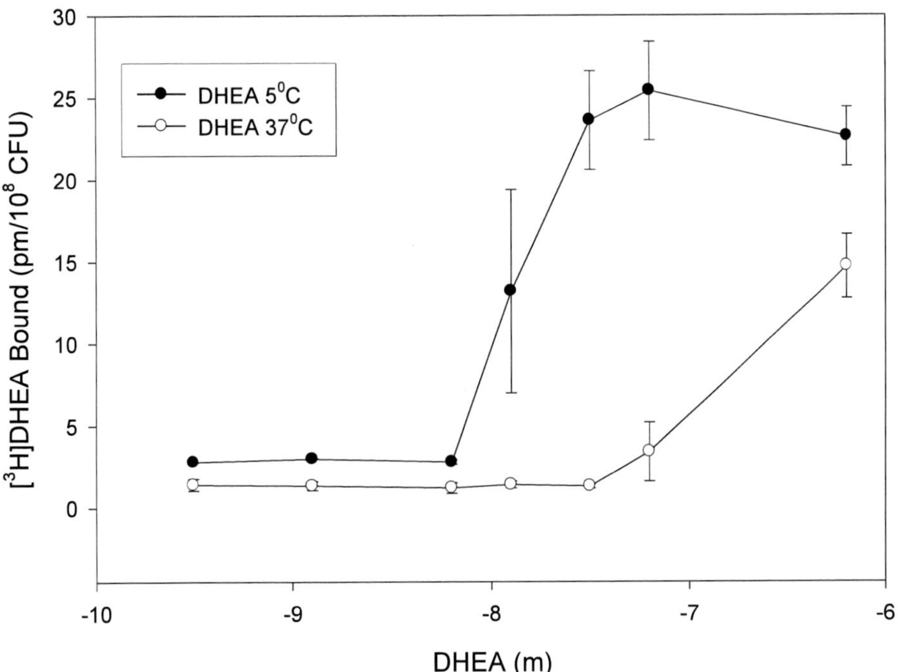

Figure 2. Binding of [^{3}H]DHEA to *S. aureus* ATCC 25923. Mid-logarithmic growth phase bacteria (10^{8}CFU/ml) were incubated (37^{0}C, 30 min; 5^{0}C, 18 hr) with a constant concentration of [^{3}H]DHEA (2.94 Ci/mmol; 20 nM) and increasing levels of cold DHEA (0.2 to 100μM in a total volume of 150μL). After incubation the cells were harvested by centrifugation, washed four times and cell-associated radioactivity determined by scintillation count. [^{3}H]DHEA associated with heat-killed cells was subtracted from that measured for live cell levels under homologous incubation conditions. All experiments were done in triplicate and repeated once. Data was analyzed and plotted using Prism/EnZ Plot (GraphPad).

3. DHEA's Effect on Cell Surface Characteristics

QSARIS analysis of DHEA's effect on antibiotic resistance indicated that DHEA affects the surface characteristics of *S. aureus,* including alterations in surface logP (hydrophobicity) and relative electronegative charge. Alterations in surface characteristics mediated by DHEA included an increase in the ability to adhere to plastic, a measure of surface hydrophobicity, and ability to form biofilm [54-55] (Table 2). Plastic adherence assays revealed that growth in the presence of DHEA resulted in a significant increase in surface hydrophobicity at low to mid-range DHEA concentrations, with the highest level of plastic adherence measured for cells grown in 0.5μM DHEA. Data shown in Figure 2 suggests that DHEA binds to *S. aureus*. DHEA-mediated alteration in cell membrane composition via signal transduction and/or DHEA intercalation into the membrane could account for the increase in hydrophobicity as measured by adherence to plastic.

Table 2. Effect of DHEA on *S.aureus* ATCC 25923 characteristics

	Relative Hydrophobicity	Relative Electronegativity	Lysozyme MIC (μg/ml)	β-1 Defensin MIC (μg/ml)
Control	21.8 $\pm$ 0.3[a,c]	0.1363 $\pm$ 0.052 [ac]	32	25
DHEA 0.1μM	34.3 $\pm$ 0.7[b]	0.1418 $\pm$ 0.052	16	>100 [b]
DHEA 0.5μM	36.7 $\pm$ 7 [b]	0.2890 $\pm$ 0.004[b]	16	>100 [b]
DHEA 1.0μM	29.9 $\pm$ 1.6 [b]	0.2233 $\pm$ 0.011[b]	4 [b]	>100 [b]
DHEA 5.0μM	24.3 $\pm$ 2.7	0.1518 $\pm$ 0.014	8 [b]	50

[a] Mean $\pm$ S.D
[b] Significantly different from control, $p \leq 0.01$
[c] Absorbance 560nm

We previously reported that in the presence of DHEA *S. aureus* exhibits colony pleomorphism with a preponderance of the small colonies similar in appearance to that described for heteroresistant *S. aureus* [56-57]. Heteroresistance in *S. aureus* correlates with increased vancomycin resistance [56-57]. However, DHEA-mediated colony variants were unstable; they are observed only in the presence of DHEA, regardless of medium NaCl content, which is reported to stabilize appearance of heteroresistant colonies [50]. To ascertain whether DHEA-mediated alterations were analogous to changes associated with those reported for heteroresistance, a systematic analysis of DHEA-mediated changes was done. This analysis tested the following characteristics: increased resistance to Triton-X 100, which is used to evaluate murein hydrolase autolytic activity [42, 58]; increased resistance to lysozyme, which cleaves the β1-4 linkage in peptidoglycan strands; increased resistance to lysostaphin, a glycylglycine endopeptidase; depressed coagulase production; decreased capsule production; and slower generation times (for review see [59]).

In contrast to phenotypes associated with heteroresistance, *S. aureus* grown in the presence of DHEA (0.5 μM) was slightly but significantly ($p < 0.05$) more sensitive to lysis by both Triton X-100, and four to eight fold more sensitive to lysis by lysozyme (Figure 3 and Table 2, respectively). The effect of DHEA was concentration specific for both Triton X-100 and lysozyme, with the optimum concentration of DHEA at 0.5 μM. In addition, sensitivity to lysostaphin lysis, levels of capsule production and coagulase production were unaffected by the presence of DHEA. With respect to growth kinetics, although heteroresistant *S. aureus* exhibit decreased rates of growth, as previously reported, *S. aureus* grown in the presence of DHEA had an increased rate of growth (1.3 fold), although at a concentration that does not correlate with DHEA levels triggering altered cell surface changes or drug MICs (5.0 μM DHEA) [10]. These results taken together indicate that the mechanism by which DHEA mediates increased resistance to vancomycin is dissimilar to that reported for heteroresistance [56-57].

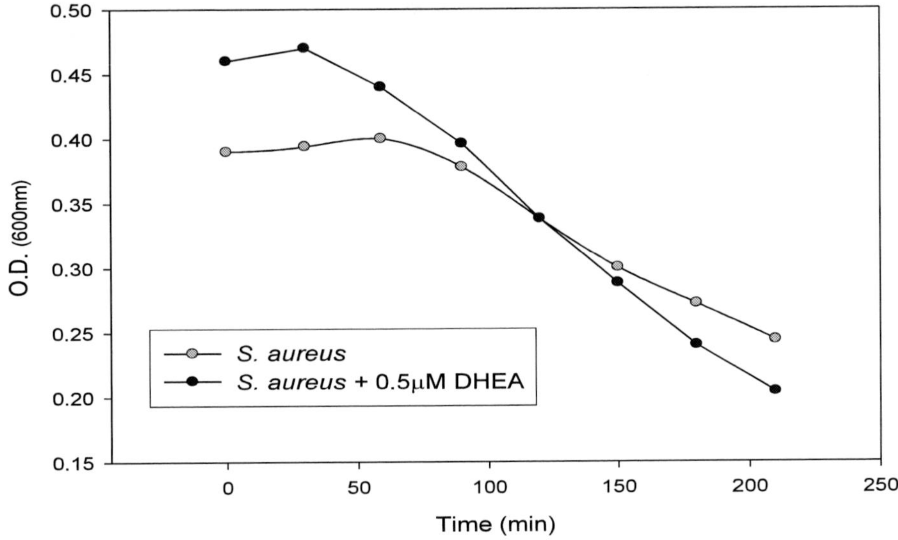

Figure 3. Effect of Triton X-100 on lysis of *S. aureus* ATCC 29523 grown in the presence (●) and absence ● of 0.5 μM DHEA; (p<0.001).

An alternative to heteroresistance, as a vancomycin resistance mechanism, is the increase in cell surface positive charge resulting from amino acid substitutions in teichoic/lipoteichoic acids and phosphatydlglycerol [12, 16, 19, 60]. These alterations are also linked to increased resistance to a cationic immune defense protein human β-1 defensin [16, 18]. DHEA (0.1, 0.5, and 1.0μM) induced an increase in resistance to human β-1 defensin (Table 2); a finding that supports QSARIS analysis of DHEA's effect on resistance to antibiotics.

To directly confirm that DHEA alters *S. aureus* surface charge, adherence to glass, a negatively charged surface, was measured (Table 2) [41]. DHEA (0.5 and 1.0 μM) induced a significant increase in cell surface positive charge as measured by adherence to glass. These findings may also explain the modest but significant effect of DHEA on Triton X-100 lysis (Figure 3) since this detergent carries a positive charge [61-62]. Both the increased ability to bind to glass and increased hydrophobicity also indicate that *S. aureus*' ability to form biofilms on a variety of surfaces, including host cell surfaces which have a negative surface charge, is enhanced by the presence of DHEA, [63-64]. The DHEA concentrations resulting in altered surface charge and response to β-1 defensin correlate with the DHEA levels that most often resulted in alteration in drug activity (Table 1). Taken together, these data support the hypothesis that the DHEA induction of resistance to cationic peptides of the innate immune system (β-1 defensin) and induction of increased resistance to weakly cationic drugs, including vancomycin, occur via a similar mechanism. A mechanism by which DHEA-may be mediating these alterations in *S. aureus* phenotype is repression of *mprF* gene transcription which has been demonstrated to result in an increase in surface charge. The negative surface charge of staphylococci is contributed to by anionic phospholipids and by the teichoic acids and lipoteichoic acids, which are highly negatively-charged polymers of the cell wall and membrane complex of staphylococci [18-19, 65]. *S. aureus* increases its surface positive charge by incorporating D-alanine into teichoic and lipoteichoic acids and introducing L-lysine-modified phosphatidylglycerol, the only phospholipid with a positive net charge into the cell membrane. *mprF* mutants exhibit increased susceptibility to vancomycin and are highly susceptible to human defensin killing. However, the signaling trigger for induction of altered surface charge has not yet been described. An essential characteristic required of such a trigger would be its consistent presence in sites where exposure to harmful substances, e.g. innate immune factors is inevitable. DHEA is a hormone present in the human host, 24hrs/day, throughout life, regardless of gender [20, 66-67] and fits the qualifications for such a gene trigger. An alternative mechanism by which DHEA could affect overall S. *aureus* surface charge is by increasing the levels of neutrally charged molecules. Molecules which would fit this parameter, and which have also been shown to correlate with increased virulence are carotenoids [15, 68-72]. To determine if cell composition has been altered by DHEA, the effect of DHEA, as well as other steroids, e.g. testosterone, dehydroxytestosterone which lack a 3 hydroxyl group and had no effect on staphylococcal response to vancomycin, were tested (Table 3).

Table 3. Effect of sterols on *S. aureus* susceptibility to vancomycin and carotenoid production

	Sterol Concentration (μM)	Vancomycin MIC (μg/ml)	Carotenoid Production (Abs_{490nm})	Carotenoid Level (Test/Control)
DHEA	0.1	8.0 (I)	6.2± 0.2	1.61
	0.5	4.0 (S)	14.0*±0.3	3.61*
	1.0	0.25	7.6±0.4	1.96
	5.0	0.25	6.1±0.2	1.68
Estradiol	0.1	0.25	5.4±0.3	1.91
	0.5	0.25	11.4±0.8	3.26*
	1.0	0.25	5.7±0.2	1.67
	5.0	0.25	7.0±0.2	1.8
Pregnenolone	0.1	0.25	7.6±0.3	1.16
	0.5	0.25	12.6±0.3	1.2
	1.0	0.25	6.4±0.1	1.0
	5.0	0.50	7.0±0.2	1.65
Progesterone	0.1	0.25	4.5±0.2	1.4
	0.5	0.25	4.6±0.3	2.95*
	1.0	0.25	4.0±0.2	0.89
	5.0	0.25	5.4±0.2	1.0
Cholesterol	0.1	1.0 (S)	5.1±0.2	1.3
	0.5	0.25	10.4±0.3	2.71*
	1.0	0.25	3.4±0.1	0.89
	5.0	0.25	3.9±0.2	1.0

* Significantly different from control, $p \leq 0.01$

Of the sterols tested, DHEA caused the maximal response both with respect to resistance to vancomycin and with regards to carotenoid synthesis. However, elevation in carotenoid synthesis was not limited to DHEA but was also affected by cholesterol, estradiol and progesterone. *S. aureus* produces carotenoids via pathways that have been demonstrated as being highly mutable [70, 73]. Carotenoids with one ring can mimic the tetracyclic sterols. The ability to regulate carotenoid production, as an intrinsic mechanism of resistance to cationic compounds, would provide *S. aureus* with a selective advantage in the host granting protection against highly cationic peptides, e.g., defensins, as well as reactive oxygen intermediates [15, 69]. The ability of *S. aureus* to convert sterol to carotenoid would hypothetically require a cyclase. *S. aureus* may contain such as cyclase as part of its ability to produce cholesterol ostensibly to promote its survival as an L-form [74]. If such a hypothetical enzyme exists, then growth in the presence of the 3-hydroxylated sterols should affect levels of carotenoid present. Growth in the presence of all 3-hydroxyl sterols tested (cholesterol, DHEA, estradiol), with the exception of pregnenolone, resulted in significantly higher ($p<0.01$) carotenoid levels at sterol concentrations (0.5 μM) which overlap those affecting antibiotic response (Table 1). Growth in the presence of sterols that lack an OH group at the 3-C position i.e., testosterone, dehydrotestosterone, progesterone, cortisol, hydrocortisone (Figure 1) had no effect on either vancomycin MIC or carotenoid levels (data not shown).

Conclusion

We propose that DHEA, and possibly other 3-hydroxylated sterols, are interkingdom signals which act as environmental triggers for *S. aureus,* warning of the potential for exposure to highly cationic host peptides that are part of the innate immune defense system. This cell surface remodeling also, incidentally, causes an increase in resistance to those antibiotics whose associated positive charge, or other chemical properties, are insufficient to overcome the charge-repulsion. The staphylococcal response is at minimum increasing carotenoid levels, but may also include altered *mprF* gene expression. This alteration in cell surface charge also incidentally causes an increase in resistance to those antibiotics whose associated positive charge is insufficient to overcome the repulsive forces to allow for interaction with their cellular targets. The DHEA concentration specificity may be indicative of tight genetic regulation, or membrane functional requirements. Previous studies have shown that L-forms of *S. aureus* can synthesize cholesterol, indicating the presence of a mevalonate-like pathway, the pathway which may be affected by statins [69]. We reported that growth in the presence of DHEA also results in alterations of colonial morphology with the production of small to petite colony types reminiscent of persisters. To survive the presence of antibiotics produced by soil organisms, or cationic peptides of the host including lysozyme and β-1 defensin requires plasticity. In the soil, microbes alter their levels of carotenoids through environmentally-induced changes in hopanoid synthesis. *S. aureus* while not having been demonstrated to produce hopanoids, appears instead to use the readily available 3'-hydroxy steroids present in the host to alter its membrane and increase its carotenoid production. Interestingly, genetic changes resulting in a broad spectrum of resistance to antimicrobials, including presence of PBP2a and efflux pumps, appear to override the organism's intrinsic ability to defend itself against deleterious environmental assaults. These data indicate that *in vivo* the cell surface architecture of *S. aureus* is fundamentally different due to the presence and utilization of 3-hydroxysterols resulting in increased levels of carotenoid synthesis. The resulting alteration in cell surface charge and hydrophobic characteristics may provide a new approach to treatment management if the patient is infected with strains responsive to host-derived sterols.

REFERENCES

[1] Catron D, Lange Y, Borensztajn J, Sylvester M, Jones B, Haldar K: *Salmonella enterica* Serovar Typhimurium Requires Nonsterol Precursors Of The Cholesterol Biosynthetic Pathway For Intracellular Proliferation. *Infect Immun* 2004, 72:1036-1042.

[2] Cheng I: The Effect of Cholesterol On The Growth Of *Mycobacterium tuberculosis*, BCG And *Mycobacterium phlei*. *Scientia Sinica* 1963, 12:1541-1552.

[3] Lowery C, Dickerson T, Janda K: Interspecies And Interkingdom Communication Mediated By Bacterial Quorum Sensing. *Chem Soc Rev* 2008, 37:1337-1346.

[4] Klosowska K, Plotkin B: Human Insulin Modulation Of *Escherichia coli* Adherence And Chemotaxis. *American Journal Of Infectious Diseases* 2006, 2:197-200.

[5] Lenard J: Mammalian Hormones In Microbial Cells. *Trends In Biochemical Sciences* 1992, 17:147-150.

[6] Ourisson G, Rohmer M: Procaryotic Polyterpenes: Phylogenetic Precursors Of Sterols. *Current Topics In Membranes And Transport* 1982, 17:153-182.

[7] Martinotti M, Savoia D: Effect Of Some Steroid Hormones On The Growth Of *Trichomonas vaginalis*. *G Batteriol Virol Immunol* 1985, 78:52-59.

[8] Meier K, Eberhard K, Ivens K: Effect Of Hormones On Bacteria Growth. *Naturwissenschaften* 1954, 41:284-285.

[9] O'Connor C, Essmann M, Larsen B: 17-ß-Estradiol Upregulates The Stress Response In Candida albicans: Implications For Microbial Virulence. *Infect Dis Obstet Gynecol* 1998, 6:176-181.

[10] Plotkin B, Erickson Q, Roose R, Viselli S: Effect Of Androgens And Glucocorticoids On Microbial Growth And Antimicrobial Susceptibility. *Current Microbiol* 2003, 47:514-520.

[11] Yotis W: Responses Of Staphylococci And Other Microorganisms To Steroids. *Adv Steroid Biochem Pharmacol* 1972, 3:193-235.

[12] Koprivnjak T, Peschel A, Gelb M, Liang N, Weiss J: Role Of Charge Properties Of Bacterial Envelope In Bactericidal Action Of Human Group Iia Phospholipase A2 Against *Staphylococcus Aureus*. *J Biol Chem* 2002, 277:47636-47644.

[13] Koprivnjak T, Weidenmaier C, Peschel A, Weiss J: Wall Teichoic Acid Deficiency In *Staphylococcus aureus* Confers Selective Resistance To Mammalian Group Iia Phospholipase A(2) And Human Beta-Defensin 3. *Infect Immun* 2008, 76:2169-2176.

[14] Kristian S, Durr M, Strijp Jv, Neumeister B, Peschel A: *mprF*-Mediated Lysinylation Of Phospholipids In *Staphylococcus aureus* Leads To Protection Against Oxygen-Independent Neutrophil Killing. *Infect Immun* 2003, 71:546-549.

[15] Liu G, Essex A, Buchanan J, Datta V, Hoffman H, Bastian J, Fierer J, Nizet V: *Staphylococcus aureus* Golden Pigment Impairs Neutrophil Killing And Promotes Virulence Through Its Antioxidant Activity. *J Exp Med* 2005, 202:209-215.

[16] Peschel A, Jack R, Otto M, Collins L, Staubitz P, Nicholson G, Kalbacher H, Nieuwenhuizen W, Jung G, Tarkowski A *Et Al*: *Staphylococcus aureus* Resistance To Human Defensins And Evasion Of Neutrophil Killing Via The Novel Virulence Factor *mprF* Is Based On Modification Of Membrane Lipids With L-Lysine. *J Exp Med* 2001, 193:1067-1076.

[17] Plotkin B, Morejon A, Laddaga R, Viselli S, Tjhio J, Schreckenberger P: Dehydroepiandosterone Induction Of Increased Resistance To Vancomycin In *Staphylococcus aureus* Clinical Isolates (MSSA, MRSA). *Lett Appl Microbiol* 2005, 40:249-254.

[18] Peschel A, Otto M, Jack R, Kalbacher H, Jung G, Gotz F: Inactivation Of The *dlt* Operon In *Staphylococcus aureus* Confers Sensitivity To Defensins, Protegrins, And Other Antimicrobial Peptides. *J Biol Chem* 1999, 274:8405-8410.

[19] Peschel A, Vuong C, Otto M, Gotz F: The D-Alanine Residues Of *Staphylococcus aureus* Teichoic Acids Alter The Susceptibility To Vancomycin And The Activity Of Autolytic Enzymes. *Antimicrob Agents Chemother* 2000, 44:2845-2847.

[20] Orentreich N, Brind J, Vogelman J, Andres R, Baldwin H: Longterm Longitudinal Measurements Of Plasma Dehydroepiandrosterone Sulfate In Normal Men. *Journal Of Clinical Endocrinology And Metabolism* 1992, 75:1002-1004.

[21] Arlt W: Dehydroepiandrosterone And Ageing. Best Practice And Research In Clinical Endocrinology And Metabolism 2004, 18:363-380.

[22] Aoki K, Taniguchi H, Ito Y, Satoh S, Nakamura S, Muramatsu K, Yamashita R, Ito S, Mori Y, Sekihara H: Dehydroepiandrosterone Decreases Elevated Hepatic Glucose Production In C57bl/Ksj-Db/Db Mice. *Life Sciences* 2004, 74:3075-3084.

[23] Herbert J: The Age Of Dehydroepiandrosterone. *Lancet* 1995, 345:1193-1194.

[24] James K, Premchand N, Skibinska A, Skibinska J, Nicol M, Mason J: Il-6, DHEA And The Aging Process. *Mech Ageing Dev* 1997, 93:15-24.

[25] Bottasso O, Bay M, Besedovsky H, Rey Ad: Immunoendocrine Alterations During Human Tuberculosis As An Integrated View Of Disease Pathology. *Neuroimmunomodulation* 2009, 16:68-77.

[26] Mege J, Capo C, Michel B, Gastaut J, Bongrand P: Phagocytic Cell Function In Aged Subjects. *Neurobiol Aging* 1988, 9:217-220.

[27] Spencer N, Norton S, Harrison L, Li G, Daynes R: Dysregulation Of IL-10 Production With Aging: Possible Linkage To The Age-Associated Decline In DHEA And Its Sulfated Derivative. *Exper Gerontol* 1996, 31:393-408.

[28] Sergio G: Exploring The Complex Relations Between Inflammation And Aging (Inflamm-Aging): Anti-Inflamm-Aging Remodelling Of Inflamm- Aging, From Robustness To Frailty. *Inflamm Res* 2008, 57:558-563.

[29] Genazzani A, Lanzoni C, Genazzani A: Might DHEA Be Considered A Beneficial Replacement Therapy In The Elderly? *Drugs Aging* 2007, 24:173-185.

[30] Krysiak R, Frysz-Naglak D, Okopien B: Current Views On The Role Of Dehydroepiandrosterone In Physiology, Pathology And Therapy. *Pol Merkur Lekarski* 2008, 24:66-71.

[31] Cameron D, Braunstein G: Androgen Replacement Therapy In Women. *Fertility And Sterility* 2004, 82:273-289.

[32] NCCLS: Methods For Dilution Antimicrobial Susceptibility Testing For Bacteria That Grew Aerobically: Approved Standard. Wayne, Pennsylvania: National Committee For Clinical Laboratory Standards; 1997.

[33] NCCLS: Performance Standards For Antimicrobial Susceptibility Testing: Eighth Informational Supplement. Villanova, PA: National Committee For Clinical Laboratory Standards; 1998.

[34] Udo E, Grubb W: Molecular And Phage Typing Of *Staphylococcus aureus* Harbouring Cryptic Conjugative Plasmids. *Eur J Epidemiol* 1996, 12:637-641.

[35] Novick R, Morse S: In Vivo Transmission Of Drug Resistance Factors Between Strains Of *Staphylococcus aureus*. *J Exp Med* 1967, 125:45-59.

[36] Urani A, Privat A, Maurice T: The Modulation By Neurosteroids Of The Scopolamine-Induced Learning Impairment In Mice Involves An Interaction With Sigma1 Receptors. *Brain Res Brain Res Rev* 1998, 799:64-77.

[37] Zinder O, Dar D: Neuroactive Steroids: Their Mechanism Of Action And Their Function In The Stress Response. *Acta Physiol Scand* 1999, 167:181-188.

[38] Plotkin B, Bemis D: Adherence Of *Bordetella bronchiseptica* To Hamster Lung Fibroblasts. *Infect Immun* 1984, 46:697-702.

[39] Morikawa K, Maruyama A, Inose Y, Higashide M, Hayashi H, Ohta T: Overexpression Of Sigma Factor B Urges *Staphylococcus aureus* To Thicken The Cell Wall And Resist Beta Lactams. *Biochem Biophys Res Commun* 2001, 288:385-389.

[40] Farzam F, Plotkin B: Effect Of Sub-MICs Of Antibiotics On The Hydrophobicity And Production Of Acidic Polysaccharide By *Vibrio vulnificus*. *Chemotherapy* 2001, 47:184-193.

[41] Gross M, Cramton S, Gotz F, Peschel A: Key Role Of Teichoic Acid Net Charge In *Staphylococcus aureus* Colonization Of Artificial Surfaces. *Infection And Immunity* 2001, 69:3423-3426.

[42] Boyle-Vavra S, Carey R, Daum R: Development Of Vancomycin And Lysostaphin Resistance In A Methicillin-Resistant *Staphylococcus aureus* Isolate. *J Antimicrob Chemother* 2001, 48:617-625.

[43] Kahl B, Belling G, Reichelt R, Herrmann M, Karam C, Mckinnon P, Neuhauser M, Rybak M: Outcome Assessment Of Minimizing Vancomycin Monitoring And Dosing Adjustments. *Pharmacotherapy* 1999, 19:257-266.

[44] Zimmermann A, Katona B, Plaisance K: Association Of Vancomycin Serum Concentrations With Outcomes In Patients With Gram-Positive Bacteremia. *Pharmacotherapy* 1995, 15:85-91.

[45] Niebergall P, Hussar D, Cressman W, Sugita E, Doluisio J: Metal Binding Tendencies Of Various Antibiotics. *J Pharm Pharmacol* 1966, 18:729-738.

[46] Hanberger H, Nilsson L, Maller R, Isaksson B: Pharmacodynamics Of Daptomycin And Vancomycin On *Enterococcus faecalis* And *Staphylococcus aureus* Demonstrated By Studies Of Initial Killing And Postantibiotic Effect And Influence Of Ca^{2+} And Albumin On These Drugs. *Antimicrob Agents Chemother* 1991, 35:1710-1716.

[47] Fietta A, Merlini C, Gialdroni G, Grassi G: Requirements For Intracellular Accumulation And Release Of Clarithromycin And Azithromycin By Human Phagocytes. *J Chemother* 1997, 9:23-31.

[48] Obaseiki-Ebor E: Enhanced *Escherichia coli* Susceptibility To Nitrofurantoin By EDTA And Multiple Aminoglycoside Antibiotics Resistance Mutation. *Chemother* 1984, 30:88-91.

[49] Mtairag E, Abdelghaffar H, Douhet C, Labro M: Role Of Extracellular Calcium In In Vitro Uptake And Intraphagocytic Location Of Macrolides. *Antimicrob Agents Chemother* 1995, 39:1676-1682.

[50] Plotkin B, Konaklieva M: Possible Role Of *sarA* In Dehydroepiandosterone (DHEA) - Mediated Increase In *Staphylococcus aureus* Resistance To Vancomycin. *Chemother* 2007, 53:181-184.

[51] Nes W: Role Of Sterols In Membranes. *Lipids* 1974, 9:596-612.

[52] Altschul S, Madden T, Shaffer A, Zhang J, Zhang Z, Miller W, Lipman D: Gapped Blast And Psi-Blast: A New Generation Of Protein Database Search Programs. *Nucleic Acids Research* 1997, 25:3389-3402.

[53] Aragno M, Cutrin J, Mastrocola R, Perrelli M-G, Restivo F, Poli G, Danni O, Boccuzzi G: Oxidative Stress And Kidney Dysfunction Due To Ischemia/Reperfusion In Rat: Attenuation By Dehydroepiandrosterone. *Kidney International* 2003, 64:836–843.

[54] Beenken K, Blevins J, Smeltzer M: Mutation Of *sarA* In *Staphylococcus aureus* Limits Biofilm Formation. *Infect Immun* 2003, 71:4206-4211.

[55] Wu J, Kusuma C, Mond J, Kokai-Kun J: Lysostaphin Disrupts *Staphylococcus aureus* And *Staphylococcus epidermidis* Biofilms On Artificial Surfaces. *Antimicrob Agents Chemother* 2003, 47:3407-3414.

[56] Plipat N, Livni G, Bertram H, Rb Thomson J: Unstable Vancomycin Heteroresistance Is Common Among Clinical Isolates Of Methiciliin-Resistant *Staphylococcus aureus. J Clin Microbiol* 2005, 43:2494–2496.

[57] Wong S, Ho P, Woo P, Ky Ky: Bacteremia Caused By Staphylococci With Inducible Vancomycin Heteroresistance. *Clin Infect Dis* 1999, 29:760-767.

[58] Tomasz A, Waks S: Mechanism Of Action Of Penicillin: Triggering Of The Pneumococcal Autolytic Enzyme By Inhibitors Of Cell Wall Synthesis. *Proc Natl Acad Sci U S A* 1975, 72:4162-4166.

[59] Walsh T, Howe R: The Prevalence And Mechanisms Of Vancomycin Resistance In *Staphylococcus aureus*. *Annual Review Of Microbiology* 2002, 56:657-675.

[60] Ruzin A, Severin A, Moghazeh S, Etienne J, Bradford P, Projan S, Shlaes D: Inactivation Of *mprF* Affects Vancomycin Susceptibility In *Staphylococcus aureus*. *Biochim Biophys Acta* 2003, 1621:117-121.

[61] Helenius A, Simons K: Charge Shift Electrophoresis: Simple Method For Distinguishing Between Amphiphilic And Hydrophilic Proteins In Detergent Solution. *Proc Natl Acad Sci Usa* 1977, 74:529-532.

[62] Retailleau P, Ducruix A, Ries-Kautt M: Importance Of The Nature Of Anions In Lysozyme Crystallisation Correlated With Protein Net Charge Variation. *Acta Crystallogr D Biol Crystallogr* 2002, 58:1576-1581.

[63] Foged C, Brodin B, Frokjaer S, Sundblad A. Particle Size And Surface Charge Affect Particle Uptake By Human Dendritic Cells In An In Vitro Model. *Internat J Pharmaceut* 2005, 298; 315-322.

[64] Stylos W, Merryman C, Maurer P: Antigenicity Of Polypeptides (Poly-Alpha-Amino Acids). Distribution Of Sheep Antibodies To Polymers Of Alpha-L-Amino Acids Of

Varying Net Electrical Charge And Lysozyme. *Int Arch Allergy Appl Immunol* 1970, 39:381-390.

[65] Fischer W, Koch H, Rosel P, Fiedler F, Schmuck L: Structural Requirements Of Lipoteichoic Acid Carrier For Recognition By The Poly(Ribitol Phosphate) Polymerase From *Staphylococcus aureus* H. A Study Of Various Lipoteichoic Acids, Derivatives, And Related Compounds. *J Biol Chem* 1980, 255:4550-4556.

[66] Lamberts S, Vandenbeld A, Vanderlely A: The Endocrinology Of Aging. *Science* 1997, 278:419-424.

[67] Hubert S, Mohammed J, Fridkin S, Gaynes R, Mcgowan J, Tenover F: Glycopeptide-Intermediate *Staphylococcus aureus*: Evaluation Of A Novel Screening Method And Results Of A Survey Of Selected U.S. Hospitals. *Journal Of Clinical Microbiology* 1999, 37:3590-3593.

[68] Xiong Z, Kapral F: Carotenoid Pigment Levels In *Staphylococcus aureus* And Sensitivity To Oleic Acid. *J Med Microbiol* 1992, 37:192-194.

[69] Liu C, Liu G, Song Y, Yin F, Hensler M, Jeng W, Nizet V, Wang A, Oldfield E: A Cholesterol Biosynthesis Inhibitor Blocks *Staphylococcus aureus* Virulence. *Science* 2008, 319:1391-1394.

[70] Umeno D, Arnold F: A C35 Carotenoid Biosynthetic Pathway. *Applied And Environmental Microbiology* 2003, 69:3573-3579.

[71] Martin Hd, Ruck C, Schmidt M, Sell S, Beutner S, Mayer B, Walsh R: Chemistry Of Carotenoid Oxidation And Free Radical Reactions. *Pure Appl Chem* 1999, Vol. 71, No. 12, Pp. 2253-2262.

[72] Marshall J, Wilmoth G: Proposed Pathway Of Trierpenoid Carotenoid Biosynthesis In *Staphylococcus aureus*: Evidence From A Study Of Mutants. *J Bacteriol* 1981, 147:914-919.

[73] Umeno D, Arnold F: Evolution Of A Pathway To Novel Long-Chain Carotenoids. *Journal Of Bacteriology* 2004, 186:1531-1536.

[74] Hayami M, Okabe A, Sasai K, Hayashi H, Kanamasa Y: Presence And Synthesis Of Cholesterol In Stable Staphylococcal L-Forms. *J Bacteriol* 1979, 140:859-863.

INDEX

A

B

C

E

F

G

H

I

J

K

L

M

N

O

P

Q

R

S

T

U

V

W

X

Y

Z